Jürgen Zierep

G. H. SCHNERR, R. BOHNING,
K. BÜHLER, AND W. FRANK (EDS.)

Fluid- and Gasdynamics

ACTA MECHANICA
SUPPLEMENTUM 4

SPRINGER-VERLAG WIEN NEW YORK

Prof. Dr.-Ing. habil. G. H. Schnerr
Prof. Dr.-Ing. habil. R. Bohning
Prof. Dr.-Ing. habil. W. Frank
Institut für Strömungslehre und Strömungsmaschinen, Universität Karlsruhe (TH),
Karlsruhe, Federal Republic of Germany

Prof. Dr.-Ing. habil. K. Bühler
Fachbereich Maschinenbau, Fachhochschule Offenburg,
Offenburg, Federal Republic of Germany

Printed on acid-free and chlorine-free bleached paper

With 205 Figures

ISSN 0939-7906
ISBN-13:978-3-211-82495-5   e-ISBN-13:978-3-7091-9310-5
DOI: 10.1007/978-3-7091-9310-5

Professor Dr.-Ing. Dr. techn. E. h. Jürgen Zierep

It is often observed that great scientists originate from distinguished teachers. Jürgen Zierep is no exception to this rule. His scientific background can be traced back to eminent scientists like H. Koschmieder, J. Ackeret, and K. Oswatitsch. Professor Koschmieder not only was the admired teacher, who raised his student's interest in meteorological applications of mathematics, he later also became Zierep's father-in-law. When Jürgen Zierep spent about one year with Professor Ackeret in Zürich, he was no older than 23 and had already received a Doctor of Engineering degree from the Technical University of Berlin. Four years later he received the *venia legendi* in applied mathematics, also at Berlin, and then joined the former DVL (now DLR) Institute of Theoretical Gasdynamics at Aachen. Polishing his scientific education under the guidance of Professor Oswatitsch, Zierep shifted his main field of interest closer to gasdynamics. In 1961, Jürgen Zierep was appointed Professor of Theoretical Fluid Mechanics at the Technical University of Karlsruhe, and just two years later, at the age of 34, he was promoted to the Chair of Fluid Mechanics. Although Professor Zierep received many honourable offers from other universities and research institutions, he decided to stay in Karlsruhe until now. Today Professor Zierep is among the leading scientists in his field, both in Germany and on an international scale. Within the limits of a preface, Zierep's scientific work can be described just briefly as follows.

With his first scientific papers, which were published at the beginning of the 1950's, Zierep already became known outside his immediate research area by introducing fluid mechanical methods into meteorology. Zierep's papers on lee waves and cellular convection were early contributions to a branch of science that is now often called "dynamic meteorology".

In those days without large electronic computers, solving the basic equations of gasdynamics was an extremely difficult problem. Zierep applied, and improved, contemporary mathematical methods for solving the equations of gasdynamics, e.g. the so-called "parabolic method", the "integral equation method", and a perturbation method for weakly nonlinear waves. Even in the age of the computer, those analytical methods have not lost importance; on the contrary, they often serve as a basis for numerical investigations.

Also of large, and certainly permanent, value are Zierep's investigations on mechanical similarity and its generalizations. After he became interested in that field many years ago, he resumed working on those problems again and again. His efforts became the source of new similarity laws, now to be found in most text books on gasdynamics. He was also successful in applying similarity methods to problems that were considered "unsolvable" before.

In co-operation with younger colleagues and students, Jürgen Zierep performed theoretical and experimental investigations on rotating systems. This work gave access to a field that has become relevant to the modern concept of bifurcating solutions. New directions of research also have been pointed out by Jürgen Zierep with respect to the development and design of experimental facilities, in particular for studying the interaction between thermodynamic and fluid mechanical processes. Flows with heat supply and condensation in high-speed flows can be given as examples of research areas that are intimately related to the name Zierep, in addition to Oswatitsch, of course. Not that Jürgen Zierep would intend to rest now — he is most successful in keeping his institute on top of recent developments, such as passive control of shock/boundary-layer interaction by ventilation.

Any attempt that briefly evaluates the scientific work of Jürgen Zierep would be incomplete if his books were not mentioned. Zierep's ability to describe, and explain, complex processes in the clearest and most convincing way is rare even among very distinguished scientists. His textbooks provide engineers and physicists, among others, with an introduction to areas of fluid mechanics and gasdynamics, which previously were of access to specialists only. Several translations into other languages (including, of course, English, but also Polish, Turkish and Japanese) prove that Zierep's books are appreciated far beyond the boundaries of the German-speaking countries. As far as the quality of presentation is concerned, Zierep's lectures match his books. Colleagues and students alike are regularly fascinated by the unusual clarity and beauty of the subjects presented by Jürgen Zierep. Over the years, his rhetoric has become legendary.

Zierep's contributions to the administration of science are in accordance with his research work. While he was in charge, his institute became one of the largest of its kind in Germany. Experimental facilities that are rarely seen at universities are available at the institute at Karlsruhe. As treasurer, president, and vice president of the Gesellschaft für Angewandte Mathematik und Mechanik (GAMM), Jürgen Zierep had a large impact on one of the most important scientific societies in Germany. In carrying out the functions of his office, he has never stressed national interests. He helped to organize international conferences taking place as far away as Beijing, and his efforts with respect to improving the relationship with former East Germany and other Eastern European countries provided quiet, but valuable, help to many colleagues in those difficult years before the Iron Curtain and the Wall were removed. Last, but not least, it should be mentioned that Jürgen Zierep has been serving as a co-editor of this journal from the very beginning. Despite the modern, yet regrettable, trend towards specialisation, *Acta Mechanica* still comprises mechanics of both solids and fluids. This is so largely thanks to Jürgen Zierep.

Given his merits, it is natural that Jürgen Zierep has received many honours. To name just a few in chronological order, he was asked to present the Ludwig Prandtl Memorial Lecture, was appointed honorary member of the Jugoslavian Society of Mechanics and honorary professor at the University of Beijing, and received an honorary doctoral degree from the Technical University of Vienna.

If asked about his greatest achievements, Professor Zierep will certainly refer to the large number and high quality of his students. He has supervised about 40 doctoral dissertations, and 23 of his former students have received a *venia legendi*. That Professor Zierep enjoys his student's high esteem was demonstrated most impressively when a torch procession was held at the University of Karlsruhe to celebrate his decision to turn down an offer from another university.

Jürgen Zierep's academic and scientific success is intertwined with a remarkable family life. Married for nearly forty years to his wife Elisabeth, he could always rely on family support should difficulties arise in his academic or professional life. In turn, Jürgen Zierep is a father and grandfather who even places his beloved science second when an important family event calls for his attention.

On the occasion of his 65th birthday, friends, colleagues and former students wish Professor Zierep, the scientist who loves to teach, that he will enjoy many more years of fruitful research and stimulating teaching.

W. Schneider, Wien

List of publications by J. Zierep

[1] Das Leewellenproblem bei geschichteter Anströmung. Dissertation TU Berlin (1951).

[2] Leewellen bei geschichteter Anströmung. Ber. dt. Wetterdienst. US-Zone **35**, 85–90 (1952).

[3] Die Bestimmung der Leewellenströmung bei beliebigem Anströmprofil. Z. Flugwiss. **1**, 9–11 (1953).

[4] Der Stand des Leewellenproblems. Jahrb. WGL, 76–78 (1953).

[5] Berechnungsmethoden von Lavaldüsen auf Grund eines Formelsystems für die Strömung in der Düsenkehle (gemeinsam mit W. Haack). Z. Flugwiss. **2**, 41–50 (1954).

[6] Zur Theorie der instationären Leewellen. Jahrb. WGL, 311–317 (1954).

[7] Instationäre Leewellen. Sc. Proc. Int. Ass. Meteorol., Rome, 260–263 (1954).

[8] Über die Kräfte und Momente, die auf einen Tropfen in scherender Strömung wirken. Z. Flugwiss. **3**, 22–25 (1955).

[9] Das Verhalten der Leewellen in der Stratosphäre. Beitr. Phys. Atmos. **29**, 10–20 (1956).

[10] Ein Charakteristikenverfahren zur angenäherten Berechnung der unsymmetrischen Überschallströmung um mehrere hintereinander angeordnete ringförmige Körper. (Habilitationsschrift). Z. Flugwiss. **4**, 290–300 (1956)

[11] Über den Auftrieb eines Ringflügels bei beschleunigtem und verzögertem Überschallflug. Z. Flugwiss. **4**, 269–272 (1956).

[12] Neue Forschungsergebnisse aus dem Gebiet der atmosphärischen Hinderniswellen. Beitr. Phys. Atmos. **29**, 143–153 (1957).

[13] Kusnezows Theorie der Schrägsicht (gemeinsam mit H. Koschmieder). Z. Flugwiss. **5**, 73–77 (1957).

[14] Auftrieb und Widerstand langer Ringflügel in Überschallströmung. Jahrb. WGL, 83–88 (1957).

[15] Verdichtungsstoß am gekrümmten Profil. ZAMP **9b**, 764–776 (1958).

[16] Der Verdichtungsstoß am gekrümmten Profil. DVL-Ber. **51**, 1–36 (1958)

[17] Eine rotationssymmetrische Zellularkonvektionsströmung. Beitr. Phys. Atmos. **30**, 215–222 (1958).

[18] Über die Bevorzugung der Sechseckzellen bei Konvektionsströmungen über einer gleichmäßig erwärmten Grundfläche. Beitr. Phys. Atmos. **31**, 31–39 (1958).

[19] Das Leewellenproblem der Meteorologie. Naturwiss. 197–200 (1958).

[20] Über rotationssymmetrische Zellularkonvektionsströmungen. ZAMM **38**, 329–333 (1958).

[21] Zur Theorie der Zellularkonvektion III. Beitr. Phys. Atmos. **32**, 23–33 (1959).

[22] Ergänzung zum Aufsatz [15]. ZAMP X, 429 (1959).

[23] The Airflow over Mountains (gemeinsam mit P. Queney, G. A. Cordy, N. Gerbier, H. Koschmieder). World Met. Organ., Genf, Techn. Note **34**, (1960).

[24] Zur Theorie der Zellularkonvektion IV. Über die Bewegungsrichtung in thermokonvektiven Zellularströmungen. Beitr. Phys. Atmos. **32**, 158–160 (1960).

[25] Das Problem des senkrechten Stoßes an einer gekrümmten Wand (gemeinsam mit K. Oswatitsch). ZAMM **40**, T 143–144 (1960).

[26] Thermokonvektive Zellularströmungen bei inkonstanter Erwärmung der Grundfläche. ZAMM **41**, 114–125 (1961).

[27] Die Überschallströmung um einen Flügel mit zwei Unterschallvorderkanten. ZAMM **41**, T 166–169 (1961).

[28] Stationäre, ebene schallnahe Strömungen (gemeinsam mit K. Oswatitsch). DVL-Bericht **189** (1962).

[29] Vorlesungen über theoretische Gasdynamik. **Buch**veröffentlichung Verlag G. Braun, Karlsruhe (1963).

[30] Zur Theorie der Zellularkonvektion V. Zellularkonvektionsströmungen in Gefäßen endlicher horizontaler Ausdehnung. Beitr. Phys. Atmos. **36**, 70–76 (1963).

[31] Schallnahe Strömungen. Beiträge zur parabolischen Methode. ZAMM **43**, T 182–187 (1963).

[32] Beiträge zur Ausbreitung von Kugelwellen (gemeinsam mit J. T. Heynatz). ZAMM **43**, T 137–143 (1963).

[33] Die Integralgleichungsmethode zur Berechnung schallnaher Strömungen. Proceedings des IUTAM-Symposiums über transsonische Strömungen, Sept. 1962, Aachen, Springer-Verlag, 92−109 (1962).

[34] Theoretische Untersuchungen über Zellularkonvektionsströmungen. Arch. Mech. Stos. **16**, 491−505 (1964).

[35] Ein analytisches Verfahren zur Berechnung der nichtlinearen Wellenausbreitung (gemeinsam mit J. T. Heynatz). ZAMM **45**, 37−46 (1965).

[36] Der Äquivalenzsatz und die parabolische Methode für schallnahe Strömungen. ZAMM **45**, 19−27 (1965).

[37] Ähnlichkeitsgesetze für Profilströmungen mit Wärmezufuhr. Acta Mech. **1**, 60−70 (1965).

[38] Profile geringsten Widerstandes bei Schallanströmung (gemeinsam mit K. Burg). Acta Mech. **1**, 91−108 (1965).

[39] Theorie der schallnahen und der Hyperschallströmungen. **Buch**veröffentlichung Verlag G. Braun, Karlsruhe (1966).

[40] Über den Einfluß der Wärmezufuhr bei Hyperschallströmungen. Acta Mech. **2**, 217−230 (1966). On the influence of the addition of heat to hypersonic flow. RAE, LTr No. 1222 (1967). (Übersetzung von [40]).

[41] Similarity laws for flows past aerofoils with heat addition. Royal Aircraft Establishment, Library Translation No. 1114 (1965). (Übersetzung von [37]).

[42] Charakteristikenverfahren zur Berechnung von Hyperschallströmungen. Bericht 7/66 des Deutsch-Französischen Forschungsinstitutes Saint Louis ISL, 469−487 (1966).

[43] Das schwach angestellte Segel bei Überschall-, bei Hyperschall- und bei Schallgeschwindigkeit (gemeinsam mit J. T. Heynatz). Acta Mech. **3**, 278−294 (1967).

[44] Bestimmung des Kondensationsbeginns bei der Expansion feuchter Luft in Überschalldüsen (gemeinsam mit S. Lin). Forsch. Ing. **33**, 169−172 (1967).

[45] Die Ackeret-Formel für Überschallströmungen bei Wärmezufuhr. ZAMM **47**, T 181−182 (1967).

[46] Ein Ähnlichkeitsgesetz für instationäre Kondensationsvorgänge in Lavaldüsen (gemeinsam mit S. Lin). Forsch. Ing. **34**, 97−99 (1968).

[47] Wie kann das Flugzeug der Zukunft aussehen? Einige wissenschaftliche und technische Probleme dieses Projekts. Ehrensenatsvortrag. Hochschulzeitschrift **2**, 22−31 (1968).

[48] Der Kopfwellenabstand bei einem spitzen, schlanken Körper in schallnaher Überschallanströmung. Acta Mech. **5**, 204−208 (1968).

[49] Ein Ähnlichkeitsgesetz für den Kanalwandeinfluß für alle Geschwindigkeitsbereiche. Acta Mech. **6**, 113−114 (1968).

[50] Theoretische und experimentelle Bestimmung des Kopfwellenabstandes bei einem spitzen, schlanken Profil in transsonischer Überschallströmung. Revue Roumaine des Sciences Techniques. Ser. Méc. Appl. **15**, 141−147 (1970).

[51] Die quergestellte Platte bei Schallanströmung. Acta Mech. **9**, 137−141 (1970).

[52] Schallnahe Strömungen mit Wärmezufuhr. Acta Mech. **8**, 126−132 (1969).

[53] Das Stromfeld im Spalt zwischen zwei konzentrischen Kugelflächen, von denen die innere rotiert (gemeinsam mit O. Sawatzki). Acta Mech. **9**, 13−35 (1970).

[54] Die Strömung im Spalt zwischen zwei konzentrischen Kugeln, von denen die innere rotiert (gemeinsam mit O. Sawatzki). ZAMM **50**, T 205−208 (1970).

[55] Theoretische und experimentelle Ergebnisse für die schallnahe Umströmung von schlanken Profilen. Ber. DGLR (1969).

[56] Similarity laws and modeling. **Buch**veröffentlichung Verlag M. Dekker, New York (1971).

[57] Ähnlichkeitsgesetze und Modellregeln der Strömungslehre. **Buch**veröffentlichung Verlag G. Braun, Karlsruhe (1972).

[58] Theorie und Experiment bei schallnahen Strömungen. In: Übersichtsbeiträge zur Gasdynamik (Leiter, E., Zierep, J.). Springer-Verlag, Wien, 117−162 (1971).

[59] Ein Ähnlichkeitsgesetz für Strömungen mit Relaxation. Acta Mech. **11**, 313−318 (1971).

[60] Three dimensional instabilities and vortices between two rotating spheres (gemeinsam mit O. Sawatzki). Proceedings '8th Symposium on Naval Hydrodynamics' Pasadena, 275−288 (1970).

[61] Die transsonische Umströmung der welligen Wand mit Verdichtungsstößen. Revue Roumaine des Sciences Techniques. Ser. Méc. Appl. **17**, 721−729 (1972).

[62] Die Strömung in der Lavaldüse mit mehreren Einschnürungen. DLR-FB 72-27, 416−423 (1972).

[63] Theoretische Gasdynamik 1 − Theorie der Strömungen kompressibler Medien. **Buch**veröffentlichung Verlag G. Braun, Karlsruhe (1972). (Rev. Taschenbuchausg. von [29]).

[64] Theoretische Gasdynamik 2 — Schallnahe und Hyperschallströmungen. **Buch**veröffentlichung Verlag G. Braun, Karlsruhe (1972). (Rev. Taschenbuchausg. von [39]).

[65] Die Überschallströmung längs einer welligen Wand mit gerader Gegenwand (gemeinsam mit K. Burg und S. Viriyabhun). Acta Mech. **16**, 271—278 (1973).

[66] Theory of compressible fluid flow with heat addition. **Buch**veröffentlichung. AGARDographie **191** (1974).

[67] Schallnahe Überschallströmung um rotationssymmetrische Körper (gemeinsam mit W. Frank). Acta Mech. **19**, 277—287 (1974).

[68] Theory and experiments in transonic flows. Rendiconti del Seminario Matematico dell'Universita e Politecnico di Torino, 32, Turin, 145—181 (1975).

[69] Aperiodische instationäre schallnahe Strömungen (gemeinsam mit S. Turbatu). Acta Mech. **21**, 165—169 (1975).

[70] Strömungen mit Energiezufuhr. **Buch**veröffentlichung Verlag G. Braun, Karlsruhe (1975).

[71] Was leistet die Ähnlichkeitsmechanik? Abh. des Aachener Aerodynamischen Institutes **22**, 29—31 (1975).

[72] Schallnahe Überschallströmung um stumpfe Rotationskörper (gemeinsam mit W. Frank). ZAMM **56**, T 178—181 (1976).

[73] Unsteady transonic supersonic flow around suddenly produced bodies (gemeinsam mit W. Frank und G. Patz). Proc. IUTAM Symp. Transsonicum II, 101—108 (1976). Bericht CO 210/75 des ISL.

[74] The normal shock at a curved wall in the viscous case (gemeinsam mit R. Bohning). Proc. IUTAM Symp. Transsonicum II, 236—243 (1976).

[75] Der senkrechte Verdichtungsstoß an der gekrümmten Wand unter Berücksichtigung der Reibung (gemeinsam mit R. Bohning). ZAMP **27**, 225—240 (1976).

[76] Theoretische Gasdynamik. **Buch**veröffentlichung Verlag G. Braun, Karlsruhe (1976).

[77] Zur instationären Lösung der transsonischen Gleichungen. DLR-FB 77-16, 299—302 (1977)

[78] Instationäre, schallnahe Überschallströmung um plötzlich erzeugte Körper (gemeinsam mit W. Frank und G. Patz). Acta Mech. **25**, 273—283 (1977).

[79] Ablösung der turbulenten Grenzschicht an der gekrümmten Wand mit senkrechtem Verdichtungsstoß (gemeinsam mit R. Bohning). ZAMM **58**, T 251—252 (1978).

[80] Separation of turbulent boundary layer at a curved wall with normal shock (gemeinsam mit R. Bohning). Arch. Mech. Stosowanej **30**, 353—358 (1978).

[81] Bedingung für das Einsetzen der Ablösung der turbulenten Grenzschicht an der gekrümmten Wand mit senkrechtem Verdichtungsstoß (gemeinsam mit R. Bohning). ZAMP **29**, 190—198 (1978).

[82] Instabilitäten in Strömungen zäher, wärmeleitender Medien. Ludwig-Prandtl-Gedächtnisvorlesung. ZFW **2**, 143—150 (1978).

[83] Kryteria podobienstwa i zasady modelowania w mechanice plynow. **Buch**veröffentlichung PWN-Verlag (1989). (Polnische Übersetzung von [57]).

[84] Theoretical gasdynamics. **Buch**veröffentlichung Springer-Verlag (1978). (Englische Übersetzung von [76]).

[85] Das Rayleigh-Stokes-Problem für die Ecke. Acta Mech. **34**, 161—165 (1979).

[86] Grundzüge der Strömungslehre. **Buch**veröffentlichung Verlag G. Braun, Karlsruhe (1979).

[87] Neuere Entwicklungen auf dem Gebiet der Radialventilatoren hoher Leistungsdichte (gemeinsam mit H. Leist, H. W. Roth, R. Schilling). HLH **30**, 443—447 (1979).

[88] Theory of flows in compressible media with heat addition. Fluid Dyn. Trans. **10**, 213—240 (1980).

[89] Elementare Betrachtungen über Görtler-Wirbel. ZAMM **61**, T 199—202 (1981).

[90] Axisymmetric and non-axisymmetric convection in a cylindrical container (gemeinsam mit K. R. Kirchartz, U. Müller und H. Oertel jr.). Acta Mech. **40**, 181— 194 (1981).

[91] Normal shock-turbulent boundary layer interaction (gemeinsam mit R. Bohning). AGARD Conf. Proc., 291, 17-1—17-5 (1981).

[92] Lineare Theorie der instationären thermischen Konvektionsströmungen bei überkritischem Temperaturgradienten. Acta Mech. **39**, 271—275 (1981).

[93] Strömungsprobleme bei Energiezufuhr. In: Neue Wege der Mechanik. (Geburtstagsband Schultz-Grunow 75 Jahre), Aachen, VDI-Verlag, 27—30 (1981).

[94] Special solutions of the Navier-Stokes equation in the case of spherical symmetry. Rev. Roum. Sc. **27**, 423—428 (1981).

[95] Analogies between thermal and viscous instabilities. In: Convective transport and instability phenomena. Verlag G. Braun, Karlsruhe, 25—37 (1982).

[96] Strömung zwischen zwei rotierenden Kugeln mit Durchfluß (gemeinsam mit K. Bühler). ZAMM **62**, 198−201 (1982).

[97] Time-dependent convection (gemeinsam mit H. Oertel jr. und K. R. Kirchartz). In: Convective transport and instability phenomena. Verlag G. Braun, Karlsruhe, 101−122 (1982).

[98] The length of the separation bubble in turbulent shock-boundary layer interaction at curved walls. Arch. Mech. Stosowanej **34**, 685−689 (1982).

[99] Moderne Fragen der Gasdynamik. (Geburtstagsband Haack 80 Jahre). Hahn-Meitner-Institut für Kernforschung, Berlin, 95−104 (1982).

[100] Ähnlichkeitsgesetze und Modellregeln der Strömungslehre. **Buch**veröffentlichung Verlag G. Braun, Karlsruhe (1982). (2. überarb. Auflage von [57]).

[101] Stoß-Grenzschichtinterferenz bei turbulenter Strömung an gekrümmten Wänden mit Ablösung (gemeinsam mit R. Bohning). ZFW **6**, 69−74 (1982).

[102] Grundzüge der Strömungslehre. **Buch**veröffentlichung Verlag G. Braun, Karlsruhe (1982). (2. Auflage von [86]).

[103] Laminare Strömung im Innern einer rotierenden Hohlkugel beim Anfahren (gemeinsam mit L. Kullmann). ZAMM **63**, T 302−304 (1983).

[104] What can be achieved by similarity laws. In: Recent Contributions to Fluid Mechanics, Springer-Verlag, 329−339 (1982).

[105] Einige moderne Aspekte der Strömungsmechanik. ZFW **7**, 357−361 (1983).

[106] Viscous potential flow. Arch. Mech. Stosowanej **36**, 127−133 (1984).

[107] Schallnahe Unterschallströmung um Profile ohne Anstellung (gemeinsam mit G. H. Schnerr). ZAMM **65**, T 240−243 (1985).

[108] Das reverse flow theorem bei Schallanströmung (gemeinsam mit G. H. Schnerr). Lecture Notes in Physics **235**, Springer-Verlag, 17−27 (1985).

[109] New secondary instabilities for high Re-number flow between two rotating spheres (gemeinsam mit K. Bühler). In: Laminar-turbulent transition. (V. V. Kozlov). Proceedings IUTAM-Symposium, Novosibirsk 1984, 677−685 (1985).

[110] Calculation of 2D turbulent shock-boundary layer interaction at curved surfaces with suction and blowing (gemeinsam mit R. Bohning). Turbulent Shear Layer/Shock Wave Interactions IUTAM Symposium, Palaisan, Paris 1985 (J. Délery). Springer-Verlag, Berlin−Heidelberg−New York−Tokio, 105−112 (1985).

[111] Flows with taylor vortices: a report on the 4th Taylor Vortex Flow Working Party Meeting (gemeinsam mit K. Bühler, J. E. R. Coney, M. Wimmer). Acta Mech. **62**, 47−61 (1986).

[112] Potential flow of viscous media (Gewidmet Prof. Schultz-Grunow zum 80. Geburtstag). Strömungs-mech. Strömungsmasch. **38**, 1−9 (1986).

[113] Japanische Übersetzung von [100]. **Buch**veröffentlichung Kanazawa-Verlag, Tokyo (1986).

[114] Dynamical instabilities and transition to turbulence in spherical gap flows (gemeinsam mit K. Bühler). In: Advances in Turbulence. European Turbulence Conf. Lyon, July 1986, Springer-Verlag, 16−26 (1987).

[115] Modern aspects of transonic research. Proceedings International Conference on Fluid Mechanics, Beijing, China, July 1987, Peking University Press, 18−27 (1987).

[116] Japanische Übersetzung von [102]. **Buch**veröffentlichung Shisei-bunko Press (1987).

[117] Grundzüge der Strömungslehre. **Buch**veröffentlichung Verlag G. Braun, Karlsruhe (1987). (3. Auflage von [86]).

[118] Japanische Übersetzung von [70]. **Buch**veröffentlichung Shisei-bunko Press (1988).

[119] An experimental investigation of the Reynolds number effect on a normal shock wave turbulent boundary layer interaction on a curved wall (gemeinsam mit P. Doerffer). Acta Mech. **73**, 77−93 (1988).

[120] Strömungsmechanik (gemeinsam mit K. Bühler). **Buch**beitrag in der ‚Hütte'. Springer-Verlag, E 120−188 (1989).

[121] Profile in Überschallquellströmung. ZAMM **69**, T 591−593 (1989).

[122] Airfoils in supersonic source and sink flows (gemeinsam mit G. H. Schnerr). ZFW **13**, 281−290 (1989).

[123] Der gerade anliegende Stoß an der Profilspitze $M_\infty \gtrless 1$ (gemeinsam mit G. H. Schnerr). ZAMM **70**, T 409−412 (1990).

[124] Grundzüge der Strömungslehre. **Buch**veröffentlichung Verlag G. Braun, Karlsruhe (1990). (4. Auflage von [86]).

[125] Förderschwerpunkt Strömungsforschung. In: „25 Jahre KSB-Stiftung", 49−73 (1990).

[126] Strömungen mit Energiezufuhr. **Buch**veröffentlichung Verlag G. Braun, Karlsruhe (1990). (2. Auflage von [70]).

[127] Instationäre Plattenströmung mit Absaugung und Ausblasen (gemeinsam mit K. Bühler). ZAMM **70**, 589 – 590 (1990).

[128] The just attached shock wave at the leading edge of a profile (gemeinsam mit G. H. Schnerr). Archives of Mechanics **42**, 617 – 622 (1990).

[129] Computation of transonic viscous flow over airfoils with control by an elastic membrane and comparison with control by passive ventilation (gemeinsam mit T. Breitling). Acta Mechanica **87**, 23 – 36 (1991).

[130] Strömungsmechanik (gemeinsam mit K. Bühler). **Buch**veröffentlichung, Springer-Lehrbuch, Springer-Verlag (1991).

[131] Ähnlichkeitsgesetze und Modellregeln der Strömungslehre. **Buch**veröffentlichung Verlag G. Braun, Karlsruhe (1991). (3. überarb. Auflage von [57]).

[132] Theoretische Gasdynamik. **Buch**veröffentlichung Verlag G. Braun, Karlsruhe (1991). (4. Auflage von [29] und [39]).

[133] The mathematical theory of thermal choking in nozzle flows (gemeinsam mit C. F. Delale und G. H. Schnerr). ZAMP 44, No. 6, 943 – 976 (1993).

[134] Beschleunigte/verzögerte Platte mit homogenem Ausblasen/Absaugen (gemeinsam mit K. Bühler). ZAMM **73**, T527 – 529 (1993).

[135] Strömungen in einem zähen Medium aufgrund einer Plattenbewegung mit Absaugung oder Ausblasen (gemeinsam mit K. Bühler). In: Turbulente Strömungen in Forschung und Praxis. (A. Leder). Verlag Shaker, 386 – 392 (1993).

[136] A local solution method for shock-boundary-layer interaction on a swept wing (gemeinsam mit C. Nellner). Acta Mechanica **101**, 45 – 57 (1993).

[137] Periodic transonic flow without and with condensation in a 2-D plane channel (gemeinsam mit P. Li). Acta Mechanica **100**, 1 – 12 (1993).

[138] Asymptotic solution of transonic nozzle flows with homogeneous condensation. I: Subcritical flows (gemeinsam mit C. F. Delale und G. H. Schnerr). Physics of Fluids A, Vol. **5**, No. 11, 2969 – 2981 (1993).

[139] Asymptotic solution of transonic nozzle flows with homogeneous condensation. II: Supercritical flows (gemeinsam mit C. F. Delale und G. H. Schnerr). Physics of Fluids A, Vol. **5**, No. 11, 2982 – 2995 (1993).

[140] Condensation phenomena in Laval nozzle (gemeinsam mit G. H. Schnerr). Erscheint in ATLAS OF VISUALIZATION, Progress in Visualization, Vol. **2**. (The Visualization Society of Japan). Pergamon Press Oxford, New York, Seoul, Tokyo (1994).

[141] The singularity in the integral equation of Oswatitsch for transonic flow. (Gewidmet Professor Dr. N. Rudraiah/Indien).

[142] Trends in transonic research. Invited Sectional Lecture. In: Theoretical and Applied Mechanics 1992 (S. R. Bodner, J. Singer, A. Solan, Z. Hashin). IUTAM Conference Haifa 1992. Elsevier Science Publisher, Amsterdam, 423 – 436 (1993).

[143] Thermal convection between two concentric spheres (gemeinsam mit T. Nakagawa, K. Bühler, K.-R. Kirchartz, M. Wimmer). Proc. 2nd JSME-KSME Thermal Engineering Conference, October 19 – 21, 1992, Kitakyushu, Japan, 1-13 – 1-18 (1993).

[144] Turbulent shock-boundary-layer interaction with passive control by normal suction and normal/tangential blowing (gemeinsam mit R. Bohning). In: Problems of Fluid-Flow Machines (Geburtstagsband Prof. Szewalski 90 Jahre), Danzig, Polen, 117 – 129 (1993).

[145] Die Singularität in der Integralgleichung von Oswatitsch für transsonische Strömungen. ZAMM, zur Veröffentlichung eingereicht.

[146] Experimental investigation on Taylor vortices in flow of viscoelastic fluid — Konjaku Aqueous Solution (gemeinsam mit Han Shifang, Wang Yubin und M. Wimmer). Proc. 2nd International Conference on Fluid Mechanics, Beijing, China, July 7 – 10, 1993 (Zhao Dagang, Zhang Zhixin). Peking University Press, Beijing, 86 – 90 (1993).

[147] A stability theory of viscoelastic fluid flow between two coaxial cylinders (gemeinsam mit Han Shifang). Proc. 2nd International Conference on Fluid Mechanics, Beijing, China, July 7 – 10, 1993 (Zhao Dagang, Zhang Zhixin). Peking University Press, Beijing, 91 – 95 (1993).

[148] Turbulent shock-boundary-layer interaction with passive control by normal suction and normal/tangential blowing (gemeinsam mit R. Bohning). Proc. 2nd International Conference on Fluid

Mechanics, Beijing, China, July 7—10, 1993 (Zhao Dagang, Zhang Zhixin). Peking University Press, Beijing, 235—258 (1993).

[149] Shock tube flows with homogeneous condensation (gemeinsam mit C. F. Delale und G. H. Schnerr). Proc. 2nd International Conference on Fluid Mechanics, Beijing, China, July 7—10, 1993 (Zhao Dagang, Zhang Zhixin). Peking University Press, Beijing, 293—298 (1993).

[150] Adiabatic and condensing periodic transonic flows in a 2-D plane channel (gemeinsam mit P. Li und G. H. Schnerr). Proc. 2nd International Conference on Fluid Mechanics, Beijing, China, July 7—10, 1993 (Zhao Dagang, Zhang Zhixin). Peking University Press, Beijing, 484—489 (1993).

[151] Numerical and experimental investigation of passive control of the shock-boundary layer interaction in a transonic compressor cascade (gemeinsam mit G. H. Schnerr, U. Dohrmann und O. Sadi). Proc. 2nd International Conference on Fluid Mechanics, Beijing, China, July 7—10, 1993 (Zhao Dagang, Zhang Zhixin). Peking University Press, Beijing, 504—510 (1993).

[152] Passive control of the shock-boundary layer interaction in a transonic compressor cascade (gemeinsam mit G. H. Schnerr und U. Dohrmann). Proc. 2nd International Symposium on Experimental and Computational Aerothermodynamics of Internal Flow, Prague, July 12—15, 1993 (R. Dvořák, J. Kvapilová). Society of Czech Mathematicians and Physicists, 491—498 (1993).

[153] Grundzüge der Strömungslehre. **Buch**veröffentlichung, Springer-Lehrbuch, Springer-Verlag (1993). (5. überarbeitete Auflage von [86]).

Acknowledgements

On behalf of all editors I gratefully acknowledge the stimulating response and support provided by Professor Dipl.-Ing. Dr. F. Ziegler as one of the editors of *Acta Mechanica*, and Springer-Verlag Wien New York, and especially Mrs. S. Schilgerius for the preparation and realization of this supplement issue. Thanks are due to all authors from 17 countries for their preparation of the 42 original manuscripts and to all referees for evaluating the contributions and for their valuable criticisms.

Finally I would like to thank the publisher for the excellent presentation of this volume and their pleasant cooperation, in making it possible to have this book available at the time of the 65th birthday of Professor Dr.-Ing. Dr. techn. E. h. Jürgen Zierep.

Günter H. Schnerr

Contents

Part 5: Fluid Machinery

Part 6: Computational Fluid Dynamics

Part 7: Miscellaneous Problems

Listed in Current Contents

List of contributors

Asai, M., Professor, Department of Aerospace Engineering, Tokyo Metropolitan Institute of Technology, Hino, Tokyo 191, Japan

Bogdanova-Ryzhova, E. V., Professor, Department of Mathematics, Rensselaer Polytechnic Institute, Troy, NY 12180-3590, U.S.A.

Bohning, R., Professor Dr.-Ing. habil., Institut für Strömungslehre und Strömungsmaschinen, Universität Karlsruhe (TH), D-76128 Karlsruhe, Federal Republic of Germany

Boričić, Z., Professor Dr., T. Roksandića 3a, YU-18000 Niš, Yugoslavia

Borissov, A. A., Professor Dr., Institute of Thermophysics of the Sibirian Branch of the Academy of Sciences, Acad. Kutateladze Street 2, 630090 Novosibirsk, GUS, Russia

Bühler, K., Professor Dr.-Ing. habil., Fachhochschule Offenburg, Fachbereich Maschinenbau, Badstrasse 24, D-77652 Offenburg, Federal Republic of Germany

Bühler, L., Dr.-Ing., Institut für Angewandte Thermo- und Fluiddynamik, Kernforschungszentrum Karlsruhe GmbH, D-76021 Karlsruhe, Federal Republic of Germany

Buschmann, M., Dr.-Ing., Technische Universität Dresden, Rudolf-Renner-Str. 26, D-01159 Dresden, Federal Republic of Germany

Busse, F. H., Professor Dr., Institut für Physik, Universität Bayreuth, D-95447 Bayreuth, Federal Republic of Germany

Candler, G., Professor Dr., Dept. of Aerospace Engineering and Mechanics, University of Minnesota, 110 Union St. S. E., Minneapolis, MN 55455, U.S.A.

Cercignani, C., Professor, Dipartimento di Matematica, Politecnico di Milano, Piazza Leonardo da Vinci 32, I-20133 Milano, Italy

Dallmann, U., Dr., DLR-Institut für Theoretische Strömungsmechanik, Bunsenstrasse 10, D-37073 Göttingen, Federal Republic of Germany

Dargel, G., Dipl.-Ing., Deutsche Aerospace Airbus GmbH, Abt. EF 11, D-28078 Bremen, Federal Republic of Germany

Delale, C. F., Professor Ph. D., Department of Mathematics, Bilkent University, 06533 Bilkent, Ankara, Turkey

Délery, J., Professor Dr., Aerodynamics Department, Office National d'Etudes et de Recherches Aérospatiales ONERA, 29 av. de la Division Leclerc, B.P. 72, F-92322 Chatillon Cedex, France

Dimanlig, A. C. B., College of Engineering, Department of Mechanical, Aeronautical and Materials Engineering, University of California, Davis, California 95616, U.S.A.

Doerffer, P., Priv.doc. Dr.-Ing. habil., Institute of Fluid Flow Machinery, Polish Academy of Sciences, ul. Gen. J. Fiszera 14, PL-80952 Gdansk, Poland

Dohrmann, U., Dr.-Ing., Institut für Strömungslehre und Strömungsmaschinen, Universität Karlsruhe (TH), D-76128 Karlsruhe, Federal Republic of Germany

Dvořák, R., Dr.-Ing. DrSc., Institute of Thermomechanics, Czech Academy of Sciences, Dolejškova 5, CS-18200 Praha 8, Czech Republic

Eliasson, P., Dr., FFA, The Aeronautical Research Institute of Sweden, S-16111 Bromma, Sweden

Felsch, K.-O., Professor Dr.-Ing. Dr. h. c., Institut für Strömungslehre und Strömungs-maschinen, Universität Karlsruhe (TH), D-76128 Karlsruhe, Federal Republic of Germany

Fernholz, H. H., Professor Dr.-Ing., Hermann-Föttinger-Institut für Thermo- und Fluiddyna-mik, Technische Universität Berlin, Sekr. HF 1, Straße des 17. Juni 135, D-10623 Berlin, Federal Republic of Germany

Ferrari, C., em. Professor, Politecnico di Torino, Corso Galileo Ferraris 146, I-10129 Torino, Italy

Fiebig, M., Professor Dr.-Ing., Institut für Thermo- und Fluiddynamik, Ruhr-Universität Bochum, Universitätsstrasse 150, D-44801 Bochum, Federal Republic of Germany

Fiszdon, W., Professor Dr., Institute of Fundamental Technological Research, Polish Academy of Sciences, ul. Swietokrzyska 21, PL 00-049 Warszawa, Poland

Fleberger, G., Dipl.-Ing., Institut für Strömungslehre und Wärmeübertragung, Technische Universität Wien, Wiedner Hauptstrasse 7, A-1040 Wien, Austria

Friedrich, R., Professor Dr.-Ing., Lehrstuhl für Fluidmechanik, Technische Universität München, Arcisstrasse 21, D-80333 München, Federal Republic of Germany

Furumoto, S., Grad. Stud., Graduate School, University of Osaka Prefecture, Sakai, Osaka 593, Japan

Gebing, H., Dipl.-Phys., DLR-Institut für Theoretische Strömungsmechanik, Bunsenstrasse 10, D-37073 Göttingen, Federal Republic of Germany

Gersten, K., Professor Dr.-Ing., Dr.-Ing. E. h., Institut für Thermo- und Fluiddynamik, Ruhr-Universität Bochum, Universitätsstrasse 150, D-44801 Bochum, Federal Republic of Germany

Hackeschmidt, M., Professor Dr.-Ing. habil., Institut für Experimentelle Strömungs- und Wärmetechnik, Technische Universität Dresden, Einsteinstrasse 10a, D-01445 Radebeul, Federal Republic of Germany

Hafez, M., Professor Ph. D., College of Engineering, Department of Mechanical, Aeronautical and Materials Engineering, University of California, Davis, California 95616, U.S.A.

Hannappel, R., Dr. rer. nat., Lehrstuhl für Fluidmechanik, Technische Universität München, Arcisstrasse 21, D-80333 München, Federal Republic of Germany

Hornung, H., Professor Dr., Graduate Aeronautical Laboratories, California Institute of Technology, Pasadena, California 91125, U.S.A.

Keck, H., Dipl.-Ing. Dr. techn., Sulzer-Escher-Wyss AG, CH-8023 Zürich, Switzerland

Kluwick, A., o. Univ. Professor Dipl.-Ing. Dr. techn., Institut für Strömungslehre und Wärmeübertragung, Technische Universität Wien, Wiedner Hauptstrasse 7, A-1040 Wien, Austria

Kost, A., Dr.-Ing., Institut für Thermo- und Fluiddynamik, Ruhr-Universität Bochum, Universitätsstrasse 150, D-44801 Bochum, Federal Republic of Germany

Kovalenko, V. M., Professor Dr. Sc., Corresponding Member Engineering, Academy of the Ukraine, Sumy Institute of Physics and Technology, 2 R-Korsakov St., 244007 Sumy 7, Ukraine

Kuibin, P. A., Institute of Thermophysics of the Sibirian Branch of the Academy of Sciences, Acad. Kutateladze Street 2, 630090 Novosibirsk, GUS, Russia

Leidner, P., Dipl.-Ing., Institut für Strömungslehre und Strömungsmaschinen, Universität Karlsruhe (TH), D-76128 Karlsruhe, Federal Republic of Germany

Li, P., Dr.-Ing., Institut für Strömungslehre und Strömungsmaschinen, Universität Karlsruhe (TH), D-76128 Karlsruhe, Federal Republic of Germany

Lin, S., Professor Dr.-Ing., Department of Mechanical Engineering, Concordia University, 1455 de Maisonneuve Blvd. West, Montreal, Quebec, Canada H3G 1M8

Merzkirch, W., Professor Dr. rer. nat., Lehrstuhl für Strömungslehre, Universität Essen, Schützenbahn 70, D-45127 Essen, Federal Republic of Germany

Mitra, K.-N., Professor Dr., Institut für Thermo- und Fluiddynamik, Ruhr-Universität Bochum, Universitätsstrasse 150, D-44801 Bochum, Federal Republic of Germany

Molton, P., Aerodynamics Department, Office National d'Etudes et de Recherches Aérospatiales ONERA, 29 av. de la Division Leclerc, B.P. 72, F-92322 Chatillon Cedex, France

Mrosewski, T., Dipl.-Ing., Lehrstuhl für Strömungslehre, Universität Essen, Schützenbahn 70, D-45127 Essen, Federal Republic of Germany

Müller, U., Professor Dr.-Ing., Institut für Angewandte Thermo- und Fluiddynamik, Kernforschungszentrum Karlsruhe GmbH, D-76021 Karlsruhe, Federal Republic of Germany

Nikodijević, D., Professor Dr., S. Mladenovića 138/21, YU-18000 Niš, Yugoslavia

Nishioka, M., Professor, Department of Aerospace Engineering, University of Osaka Prefecture, 1-1 Gakuen-Cho, Sakai, Osaka 593, Japan

Noack, B., Dr., Max-Planck-Institut für Strömungsforschung, Bunsenstrasse 10, D-37073 Göttingen, Federal Republic of Germany

Oertel jr., H., Professor Dr.-Ing., Institut für Strömungstechnik, Technische Universität Braunschweig, Bienroder Weg 3, D-38106 Braunschweig, Federal Republic of Germany

Okulov, V. L., Institute of Thermophysics of the Sibirian Branch of the Academy of Sciences, Acad. Kutateladze Street 2, 630090 Novosibirsk, GUS, Russia

Olszok, T., Dipl.-Phys., Max-Planck-Institut für Strömungsforschung, Bunsenstrasse 10, D-37073 Göttingen, Federal Republic of Germany

Orth, A., Dr.-Ing., Institut für Hydromechanik, Universität Karlsruhe (TH), D-76128 Karlsruhe, Federal Republic of Germany

Peters, N., Professor Dr.-Ing., Institut für Technische Mechanik, Rheinisch-Westfälische Technische Hochschule, Templergraben 64, D-52062 Aachen, Federal Republic of Germany

Piechna, J., Dr., Wydz. MEIL, P. W., ul. Mowowiejska 24, PL 00-665 Warszawa, Poland

Rautmann, R., Professor Dr. rer. nat., Universität GHS Paderborn, FB 17, Abt. Mathematik und Informatik, Warburger Strasse 100, D-33098 Paderborn, Federal Republic of Germany

Rizzi, A., Professor Dr., FFA, The Aeronautical Research Institute of Sweden, S-16111 Bromma, Sweden

Rocklage, B., Dipl.-Ing., Institut für Thermo- und Fluiddynamik, Ruhr-Universität Bochum, Universitätsstrasse 150, D-44801 Bochum, Federal Republic of Germany

Rodi, W., Professor Dr., Institut für Hydromechanik, Universität Karlsruhe (TH), D-76128 Karlsruhe, Federal Republic of Germany

Ryzhov, O. S., Professor, Department of Mathematics, Rensselaer Polytechnic Institute, Troy, NY 12180-3590, U.S.A.

Sadi, O., Dipl.-Ing., Institut für Strömungslehre und Strömungsmaschinen, Universität Karlsruhe (TH), D-76128 Karlsruhe, Federal Republic of Germany

Saljnikov, N. V., Professor Dr., Nevesinjska 17, YU-11000 Belgrade, Yugoslavia

Schneider, W., o. Univ. Professor Dipl.-Ing. Dr. techn., Institut für Strömungslehre und Wärmeübertragung, Technische Universität Wien, Wiedner Hauptstrasse 7, A-1040 Wien, Austria

Schnerr, G. H., Professor Dr.-Ing. habil., Institut für Strömungslehre und Strömungsmaschinen, Universität Karlsruhe (TH), D-76128 Karlsruhe, Federal Republic of Germany

Schröder, E., Dipl.-Ing., Institut für Strömungslehre und Strömungsmaschinen, Universität Karlsruhe (TH), D-76128 Karlsruhe, Federal Republic of Germany

Schumann, U., Professor Dr.-Ing. habil., DLR-Institut für Physik der Atmosphäre, D-82234 Oberpfaffenhofen, Post Weßling, Federal Republic of Germany

Sobieczky, H., Professor Dr. habil., DLR-Institut für Theoretische Strömungsmechanik, Bunsenstrasse 10, D-37073 Göttingen, Federal Republic of Germany

Sreenivasan, K. R., Professor Dr., Department of Engineering and Applied Sciences, Mason Laboratory, Yale University, P. O. Box 2159, New Haven, Connecticut 06520, U.S.A.

Stamm, G., Dr., Max-Planck-Institut für Strömungsforschung, Bunsenstrasse 10, D-37073 Göttingen, Federal Republic of Germany

Stolovitzky, G., Dr., Department of Engineering and Applied Sciences, Mason Laboratory, Yale University, P.O. Box 2159, New Haven, Connecticut 06520, U.S.A.

Stütz, W., Dr.-Ing., Ziehl Abegg GmbH & Co. KG, D-74653 Künzelsau, Federal Republic of Germany

Thiede, P., Professor Dr., Deutsche Aerospace Airbus GmbH, Abt. EF 11, D-28078 Bremen, Federal Republic of Germany

Turbatu, S., Professor Dr., University of Bucharest, Str. Bocsa Nr. 3, et. I, ap. 5, Sector 2, Cod 70224, Bucharest, Romania

Wen, Ch., Dr., Graduate Aeronautical Laboratories, California Institute of Technology, Pasadena, California 91125, U.S.A.

Wimmer, M., Dr.-Ing., Institut für Strömungslehre und Strömungsmaschinen, Universität Karlsruhe (TH), D-76128 Karlsruhe, Federal Republic of Germany

Wintrich, H., Dipl.-Ing., Lehrstuhl für Strömungslehre, Universität Essen, Schützenbahn 70, D-45127 Essen, Federal Republic of Germany

Wu, L. Y., Professor, Institute of Fluid Mechanics, Beijing University of Aeronautics and Astronautics, Beijing 100083, People's Republic of China

Wuest, W., Professor Dr. rer. nat., DLR-Institut für Experimentelle Strömungsmechanik, Bunsenstrasse 10, D-37073 Göttingen, Federal Republic of Germany

Xia, Z. X., Ph. D., Institute of Fluid Mechanics, Beijing University of Aeronautics and Astronautics, Beijing 100083, People's Republic of China

Yu, S., Professor, Institute of Engineering Thermophysics, Chinese Academy of Sciences, P. O. Box 2706, Beijing 100080, People's Republic of China

Zhu, Z. Q., Professor, Institute of Fluid Mechanics, Beijing University of Aeronautics and Astronautics, Beijing 100083, People's Republic of China

Acta Mechanica (1994) [Suppl] 4: 1–9

Part 1: Stability Phenomena

Dynamics of flow instabilities in thermal and hydrodynamic systems

K. Bühler, Offenburg, **E. Schröder,** and **M. Wimmer**, Karlsruhe, Federal Republic of Germany

Summary. The stability behaviour and time-dependence of vortex patterns in different physical systems are considered. The classical stability problems of heat transfer in horizontal fluid layers and momentum transfer between rotating cylinders are generalized by the extension from closed flow systems to open flow systems and variations of the geometry. For both cases, for convective and centrifugal instabilities, the superimposed throughflow effects phenomena with respect to the onset of time-dependent motions. The special time-behaviour of the different physical systems is described as well as their similarities and differences.

1 Introduction

Starting points are the classical stability problems of Rayleigh-Bénard and Taylor. Both different physical systems have a lot of common properties [1], which are described in the literature [2]. The present paper considers some generalizations with respect to a superimposed mass flux and the variation of the geometry.

Figure 1 shows the two systems to be studied: the heat transfer in a horizontal fluid layer and the momentum transfer between two concentric rotating cylinders.

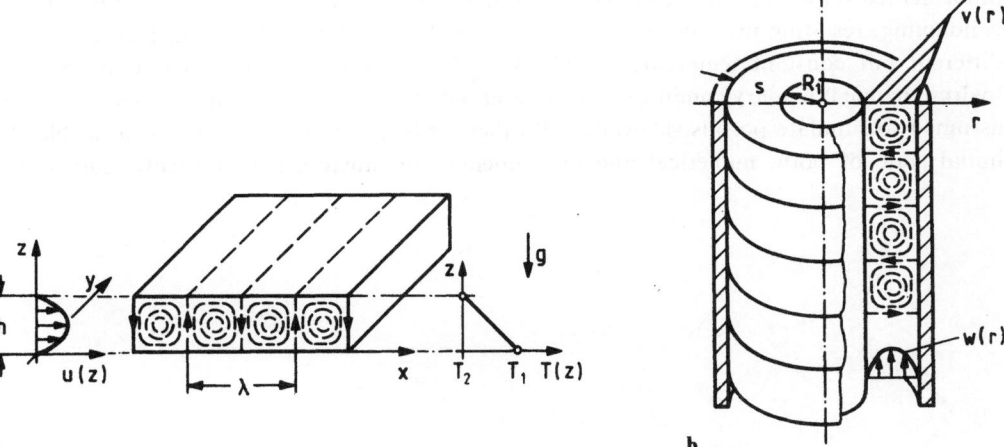

Fig. 1. Sketch of the thermal and hydrodynamic systems with massflux. **a** Rayleigh-Bénard problem
$\mathrm{Ra} = \dfrac{\alpha \cdot g \cdot \Delta T \cdot h^3}{v \cdot a}$; **b** Taylor-Problem $\mathrm{Ta}^2 = \dfrac{R_1 \cdot \omega^2 \cdot s^3}{v^2}$

In both cases we obtain disturbances of the basic state, displaying the form of defined, well organized cells. In the case of the Rayleigh-Bénard problem the state of heat conduction changes into a convective motion. The circular Couette flow between concentric cyclinders, with the inner one rotating and the outer one at rest, changes into a flow with toroidal vortices. The onset of instabilities is found by the linear stability theory at the same value for the characteristic critical parameters $Ra_c = Ta_c^2 = 1\,708$, provided the non-slip boundary condition is introduced in both cases.

By superimposing an axial mass flux the closed flows are extended to open flow systems. The influence of the axial flow on the stability and the time-behaviour of supercritical flow patterns is of special interest. Furthermore, we change the geometry from circular to conical cylinders to study the Taylor vortex motion in a gap between two cones, showing a rich variety of possible flow patterns [3].

2   Three-dimensional thermal convection with throughflow

The classical Rayleigh-Bénard convection examines a fluid layer between two infinite horizontal plates. For more technical relevance the problem is studied with vertical sidewalls. The investigation of thermal convection in boxes is much more difficult, than that of infinite fluid layers, since the variables of the Boussinesq equations can no longer be separated [4]. Therefore, it is necessary to use numerical methods for the calculation of the three-dimensional convection in boxes. In rectangular boxes the convection rolls are orientated parallel to the shorter side, called transversal rolls [5].

A further extension of the Rayleigh-Bénard problem is the mixed convection, i.e. a combination of the thermally driven convective motion and a superimposed horizontal shear flow. Our investigation concerns the three-dimensional investigations of mixed convection in a box with aspect ratios $l_y/h = 4.1$ and $l_x/h = 7.4$. The measuring techniques used for the experiments are the whole-field and the laser differentialinterferometer to visualize the flow and to obtain the dynamical behaviour, respectively.

For the numerical simulations the MAC method (Marker and Cell Method) [6] is implemented on a staggered grid. The vertical boundaries are considered to be ideally heat conducting, resulting in a linear temperature profile. The top and bottom plates are set at different but constant temperatures. The non-slip condition is employed to describe the hydrodynamic boundary conditions. As in- and outlet conditions a constant velocity profile is assumed to simulate porous sidewalls. All other walls are regarded to be impermeable. The initial state of both, numerical and experimental investigations, is a steady, supercritical

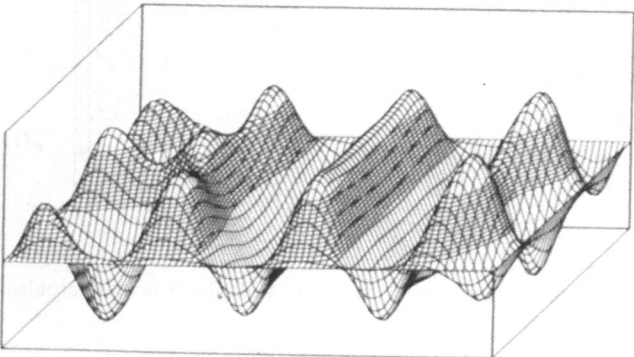

Fig. 2. Vertical velocity in the midplane, $Ra = 4\,700$, $Pr = 530$, $Re = 7.5 \cdot 10^{-5}$

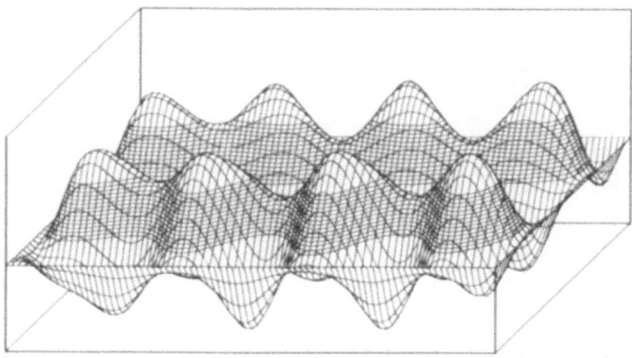

Fig. 3. Vertical velocity in the mid-plane, $Ra = 4700$, $Pr = 530$, $Re = 3 \cdot 10^{-3}$

convection state without throughflow. It should be mentioned that the rolls are deformed near the sidewalls and get a three-dimensional character. After the onset of the superimposed throughflow, perpendicular to the axes of the rolls, three different types of convection can be distinguished:

1. If the throughflow, characterised by the Reynolds number Re, does not exceed a certain value, the transversal rolls are displaced downstream and deformed (Fig. 2). At the inlet, the rolls change their orientation and become longitudinal. This situation remaines unchanged even during thirty hours.
2. If the Reynolds number is increased, the rolls start to move downstream. In the first half of the box, the rolls become longitudinal, in the second half they are still transversal, moving downstream with a constant velocity. The two parts of rolls with different orientations are separated by a mixed flow region, where new transversal rolls are formed which move towards the outlet and disappear.
3. If the throughflow is strong enough, the whole box is filled with longitudinal rolls. Near the sidewalls the longitudinal rolls are deformed and get a transversal character (Fig. 3).

The following pictures show a comparison between numerical and experimental results of the oscillation frequency of the vertical velocity component.

In Fig. 4 the y-axis represents the oscillation frequency, the x-axis shows the Reynolds number. In the considered range of parameters the frequency is a linear function of the Reynolds number, so that the velocity of the travelling rolls grows with increasing strength of the throughflow. The curve does not reach the x-axis at $Re = 0$, but at a higher Reynolds number, proving that below this value a steady state can be found (convection form 1).

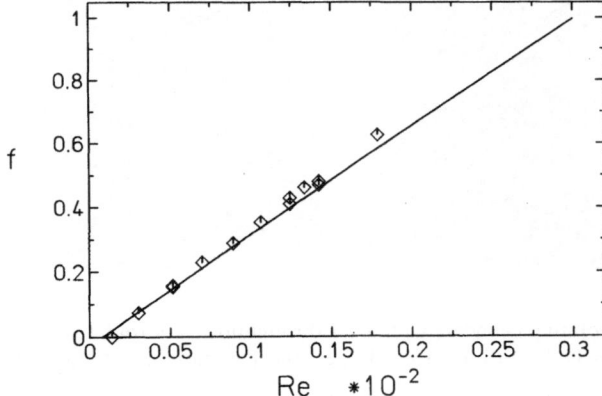

Fig. 4. Dependence of the oscillation frequency on the Re-number, comparison between numerical results (straight line) and experiments (rhomb), $Ra = 4700$, $Pr = 530$

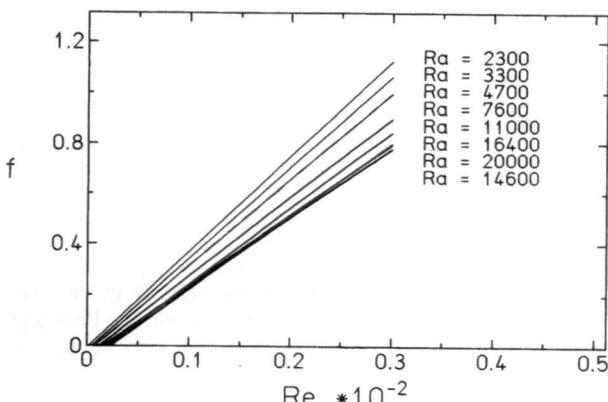

Fig. 5. Numerical results for different Rayleigh numbers and Pr = 530

Figure 5 shows the decreasing effect of the throughflow on the oscillation frequency at higher Rayleigh numbers with the result, that high Rayleigh numbers have a stabilizing effect on the transversal cell patterns.

3 Cylindrical gap flow with a superimposed axial mass flux

The superposition of an axial mass flux on the flow between two concentric rotating cylinders exhibits interesting phenomena with respect to the pattern formation and the dynamical behaviour of the vortices.

Experimentally observed vortex configurations are shown in Fig. 6a, which displays rotationally symmetric ring vortices travelling in the axial direction of the throughflow. The size of the vortices depends on their sense of rotation. At higher throughflow rates helical vortices are formed with different, but discrete inclinations. The configuration with the smallest inclination

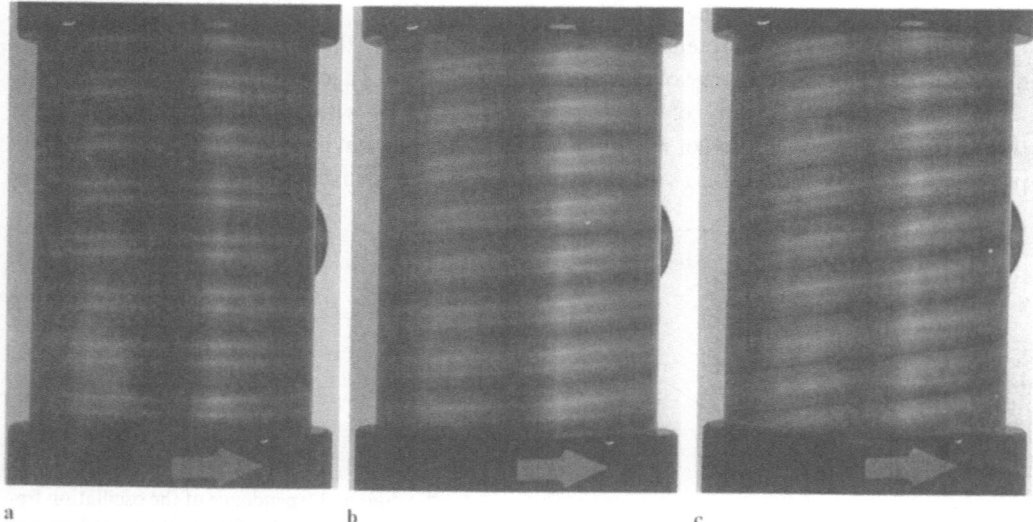

Fig. 6. Taylor vortices between concentric rotating cylinders with axial throughflow: **a** toroidal vortices Ta = 67.6, $Re_D = 1.3$; **b** helical vortices n = 1, Ta = 61.5, $Re_D = 3.5$; **c** helical vortices n = 2, Ta = 66, $Re_D = 8$

value $n = 1$ is shown in Fig. 6b. This inclination is proportional to one wavelength over the circumferential length. The next higher inclination with $n = 2$ can be seen in Fig. 6c.

The discrete inclination of the helical vortices leads to the following time-behaviour. Toroidal vortices with $n = 0$ can only move in the direction of the throughflow, while helical vortices can travel *in* the direction of the throughflow or in *opposite* direction. A steady state is possible for special combinations of the Taylor number and the throughflow Reynolds number. A similar behaviour is also true for Taylor vortices in spherical gaps with superimposed throughflow [7].

The regimes of existence of these flow configurations are given in the stability diagram in Fig. 7. The slightly increasing critical Taylor numbers display the stabilizing effect of the throughflow on the onset of Taylor vortices with increasing Re_D.

The three solid lines through the origin of the coordinates are crossing the regions of steady states with $n = 1$, $n = 2$ and $n = 3$ as inclination values. These lines correspond to constant Ta/Re_D-ratios and confirm that steady states occur, when the vector of the mean axial and circumferential velocity is parallel to the inclination of the helical vortices. Outside the vortices are travelling. The direction of the movement in the direction of the throughflow or opposite to it is marked by the arrows in Fig. 7.

The time-behaviour of these flows can be explained by the linear stability theory [8]. The axial velocity w_c of the travelling vortices is proportional to the eigenvalue $-\gamma$, which depends on the throughflow Reynolds number and on the inclination values n, shown in Fig. 8. It follows that the toroidal ring vortices with $n = 0$ have a positive axial velocity, $(w_c \sim -\gamma \sim Re_D)$.

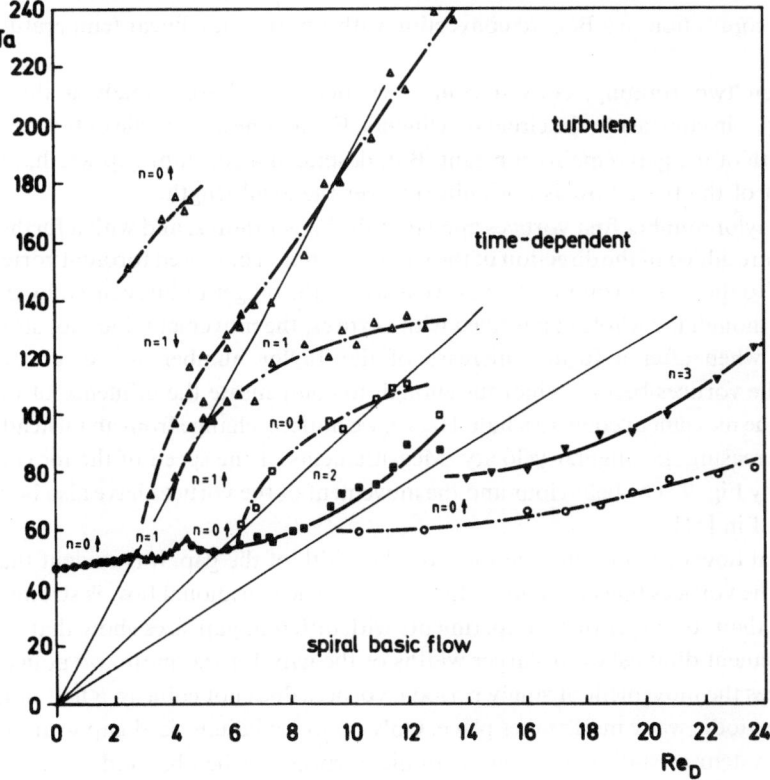

Fig. 7. Stability diagram for helical flows with regimes of existence, gap width $\sigma = s/R_1 = 0.25$, $Re_D = \dfrac{W \cdot s}{\nu}$

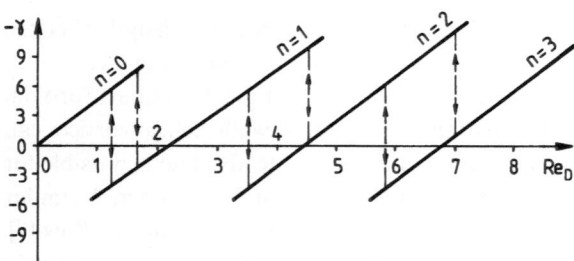

Fig. 8. Coefficient $-\gamma$ for the axial velocity of the patterns as function of Re_D and n, possible transitions are marked by arrows

Starting with steady helical vortices $n = 1$ and a corresponding Re_{D1} we find with increasing $\mathrm{Re}_D > \mathrm{Re}_{D1}$ helical vortices travelling in the direction of throughflow with $w_c > 0$. If we decrease the throughflow $\mathrm{Re}_D < \mathrm{Re}_{D1}$, then the helical vortices travel in opposite direction of the throughflow with $w_c < 0$. The same behaviour is possible for all helical vortices with other inclinations and is in qualitative agreement with experiments.

Transitions between the different modes are possible and marked by the arrows in Fig. 8. The values of Re_D, at which the transition occurs, must be determined by a non-linear stability theory.

4   Taylor vortex flow between conical cylinders

Moving closed vortex cells can also be detected in a closed system in a gap between conical cylinders. For rotating cones one obtains a linear distribution of the centrifugal forces. Here, too, we have again the connection to a Bénard convection with a horizontal, linear temperature gradient [9, 10].

In the gap between two rotating, coaxial conical cylinders we have already a three-dimensional basic flow — in contrast to the circular cylinders. For a same apex angle of the inner and outer cone the pidth of the gap remains constant. But, despite of a constant gap width, the meridional component of the basic flow is not uniform over the axial length.

By increasing the Taylor number first vortices appear at the bigger radius, and with a further increase more vortices are added in the direction of the smaller radius. The closed toroidal vortex cells are perpendicular to the axis of rotation and move towards the bigger radius caused by the meridional flow. Even though the whole gap is filled with vortices, the movement does not stop, but diminishes. Only when after a further increase of the Taylor number — i.e. greater centrifugal forces — the vortices become vigorous enough to compensate the influence of the meridional flow, does the movement come to a halt. Hence, we notice a change from an unsteady to a steady flow by increasing the angular velocity. Measurements of the speed of the moving vortices are displayed by Fig. 9. The behaviour and the movement of the vortices have also been calculated and reported in [11].

Since the meridional flow depends among others on the width of the gap, it is evident that also the movement of the vortices must depend on the gap size. The meridional flow is stronger for smaller gap widths than for larger ones. Experiments with different gap sizes show that the magnitude of the movement diminishes for larger widths of the gap. Furthermore, we noticed that for smaller gap sizes the movement of singly periodic vortices does not come to a halt until a change to a doubly periodic wavy mode takes place. Only up to a medium sized gap width of $\sigma = s/R_{\mathrm{max}} = 0.16$, a system of stationary singly periodic vortices can be observed.

If the strength of the meridional flow is no longer sufficient to force the vortices to move, then the influence of the meridional flow is reduced to a vortex cell itself. This results in a deformation

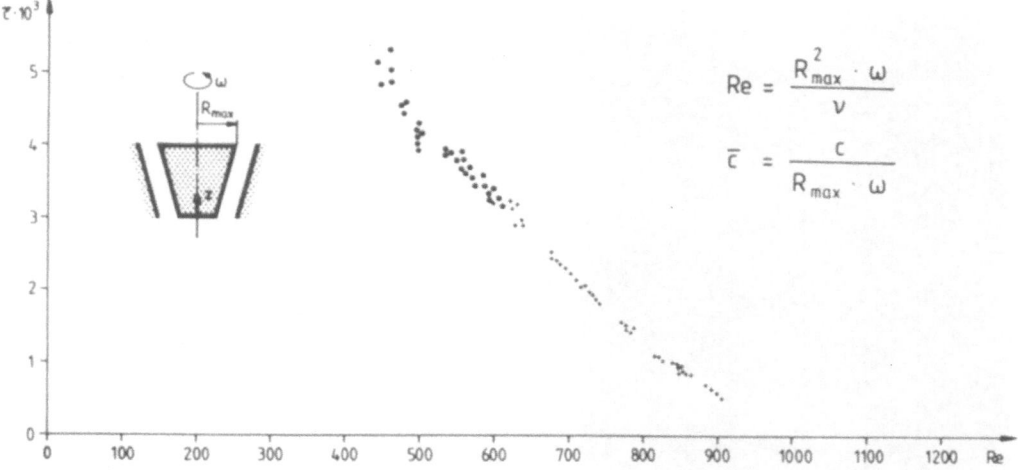

Fig. 9. Speed of travelling vortices as function of the Reynolds number, $\sigma = 0.25$

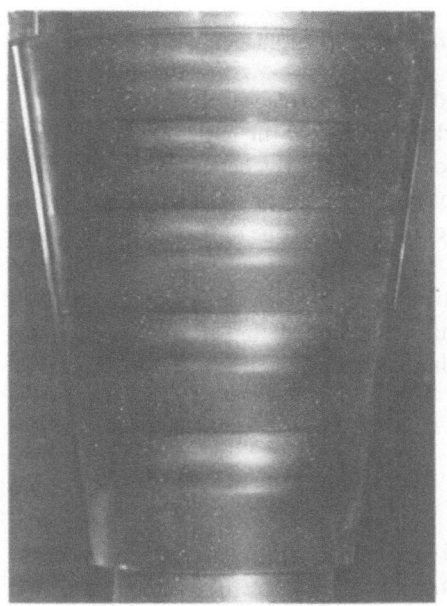

Fig. 10. Rotationally symmetric steady vortices of different sizes, $\sigma = 0.25$

of single cells. Thus, we obtain e.g. larger vortex cells if they rotate *in* the direction of the meridional flow and smaller ones for those rotating *opposite* to it, resulting in alternating big and small neighbouring vortices as it is displayed in Fig. 10.

The described steady state with stationary vortices is however not uniform, but a function of initial conditions. Depending on the rate of acceleration of the inner body, one obtains in the same gap more or less vortices with a corresponding different wavelength. There is again an analogue to the Bénard convection, where different cell sizes are possible for different heating rates.

Another interesting unsteady flow exists between conical cylinders. It is generated by a disturbance asymmetric to the axis of rotation. The vortices take now the form of a helix, coiled around the rotating cone. The whole arrangement moves to the smaller radius only depending on the angular velocity. Again as a function of initial conditions we may obtain a case that both

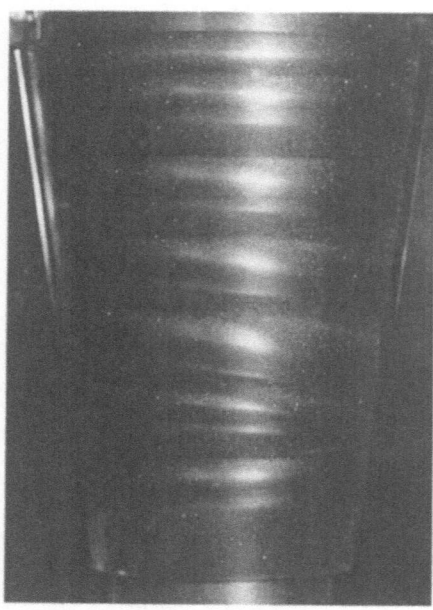

Fig. 11. Steady and time-dependent flow regimes, $\sigma = 0.25$

modes of stationary, toroidal vortex rings and unsteady helical vortices exist side by side. This can happen in a way that the stationary part is located at one end of the gap and the unsteady helix at the other end; or — again for another rate of acceleration — the helical unsteady part is imprisoned between the two stationary perpendicular ones, as it is demonstrated by Fig. 11. The simultaneous existence of steady and unsteady flows in conical gaps is one of the rare examples in fluid mechanics.

Similar, but still more complex, flow patterns can be obtained by a consequent variation of the geometry, as it was described in [12]. The combination of cylinders and cones demonstrates the dominant influence of the gap width and illuminates the rich variety of possible flow patterns.

5 Conclusions

The generalization of the original Rayleigh-Bénard and Taylor problem with respect to a superimposed throughflow and a change of geometry leads to interesting behaviours. It is remarkable that in all investigated cases a direct transition from the steady basic state into a time-dependent supercritical flow state can be observed. That means that the "principle of exchange of stability" is violated in these cases.

An onset of steady and unsteady instabilities are observed for the different physical systems. The dynamical behaviour of the moving vortex patterns is, however, different for the three investigated cases. The time-dependent motion of the different cells can be determined by their characteristic parameters, showing the influence of the geometry and boundary conditions.

References

[1] Zierep, J.: Instabilitäten in Strömungen zäher, wärmeleitender Medien. ZFW **2**, 143−150 (1978).
[2] Bühler, K., Kirchartz, K. R., Wimmer, M.: Strömungsmechanische Instabilitäten. Stroemungsmech. Stroemungsmasch. **40**, 99−126 (1989).

[3] Wimmer, M.: Viscous flows and instabilities near rotating bodies. Prog. Aerospace Sci. **25**, 43—103 (1988).

[4] Zierep, J.: Zur Theorie der Zellularkonvektionsströmungen in Gefäßen endlicher horizontaler Ausdehnung. Beitrag. Phys. Atmos. **36**, 70—76 (1963).

[5] Stork, K., Müller, U.: Convection in boxes: experiments. J. Fluid Mech. **54**, 599—611 (1972).

[6] Kirchartz, K. R.: Dreidimensionale Konvektion in quaderförmigen Behältern. Habilitationsschrift, Karlsruhe 1988.

[7] Bühler, K.: Strömungsmechanische Instabilitäten zäher Medien im Kugelspalt. Fortschritt-Berichte VDI, Reihe 7, **96**, Düsseldorf 1985.

[8] Bühler, K., Polifke, N.: Dynamical behaviour of Taylor vortices with superimposed axial flow. Nonlinear evolution of spatio-temporal structures in dissipative continuous systems (Busse, F. H., Kramer, L., eds.), pp. 21—29. New York: Plenum Press 1990.

[9] Srulijes, J.: Zellularkonvektion in Behältern mit horizontalen Temperaturgradienten. Dissertation, Karlsruhe 1979.

[10] Wimmer, M.: Die viskose Strömung zwischen rotierenden Kegelflächen. ZAMM **63**, T299—T301 (1983).

[11] Abboud, M.: Ein Beitrag zur Untersuchung von Taylor Wirbeln in Spalt zwischen Zylinder/Kegel-Konfigurationen. ZAMM **68**, T275—T277 (1988).

[12] Wimmer, M.: Wirbelbehaftete Strömungen im Spalt zwischen Zylinder-Kegel-Kombinationen. Stroemungsmech. Stroemungsmasch. **44**, 59—83 (1992).

Authors' addresses: Professor Dr.-Ing. habil. K. Bühler, Fachhochschule Offenburg, Fachbereich Maschinenbau, Badstrasse 24, D-77652 Offenburg; Dipl.-Ing. E. Schröder and Dr.-Ing. M. Wimmer, Institut für Strömungslehre und Strömungsmaschinen, Universität Karlsruhe (TH), D-76128 Karlsruhe, Federal Republic of Germany

Acta Mechanica (1994) [Suppl] 4: 11−17

Spoke pattern convection

F. H. Busse, Bayreuth, Federal Republic of Germany

Summary. An introduction to spoke pattern convection in a horizontal fluid layer heated from below is given. Spoke pattern convection is an ubiquitous form of turbulent convection in fluids with moderate or large Prandtl numbers. A large scale square structure can often be recognized within which the small scale fluctuating velocity field is organized. Experimental measurements and theoretical considerations are presented and possibilities for future analysis are outlined.

1 Introduction

The term spoke pattern convection was introduced by Busse and Whitehead [1] to describe the time dependent form of convection that is observed in fluid layers heated from below when the Rayleigh number exceeds a value of the order 3×10^4 for Prandtl numbers above the order unity. The thermal boundary layers at the rigid upper and lower boundaries of the fluid layer are typically unstable and the erupting sheets of hot and cold fluid tend to move radially inward towards central plumes which carry fluid to the opposite boundary. In visualizations of convection with the shadowgraph method the radial spokes together with the plumes represent a characteristic feature of this type of convection. Of course, the appearance of spoke pattern convection changes with varying Rayleigh and Prandtl numbers. The spokes tend to be nearly stationary at low Rayleigh numbers close to the onset of this type of convection. But at high Rayleigh numbers of the order 10^5 or 10^6 spokes fluctuate strongly in time and show a chaotic appearance. Inspite of the small scale turbulence the convection pattern retains a large scale nearly steady structure in which rising and descending plumes are arranged at well defined distances. The wavelength of these spoke pattern cells varies considerably with the Prandtl number. In this paper some measurements of the wavelength are presented and various features of spoke pattern convection are discussed.

2 Phenomenological description

Some of the characteristic features of spoke pattern convection, namely the eruption of thermal blobs form the boundary layer and their transport in the form of plumes to the opposite boundary, can already be seen in steady knot convection which arises as an instability of two-dimensional convection rolls [2] and exhibits the typical knot like structure along the rising and descending fluid sheets of the rolls as shown in Fig. 1. For details on the computations on which the plots are based we refer to [3]. Because of the acceleration of motion in the buoyant plumes, the upper thermal boundary layer becomes rather thin at the stagnation point at which the plume impinges onto the boundary. This process can also be seen in the left pictures of Fig. 1 at the place where the cold plume meets the hot boundary. Because of the symmetric boundary conditions the solution exhibits a symmetry between hot and cold plumes.

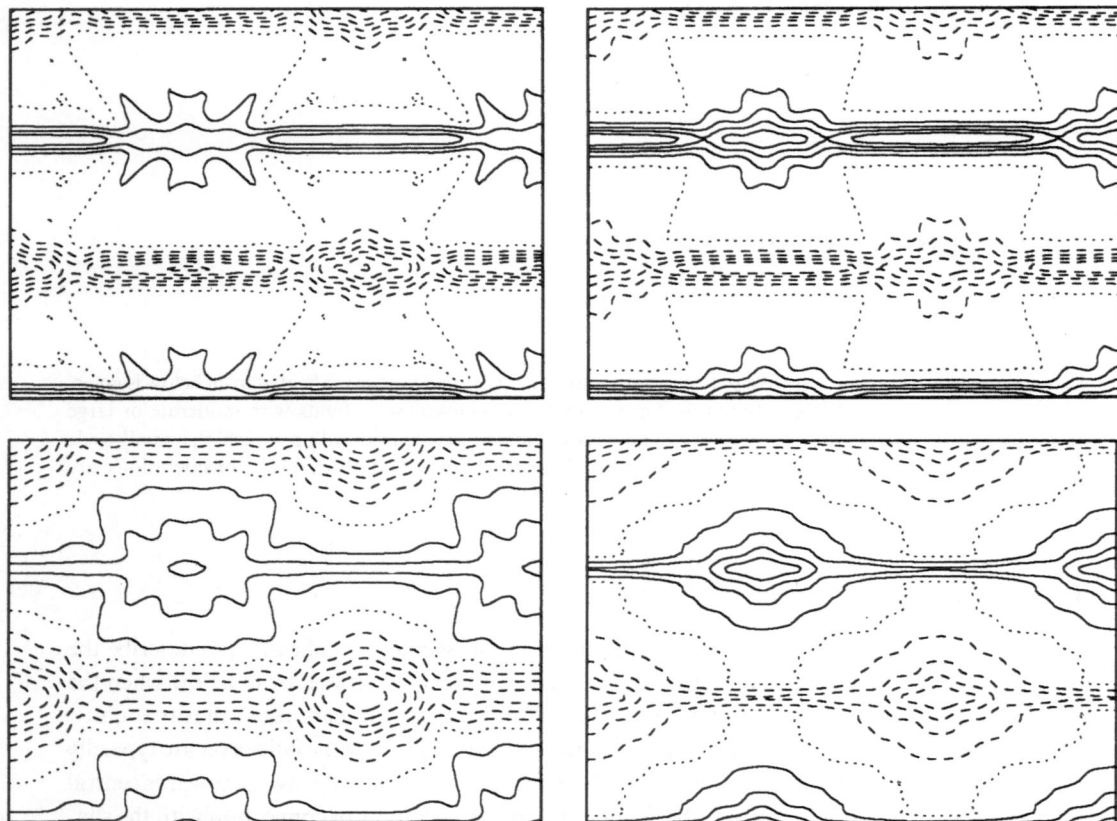

Fig. 1. Lines of constant vertical velocity in the horizontal planes $z = -.3$ (upper left) and $z = 0$ (upper right), lines of constant $\partial(\theta - \bar{\theta})/\partial z$ at $z = -.5$ (lower left), and isotherms in the midplane, $z = 0$, (lower right) are shown for steady knot convection for $R = 4 \cdot 10^4$, $P = 2.5$ with wavenumbers $\alpha_x = 1.1$ and $\alpha_y = 1.5$. Solid (dashed) lines indicate positive (negative) values and the dotted lines correspond to zero. The layer thickness d is used as length scale such that the rigid boundaries are located at $z = \pm .5$ $\bar{\theta}$ indicates the horizontal average of the deviation θ of the temperature from the static solution of the problem

From observations [2] and computations [3] it is known that steady knot convection is stable only in a small region of the parameter space and that typically the blobs emerging from the thermal boundary layers become time dependent. In the form of oscillatory knot convection [3] the large scale structure in the form of plumes and sheets remains steady, while blobs of hot and cold fluid circulate in time periodic fashion. Oscillatory knot convection can also be reached through a transition from steady bimodal cells as shown in Fig. 6 of [1]. Because of the spoke-like structures arranged around the plumes, this type of convection was called spoke pattern convection in [1]. But after the term knot convection has been introduced for the pattern in which the original roll structure is still apparent it seems appropriate to reserve the term spoke pattern to the polygonal pattern in which all memory of the original roll structure has disappeared. In a similar vein the collective instability described in [1] can be identified with the knot instability even though the former occurs as the instability of bimodal convection instead of convection rolls.

The property that the original roll structure loses importance, once the plumes become a dominant feature, is demonstrated by the fact that knot convection with a square like pattern can easily be generated in numerical simulations. The steady solution shown in Fig. 2 exhibits the

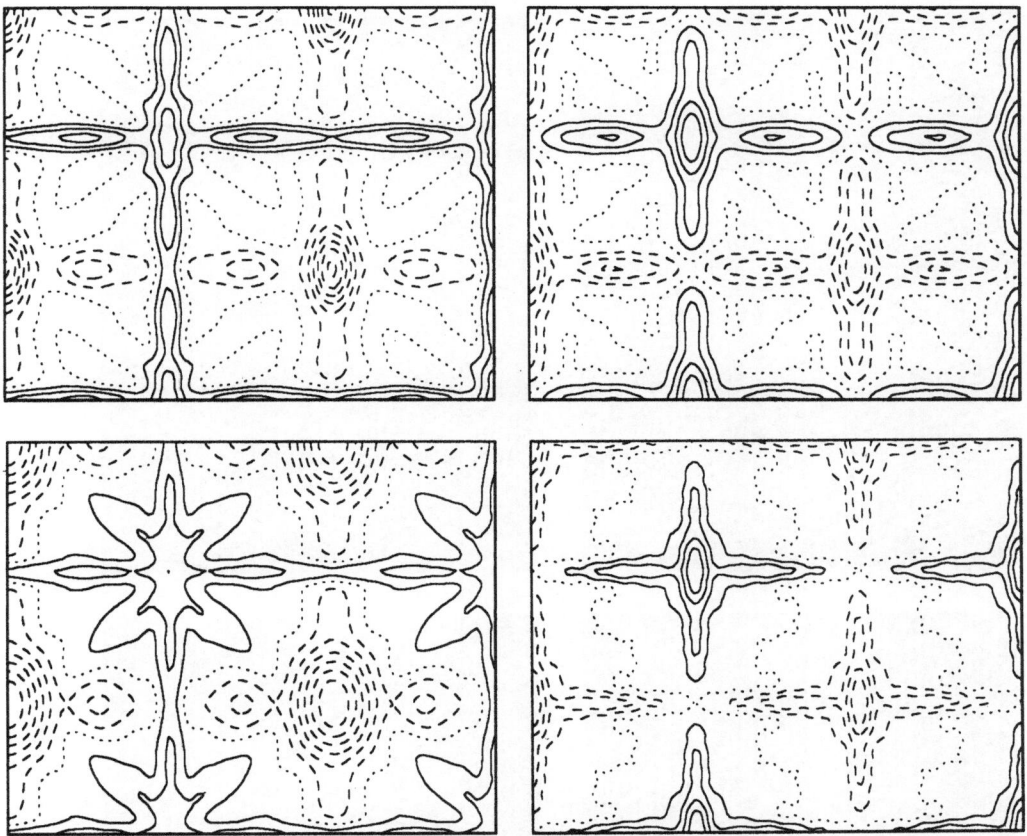

Fig. 2. Same as Fig. 1, but for square type convection with $R = 6 \cdot 10^4$, $P = 7$, $\alpha_x = 1.6$, $\alpha_y = 2.0$

typical combination of sheet and plume like rising and descending motions, but without a clear preference for a single horizontal direction.

Experimental observations of spoke pattern convection are shown in Fig. 3. Two cases with two pictures each are shown in this figure. The shadowgraph method is used for visualization, and dark (white) lines thus indicate rising (descending) motion. Details on the experimental techniques can be found in [1]. The two photographs of each case were taken a minute or two apart in order to demonstrate that the large scale cells are stationary while the small scale motions are strongly fluctuating. In the case of the lower Rayleigh number, cells can be distinguished bounded by black or white lines which seems to be typical for a Prandtl number of the order 7. In the high Rayleigh number case the locations of hot and cold plumes are not clearly evident and are only identified as the maxima in the density of black and white lines, respectively. But, when the pattern is seen in a movie, the plume locations become clearly evident as sinks of the motions of black white streamers. In this way highly turbulent spoke pattern convection may be observed up to Rayleigh numbers of at least $2 \cdot 10^6$.

3 Prandtl number dependence

While the degree of turbulence increases strongly with the Rayleigh number R the large scale wavelength of spoke pattern convection changes relatively little with R. But there is a strong influence of the Prandtl number P on the large scale wavelength as is evident from the four

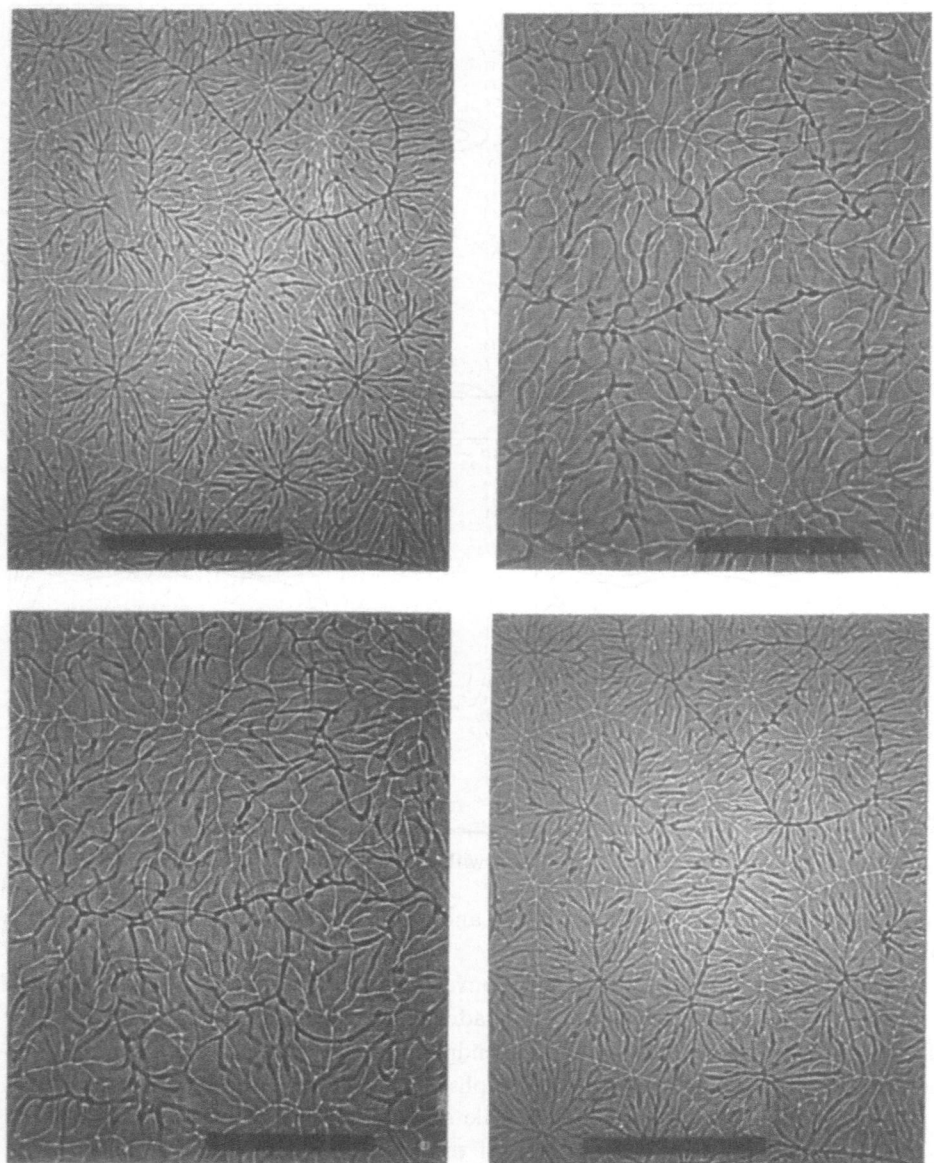

Fig. 3. Shadowgraph pictures of spoke pattern convection in a convection layer of depth d heated from below. The upper pictures were taken one minute apart in the case of $d = 0.84$ cm, $R \cong 1.6 \cdot 10^5$, $P = 7$ with methyl alcohol as fluid. The lower pictures were taken 2 min apart in the case of a water layer with $d = 2$ cm, $R \cong 7 \cdot 10^5$, $P = 5.7$. The black bar indicates a length of 10 cm

photographs of Fig. 4 which show shadowgraph pictures for four different convection fluids. The spoke pattern cells often exhibit the approximate symmetry of a square lattice in which upward and downward plumes alternate. Only in the case of the first panel of Fig. 4 the symmetry between upward and downward motion is broken locally as in hexagonal convection cells. But this asymmetry does not imply an asymmetry of the fluid layer since both types of cells appear with equal probability. Assuming a square lattice we identify the wavelength of the pattern with the mean distance between spoke centers of the same sign. By measuring the mean distance of

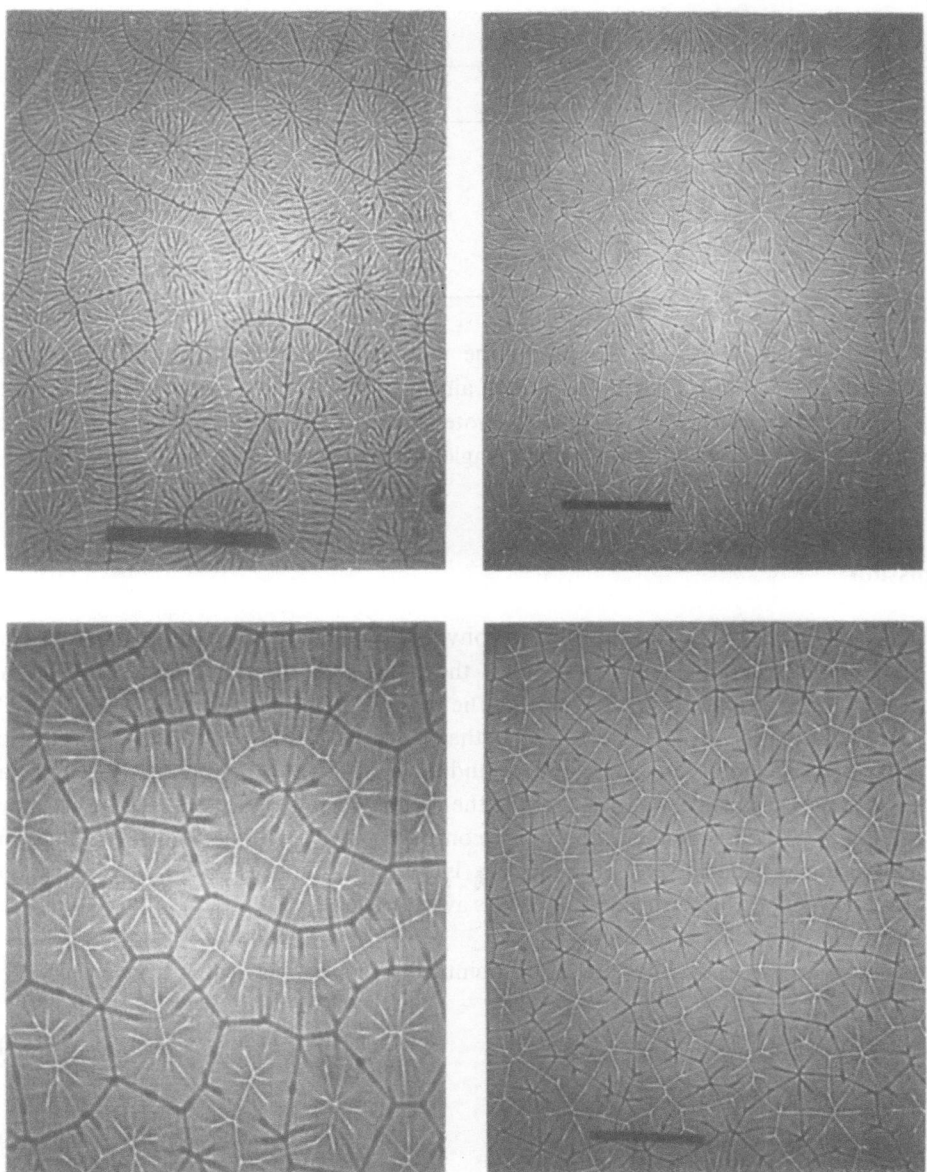

Fig. 4. Shadowgraph pictures of spoke pattern convection in the cases $d = 0.56$ cm, $R \cong 5 \cdot 10^4$, $P = 7$ with methyl alcohol as fluid (upper left), $d = 2$ cm, $R \cong 8 \cdot 10^5$, $P = 48$ with 3.23 cSt. silicone oil (upper right), $d = 1.3$ cm, $R \cong 9 \cdot 10^4$, $P = 64$ with 4.4 cSt. silicone oil (lower left) and $d = 2$ cm, $R \cong 2.3 \cdot 10^5$, $P = 170$ with 15 cSt. silicone oil (lower right). The length of the black bar is 10 cm

spoke centers from photographs of shadowgraph pictures we have evaluated quantitatively the variation of the wavelength with Prandtl number. Although the Rayleigh number usually was varied by at least a factor 10 no significant variation was found. Hence the results shown in Table 1 represent typical results for values of R of a few times 10^5. Although the table exhibits a strong variation of the wavelength of spoke pattern convection, little variation occurs outside the Prandtl number range that has been investigated. At high Prandtl numbers the wavelength tends to stay around 3 which is less than 50% larger than the critical wavelength for the onset of

Table 1. Large scale wavelength λ of spoke pattern convection
(the silicone oils are Dow Corning® 200 fluids)

P	Fluid	Depth d	λ in units of d
5.7	water	2 cm	10.6
7	methyl-alcohol	0.84 cm	9.2
48	3.23 cSt. silicone oil	2 cm	6.5
64	4.4 cSt. silicone oil	1.3 cm	5.4
170	15 cSt. silicone oil	2 cm	3.1

convection. Towards lower Prandtl numbers the wavelength does not vary much either. Numerical simulations of turbulent convection in air [4] and for $P = 6$ [5] show rather similar scales of the spoke patterns, although it should be noted that these simulations correspond to the internally heated case and thus involve only a single thermal boundary layer.

4 Discussion

The bottleneck of the heat transport in turbulent convection with rigid boundaries is the transfer of fluid from the thermal boundary layers into the interior of the fluid layer. For steady convection rolls this transfer is so ineffective that the Nusselt number grows only with the $R^{1/5}$ power [7]. Knot convection is more effective [2] in that plumes are generated which enhance the asymmetry between the flow from the boundary and the flow towards the boundary. Through the concentrated positive or negative buoyancy in the plumes the fluid is accelerated towards the opposite boundary where its momentum causes a compression of the thermal layer and thus an effective heat transport. Since only a small area is required for the heat transfer between boundary and impinging plume, most of the area is available for the formation of the spoke-like eruptions form the thermal boundary layer.

The normal velocity w near the stagnation point of the impinging plume is given by

$$w = W\zeta 2/d \quad \text{for} \quad \zeta \gg \left(\frac{vd}{W}\right)^{1/2} \tag{1.1}$$

$$w = 2\alpha_2\zeta^2 \left(\frac{W^3}{d^3 v}\right)^{1/2} \quad \text{for} \quad \zeta \ll \left(\frac{vd}{W}\right)^{1/2} \tag{1.2}$$

where ζ is the distance from the boundary, v is the kinematic viscosity and d is the layer thickness. W denotes a characteristic value of the vertical velocity which is assumed in the middle of the layer according to expression (1.1). The coefficient α_2 assumes a value of 0.66 at an axisymmetric stagation point according to Homann [6]. Since for $P > 1$ the thermal boundary layer is thinner than the viscous one, the heat equation

$$\varkappa \frac{\partial^2}{\partial \zeta^2} T = w \frac{\partial}{\partial \zeta} T \tag{2}$$

can be solved by

$$\frac{\partial}{\partial \zeta} T = \frac{\Delta T_p}{\eta \Gamma(1/3)} 3 \exp\left\{-(\zeta/\eta)^3\right\} \quad \text{with} \quad \eta \equiv \left(\frac{dv}{W}\right)^{1/2} (2P\alpha_2/3)^{-1/3} \tag{3}$$

where ΔT_p denotes the temperature difference between boundary and plume. Because the typical convection velocity W is given by $W \cong f(R) \, \varkappa/d$ where $f(R)$ is independent of the Prandtl number P for $P \gtrsim 1$ [8, 9], expression (3) indicates a growth of the temperature gradient in proportion to $P^{1/6}$. This growth is limited, of course, by the fact that the assumption (1) of a viscous boundary layer requires that the Reynolds number, Wd/v, is large compared with unity. Since the function $f(R)$ is approximately given by $\gamma(R - R_c)^{1/2}$, where γ is a numerical factor of the order 0.1, we conclude that the momentum effect and thus the tendency towards a large scale spoke pattern convection occurs when the inequalities

$$\gamma(R - R_c)^{1/2} \, P^{-1} \gg 1, P \gtrsim 1 \tag{4}$$

are satisfied. A more detailed boundary layer theory for a quantitative explanation of the results of Table 1 becomes rather complicated. But, as recent advances have demonstrated [4, 5], numerical simulations of spoke pattern convection are feasible and it is likely that a more detailed understanding of its physical properties as a function of Rayleigh and Prandtl numbers will soon become possible.

Acknowledgements

The author is grateful to Dr. R. M. Clever for his help in producing Figs. 1 and 2. The research reported in the paper has been supported in part by the Atmospheric Sciences Section of the U. S. National Science Foundation.

References

[1] Busse, F. H., Whitehead, J. A.: Oscillatory and collective instabilities in large Prandtl number convection. J. Fluid. Mech. **66**, 67–79 (1974).

[2] Clever, R. M., Busse, F. H.: Three-dimensional knot convection in a layer heated from below. J. Fluid Mech. **198**, 345–363 (1989).

[3] Busse, F. H., Clever, R. M.: Instabilities of convection rolls in a fluid of moderate Prandtl number. J. Fluid Mech. **91**, 319–335 (1979).

[4] Schmidt, H., Schumann, U.: Coherent structure of the convective boundary layer derived from large-eddy simulations. J. Fluid Mech. **200**, 511–562 (1989).

[5] Grötzbach, G.: Direct numerical and large eddy simulation of turbulent channel flows. In: Encyclopeadia of Fluid Mechanics, vol. 6 (Cheremisinoff, N. P., ed.), pp. 1337–1391. Houston: Gulf Publ. Comp. 1987.

[6] Homann, F.: Der Einfluß großer Zähigkeit bei der Strömung um den Zylinder und um die Kugel. ZAMM **16**, 153–164 (1936).

[7] Roberts, G. O.: Fast viscous Bénard convection. Geophys. Astrophys. Fluid Dyn. **12**, 235–272 (1979).

[8] Malkus, W. V. R.: The heat transport and spectrum of thermal turbulence. Proc. Roy. Soc. London Ser. A **225**, 196–212 (1954).

Author's address: Prof. Dr. F. H. Busse, Institute of Physics, University of Bayreuth, D-95447 Bayreuth, Federal Republic of Germany

Acta Mechanica (1994) [Suppl] 4: 19 — 26

Digital particle image velocimetry applied to a natural convective flow

W. Merzkirch, T. Mrosewski, and **H. Wintrich,** Essen, Federal Republic of Germany

Summary. The method of digital particle image velocimetry (DPIV) is applied to the study of a time-dependent natural convective flow. The particle image patterns are recorded with a CCD camera and stored on video tape. The velocity distributions are obtained by cross-correlating the information of sucessive video frames. The experiments verify the possibility of taking "cinematographic" velocity measurements of an unsteady flow with a relatively long characteristic time scale.

1 Introduction

During the past few years there has been a rapidly growing interest in using optical whole-field methods for planar flow measurements. One of the novel methods, particle image velocimetry (PIV), relies on the use of tracer particles whose motion in a plane of the flow field is recorded with a camera (e.g. Adrian [1]; Merzkirch [3]). In the majority of the applications, the particle motion is recorded on photographic film and, after development, the photograph is interrogated, point by point, with a thin laser beam, that produces an optical signal in the form of interference fringes as a measure of the plane velocity vector at the interrogation point. Inconvenient for the experimentation is the photographic process that interrupts the measuring procedure. More advantageous is the direct recording of the particle motion with a CCD camera and the subsequent evaluation of the electronically recorded image pattern with an image processing procedure. This may provide the means for an on-line measurement of the flow velocity in the field under study.

Willert and Gharib [7] verified that direct digital recording and subsequent computer-based evaluation of the velocity distribution is possible for certain flow situations ("digital particle image velocimetry" = DPIV). The tracer particle patterns recorded on successive video frames are compared with one another, and the displacements of the particle images during the time interval separating the video frames is determined from a cross-correlation of two patterns. The limited spatial resolution of CCD recordings and the available video frequency restrict the application to low-speed flows in liquids where relatively large tracer particles may be used.

In this paper we describe the use of DPIV for measuring the velocity distribution of the natural convective flow developing in a tank with heated side walls. One aim of the experiments was to explore the usefulness of the method for recording a time-dependent flow whose characteristic time scale is such that it can be resolved with the given video frequency. Therefore, the heating of the walls was controlled in a manner that the wall temperature gradually raised from the ambient temperature to approach a constant value, about 15 °C above ambient temperature, after more than one hour. This way an unsteady flow pattern was produced that was viewed and recorded with a CCD camera. In contrast to Willert and Gharib [7], who stored the whole information of an experiment on a real time disk in digital form, we used an analog recording of the frames on video tape. It is known that this analog storage on tape may cause problems in linearity of gray levels and geometrical distortions. But we believed that, for the present experimental conditions, this disadvantage was compensated by the relatively low costs of recording an experiment for which a long measuring time is required.

In the following we describe the experimental arrangement, the evaluation procedure, and we demonstrate the practicability of the system with a number of examples.

2 Experimental procedure

The experimental set-up is shown in Fig. 1. The test tank of size $40 \times 40 \times 40$ mm^3 is filled with water. Two opposite side walls can be heated by means of Peltier elements that are in close contact with plane copper plates in order to provide a uniform heat transfer from the wall to the fluid. Several Ni — CrNi thermoelements serve to measure the wall temperature. Their reading is used to control the heating rate with a PC such that the wall temperature follows a prescribed pattern with time.

Two high quality optical glass windows form the second pair of side walls and allow the flow to be viewed by the CCD camera. The water in the tank is seeded with tracer particles ("Polyamid 12") whose density $\rho_p = 1\,012$ kg/m^3 is only slightly different from the density of water. The particles have a mean diameter of 25 µm, and the seeding is 0.07 g per liter. Illumination of the tracer particles is provided in form of a plane light sheet, parallel to the viewing windows and generated by expanding the beam from a 35 mW He — Ne laser with a cylindrical lens. The visible particle patterns are recorded by the CCD camera on a format of 756×581 pixels and stored on a video tape with the given video frequency of 25 frames per second; i.e., the time intervall between successive frames is 40 ms.

The signal from the video tape is given to a frame grabber of a 80 386 personal computer. The frame grabber can store the information of 24 video frames in digital form with a resolution of 256 gray values per pixel. The further processing of the digitized particle image patterns by applying the cross-correlation algorithm (see next Section) is performed with the PC. A monitor allows to observe on-line the motion of the tracer particles in the flow.

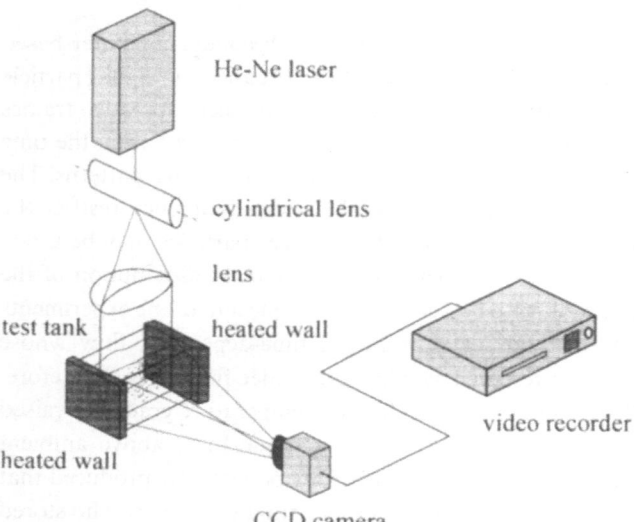

He-Ne laser

cylindrical lens

lens

test tank   heated wall

heated wall

video recorder

CCD camera   **Fig. 1.** Experimental set-up

3   Evaluation by cross-correlation

The displacement that the tracer particles experience within the time interval $\Delta t = 40$ ms (or multiples of Δt) is found from the patterns recorded in successive video frames. The velocity follows by dividing the displacement by Δt. The displacement (or the velocity) is determined as a vector in the plane of the figure at discrete positions of the field of view. This evaluation consists of cross-correlating the patterns of small sections ("windows") of digitized successive frames. A description of applying this cross-correlation technique to the measurements of tracer particle velocities has been given by Utami et al. [6] and Willert and Gharib [7].

A window of size 32×32 pixels is used for the evaluation procedure. The particle image pattern included in the window at a specific position of frame no. 1 taken at time t is cross-correlated with the pattern contained in the window at the same position in frame no. 2 taken at $t + \Delta t$ (Fig. 2). The information of the two windows (i.e., the intensity distributions stored in the 32×32 pixels) is Fourier transformed into the spatial frequency domain, and the conjugated complex product is formed. The two-dimensional (plane) correlation function $\phi(\Delta x, \Delta y)$ is obtained by a Fourier transformation back into the space domain. The position of the maximum of the correlation function, $(\Delta x, \Delta y)_{max}$, determines the most frequent displacement experienced by the particle image pattern of the window within the time interval Δt (Fig. 3). The position of the maximum can be determined with sub-pixel accuracy.

The accuracy with which the maximum of the correlation function can be identified depends on the degree of uniformity of the particle velocity in the area covered by the test

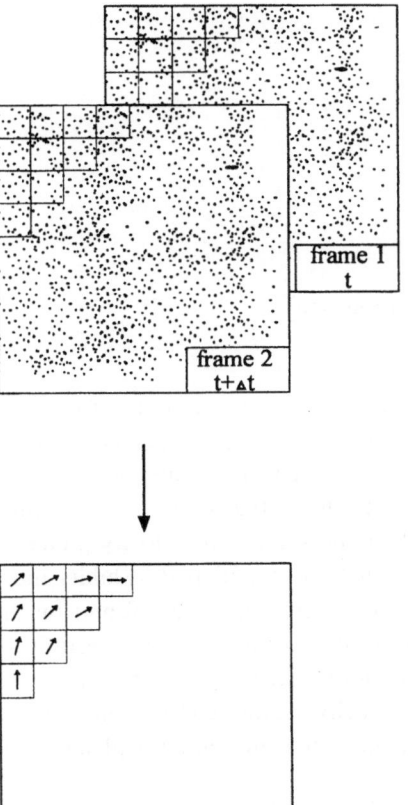

frame 1
t

frame 2
t+$_\Delta$t

Fig. 2. Scheme of scanning two successive video frames with the test window and determination of the velocity vector averaged across the window

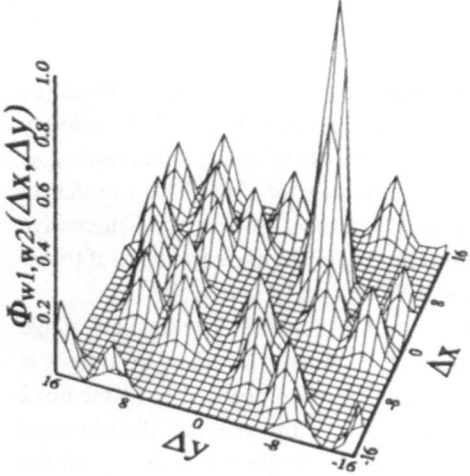

Fig. 3. Example of a two-dimensional (plane) correlation function with a clear maximum as a measure of the displacement vector $(\Delta t, \Delta y)$

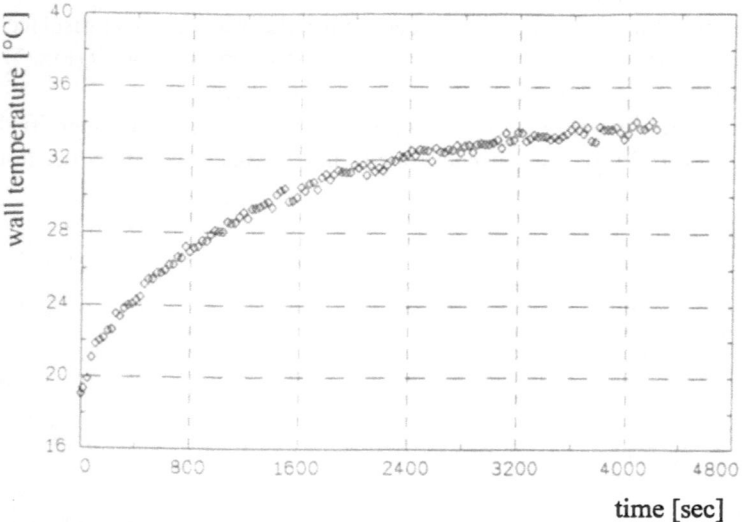

Fig. 4. Change of wall temperature with time during heating-up

window. Keane and Adrian [2] have developed criteria to determine the validity of the used approximation by which a locally constant velocity is assumed in the measurement volume. The present experimental conditions are such that "weak" velocity gradients, according to Keane and Adrian's terminology, occur in the flow, and it follows that flow non-uniformity can be neglected with good accuracy. An indication of the occurrence of velocity gradients in the measurement volume (interrogation window) is that the function ϕ exhibits more than one peak. We use a specific criterion for deciding whether a maximum found by the search algorithm can be accepted as a valid result. According to this criterion the ratio of the amplitudes of the first (highest) and second peak must exceed a certain value. The displacement determined in this way or the respective velocity vector (u,v) is then assigned to the central position of the i-th window (x_i, y_i), in the measurement plane (or plane of the video frame).

The test window is moved step by step across the digitized particle image patterns of two successive frames. This scanning of the two frames is shown in Fig. 2 without any overlap of the windows. For a practical evaluation, however, we use a 50% or 75% overlap in order to smooth the data. With 50% overlap we obtain 981 data points (velocity vectors) per frame, and 3721 vectors can be obtained with 75% overlap. The whole evaluation procedure is automated. The procedure includes the presentation of the results in form of vector diagrams. Details of both the data processing and the experimental set-up have been described by Mrosewski [4].

4 Results

The heating of the two side walls by the Peltier elements is chosen in a manner that the wall temperature approaches a constant value of approximately 35 °C more than one hour after the heating had started at the room temperature of 20 °C (Fig. 4). A number of results of the velocity

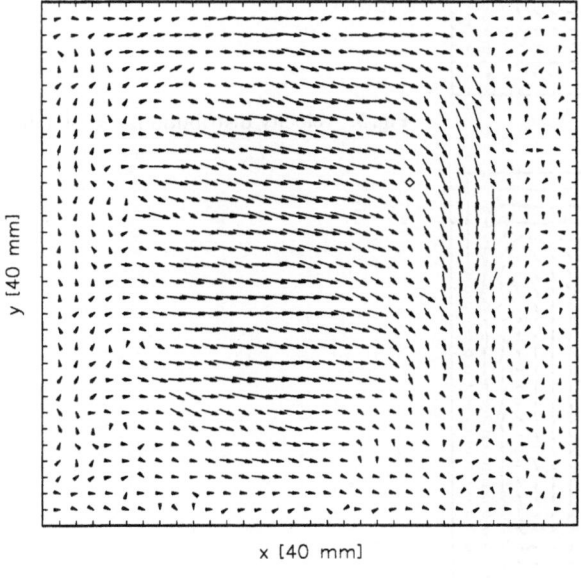

Fig. 5. Velocity distribution at time $t = 5$ min after start of heating

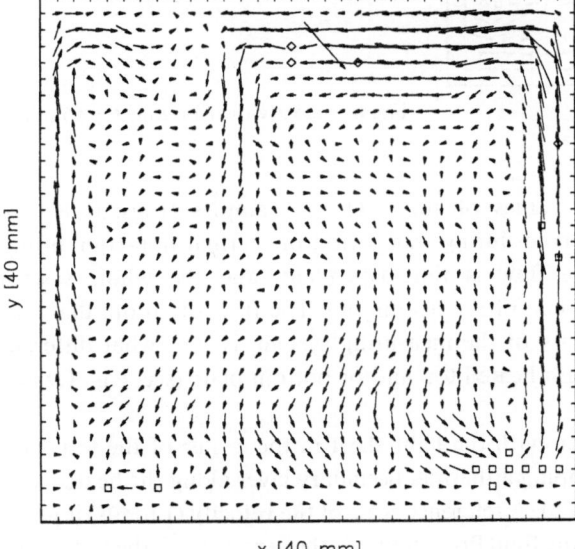

Fig. 6. Velocity distribution, $t = 15$ min

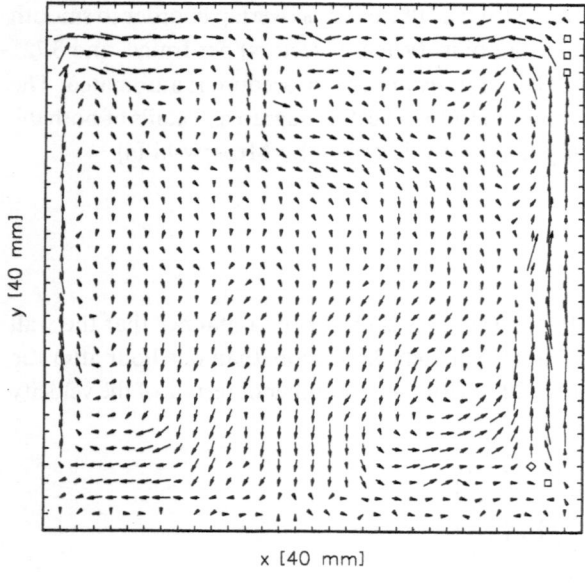

Fig. 7. Velocity distribution, $t = 30$ min

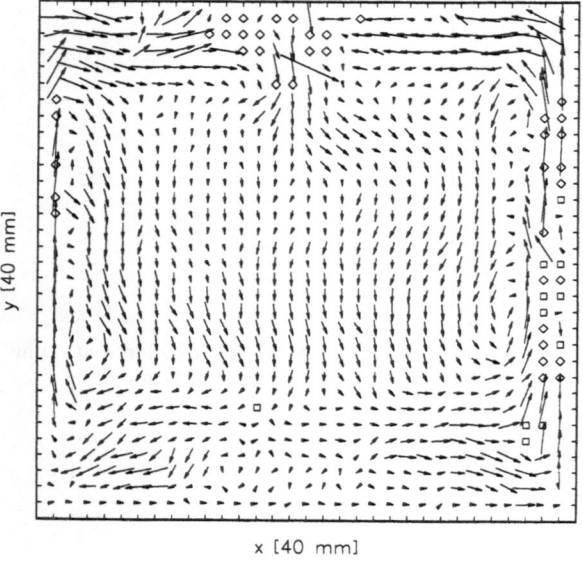

Fig. 8. Velocity distribution, $t = 55$ min

measurements are shown for different values of time. These examples (Figs. 5–8) are representative for the flow patterns observed during this period of thermal non-equilibrium in the fluid. The developing convective flow is time-dependent, and the whole system approaches an equilibrium state near the end of the heating-up. The flow situation, that was primarily chosen to demonstrate the ability of the measuring system, is somewhat different from the more frequently studied cases with heated top and/or bottom plate (see, e.g., Ozawa et al. [5]).

The velocity vector diagrams are taken with the light sheet in the center of the test tank (Fig. 1). The convection starts with developing a thermal boundary layer along the heated side walls and upward flow in these layers (Fig. 5). A reference scale of the velocity (1 mm/s) is given in the figures. Near the upper free surface the fluid flows towards the center. Near the bottom of the tank the fluid flows from the center to the walls where it is entrained into the thermal boundary layer. The thickness of the boundary layers increases with time, and a reverse flow system develops in the tank with strong downflow near the center (Fig. 6). The flow system is not symmetric; vortex type rolls appear in two corners of the square tank (Figs. 6, 7). From the length of the velocity vectors it is apparent that the mean velocity magnitude increases with time due to the increasing temperature differences. A comparison of Figs. 5, 6, 7, and 8 shows that the flow is unsteady, but with a relatively long characteristic time scale (as it was desired; see Section 1). Even after 55 min of heating (Fig. 8) the flow is far from approaching a steady, symmetric equilibrium state, that will consist of two symmetrical, counter-rotating cells.

5   Discussion

The experiments verify that DPIV is a powerful tool for measuring planar distributions of the velocity in certain time-dependent flows. The method avoids the photographic recording of particle image patterns; instead, these patterns are recorded with a CCD camera so that an online measurement becomes possible. The used experimental arrangement differs from the system described by Willert and Gharib [7] by the use of a commercial video tape for storing the optical information. An advantage of this system is the possibility of storing, at relatively low cost, the experimental information recorded during a long measurement time ($\sim$ 60 minutes). Furthermore, it is possible to select the time interval Δt between frames that are cross-correlated in multiples of $\Delta t_0 = 40$ ms (video norm) if this is appropriate for resolving the very low velocities as experienced in the present flow situation; i.e., the value of Δt can be adapted to the velocity range *after* the experiment. The present technology of electronic recording systems restricts the application of this "cinematographic" method for measuring the flow velocity to particular flow situations. The main resctrictions are:

— The characteristic time scale of the unsteady phenomena must be long enough for being resolved with the given video frequency.
— The velocities to be measured must be in the range of only a few mm/s, due to the relatively long time interval $\Delta t = 40$ ms between successive video frames.
— Relatively large tracer particles ($\sim$ 25 μm diameter) must be tolerated because of the limited spatial resolution (number of pixels) of CCD cameras.

However, it must be expected that, with further progress of the video and CCD camera technology, this DPIV method will become more attractive for general use in flow velocimetry.

References

[1] Adrian, R. J.: Particle imaging techniques for experimental fluid mechanics. Ann. Rev. Fluid Mech. **23**, 261−304 (1991).
[2] Keane, R. D., Adrian, R. J.: Optimization of particle image velocimeters. Part I: Double pulsed systems. Meas. Sci. Technol. **1**, 1202−1215 (1990).
[3] Merzkirch, W.: Laser-Speckle-Velocimetrie. In: Lasermethoden in der Strömungsmeßtechnik (Ruck, B. ed.), pp. 71−97. Stuttgart: AT-Verlag 1990.

[4] Mrosewski, T.: Anwendung der Digital Particle Image Velocimetry auf eine instationäre Konvektions-strömung. Diplomarbeit, Lehrstuhl für Strömungslehre, Universität Essen 1992.

[5] Ozawa, M., Müller, U., Kimura, J., Takamori, T.: Flow and temperature measurement of natural convection in a Hele-Shaw cell using a thermo-sensitive liquid-crystal tracer. Exp. Fluids 12, 213−222 (1992).

[6] Utami, T., Blackwelder, R. F., Ueno, T.: A cross-correlation technique for velocity field extraction from particulate visualization. Exp. Fluids 10, 213−223 (1991).

[7] Willert, C. E., Gharib, M.: Digital particle image velocimetry. Exp. Fluids 10, 181−193 (1991).

Authors' address: Prof. Dr. rer. nat. W. Merzkirch, Dipl.-Ing. T. Mrosewski, and Dipl.-Ing. H. Wintrich, Lehrstuhl für Strömungslehre, Universität Essen, D-45127 Essen, Federal Republic of Germany

Acta Mechanica (1994) [Suppl] 4: 27 – 37

Part 2: Transition, Turbulence and Separation

On singular solutions of the incompressible boundary-layer equation including a point of vanishing skin friction

E. V. Bogdanova-Ryzhova and **O. S. Ryzhov**, Troy, New York

Summary. Three different types of singularities of the Prandtl equation in the vicinity of a point where the skin friction becomes zero are under discussion. According to the weakest type singularity the skin friction varies as the square root of the cube of the local distance. The next type involves a sudden change in the skin friction derivative with respect to coordinate along the rigid surface. It is superseded by the famous Landau-Goldstein singularity with the skin friction being proportional to the square root of the distance. Then, a still more complicated flow pattern may be composed of the singularity with the sudden change in the skin friction derivative followed by the Landau-Goldstein singularity at some small distance downstream.

1 Limit regime

To perceive physical background behind the mathematical problem to be posed below let us consider the incompressible boundary layer on a thin airfoil at incidence. The skin friction distribution over the body contour displays a local minimum which remains positive until the angle of attack is below some limit value. In the limit regime the minimum of the skin friction first becomes zero. The further increase of incidence leads to the boundary-layer separation giving rise to a small bubble with reverse flow.

To describe the phenomenon we employ the classical Prandtl equation

$$\frac{\partial \psi}{\partial y} \frac{\partial^2 \psi}{\partial x \, \partial y} - \frac{\partial \psi}{\partial x} \frac{\partial^2 \psi}{\partial y^2} = -\frac{dp}{dx} + \frac{\partial^3 \psi}{\partial y^3} \tag{1}$$

for the stream-function ψ depending on orthogonal curvelinear coordinates x, y of the boundary layer. The pressure gradient on the right-hand side of (1) is supposed to be given by the Taylor series

$$\frac{dp}{dx} = c_0 + c_1 x + c_2 x^2 + \dots \tag{2}$$

On the body surface the no-slip conditions

$$\psi = \frac{\partial \psi}{\partial y} = 0 \quad \text{at} \quad y = 0 \tag{3}$$

take place. The limit regime of flow is fixed, according to our basic assumption, by

$$\frac{\partial^2 \psi}{\partial y^2} = \frac{\partial^3 \psi}{\partial x \, \partial y^2} = 0, \quad \frac{\partial^4 \psi}{\partial x^2 \, \partial y^2} > 0 \quad \text{at} \quad x = y = 0 \tag{4}$$

Thus, the velocity field includes no singularities at the point $x = 0$ of the vanishing skin friction $\tau_w = \partial^2 \psi(x, 0)/\partial y^2$. This regime plays a key role in understanding of how the incipient separation originates on the blunt nose of a body.

Contrary to (4), on the basis of simple physical reasoning Landau [3] arrived at the conclusion that $\tau_w \sim (-x)^{1/2}$ whenever it approaches zero. Using a refined analysis Goldstein [2] independently revealed the same type of singularity at the position of separation that seemed to be not removable for many years (Stewartson [6]). Later on, Ruban [4, 5] and Stewartson et al. [7] postulated a finite sudden change in the derivative $d\tau_w/dx = \partial^3 \psi(x, 0)/\partial x \, \partial y^2$, as $x \to 0$ from both sides, to be inherent in the limit regime. However, the Landau-Goldstein singularity must arise in supercritical regimes if the pressure gradient is given in advance in the form (2), as it follows from the above statements.

2 Viscous wall sublayer

Let us split the boundary layer into two sublayers with different properties. In the viscous wall sublayer we introduce for the stream function an asymptotic sequence

$$\psi = \sum_{i=0}^{\infty} (-x)^{\alpha_i} f_i(\eta), \qquad \eta = \frac{y}{(-x)^{1/4}} \tag{5}$$

with exponents $\alpha_i \geq \dfrac{3}{4}$, $i = 0, 1, 2, \ldots$ increasing along with the function numbers. The leading term in (5) with $\alpha_0 = \dfrac{3}{4}$ is given by

$$f_0 = \frac{1}{6} c_0 (-x)^{3/4} \eta = \frac{1}{6} c_0 y^3. \tag{6}$$

In the original paper by Goldstein [2] all the exponents were from the very beginning supposed to be defined as $\alpha_i = (3 + i)/4$. This choice resulted in some difficulties that will not be discussed here for lack of space. According to Ruban [4] the exponents at hand may be thought as being set up by solving an eigen-value problem for the functions f_i with $\alpha_i > \dfrac{3}{4}$. In our analysis we follow the same approach. However, the eigen-value problem

$$\frac{d^3 f_i}{d\eta^3} - \frac{1}{8} c_0 \eta^3 \frac{d^2 f_i}{d\eta^2} + \frac{1}{2} c_0 \left(\alpha_i + \frac{1}{4} \right) \eta^2 \frac{df_i}{d\eta} - c_0 \alpha_i \eta f_i = 0 \tag{7}$$

$$f_i = \frac{df_i}{d\eta} = 0 \quad \text{at} \quad \eta = 0 \tag{8}$$

f_i does not exponentially grow   as $\eta \to \infty$ $\tag{9}$

is marked by some specific properties which are beyond the scope of the general theory. Really, it is easily seen that

$$f_i = \frac{1}{2} a_i \eta^2, \qquad a_i = \text{const} \tag{10}$$

satisfies Eq. (7) as well as the boundary conditions (8) at the solid surface and the limit condition (9) at the outer reaches of the viscous wall sublayer for any value of α_i. So, we need to proceed further on and consider the second and third order approximations obeying nonhomogeneous equations subject to constraints similar to (8) and (9). Mathematical treatment of corresponding boundary value problems is based on a transformation

$$\zeta = \frac{c_0}{32}\eta^4, \quad f = \eta^2 F, \quad \frac{dF}{d\eta} = \zeta^{-1/2}\Phi(\zeta) \tag{11}$$

that allows to lower the order of governing equations by one and reduce them to nonhomogeneous equations for confluent hypergeometric functions. As a result, the eigen-value spectrum must be confined to $\alpha_i = (1 + i)/2$, $i = 1, 2, \ldots$ according to the analysis of the second approximation (Ruban [4]), whereas another requirement $\alpha_i = (3 + 2i)/6$, $i = 1, 2, \ldots$ stems from inquiring into the third approximation. The common part

$$\alpha_i = \frac{1}{2} + i, \quad i = 1, 2, \ldots \tag{12}$$

of both spectra consisting of half-integer numbers determines the full set of eigen-values α_i. Consideration of higher order approximations brings no additional limitations.

Hence we may conclude on account of (11) that in original variables all the eigen-functions f_i are polynomials in η. In order to meet the definition (4) of the limit flow regime a coefficient a_1 of the first eigen-function with $\alpha_1 = 3/2$ must be put to zero. Incorporating into consideration terms induced by the pressure gradient the final expression for the stream-function in the viscous wall sublayer becomes

$$\psi = (-x)^{3/4}\frac{1}{6}c_0\eta^3 + (-x)^{7/4}\left(\frac{2c_0c_1}{7!}\eta^7 - \frac{c_1}{3!}\eta^3\right) + (-x)^{10/4}\frac{1}{2}a_2\eta^2$$

$$+ (-x)^{11/4}\left(\frac{-16c_0c_1^2 + 80c_0^2c_2}{11!}\eta^{11} - \frac{2c_1^2 + 4c_0c_2}{7!}\eta^7 + \frac{c_2}{3!}\eta^3\right)$$

$$+ (-x)^{14/4}\left(\frac{2a_2c_1}{6!}\eta^6 + \frac{a^3}{2!}\eta^2\right)$$

$$+ (-x)^{15/4}\left(16\frac{133c_0c_1^3 - 380c_0^2c_1c_2 + 810c_0^3c_3}{15!}\eta^{15} - 16\frac{c_1^3 - 6c_0c_1c_2 - 15c_0^2c_3}{11!}\eta^{11}\right.$$

$$\left. + 6\frac{c_1c_2 + c_0c_3}{7!}\eta^7 - \frac{c_3}{3!}\eta^3\right)$$

$$+ (-x)^{17/4}\left(\frac{36 \cdot 77 \cdot 105c_0^3a_2^2}{17!}\eta^{17} - \frac{36 \cdot 105c_0^2a_2^2}{13!}\eta^{13} + \frac{3 \cdot 18c_0a_2^2}{9!}\eta^9 - \frac{2a_2^2}{5!}\eta^5\right) + \ldots \tag{13}$$

This is nothing else than the analytic solution, first mentioned by Goldstein [2], with arbitrary constants $a_2, a_3, \ldots$ being coefficients of eigen-functions. The very existence of an analytic solution to Prandtl equation in the vicinity of the y-axis is a direct consequence of the general Cauchy-Kovalevsky theorem.

3 Small disturbances

We turn now to flows with velocity fields slightly deviating from that intrinsic to the limit regime. As before, the pressure gradient dp/dx is assumed to be given by (2) with somewhat disturbed constants $c_i \rightarrow c_i + \Delta c_i'$ where Δ is a small parameter. In supercritical flow regimes the wall shear stress can remain positive everywhere, nevertheless singularities come about as a rule from its becoming zero.

Let us write down an expression for the stream-function as

$$\psi = \tilde{\psi}(x, y) + \Delta \psi_c(x, y) \tag{14}$$

with $\tilde{\psi}$ implying the solution (13) for the limit regime. The correction term ψ_c satisfies the linearized Prandtl equation

$$\frac{\partial \tilde{\psi}}{\partial y} \frac{\partial^2 \psi_c}{\partial x \partial y} - \frac{\partial \tilde{\psi}}{\partial x} \frac{\partial^2 \psi_c}{\partial y^2} + \frac{\partial^2 \tilde{\psi}}{\partial x \partial y} \frac{\partial \psi_c}{\partial y} - \frac{\partial^2 \tilde{\psi}}{\partial y^2} \frac{\partial \psi_c}{\partial x} = -\frac{dp_c}{dx} + \frac{\partial^3 \psi_c}{\partial y^3} \tag{15}$$

the pressure gradient dp_c/dx arising from the aforementioned shift in values of c_i. The no-slip conditions for (15) on the body surface $y = 0$ are similar to (3) whereas no constraints as in (4) bear on supercritical regimes.

In the viscous wall sublayer to the left of the characteristic line $x = 0$ the correction term ψ_c is expanded into an asymptotic sequence

$$\psi_c = \sum_{i=0}^{\infty} (-x)^{\beta_i} g_i(\eta) \tag{16}$$

With the leading term in (5) given by (6) the eigen-value problem for g_i reads

$$\frac{d^3 g_i}{d\eta^3} - \frac{1}{8} c_0 \eta^3 \frac{d^2 g_i}{d\eta^2} + \frac{1}{2} c_0 \left(\beta_i + \frac{1}{4} \right) \eta^2 \frac{dg_i}{d\eta} - c_0 \beta_i \eta g_i = 0 \tag{17}$$

$$g_i = \frac{dg_i}{d\eta} = 0 \quad \text{at} \quad \eta = 0 \tag{18}$$

$$g_i \text{ does not exponentially grow} \quad \text{as} \quad \eta \rightarrow \infty. \tag{19}$$

Within designations, the problem (17)–(19) duplicates formulation of the boundary value problem (7)–(9) except for the exponents β_i are allowed to take both positive and negative values. Hence a solution of the type (10) with a coefficient a_{ci} substituted for a_i holds for any β_i. So, to find the eigen-value spectrum a term of the next order in x from (16) should be considered again. It is derivable from a nonhomogeneous equation subject to constraints similar to (18) and (19).

The further analysis follows along the same lines that have been sketched above in connection with the eigen-value problem (7)–(9). The transformation (11) serves to lower the order of a governing equation by one reducing it to a nonhomogeneous equation for confluent hypergeometric functions. As a final result we have

$$\beta_i = i - \frac{3}{2}, \quad i = 0, 1, 2, \ldots \tag{20}$$

and no higher order approximation needs to be examined. Like (12), the spectrum of β_i consists of half-interger numbers also, nevertheless both spectra are different in that three additional eigen-values $\beta_0 = -\dfrac{3}{2}$, $\beta_1 = -\dfrac{1}{2}$, and $\beta_2 = \dfrac{1}{2}$ are included into the latter. The lowest of these does not enter the set pointed out by Ruban [4] whose analysis was based on the assumption that the skin friction distribution in the limit regime of flow experiences a weak singularity of the type $\tau_w \sim |x|$, as $x \to 0$ from both sides, rather than varying smoothly.

The eigen-functions corresponding to (20) are expressed in terms of polynomials. Reverting to the initial variable η yields

$$\psi_c = \sum_{i=0}^{\infty} (-x)^{-(3/2)+i} \frac{1}{2} a_{ci}\eta^2 + (-x)^{-1/2} \frac{2c_1 a_{c0}}{6!} \eta^6$$

$$+ (-x)^{1/2} \left(\frac{-16c_1^2 + 80c_0 c_2}{10!} a_{c0}\eta^{10} + \frac{2c_1 a_{c1} + 4c_2 a_{c0}}{6!} \eta^6 \right) + \dots \tag{21}$$

All terms proportional to perturbations c_i' of the constants determining the pressure gradient are omitted here because they are easily accounted for by means of the solution for the limit regime. Upon augmenting (13) by (21) the wall shear stress becomes

$$\tau_w = \Delta \sum_{i=0}^{3} a_{ci}(-x)^{2+i} + \sum_{i=2}^{\infty} (a_i + \Delta a_{c,i+2}) (-x)^i \tag{22}$$

Thus, the extension of the eigen-function system leads to occurrence of a rather strong singularity in the skin friction as $x \to 0-$.

4   A neighborhood of the singular characteristic

The simplest way to estimate the sizes of a neighborhood centered about the characteristic line $x = 0$ where the uniform validity of the above solution breaks down is to address the expression (22) for the wall shear stress. Its first correction term becomes of the same order of magnitude as the leading one if $|x| = O(\Delta^{1/4})$. Within such distances the velocity field is to be analyzed separately.

The affine transformation

$$x = \Delta_0 s, \quad y = \Delta_0^{1/4} n, \quad \Delta_0 = \Delta^{1/4} \tag{23}$$

of independent variables preserves the self-similar variable $\eta = y/(-x)^{1/4} = n/(-s)^{1/4}$ invariant. Being reduced to new variables the Prandtl equation reads

$$\frac{\partial \psi}{\partial n} \frac{\partial^2 \psi}{\partial s \, \partial n} - \frac{\partial \psi}{\partial s} \frac{\partial^2 \psi}{\partial n^2} = -\Delta_0^{1/2} \frac{dp}{ds} + \Delta_0^{3/4} \frac{\partial^3 \psi}{\partial n^3} \tag{24}$$

In keeping with (13) and (21) the stream-function may be expanded into an asymptotic sequence

$$\psi = \Delta_0^{3/4} \frac{1}{6} c_0 n^3 + \Delta_0^{7/4} h_1(s, n) + \Delta_0^{10/4} h_2(s, n)$$

$$+ \Delta_0^{11/4} h_3(s, n) + \Delta_0^{14/4} h_4(s, n) + \Delta_0^{15/4} h_5(s, n) + \Delta_0^{17/4} h_6(s, n) + \dots \tag{25}$$

Substituting (23) and (25) into (24) gives a chain of equations

$$\frac{1}{2} c_0 n^2 \frac{\partial^2 h_i}{\partial s \, \partial n} - c_0 n \frac{\partial h_i}{\partial s} = \frac{\partial^3 h_i}{\partial n^3} + H_i(s, n) \tag{26}$$

The right-hand sides H_i with $i = 1, 3, 5$ depend only on the external pressure distribution, hence the form of h_i with the same odd subscripts may be easily obtained from (13). Then, $H_2 = 0$ and the equation determining h_2 is homogeneous. The remaining H_i with the even $i = 4, 6$ are

$$H_4 = -\left(\frac{\partial h_2}{\partial n} \frac{\partial^2 h_1}{\partial s \, \partial n} + \frac{\partial h_1}{\partial n} \frac{\partial^2 h_2}{\partial s \, \partial n} - \frac{\partial h_2}{\partial s} \frac{\partial^2 h_1}{\partial n^2} - \frac{\partial h_1}{\partial s} \frac{\partial^2 h_2}{\partial n^2} \right) \tag{27}$$

$$H_6 = -\left(\frac{\partial h_2}{\partial n} \frac{\partial^2 h_2}{\partial s \, \partial n} - \frac{\partial h_2}{\partial s} \frac{\partial^2 h_2}{\partial n^2} \right) \tag{28}$$

The no-slip conditions to be met by h_i may be written in the usual way

$$h_i = \frac{\partial h_i}{\partial n} = 0 \quad \text{at} \quad n = 0 \tag{29}$$

As for a boundary condition at the entry into the viscous wall sublayer centered about the singular characteristic, they come from matching (25) to the above solution for the global-scaled region of the same sublayer extending upstream. This limit process is fixed by $x \to 0-$, $s \to -\infty$; $\eta = \text{const}$ due to aforementioned invariance of the self-similar variable under the affine transformation (23). In order to specify boundary conditions at the outer reaches of the viscous sublayer under discussion one has to examine beforehand the velocity field in the main part of the boundary layer. Here, independent variables are appropriately chosen as $x = \Delta_0 s$, y. Being reduced to these variables the Prandtl equation is cast in the form

$$\frac{\partial \psi}{\partial y} \frac{\partial^2 \psi}{\partial s \, \partial y} - \frac{\partial \psi}{\partial s} \frac{\partial^2 \psi}{\partial y^2} = - \frac{dp}{ds} + \Delta_0 \frac{\partial^3 \psi}{\partial y^3} \tag{30}$$

Analysis of (30) is rather tedious albeit not conceptually sophisticated. The following matching to the viscous sublayer is achieved when $y \to 0$, $n \to \infty$, $s = \text{const}$. As a first consequence we have

$h_2 \to \dfrac{1}{2} A_2(s) \, n^2$, as $n \to \infty$, where A_2 is an arbitrary function except for the asymptotic

behaviour as $s \to -\infty$. However, results of the passage to the last limit strongly depend on how small singular correction terms are supposed to be in the global-scaled region.

5   A remark on matched asymptotic expansions

A brief comment on the nature of asymptotic expansions in the close vicinity of the characteristic line $x = 0$ is relevant at this point. If, in accordance with (14), all the singular correction functions $g_i(\eta)$, $i = 0, 1, 2, 3$ from (16) are multiplied by the same small parameter $\Delta = \Delta_0^4$ then the asymptotics

$$A_2 \to a_2(-s)^2 + a_{c0}(-s)^{-2} \quad \text{as} \quad s \to -\infty \tag{31}$$

following from the limit passage $x \to 0-$, $s \to -\infty$; $y = \text{const}$ in solutions for the main part of the boundary layer involves, along with a_2, the only coefficient a_{c0} that specifies the function $g_0(\eta)$

with the strongest singularity. In order to take into account the contributions from other correction functions $g_i(\eta)$ into asymptotics of A_2 and, hence, into the skin friction in the vicinity of the characteristic $s = 0$, we need to rescale the coefficients a_{ci}, $i = 1, 2, 3$ by means of $a_{ci} \to \Delta_0^{-i} a_{ci}$. Certainly, this renormalization alters relative orders of different correction terms over distances $-x = O(1)$ in the viscous wall sublayer extending upstream, but the analytic solution still persists to be the leading term in (14). Therefore, in what follows the substitution $a_{ci} \to \Delta_0^{-i} a_{ci}$ is assumed to be performed. Passing to the limit $x \to 0-$, $s \to -\infty$; $y = \text{const}$ in solutions for most of the boundary layer yields in this case

$$A_2 \to a_2(-s)^2 + \sum_{i=0}^{3} a_{ci}(-s)^{-2+i} \quad \text{as } s \to -\infty \tag{32}$$

Unlike (31), the quantity A_2 here is a function of all four coefficients a_{ci}, $i = 0, 1, 2, 3$.

6 Formulation of a key boundary value problem

As has been mentioned above, the equation to define h_2 is homogeneous. Hence

$$h_2 = \frac{1}{2} A_2(s) n^2 \tag{33}$$

provides a solution that satisfies the boundary conditions at $n = 0$ and $n \to \infty$, the asymptotic behaviour of A_2 as $s \to -\infty$ being prescribed by (32). Upon introducing (33) into (27) we get

$$h_4 = \frac{1}{2} A_4(s) n^2 + \frac{2c_1}{6!} A_2(s) n^6 \tag{34}$$

with a new arbitrary function $A_4 = A_4(s)$. The asymptotic behaviour of the latter as $s \to -\infty$ may be dropped out because the issue of working out an expansion for the viscous wall sublayer rests on deriving a solvability condition for the boundary value problem to be posed for $h_6(s, n)$. However, the right-hand side (28) of the equation for h_6 does not depend on A_4 from (34) and is expressed solely in terms of A_2 through

$$H_6 = -\frac{1}{2} A_2 \frac{dA_2}{ds} n^2 \tag{35}$$

Substituting (35) into (26) leads to

$$\frac{1}{2} c_0 n^2 \frac{\partial^2 h_6}{\partial s\, \partial n} - c_0 n \frac{\partial h_6}{\partial s} = \frac{\partial^3 h_6}{\partial n^3} - \frac{1}{2} A_2 \frac{dA_2}{ds} n^2 \tag{36}$$

with the boundary conditions (29) at $n = 0$ which should be supplemented by the limit conditions as $s \to -\infty$ and $n \to \infty$. Before indicating their forms it is instrumental at this point to think back to the foregoing analysis of a similar approximation for the global-scaled region upstream, the result has been cast in the first constraint $\alpha_i = (1 + i)/2$, $i = 1, 2, \ldots$ on the eigen-value spectrum. Therefore, solvability of the boundary value problem for (36) may be expected to hold only under certain restrictions to be imposed on $A_2(s)$.

The way to find the limit condition at the entry into the region at hand has been briefly outlined in connection with the boundary value problem for $h_2(s, n)$. As before, leaving aside details of a tedious analysis let us draw on its results to write down the limit condition under discussion. The latter is represented by a sum of terms in fractional powers of s with coefficients depending on the self-similar variable η. The leading term in the sum originates from the analytic solution whereas the remaining terms ensue from the expansion for ψ_c. Reverting to the initial variables, we have

$$h_6 \rightarrow B_1 n^{17} + B_2(s)\, n^{13} + B_3(s)\, n^9 + B_4(s)\, n^5 \quad \text{as} \quad s \rightarrow -\infty \tag{37}$$

where $B_1 = \text{const}$ and B_2, B_3 and B_4 are polynomials in s of the first, second and third degree, respectively. For brevity, the explit form of the latter polynomials is omitted here.

What still remains to be done is to derive the limit condition at the outer edge of the viscous wall sublayer. For this purpose the expansion (25) for the region mentioned should be matched to the solution in the main part of the boundary layer. From the limit process we see that the asymptotics of h_6 is established by (37) where now $n \rightarrow \infty$, $s = \text{const}$. Thus, the boundary value problem for (36) is completely posed.

7   Result and conclusions

In order to simplify solving this key problem let us introduce a new desired function

$$\varphi = h_6 - B_1(s)\, n^{17} - B_2(s)\, n^{13} - B_3(s)\, n^9 - B_4(s)\, n^5 - \frac{1}{2c_0}\, G(s)\, n \tag{38.1}$$

$$G = A_2^2 - [a_2^2(-s)^4 + b_2(-s)^3 + b_3(-s)^2 + b_4(-s) + b_5] \tag{38.2}$$

where values of the constants b_2, b_3, b_4 and b_5 can be evaluated simply by squaring the right-hand side of (32). We deduce hence that $G \rightarrow 0$ as $s \rightarrow -\infty$. The term $\frac{1}{2}\, A_6(s)\, n^2$ is missing from (38.1) for the third eigen-function entering the analytic solution with the coefficient a_3 contributes to order of $\Delta_0^{18/4}$ giving rise to the next approximation in (25). Thus, the boundary value problem for φ reduces to

$$\frac{1}{2}\, c_0 n^2 \frac{\partial^2 \varphi}{\partial s\, \partial n} - c_0 n \frac{\partial \varphi}{\partial n} = \frac{\partial^3 \varphi}{\partial n^3} \tag{39.1}$$

$$\varphi = 0, \quad \frac{\partial \varphi}{\partial n} = -\frac{1}{2c_0}\, G(s) \quad \text{at} \quad n = 0 \tag{39.2, 3}$$

$$\varphi \rightarrow 0 \quad \text{as} \quad s \rightarrow -\infty \tag{39.4}$$

$$\varphi \rightarrow -\frac{1}{2c_0}\, G(s)\, n \quad \text{as} \quad n \rightarrow \infty. \tag{39.5}$$

Within definition (38.2) of the auxiliary function G appearing in the boundary conditions (39.3) and (39.5), the problem at hand coincides with that studied by Stewartson [6] whose analysis later on formed the basis of Ruban [4, 5] and Stewartson et al. [7]. In keeping with results set forth in all these works we need to put $G(s) \equiv 0$ in order to surpress the exponential growth of φ as $n \rightarrow \infty$. Since Eq. (39.1) as well as remaining boundary conditions (39.2) and (39.4) are homogeneous, we are led to the conclusion that $\varphi(s, n) \equiv 0$ must be a solution to the problem and this solution is unique. Thus

$$A_2 = [a_2{}^2(-s)^4 + b_2(-s)^3 + b_3(-s)^2 + b_4(-s) + b_5]^{1/2}. \qquad (40)$$

Insofar as

$$\frac{\partial^2 \psi}{\partial y^2} = \Delta_0{}^2 \frac{\partial^2 h_2}{\partial n^2} = \Delta_0{}^2 A_2 \qquad (41)$$

to the leading order, (40) provides a groundwork for concluding remarks on different types of the singularity that might be expected to occur at the point of vanishing skin friction. First, let the coefficients a_{c0} and a_{c1} of the correction eigen-functions g_0 and g_1 be zero, then

$$A_2 = a_2(-s)^2 + a_{c3}(-s) + a_{c2}$$

by virtue of the aforementioned definition of the constants b_2, b_3, b_4 and b_5. The lack of singularities in the skin friction retains the solution invariant when its being continued into the neighborhood of the characteristic $s = 0$. If the next coefficient $a_{c2} = 0$, the asymptotics of A_2 becomes

$$A_2 \to a_{c3}(-s) + \dots \qquad \text{as } s \to 0- \qquad (42)$$

in line with the basic assumption in Ruban [4, 5] and Stewartson et al. [7] that the skin friction should approach zero according to linear law. However, reverting to the initial coordinate x yields

$$\tau_w = a_2(-x)^2 + \Delta_0 a_{c3}(-x) \qquad (43)$$

in view of (41). So, there is nevertheless the outstanding distinction between results exposed here and those in the papers cited. By virtue of (42), (43) the linear dependence of the wall shear stress on the longitudinal coordinate takes place locally within distances of order $-s = O(1)$, whereas on the strength of the postulate put forward by Ruban [4, 5] and Stewartson et al. [7] the same dependence must hold globally on scale $-x = O(1)$. It is obvious that interpreting numerical data from Werle and Davis [9] and Ruban [4] calls for extreme care because in both approaches under discussion the wall shear stress is allowed to tend to zero according to linear law. Nevertheless, computed results lend credence to the theoretical prediction as in (43), at least for the problem on a parabola at incidence in a uniform stream that is typical of incompressible aerodynamics.

Second, let us consider the general case with $a_{c0} \neq 0$ and $a_{c1} \neq 0$. Then the properties of the solution are established by the number and order of the real roots of the fourth degree polynomial from the right-hand side of (40). The following possibilities are of importance for making a distinction between flow patterns.

(a) There exist two real roots $s = s_1$ and $s = s_2$. Here

$$A_2 = (-s + s_1)^{1/2} (-s + s_2)^{1/2} [a_2{}^2(-s)^2 + k_1(-s) + k_2]$$

with $a_2{}^2(-s)^2 + k_1(-s) + k_2$ being positive. If $s_2 < s_1$ the Landau-Goldstein singularity appears at the point $s = s_1$ and the velocity field does not admit continuation downstream of it. When $s_1 = s_2$ the skin friction tends to zero according to the linear law although the solution upstream involves the correction eigen-functions g_0 and g_1.

(b) All four roots are real and different. Then

$$A_2 = a_2 \prod_{i=1}^{4} (-s + s_i)^{1/2}.$$

The lowest of these roots (let them be $s_1 < s_2 < s_3 < s_4$) determines location of the Landau-Goldstein singularity through which the flow field can not be continued.

(c) The first two of four real roots coincide: $s_1 = s_2 = s_0$. The corresponding flow patterns become complicated and involve the weak singularity with a sudden change in the magnitude of the skin friction derivative and the Landau-Goldstein singularity shifted by a small distance further downstream. In the region between both singularities the skin friction remains positive. The weaker singularity may be treated as a prerequisite for incipient separation giving rise to a short bubble confined by flow reattachment to the body surface, the Landau-Goldstein singularity marks massive reseparation of the reattached boundary layer. The flow patterns of such a kind have been clearly observed experimentally, a comprehensive discussion of pertinent data of wind-tunnel tests is due to Tani [8]. Unfortunately, the last review does not include results regarding direct measurements of the skin friction distributions along the body surface at different Reynolds numbers. However, several curves showing variation of the frictional intensity over the 5.89-inch cylinder are reproduced from observations by Fage and Falkner in Figs. 156, 157 of the Volume II on "Modern Development in Fluid Dynamics" edited by Goldstein [1] as far back as 1938. The general picture is apparently in favour of our statements: the positive local minimum is clearly discernable upstream of the massive separation at Reynolds numbers in transition range. The widespread explanation of reattachment was related to the onset of turbulence in the separated flow inside the short bubble. As we see here, the same phenomenon can be predicted theoretically to occur in a laminar boundary layer suffering no turbulent pulsations. Feasibility for the local minimum (or weak singularity) to be formed in the skin friction distribution just upstream of the much stronger Landau-Goldstein singularity is intrinsic even to laminar motion.

Acknowledgements

It is a great honour for both authors to present this work as a token of their high esteem of Prof. J. Zierep's scientific activities. The authors would like to express also their gratitude to Prof. J. D. Cole for many discussions and helpful comments. The final stage of this study was carried out with support of the U.S. Air Force Office of Scientific Research to O.S.R. under Grant AFOSR 88-0037.

References

[1] Goldstein, S. (ed.): Modern developments in fluid dynamics. Oxford: Clarendon Press 1938.
[2] Goldstein, S.: On laminar boundary-layer flow near a position of separation. Q. J. Mech. Appl. Math. **1**, 43–69 (1948).
[3] Landau, L. D., Lifshitz, E. M.: Mechanics of continua. Moskwa: Gostekhizdat 1944 (in Russian).
[4] Ruban, A. I.: A singular solution of the boundary-layer equations that admits continuation past the point of vanishing skin friction. Izv. Akad. Nauk SSSR, Mekh. Zhidk. Gaza **6**, 42–52 (1981) (in Russian).
[5] Ruban, A. I.: An asymptotic theory of short separation bubbles on the leading edge of a thin airfoil. Izv. Akad. Nauk SSSR, Mekh. Zhidk. Gaza **1**, 42–51 (1982) (in Russian).
[6] Stewartson, K.: Is the singularity at separation removable? J. Fluid. Mech. **44**, 347–364 (1970).
[7] Stewartson, K., Smith, F. T., Kaups, K.: Marginal separation. Stud. Appl. Math. **67**, 45–61 (1982).

[8] Tani, I.: Low-speed flows involving bubble separations. Progr. Aero. Sci., vol. 5, (Küchemann, D., Sterne, L. H. G. eds.), pp. 70–103. Oxford: Pergamon 1964.

[9] Werle, M. J., Davis, R. T.: Incompressible laminar boundary layers on a parabola at angle of attack: a study of the separation point. Trans. ASMEE J. Appl. Mech. **39**, 7–12 (1972).

Authors' address: Professor E. V. Bogdanova-Ryzhova and Professor O. S. Ryzhov, Department of Mathematics, Rensselaer Polytechnic Institute, Troy, NY 12180-3590, U.S.A.

Acta Mechanica (1994) [Suppl] 4: 39 – 46

Turbulence and rarefied gasdynamics

C. Cercignani, Milano, Italy

Summary. The problem of the relation between turbulence and rarefied gasdynamics is considered, with particular attention to recent computer experiments on rarefied gases that seem to exhibit, in addition to the onset of physical instabilities, also features of transition to chaotic bulk motion.

1 Introduction

A common statement that one can read in the books on turbulence is that the latter has nothing to do with the molecular constitution of matter, because the scales of turbulence and molecular motions are widely separated.

This may always be true for a liquid, but what should we say for a gas? Let us consider the ratio between the mean free path l and the dissipative scale of turbulence l_D:

$$\frac{l}{l_D} = \frac{l}{L} (\mathrm{Re})^{3/4} = \mathrm{Kn}(\mathrm{Re})^{3/4} = (\mathrm{Ma})^{3/4}(\mathrm{Kn})^{1/4} \tag{1}$$

where L is a macroscopic scale and the relation $\mathrm{Re} = \mathrm{Ma}/\mathrm{Kn}$ has been taken into account. If we take $\mathrm{Re} = 10^4$ for fully developed turbulence, we obtain that l and l_D are of the same order for $\mathrm{Kn} = 10^{-3}$ and $\mathrm{Ma} = 10$. Thus in hypersonic flow for moderate rarefaction the turbulence dissipative scale and the mean free path are of the same order of magnitude!

Estimates of this kind must, of course, always be taken with caution because of the precence of possibly sizable factors "of order unity". But the above considerations should not be ignored when considering high Mach number flows.

The study of the role of the discrete structure of matter in the turbulence of a gas at high Mach number is a completely unexplored field. Whence should one start? Perhaps from the stability of steady motions, since they are believed to be the basic phenomenon from which turbulence develops. The difficulty here is that we do not know any closed form solution comparable to the Couette or Poiseuille flows for Navier-Stokes equations. Even the basic theory of existence of such solutions is in its infancy (for a recent, interesting, but still inadequate result see [1]). In this situation computer simulations appear to be the only available tool. The algorithms should be sufficiently robust to be insensitive to numerical instabilities and to exhibit only physical instabilities. Direct Monte Carlo Simulation methods appear to be sufficiently adequate for this purpose. That is why the author has started recently, in collaboration with S. Stefanov [2 – 5], a line of research on the study of physical instabilities that will be reviewed in the next sections.

2 Physical instability in a rarefied gas heated from below

While the problems related to the instability of laminar flows and their transition to turbulence have been studied for a long time in classical hydrodynamics, the corresponding problems in kinetic theory have been paid attention only recently. This circumstance is clearly related to the extremely complex character of such problems. It is clear, however, that the study of such problems might be of great importance for the purpose of understanding fundamental phenomena of instability and self-organization in molecular dynamics. One of the typical examples in this area is Bénard's instability.

The Bénard instability is a well known phenomenon in fluid dynamics. The earliest experiments with a horizontal fluid layer heated from below were made by Bénard himself [6]. Rayleigh [7] analysed the stability of the pure heat conduction solution of the Navier-Stokes equations in the Boussinesq approximation and introduced a non-dimensional parameter (nowadays called the Rayleigh number and denoted by Ra), which is of paramount importance for the stability analysis. The latter was carried out by Rayleigh through the linearization method; in this way he found a critical value of Ra, which turns out to be the actual theoretical value for the pure conduction solution to become unstable, due to the fact that the threshold for the onset of linear instability is also the value of Ra under which one can rigorously show that the perturbation energy decays in time [8].

It seems that the corresponding problem for a rarefied gas governed by the Boltzmann equation was not studied until recently, when S. Stefanov and the present author attacked it using the Direct Simulation Monte Carlo (DSMC) method [2, 3]. Since this method is based on a finite number of particles with stochastic dynamics, to show that it exhibits the Bénard instability means to show the ability of a stochastic system to get organized in a pattern of vortices.

The results presented in those papers [2, 3] have a preliminary character, but clearly indicate that the formation of such vortex patterns arising from the instability of the purely conducting state are possible. The calculations refer to Knudsen numbers of order 10^{-2} and different temperature ratios.

In order to define the problem, let us consider a monatomic rarefied gas with average number density n_0 in a steady state between two parallel plates with different temperatures T_h (at $x = 0$) and T_c (at $x = L$). The gas is assumed to be heated from below, so that $T_h > T_c$. An external constant force $\underline{F} = m\underline{g}$, directed vertically downwards (gravity), acts on each molecule of the gas at each point of the gap. We introduce Cartesian coordinates and consider a two-dimensional situation; then the Boltzmann equation reads as follows [9 − 11]:

$$\xi_1 \frac{\partial f}{\partial x} + \xi_2 \frac{\partial f}{\partial y} - g \frac{\partial f}{\partial \xi_2} = Q(f,f) \tag{2}$$

where $\underline{x} = (x, y)$ and $\underline{\xi} = (\xi_1, \xi_2, \xi_3)$ are the position and velocity vectors of a molecule while

$$Q(f,f) = \frac{1}{m} \iint (f'f_*' - ff_*) \, B(\theta, |\underline{\xi} - \underline{\xi}_*|) \, d\underline{\xi}_* \, d\theta \, d\varepsilon \tag{3}$$

is the collision operator. Here $B(\theta, |\underline{\xi} - \underline{\xi}_*|)$ is a kernel describing the details of molecular interaction, m the molecular mass, f', f_*', f_* same as f, except for the fact that $\underline{\xi}$ is replaced by $\underline{\xi}'$, $\underline{\xi}_*', \underline{\xi}_*$. $\underline{\xi}_*$ is an integration variable (the velocity of any molecule colliding with a molecule of velocity $\underline{\xi}$) while $\underline{\xi}'$ and $\underline{\xi}_*'$ are the velocities of two molecules entering a collision which brings them to a state with velocities $\underline{\xi}$ and $\underline{\xi}_*$. Finally, θ and ε give the direction along which the same molecules approach each other. We shall assume hard sphere molecules, in which case $B(\theta, |\underline{\xi} - \underline{\xi}_*|) = \sigma^2|\underline{\xi} - \underline{\xi}_*| \cos \theta \sin \theta$, where σ is the sphere diameter.

We note that from the data of the problem we can obtain three nondimensional parameters, the Knudsen number Kn based on the mean free path $\lambda_0 = (\sqrt{2}\,\pi\sigma^2 n_0)^{-1}$, the Froude number Fr based on the thermal speed $v_h = \sqrt{2RT_h}$, and the temperature ratio r:

$$\text{Kn} = \lambda_0/L, \quad \text{Fr} = v_h{}^2/gL, \quad r = T_c/T_h \tag{4}$$

The Rayleigh number in the usual theory of the Bénard instability is given by

$$\text{Ra} = \frac{gL^3(1-r)/r}{\chi\nu} \tag{5}$$

where we assumed the perfect gas law and χ and ν denote the heat diffusivity and the kinematic viscosity, respectively. According to the Chapman-Enskog method [12], we have for hard spheres (first approximations are sufficient for this discussion):

$$\nu = (5/16)\,\lambda_0 v_h \sqrt{\pi}; \quad \chi = (15/32)\,\lambda_0 v_h \sqrt{\pi} \tag{6}$$

Hence the Rayleigh number can be expressed through the parameters Kn, Fr and r in the following way:

$$\text{Ra} = \frac{512}{75\,\pi}\frac{(1-r)/r}{\text{Fr Kn}^2} \tag{7}$$

If we use

$$\text{Fr}^* = v_h{}^2/g\lambda_0 = \text{Fr}/\text{Kn}, \tag{8}$$

then:

$$\text{Ra} = \frac{512}{75\,\pi}\frac{(1-r)/r}{\text{Fr}^*\,\text{Kn}^3} \tag{9}$$

If we are close to the continuum regime, we can then conjecture that the combination $(1-r)/(r\,\text{Fr}\,\text{Kn}^2)$ is the basic parameter when investigating stability. The critical value of the Rayleigh number is about 1,710: hence if we take $\text{Fr}^* = 200$ and $\text{Kn} = 10^{-2}$, we can expect instability for

$$\frac{1-r}{r} \geq 0.16 \tag{10}$$

or

$$r \leq 0.86 \tag{11}$$

provided we are able to excite the lowest mode of the instability.

In the case under consideration, however, the accuracy of this estimate must be taken with some suspicion because the resulting temperatures are widely different and temperature jumps occur at the wall, while Rayleigh's analysis refers to constant values of χ and ν and small temperature differences and no jumps.

The formulation is completed with the following boundary conditions:

a) at the plates one assumes diffuse reflection of the molecules and hence writes:

$$f(x, 0, \underline{\xi}) = n_h \pi^{-3/2} v_h^{-3} \exp\left(-\xi^2/v_h^2\right) \qquad (\xi_2 \gcurvearrowright 0) \tag{12}$$

$$f(x, L, \underline{\xi}) = n_c \pi^{-3/2} v_c^{-3} \exp\left(-\xi^2/v_c^2\right) \qquad (\xi_2 \lcurvearrowright 0) \tag{13}$$

where n_h and n_c are determined by mass conservation at the walls and hence are given by

$$n_h = \left(2\sqrt{\pi}/v_h\right) \int\limits_{\xi_2 > 0} f(x, 0, \underline{\xi}) \, \xi_2 \, d^3\underline{\xi}; \quad n_c = \left(2\sqrt{\pi}/v_c\right) \int\limits_{\xi_2 < 0} f(x, L, \underline{\xi}) \, \xi_2 \, d^3\underline{\xi} \tag{14}$$

while v_h has been already defined and v_c is similarly defined by $v_c = \sqrt{2RT_c}$.

b) to solve numerically the problem one assumes that the solution possesses a periodic structure in the direction of the x-axis. In order to spare computer time and storage capacity, the best way to simulate periodicity is to assume that the molecules are specularly reflected at $x = 0$ and $x = \hat{L}$, where $2\hat{L}$ is the period in the x-direction.

The Monte Carlo simulation was devised in agreement with the above formulation according to standard procedures [13, 14] and is described in the above [2, 3].

As noted above, the effect of the instability appears for certain values of the parameters Kn, Fr* and r. The numerical solution is, of course, also sensitive to the parameters of the numerical scheme. The results show that in order to discover whether an instability occurs or not, one can use rather coarse meshes with cell sizes Δx and Δy of the order of 2 or 3 mean free paths and a total number of particles N_0 between 10000 and 15000. But, by varying Δx, Δy and N_0, one obtains different flow pictures with various numbers and sizes of vortices. Another important parameter, $\hat{L}$, fixes the number of vortices in the domain; in the calculations of [2] it was chosen to be 2 or 3 times the gap height L.

Two types of grids were used in [2, 3]. The first was a large scale grid of 120×40 cells, for all the calculations with Kn = 0.05 and, just to obtain preliminary results of a qualitative nature, for Kn between 0.01 and 0.02 as well; in the latter range of Knudsen numbers the final results were, however, obtained by means of a smaller scale grid $(200 \text{ or } 300) \times 100$ cells. In all the cases the average number of particles per cell, $\bar{N}$, was kept equal to 3. This means that the total number of particles varied from 14400 to 90000.

If we accept the idea, which is borne out by the previous discussion, that the relevant parameter for stability is

$$S = \frac{(1 - r)/r}{\text{Fr*} \, \text{Kn}^3} \tag{15}$$

(which is about 0.46 times the Rayleigh number), the calculations seem to indicate a critical value S_c between 7500 and 20000 or a Ra_c between 3450 and 9200 (see Table 1). This is higher than expected on the basis of analogy with the Navier-Stokes equations (with no slip); this might be due to rarefaction effects, to the fact that a higher mode is excited for the particular values of $\hat{L}/L$ (2 or 3), which have been chosen, or, more likely, to the high temperature ratios involved. We remark, however, that stability seems to be gained again for a value of S between 45000 and 90000 or a value of Ra between 20700 and 41400. This is , of course, at variance with experience. It is a known experimental circumstance [15], in fact, that a Navier-Stokes fluid exhibits a transition to turbulence when the Grashof number Gr = Ra/Pr (where Pr

Table 1. Values of the stability parameter S in Bénard's problem

Kn	r	Fr*	Stable	S
0.05	0.1	$20 \div \infty$	Yes	$0 \div 3600$
0.02	0.1	$100 \div \infty$	Yes	$0 \div 11250$
0.01	0.1	40	Yes	225000
0.01	0.1	100	Yes	90000
0.01	0.1	200	No	45000
0.01	0.2	200	No	20000
0.01	0.4	200	Yes	7500

is the Prandtl number) is about 50000. Since the Prandtl number is about 2/3 for a monatomic gas (this is the first Chapman-Enskog approximation obtained from Eqs. (2.5); the fourth approximation delivers $Pr = 0.6608$ for a gas of rigid spheres), the transition of a monatomic gas to turbulence for high values of the Knudsen number should occur for a Rayleigh number of about 33000, which is in the range indicated above. Of course, it is impossible to simulate a transition of the system to an unsteady turbulent motion with a completely two-dimensional code. In order to confirm the conjecture that one is close to exhibiting turbulence in a rarefied gas, one should investigate time sequences, use perhaps a more detailed mesh and consider three-dimensional calculations; a further study of this exciting possibility is, however, not available yet.

3 Flow in a channel

Encouraged by the positive result of their previous investigation of Bénard's instability by means of the Direct Simulation Monte Carlo Method (DSMC), S. Stefanov and the present author [4] considered the fluctuations of the macroscopic quantities in a rarefied gas flowing in a channel under the action of a constant external force in a direction parallel to the walls (which are assumed to be at rest with the same temperature). Their final aim was the study of the transition to turbulence by means of the Boltzmann equation, but they remained far from achieving this goal, since their calculations were restricted to a two-dimensional geometry. In addition, they considered a channel of limited length in the direction of the force and simulated a long tube by appropriate boundary conditions at the two ends. Actually they also performed a couple of very rough preliminary calculations for the case of a long channel and compared them with those for a short channel; the comparison showed a sufficiently good agreement between the two sets of calculations.

The main aim of the calculations performed by S. Stefanov and C. Cercignani [4] was to investigate the time evolution of both the macroscopic quantities and their fluctuations. Their numerical experiments refer to Knudsen numbers between 0.005 and 0.1 and three different values of the body force. The analysis of their results indicates an increase in the macroscopic fluctuation for $Kn \leq 0.05$ and certain magnitudes of the force (Table 2). In order to recognize the possible formation of vortex patterns and to estimate the macroscopic fluctuations, a data analysis was performed. The results of this analysis do not show clearly formed patterns on a sufficiently large scale (larger, say, than five mean free paths). If one takes into account the increase in the macroscopic fluctuations, he is tempted to conclude that he observes a transition from laminar to two-dimensional small-scale turbulence.

Table 2. Stability of channel flow (No*
means that instability is not so clear
from numerical results)

Kn	Fr*	Stable	Re
0.1	20	Yes	13.5
0.05	20	Yes	108
0.01	10^3	Yes	270
0.02	20	No*	1687.5
0.01	100	No	2700
0.005	100	No	21600
0.005	20	No	108000

In this problem the Reynolds number can be related to the force per unit mass g in the following way

$$Re = gL^3/(12v^2) \tag{16}$$

where v is given by $v = (5/16) \lambda_0 v_h \sqrt{\pi}$, according to Eq. (6). The intensity of the force is expressed in nondimensional form by means of the Froude number Fr* based on the mean free path λ_0, according to Eq. (8). Table 2 summarizes the results of the calculations.

As is well-known [8], experiments indicate that instabilities arise in a channel for Re = 1,300 in conditions under which the Navier-Stokes equations with no slip conditions are valid; thus the above results seem to indicate that a critical Reynolds number of the same order holds for the Boltzmann equation (although detecting a possible slight dependence on the Knudsen number would require more calculations).

4 Taylor vortices in a rarefied gas

Taylor's instability of Couette flow is a well-known phenomenon in hydrodynamics. The classical case of this flow occurs when we consider two coaxial cylinders of infinite length and the inner cylinder rotates while the outer one is at rest. At a certain critical value of the angular velocity Ω of the rotating cylinder, the Couette flow becomes unstable and transforms into the so-called Taylor-Couette flow. This flow is characterized by a system of Taylor cells in the form of toroidal vortices, two neighbouring vortices rotating in opposite directions. A study of the same phenomenon in the framework of kinetic theory was never attempted before the recent numerical experiments by S. Stefanov and the present author [5]. Their results refer to Knudsen numbers of order 10^{-3} and different Mach numbers, based on the velocity of the inner cylinder. These ranges of these numbers are chosen in such a way as to indicate the onset of instability in the cylindrical Couette flow of a rarefied gas. When the Taylor-Couette flow appears to be fully developed the authors of [5] extend their study to higher Mach numbers and exhibit results that seem to indicate transition to a chaotic behavior. Again the real physics of the transition to turbulence escapes these computations because of their two-dimensional nature. In fact in [5] the calculation is performed in cylindrical coordinates (x, r, φ) and the solution is assumed to be independent of φ; the method used there is the same described by Bird in [16].

The nondimensional parameters here are the Knudsen number Kn_L based on the thickness of the gap $L = R_2 - R_1$, the speed ratio S related to the Mach number Ma by $Ma = (6/5)^{1/2}S = 1.095 S$ and the Knudsen number Kn_1 based on the inner cylinder radius R_1.

Table 3. Stability of cylindrical Couette flow (No^c means transition to chaos and No* indicates no clear tendency to form a stable vortex)

Kn$_L$	Kn$_1$	S	Stable	T
0.05	0.02	2	Yes	42 597
0.02	0.02	1	Yes	43 467
0.02	0.02	1.5	No*	97 800
0.02	0.02	2	No	173 867
0.02	0.02	3	No	391 200
0.02	0.02	6	No	1 564 800
0.02	0.02	8	No	2 781 867
0.02	0.02	10	No	4 346 667
0.02	0.02	12	No^c	6 259 200
0.02	0.02	15	No^c	9 780 000
0.02	0.01	2	No	234 720
0.02	0.005	2	No*	362 222
0.01	0.02	2	No	586 800

The basic parameter governing stability should be the Taylor number:

$$T = 4\Omega^2 R_1^4 / \{v^2[(1 - (R_1/R_2)^2]^2\} = (16/5)^2 4S^2 L^4 / (\lambda_0^2 R_1^2 \pi)$$

$$= (13.04) S^2 (\text{Kn}_1)^{-2} (1 + \text{Kn}_L/\text{Kn}_1)^2 (2 + \text{Kn}_L/\text{Kn}_1) \tag{17}$$

The results of the calculations of [5] are summarized in Table 3, where the cases marked with No correspond to the formation of Taylor vortices, while No^c indicates that even the Taylor cells are not stable and one witnesses the transition to a chaotic motion. In the cases indicated with No* there are small vortices with no clear tendency to form a stable Taylor vortex. We briefly discuss here only the results for Kn$_L$ = Kn$_1$ (i.e. when outer radius is twice the inner radius). In this case the critical value of the Taylor number, according to the Navier-Stokes equation (with no slip) is 33,110 [17]. The results in Table 3 seem to indicate a value higher than 43,467; as in the case of Bénard's instability, rarefaction (with the associated phenomenon of slip at the boundary) seems to increase the critical value.

References

[1] Arkeryd, L., Cercignani, C., Illner, R.: Measure solutions of the steady Boltzmann equation in a slab. Commun. Math. Phys. **142**, 285–296 (1991).
[2] Stefanov, S., Cercignani, C.: Monte Carlo simulation of Bénard's instability in a rarefied gas. Euro. J. Mech. B **5**, 543–552 (1992).
[3] Cercignani, C., Stefanov, S.: Bénard's instability in kinetic theory. Transp. Theory Stat. Phys. **21**, 371–381 (1992).
[4] Stefanov, S., Cercignani, C.: Monte Carlo simulation of a channel flow of a rarefied gas. Euro J. Mech. B (to appear).
[5] Stefanov, S., Cercignani, C.: Monte Carlo Simulation of the Taylor-Couette flow of a rarefied gas. J. Fluid Mech. (to appear).
[6] Bénard, H.: Les tourbillons cellulaires dans une nappe liquide transportant de la chaleur par convection en régime permanent. Ann. Chim. Phys. **23**, 62–144 (1901).
[7] Lord Rayleigh: On convective currents in a horizontal layer of fluid when the higher temperature is on the under side. Phil. Mag. **32,** 529–546 (1916).

[8] Joseph, D. D.: Stability of fluid motions I and II. Berlin Heidelberg New York: Springer 1976.
[9] Cercignani, C.: The Boltzmann equation and its applications. Wien New York: Springer 1988.
[10] Kogan, M. N.: Rarefied gasdynamics. New York: Plenum Press 1969.
[11] Cercignani, C.: Mathematical methods in kinetic theory. New York: Plenum Press 1990.
[12] Chapman, S., Cowling, T. G.: The mathematical theory of non-uniform gases. London: Cambridge University Press 1952.
[13] Bird, G.: Perception of numerical methods in rarefied gasdynamics. In: Rarefied gasdynamics: theoretical and computational techniques (Muntz, E. P., Weaver, D. P., Campbell, D. H., eds.), pp. 212−226. Washington: AIAA 1989.
[14] Yanitsky, V. E.: Operator approach to direct Monte Carlo simulation theory. In: Rarefied gasdynamics (Beylich, A., ed.), pp. 770−777. New York: VCH Publishers 1991.
[15] Landau, L. D., Lifshitz, E. M.: Fluid mechanics. New York: Addison Wesley 1959.
[16] Bird, G.: Comment on "False collisions in the direct simulation Monte Carlo method" [Phys. Fluids 31, 2047 (1988)]. Phys. Fluids A 1, 897−897 (1989).
[17] Chandrasekhar, S.: Hydrodynamic and hydromagnetic stability, p. 323. London: Oxford University Press 1961.

Author's address: Professor C. Cercignani, Dipartimento di Matematica, Politecnico di Milano, Piazza Leonardo da Vinci 32, I-20133 Milano, Italy

Acta Mechanica (1994) [Suppl] 4: 47 – 56

Flow attachment at flow separation lines

On uniqueness problems between wall-flows and off-wall flow fields

U. Dallmann and **H. Gebing**, Göttingen, Federal Republic of Germany

Summary. Evolving flow structures leave definite but not unique "footprints" of the outer (off-wall, mid-air) flow on the wall. Bifurcating flows, characterized by dynamical systems for the velocity field reveal that two-dimensional separation bubbles together with their three-dimensional bifurcations are "embedded" between fully attaching and completely separating three-dimensional flow structures with a possibility that local flow attachment occurs at a separation line. In this respect we show: 1. An unsteady or steady, incompressible flow field can be completely determined by the knowledge of the wall-shear stress and the wall-pressure field without specifying any outer (farfield) boundary conditions. 2. Nonlinearity or time-dependence or an explicit Reynolds number/viscosity dependence which would reflect properties of the Navier-Stokes equations do not affect the structural changes associated with local flow bifurcations which lead to separation bubbles or near-wake vortex flow separations. Only the continuity equation is fulfilled in every case. Therefore, 3. the question is considered how far away from the bifurcation set of parameters the kinematically possible structures remain structurally stable against Navier-Stokes perturbations. This is studied by direct numerical simulations of the structural changes of separated flows around an ellipsoid at angles of attack. It is shown that flow attachment in a region where a separation line forms is possible in a wide range of angles of attack. The dilemma of defining or locating steady streamwise vortices due to open flow separation is considered.

1 Introduction

Flow fields undergo structural (topological) changes in the set of dependent variables due to the onset of flow separations or the onset of flow unsteadiness, due to large or small scale vortex formations and breakdowns. Such changes occur at critical sets of flow parameters like Reynolds number, Mach number, angle of attack, body shape parameters, etc. and are also influenced by changes of boundary or initial conditions. Between two critical sets no topological change of the flow takes place and the resulting flows are called structurally stable with respect to certain classes of disturbances.

The concept of topological flow analysis as introduced by Perry et al. and others (see [1] with refs. therein) considers *structurally stable*, non-degenerate, hyperbolic critical points of the velocity field or structurally stable flow structures. Dallmann [3], [4] suggested to describe the *structural changes*, i.e. the changes of the flow topology of velocity fields, etc. and he derived a set of bifurcating three-dimensional elementary topological flow structures which can form locally on a rigid wall by formulating the velocity and vorticity fields as a dynamical system which is locally equivalent to the equations of motion for incompressible flows. Bakker et al. [5], [6] have analyzed several more elementary structures which also evolve locally from degenerate, non-hyperbolic critical points in the wall-shear stress/wall-pressure distribution. The velocity vector field $Q^i = (U, V, W)^T$ is represented by a locally valid Taylor-series expansion:

$$\dot{x}^i = Q^i = \sum_\alpha \sum_\beta \sum_\gamma A^i_{\alpha\beta\gamma} x^\alpha y^\beta z^\gamma + 0(N + 1) \tag{1}$$

where $\alpha > 0, \beta > 0, \gamma > 0, \alpha + \beta + \gamma \leq N$ and $0(N + 1)$ denotes higher-order terms. If we require that Q^i is a local solution of certain equations of motion there are intrinsic relationships among some of the coefficients of $A^i_{\alpha\beta\gamma}$. In addition, the no-slip boundary condition has to be fulfilled at the wall ($z = 0$).

The evolving flow structures leave definite but not unique "footprints" of the outer (off-wall, mid-air) flow on the wall. Even in special cases, where vortices are created above the wall bifurcating flows have been characterized by a set of dynamical equations for the velocity field wherein all the coefficients are quantities measured along the wall. In a case considered by Bakker et al. [6] a so-called open flow separation (see Wang [7] for further refs.) results, with no definite starting point for wall-streamline convergence. This is caused by a vortex placed above the wall with a critical point within the velocity field which is shifted off the wall. There may be different ways to create open separation lines via off-wall vortices. On the other hand, the investigations of Dallmann [3], [4], [8] have revealed that two-dimensional separation bubbles together with their three-dimensional bifurcations are "embedded" between fully attaching and completely separating three-dimensional flow structures (Fig. 1a, b). Since wall-streamline convergence or divergence do not render a topological property there is a possibility that local flow attachment occurs at a separation line. We pursue this (non-) unique feature via numerical Navier-Stokes flow simulations. If it actually appears as a structurally stable flow then it would proof the existence of a non-unique relationship between wall-streamline patterns and off-wall flow structures as stated earlier by the first author [3].

The present paper is concerned with the following questions:

1. Is it always possible to detect or describe a topological change in the outer flow field by use of wall-flow information? What kind of information is necessary and sufficient to describe a flow field quantitatively and what is necessary to describe its topological structures?

2. Do the Navier-Stokes equations influence or constrain the kinematically possible velocity structures which form locally at rigid walls due to local flow degeneracies and how do these structures evolve in parameter regions far away from the critical sets of parameters?

3. Is the creation of a "vortex" in three-dimensional steady flows detectable when there is no structural change in the wall shear stress field (wall streamlines) and no critical point within the outer, off-wall flow field?

2   Wall-flow information for off-wall flow analysis

An interesting feature of the analysis of dynamical systems of the type (1) is that the set of equations describing the structural change (rather than the quantitative evolution of the velocity field) does not necessarily have to satisfy all the properties of the equations of motion. The situation will be described in the following by consideration of two-dimensional, incompressible flows. A streamfunction Ψ is designed to fulfill the no-slip boundary condition for the velocity

---

Fig. 1. a Bifurcation diagram with instantaneous streamlines and plane-of-symmetry streamlines for $(\tau, \sigma) \to 0, (\tau_x, \sigma_y) \to 0$. τ, σ are the wall-shear stresses. Changes of the local values of μ and λ cause structural changes when crossing bifurcation lines (solid lines). Flow A is fully attaching and flow B_1 is separating and reattaching. Both have topologically equivalent wall-flow patterns. **b** Structural changes between (secondary) separation bubble and an outer flow field at vanishing local wall-pressure gradient p_x. The dashed lines indicate a possible outer flow structure, namely a primary separation bubble with two cores. **c** Topological structures of bifurcations between sets of stable flows corresponding to Ψ_1 or Eq. (5), respectively

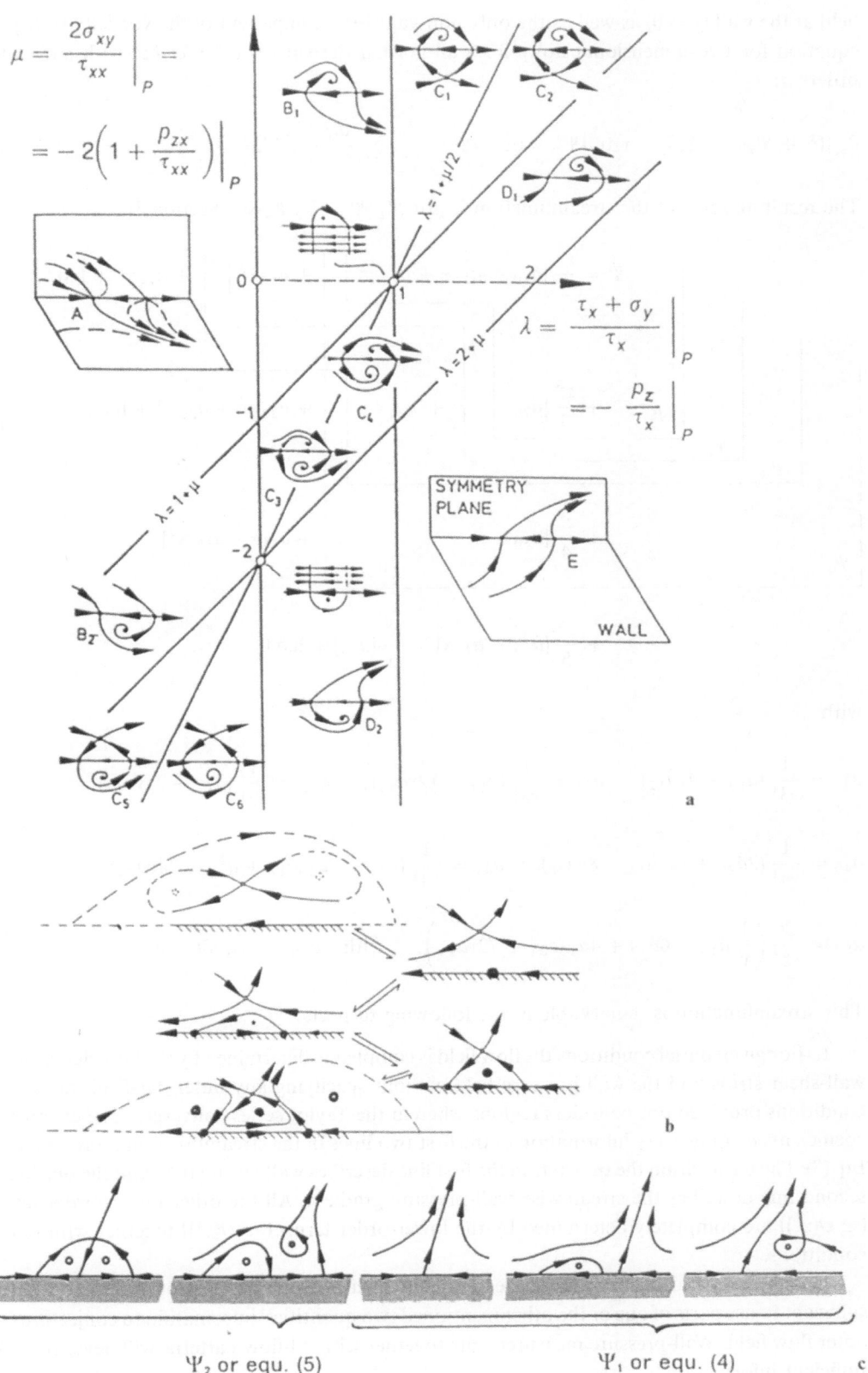

$$\mu = \frac{2\sigma_{xy}}{\tau_{xx}}\bigg|_P =$$

$$= -2\left(1 + \frac{p_{zx}}{\tau_{xx}}\right)\bigg|_P$$

$$\lambda = \frac{\tau_x + \sigma_y}{\tau_x}\bigg|_P$$

$$= -\frac{p_z}{\tau_x}\bigg|_P$$

$\lambda = 1 + \mu/2$

$\lambda = 2 + \mu$

$\lambda = 1 + \mu$

B_1 C_1 C_2 D_1

A

0 1 2

-1 C_4

C_3

-2

B_I

C_5 C_6 D_2

SYMMETRY PLANE

E

WALL

a

b

Ψ_2 or equ. (5) Ψ_1 or equ. (4)   c

field at the wall ($z = 0$) as well as the only non-vanishing component of the vorticity transport equation for two-dimensional flows for wall-normal derivatives $\partial_z{}^n = \partial^n/\partial z^n$ with successive orders n:

$$\partial_z{}^n\{(\partial_t + \Psi_z\partial_x - Y_x\partial_z - \nu\Delta)\,\Delta\Psi\} = 0, \qquad \Psi = \sum_{i=1}^{\infty}\sum_{j=0}^{\infty}\frac{a_{ij}(t)}{i+1}z^{i+1}x^j \tag{2}$$

The resulting form of the streamfunction is (for $\Psi_1, \Psi_2, \Psi_3, \Psi_4$ see Section 3):

$$\Psi = \frac{z^2}{2}[a_{10} + a_{11}x + a_{12}x^2 \;\big|\; + a_{13}x^3 \;\big|\; + a_{14}x^4 + \text{h.o.t.}]$$

$$+ \frac{z^3}{3}[a_{20} \quad + a_{21}x \;\big|\; + a_{22}x^2 + a_{23}x^3 + \text{h.o.t.}] \tag{3}$$

$$+ \frac{z^4}{4}[a_{30} \qquad\qquad\;\big|\; + a_{31}x + a_{32}x^2]$$

$$+ \frac{z^5}{5}[a_{40} + a_{41}x] + \frac{z^6}{6}[a_{50}] + \text{h.o.t.}$$

with

$$a_{30} = \frac{1}{\nu 3!}(\dot{a}_{10} - 4\nu a_{12}), \qquad a_{31} = \frac{1}{\nu 3!}(\dot{a}_{11} - 12\nu a_{13}), \qquad a_{32} = \frac{1}{\nu 3!}(\dot{a}_{12} - 24\nu a_{14})$$

$$a_{40} = \frac{1}{\nu 4!}(2\dot{a}_{20} + a_{10}a_{11} - 8\nu a_{22}), \qquad a_{41} = \frac{1}{\nu 4!}(\dot{a}_{21} + 2a_{10}a_{12} + a_{11}^2 - 24\nu a_{23})$$

$$a_{50} = \frac{1}{\nu 5!}\left(\frac{1}{\nu}\ddot{a}_{10} - 6\dot{a}_{12} + 4a_{10}a_{21} + 72\nu a_{14}\right), \qquad \text{with} \quad \dot{a}_{10} = \partial a_{10}/\partial t, \quad \text{etc.}$$

This streamfunction is remarkable in the following respects:

1. For given initial conditions the flow field is completely determined by the knowledge of the wall-shear stress and the wall-pressure field without specifying any outer (far-field) boundary conditions provided one considers regions wherein the Taylor series converges. The expansion coefficients a_{ij} only carry information of the first two lines in the streamfunction representation Eq. (3). The term within the brackets in the first line describes wall-shear stress and the one in the second line describes the streamwise wall-pressure gradient. All the other higher-order terms $(\geq O(z^4))$ are completely determined by the lower-order terms $(< O(z^4))$ together with initial conditions.

This finding is also important for experimental fluid dynamics. It proves that it is never sufficient to use parts of one or the other mentioned two quantities' information to conjecture the outer flow field. Wall-pressure measurements together with oil-flow patterns will never provide sufficient information.

However, the statement given above can also be reversed to the following:

The outer boundary conditions determine all the coefficients of the wall-shear stress and wall-pressure. This corresponds to the usual way of obtaining solutions of the equations of motion.

2. Another interesting feature of the streamfunction given by Eq. (3) is the fact that nonlinearity or time-dependence or an explicit Reynolds number/viscosity dependence which would reflect properties of the Navier-Stokes equations do only appear at sufficiently high-orders of flow bifurcations. Otherwise, these effects have no direct influence on the locally evolving flow structures.

3   Flow degeneracies and their unfoldings at rigid walls.
Influence of Navier-Stokes equations

Consider now a situation in which a degenerate critical point is instantaneously created on the wall in such ways that the local wall-shear stress and its first and second streamwise derivative vanish simultaneously, i.e. $(a_{10} \sim \mu_1, a_{11} \sim \mu_2, a_{12} \sim \mu_3) \to 0$ and that the local wall-pressure gradient vanishes or not, i.e. $(a_{20} \sim \mu_4) \to 0$ or $a_{20} \neq 0$, respectively. Hence, system (1) with Eq. (2) has multiple degeneracies in its eigenvalues. Following Bakker [5] the unfolded flows are described locally (in the vicinity of the degeneracies defined by the zeros of the a_{ij}) for non-vanishing local pressure gradient by

$$\dot{x} = \Psi_z z^{-1} = U z^{-1} = \mu_1 + \mu_2 x + z + \mu_3 x^2 - x^3$$

$$\dot{z} = -\Psi_x z^{-1} = W z^{-1} = -\frac{1}{2} \mu_2 z - \mu_3 z x + \frac{3}{2} z x^2$$

(4)

and for vanishing local pressure gradient by

$$\dot{x} = \mu_1 + \mu_2 x + \mu_4 z - z x - x^3$$

$$\dot{z} = -\frac{1}{2} \mu_2 z + \frac{1}{3} z^2 + \frac{3}{2} z x^2.$$

(5)

These flows are equivalently described by the truncated local streamfunction representations Ψ_1 (for $a_{20} \neq 0$) and Ψ_2 (for $a_{20} \to 0$), respectively, as shown in Eq. (3) where proper normalization of x and z coordinates has to be imposed to match Ψ_1 and Ψ_2 with the Eqs. (4) and (5), respectively. Ψ_1 or Eq. (4) and Ψ_2 or Eq. (5) describe bifurcations between sets of stable flows with the topological structures sketched in Fig. 1c. In both cases considered there neither nonlinearity nor time-dependence in the Navier-Stokes equations affects explicitly the set of topological flow structures. The same conclusion is derived for those local flow structures which evolve at $(a_{10} = a_{11}) \to 0$ and $a_{12} \neq 0$, $a_{20} \neq 0$. In this case the a_{ij} conditions indicate flow bifurcations at vanishing wall-shear stress and finite (adverse) pressure gradient where either a two-dimensional separation bubble evolves from this degeneracy or (via three-dimensional perturbations) the three-dimensional elementary topological flow structures of Fig. 1a evolve. The local streamfunction representation for this bubble formation is Ψ_3, shown in Eq. (3). In the case $(a_{10} = a_{11}) \to 0$ and $a_{12} \neq 0$, $a_{20} \to 0$, where the interaction of a separation bubble with reverse outer flow at zero local pressure gradient is considered (Fig. 1b) the normal form is $\Psi = \Psi_4$. Such a situation may occur during the formation of a secondary separation bubble within an already existing primary double bubble and its interactions leading to (unsteady) bubble break-up.

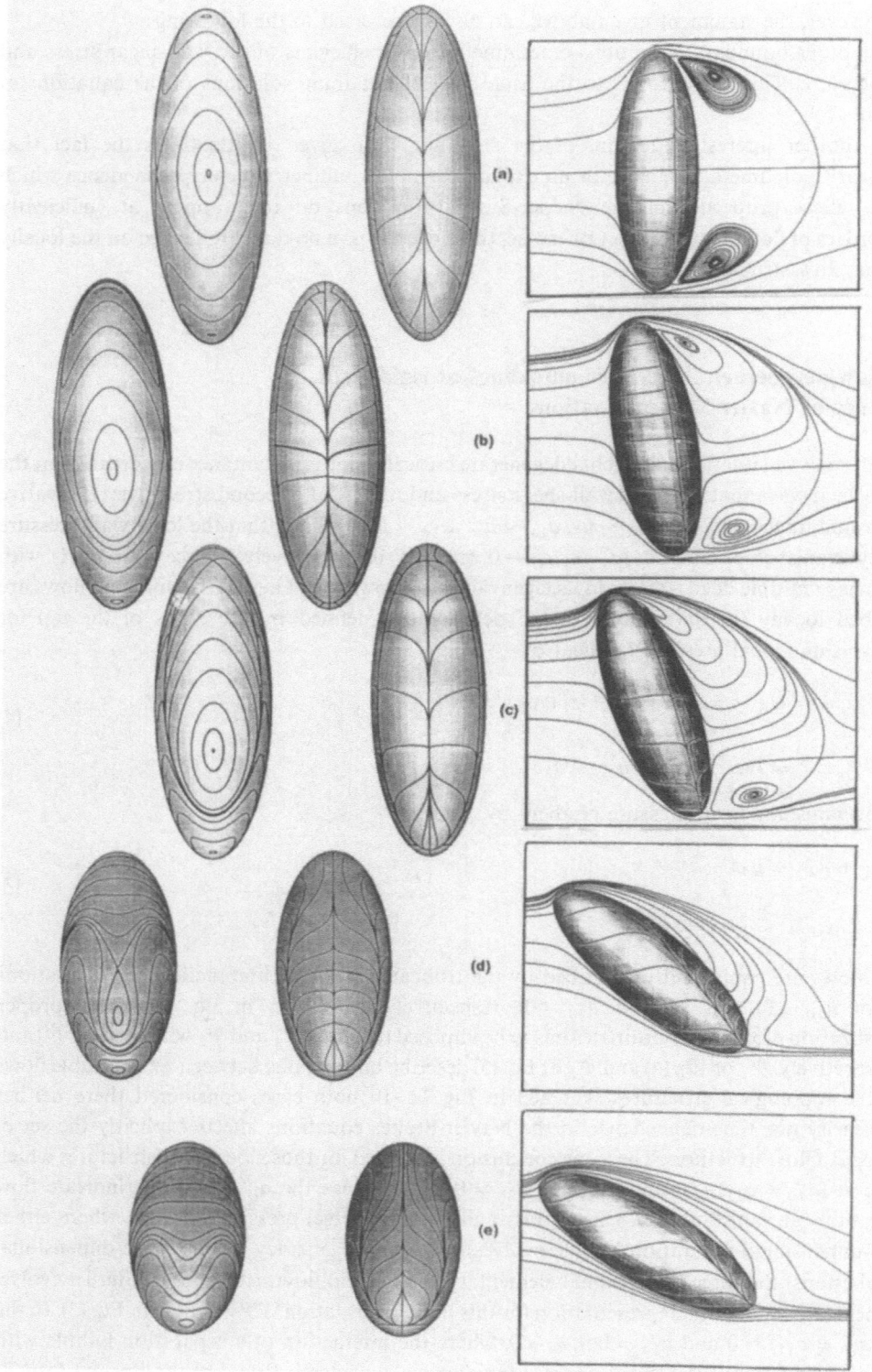

Fig. 2. Wall-vortex lines, wall streamlines and streamlines in plane of flow symmetry around an ellipsoid at Re = 100, Ma = 0.4, α = 90° **a**, 80° **b**, 70° **c**, 40° **d**, and 30° **e**. The perspective views of **b** and **c** are given in Figs. 2 bb and 2 cc, respectively

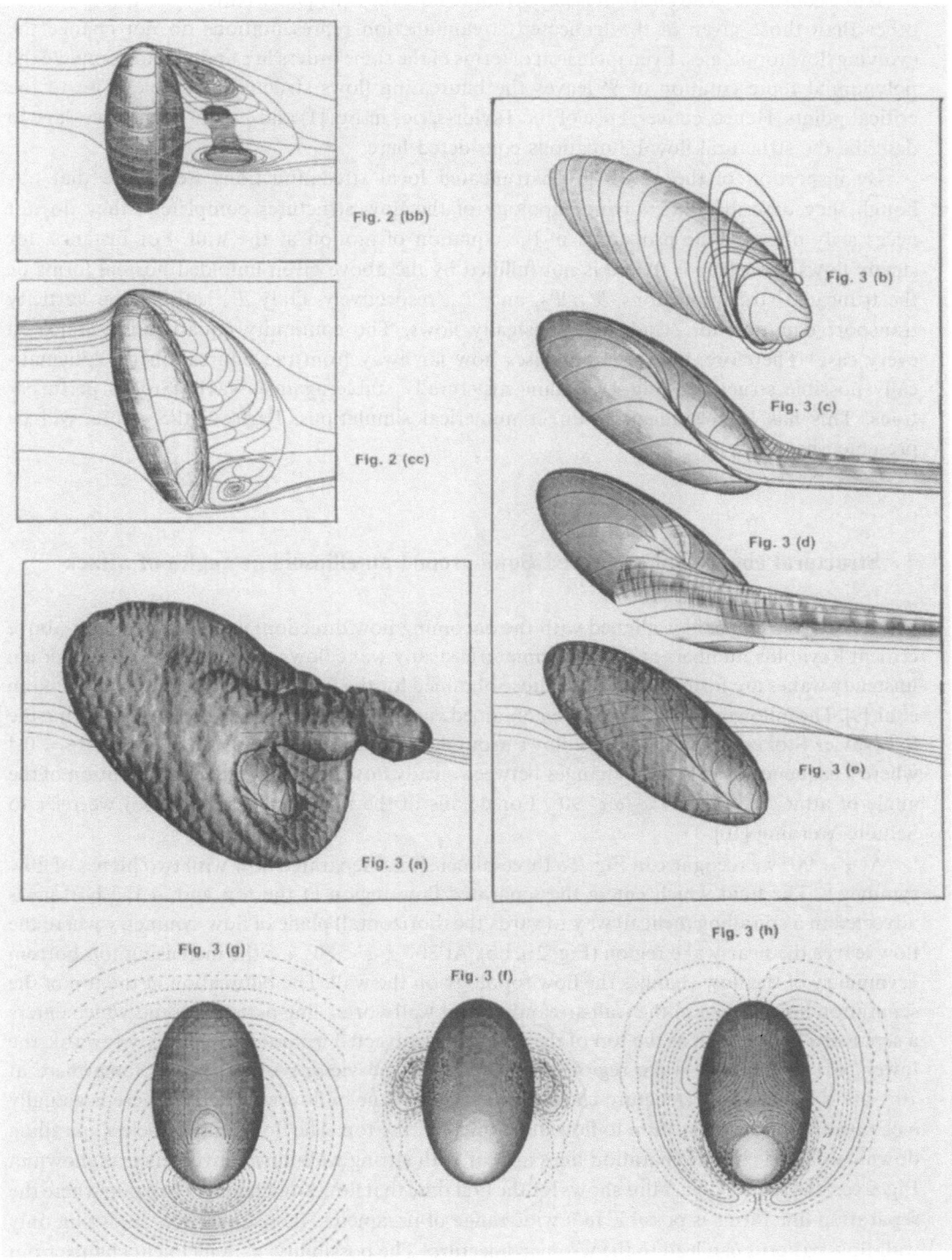

Fig. 3. Steady flow around the ellipsoid at $\alpha = 30°$, Re = 100, Ma = 0.4. **a** Region of complex eigenvalues of ∇Q^i indicating regions of vortical flow, **b** vortex lines, **c, d, e** selected streamlines, **f** region of complex eigenvalues of ∇Q^i, **g** isobars, and **h** x-component of vorticity in a vertical cross section normal to x-direction (oncoming flow direction)

If one considers steady flows it is important to note that any perturbations of Eq. (3) by terms other than those given in the truncated streamfunction representations do not change the evolving flow topologies. Even inclusion of terms of the same orders in z and x (!) to complete the polynomial representation of Ψ leaves the bifurcating flows structurally stable close to the critical points. Hence, convergence of the Taylor-series in Eq. (1) and Eq. (3) is not necessary to describe the structural flow bifurcations considered here.

By inspection of the above given truncated local streamfunctions we realize that although they describe the evolving topology of the flow structures completely, they do not necessarily obey all the properties of the equation of motion at the wall. For instance, for steady flows $\nu\Delta\Delta\phi(x, z = 0) = 0$ is not fulfilled by the above given unfolded normal forms or the truncated streamfunctions, Ψ_1, Ψ_2, and Ψ_3, respectively. Only Ψ_4 satisfies the vorticity transport equation for steady and unsteady flows. The continuity equation is fulfilled in every case! Therefore, the question arises how far away from the bifurcation the kinematically possible structures (Fig. 1) remain structurally stable against Navier-Stokes perturbations. This has been studied by direct numerical simulations. Parts of the results will be presented next.

4   Structural changes of separated flows around an ellipsoid at angles of attack

For $\alpha = 0°$ (the ellipsoid is aligned with the oncoming flow direction) we can expect that above critical Reynolds numbers at first axisymmetric steady wake flows and later three-dimensional unsteady wakes are formed, similar to those obtained for the flow around a sphere by Dallmann et al. [9]. The following results have been obtained via direct numerical simulations by solving the full Navier-Stokes equations for the flows around an ellipsoid ($a/b = 3$) at Re = 100, Ma = 0.4 where a sequence of structural changes between steady flows appears under the variation of the angle of attack between $0° \leqq \alpha \leqq 90°$. For details of the numerical method used we refer to Schulte-Werning [10].

At $\alpha = 90°$ we recognize in Fig. 2 a three-dimensional separated flow with two planes of flow symmetry. The fluid which enters the separated flow region at the top and at the bottom is advected in a spiralling (helical) way towards the (horizontal) plane of flow symmetry where the flow leaves the near-wake region (Fig. 2b, bb). At $80° < \alpha < 80° + \varepsilon$ the increasing top-bottom asymmetry of the flow changes the flow topology on the wall. The bifurcation at the top of the separation lines is seen in the wall-streamline and wall-vortex line patterns. Fluid which enters a separated flow region at the top of the ellipsoid is advected from the top directly towards the lower part of the near wake region before it leaves the vicinity of the body. Somewhere at $70° < \alpha < 80°$ the flow structure changes within the plane of flow symmetry. There is actually a change from flow separation to flow attachment at the top. The flow is nevertheless spiralling downwards and closed separation lines appear with strong wall-flow convergence as shown in Fig. 2 cc. As a matter of fact this shows for the first time that flow attachment in a region where the separation line forms is possible in a wide range of parameters (here α). Hence, analyzing only wall-flow patterns can lead to the wrong conjecture! The possibility of an incipient change from three-dimensional flow separation to flow attachment (with preservation of separation lines!) has been earlier revealed by the *local* high-order bifurcations of Dallmann [3], [4], sketched in Fig. 1. However, we have now shown that the locally evolving flow can remain structurally *stable far away* from a bifurcation set, namely in a wide range of angles of attack, i.e. for $35° \lesssim \alpha \lesssim 80°$ at Re = 100.

Decreasing the angle of attack further one observes that the two attachment points move downwards within the plane of flow symmetry. For $\alpha = 40°$ the single free saddle-focus pattern, where the fluid leaves the near wake, has disappeared completely. For $30° < \alpha < 40°$ a local flow bifurcation can be identified in the plane of flow symmetry which for $\alpha \approx 35°$ creates critical points of flow attachment and not of flow separation (!) on the wall. This is somehow in contrast to what one usually expects to happen at a finite angle of attack. At $\alpha \leqq 30°$ only so called open flow separation lines exist which do not start at critical points of the wall vorticity. A close inspection of the wall-flow patterns shows, that open separation exists, merging with the closed one, even for $\alpha > 30°$.

Now the following question arises: Is there a critical point above the wall as suggested by special flow bifurcations studied in Bakker et al. [6]. The answer is: No, we cannot find any off-wall critical point within the velocity field. We therefore asked ourselves: Is such an open flow separation, eventually, associated with any structural change in any wall-flow quantity other than the wall-shear (wall-vorticity)? How about the wall-pressure gradient? In Dallmann et al. [11] and Herberg et al. [12] we have already shown that the critical point of the wall-pressure gradient (a local minimum of the wall pressure) actually does not appear *on* the open separation line and, even more confusing, it appears *downstream* of the formation region of the open separation line. Is there, perhaps, a critical point of the (outer) vorticity field associated with open flow separation? We have not been able to identify one. Although the vortex lines are stretched and folded and locally twisted (Fig. 3) so far, we could not detect any vortex-line reconnection.

Therefore, we have checked those regions in space where the velocity gradient tensor exhibits complex eigenvalues. According to Dallmann [13] and Vollmers [14], and later reinvented by Chong et al. [2], only these regions can exhibit vortical (helical, spiralling or recirculating) instantaneous fluid motion to any observer moving in Galileian invariant frames of reference. Figure 3a provides the three-dimensional representation of this vortical flow region around the ellipsoid at $\alpha = 30°$. Vortices can be detected within the shaded region shown around the ellipsoid. The horn-like regions indicate that streamwise vortices exist. It is remarkable that no vortices can appear immediately on top of the open separation lines since the downstream parts of the open separation lines are not covered by the region of complex eigenvalues.

Iso-pressure surfaces have been frequently used for "vortex identification" amongst several other quantities [15]. However, neither the iso-pressure surface (Fig. 3g) nor the normalized helicity density (indicating the angle between streamlines and vortex lines) nor the twist of neighbouring streamlines (Herberg et al. [12]), nor the enstrophy exhibit an extremum at the position of those horns (Fig. 3a, f) in our low-Re case. So far, we have been only able to identify that the x-component of vorticity (Fig. 3h) has a local maximum in the horn region. The dilemma of defining or locating steady streamwise vortices due to open flow separation remains unsolved.

References

[1] Perry, A. E., Chong, M. S.: A description of eddying motions and flow patterns using critical point concepts. Ann. Rev. Fluid Mech. **19**, 125–155 (1987).

[2] Chong, M. S., Perry, A. E., Cantwell, B. J.: A general classification of three-dimensional flow fields. In: Topological fluid mechanics. Proc. of the IUTAM Symposium, Cambridge, August 13–18, 1989 (Moffatt, H. K., Tsinober, A. eds.), pp. 408–420. Cambridge: Cambridge University Press 1990.

[3] Dallmann, U.: Topological structures of three-dimensional vortex flow separation. AIAA-83-1735 (1983) with further details in: On the formation of three-dimensional vortex flow structures DFVLR-IB 221-85 A13 (1985).

[4] Dallmann, U.: Three-dimensional vortex structures and vorticity topology. In: Proc. of the IUTAM Symposium on Fundamental Aspects of Vortex Motion, Tokyo, Japan, 1987 (Hasimoto, H., Kambe, T., eds.), pp. 183–189. Amsterdam: North Holland 1988.

[5] Bakker, P. G.: Bifurcations in flow patterns. PhD Thesis, Delft University of Technology 1988.

[6] Bakker, P. G., de Winkel, M. E. M.: On the topology of three-dimensional separated flow structures and local solutions of the Navier-Stokes equations. In: Topological Fluid Mechanics, Proc. of the IUTAM Symposium, Cambridge, August 13–18, 1989 (Moffatt, H. K., Tsinober, A. eds.), pp. 384–394. Cambridge: Cambridge University Press 1990.

[7] Wang, K. C., Zhou, H. C., Hu, C. H., Harrington, S.: Three-dimensional separated flow structure over prolate spheroids. Proc. R. Soc. London Ser. A **421**, 73–90 (1990).

[8] Dallmann, U., Kordulla, W., Vollmers, H., Schulte-Werning, B.: Analysis of the changing topological structures of three-dimensional separated flows. DLR-IB 221-92 A12 and in: Physics of separated flows (K. Gersten, ed.), pp. 249–256. Braunschweig/Wiesbaden: Vieweg 1993 (Notes on Numerical Fluid Mechanics, Vol. 40).

[9] Dallmann, U., Gebing, H., Vollmers, H.: Unsteady three-dimensional separated flows around a sphere – analysis of vortex chain formation. In: Proc. IUTAM Symposium on Bluff-Body Wakes, Dynamics and Instabilities, 7.–11. Sept. 1992, Göttingen (Eckelmann, H. ed.), pp. 27–30. Berlin Heidelberg New York Tokyo: Springer 1993.

[10] Schulte-Werning, B.: Numerische Simulation und topologische Analyse der abgelösten Strömung an einer Kugel. DLR-FB 90-43, 1990 (Dissertation Univ. München, 1990).

[11] Dallmann, U., Hilgenstock, A., Riedelbauch, S., Schulte-Werning, B., Vollmers, H.: On the footprints of three-dimensional separated vortex flows around blunt bodies. Attempts of defining and analyzing complex flow structures. AGARD CP 494, 9.1–9.13 (1991).

[12] Herberg, T., Dallmann, U.: Untersuchung des dreidimensionalen Wirbelstärkefeldes und der Entstehung von Wirbeln am Deltaflügel. In: Proc. 8. DGLR-Symposium „Strömungen mit Ablösung", Köln, 10.–12. Nov. 1992, pp. 139–143. DGLR-Bericht 92-07, Bonn: DGLR 1992.

[13] Dallmann, U.: Topological structures of three-dimensional flow separations. DFVLR-IB 221-82 A07 (1982).

[14] Vollmers, H., Kreplin, H.-P., Meier, H. U.: Separation and vortical-type flow around a prolate spheroid – evaluation of relevant parameters. AGARD-CP-342, 14.1–14.14 (1983).

[15] Wray, A. A., Hunt, J. C. R.: Algorithms for classification of turbulent structures. In: Topological fluid mechanics, Proc. of the IUTAM Symposium, Cambridge, August 13–18, 1989, (Moffatt, H. K., Tsinober, M. A., eds.), pp. 95–104. Cambridge: Cambridge University Press 1990.

Authors' address: Dr. U. Dallmann and Dipl.-Phys. H. Gebing, DLR – Institute of Theoretical Fluid Mechanics, D–37037 Göttingen, Federal Republic of Germany

Acta Mechanica (1994) [Suppl] 4: 57 – 67

Near-wall phenomena in turbulent separated flows

H. H. Fernholz, Berlin, Federal Republic of Germany

Summary. A comparison is presented of experimental results such as pressure, skin friction (mean and higher moments), and reverse-flow factor to demonstrate the differences between weak and strong closed reverse-flow regions. The driving mechanisms are shear stress or pressure combined with shear stress, causing clearly distinctive distributions of the mean skin friction and its intensity. The "failure" of the reverse-flow parameter reaching 100% in the centre of the separation bubble can be explained by lumps of the free shear layer penetrating to the wall from above and flowing partly downstream near the wall. The outer layer flow structures interact with near-wall streaky structures moving upstream which differ considerably in width and length from those observed in boundary layers. These structures are shown by spanwise correlation measurements and by flow visualization. Suggestions are made for model flows covering weak and strong reverse-flow regions, respectively.

1 Introduction

Flows with separation and reattachment are so complex that it may be worthwhile to discuss their properties with the aim of finding two or more types of separation regions which share certain flow properties and which are therefore more amenable to modeling or to calculation. This discussion will center around near-wall phenomena because of their importance for momentum, heat, and mass transfer and because so much more information is already available about the outer shear layer.

We shall confine ourselves here to incompressible turbulent shear layers bounded by a wall which separate from and reattach again at the wall, forming a closed reverse-flow region or bubble. Separation may be fixed, as at a sharp edge, or free, as in a boundary layer on which an adverse pressure gradient is imposed. The flow is nominally two-dimensional up to separation and upstream of reattachment and steady in the mean.

Experimental investigations show that there are at least two distinct types of closed reverse-flow regions:

(1) The weak reverse flow region (WRF) and (2) the strong reverse-flow region (SRF).

In the first case (WRF) the turbulent boundary separates from the surface due to a sufficiently strong adverse pressure gradient. The separation process depends on the upstream history of the flow and the boundary conditions in the streamwise direction. In the second case (SRF), a boundary layer separates from a sharp edge forming a separated shear layer — often curved — which covers a cavity-like flow, largely independent of its upstream history, if the upstream boundary layer is turbulent.

Characteristic features of these two types of reverse-flow region will be exemplified by measurements (a) in a boundary layer with a long and shallow separation bubble (e.g. [1]) and (b) in the closed reverse-flow region downstream of a normal plate with a splitter plate in flow direction (Fig. 1a) as investigated by [2], [3], and [4], for example. In the following the

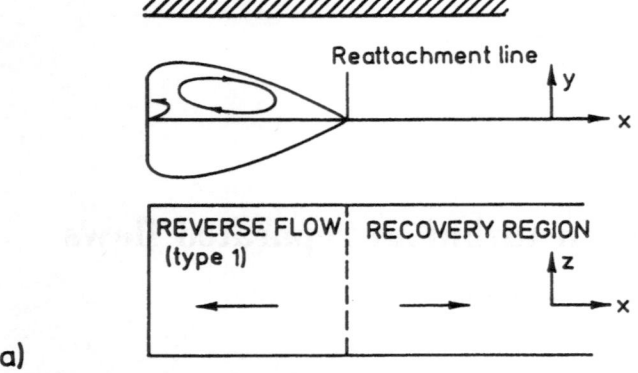

a)

Flow configuration : Normal plate / splitter plate (Re$_H$≈500)

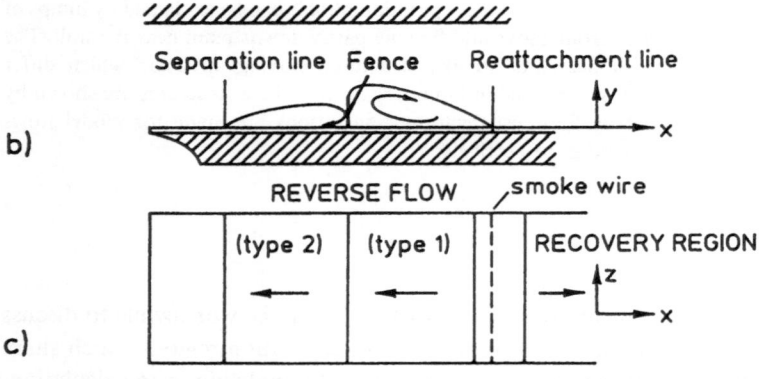

b)

c)

Flow configuration : b) Fence in a thin boundary layer (Re$_H$ = 1650)
                         and c) ZPG turbulent boundary layer (fence removed)
Flow visualization : Green (0≤y≤4mm) and blue(4≤ y≤8mm) Laser-light
                          sheet/smoke wire. Air(u$_∞$ =0.75 m/s).
Flow configuration: c) Fence removed so that the turbulent boundary
                         is undisturbed.

Fig. 1. Schematic diagram of flow configurations showing locations of streak patterns in reverse-flow regions

development in the streamwise direction of some near-wall properties will be described for these two types of flow and physical explanations will be presented in an attempt to provide a basis for the modeling of closed reverse-flow regions.

2 Flow properties in the near-wall region of a closed reverse flow

(1) The weak reverse-flow region (WRF) is characterized by distributions of a weakly rising static pressure, almost constant negative mean skin friction, and a low almost constant r.m.s. skin-friction fluctuation in the streamwise direction. This often shallow bubble is embedded in a highly disturbed boundary layer.[1]

---

[1] Highly disturbed means here that it will take many boundary layer thicknesses, before the downstream boundary layer is back to local equilibrium in the near-wall region.

Figure 2a displays the dimensionless distributions of the static pressure c_p in the mainstream direction, the mean skin friction $\bar{c}_f$, and the reverse-flow parameter at the wall χ_w in a typical WRF plotted against x/x_R, with x_R being the length of the reverse-flow region [1]. The reverse-flow parameter is defined as the ratio of the upstream events to the total number of events as measured by the sensor wires of a pulsed-wire probe. Separation occurs at $x/x_R = 0$ and reattachment at $x/x_R = 1$ (also in Fig. 2b). Since there is only a small pressure gradient, the flow in the reverse-flow region is energetically weak near the wall, resulting in a low negative mean skin-friction distribution.

The maximum value of χ_w is at most 70% which indicates that upstream events are relatively frequent. One should also note that upstream of separation ($\bar{\tau}_w = 0$) and downstream of reattachment ($\bar{\tau}_w = 0$), respectively, χ_w is well above zero (see also [5]). This means that effects caused by instantaneous reverse flow influence both the boundary layer upstream of separation and downstream of reattachment. The distribution of the r.m.s. value of the fluctuating skin friction $\tau'_{wr.m.s}$ decreases sharply with increasing pressure and then remains almost constant along the length of the reverse-flow region (see [6], Figs. 4, 5).

(2) The strong reverse-flow region (SRF) is characterized by large changes in the static pressure and the negative mean skin friction and by fluctuations of the skin friction on a high level (higher than in a zero pressure gradient boundary layer) in the streamwise direction. The SRF is pressure and shear driven and often bounded by a curved shear layer.

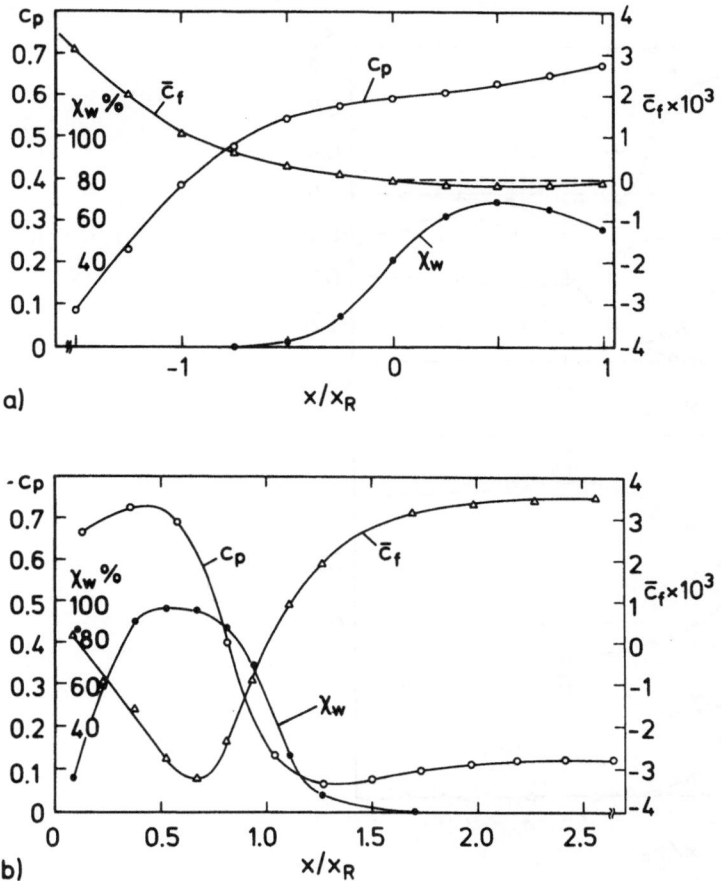

Fig. 2. Typical distributions of the static pressure $\bar{c}_p$, the reverse-flow parameter χ_w, and the skin-friction coefficient $\bar{c}_f$ in a weak **(a)** and a strong **(b)** closed reverse-flow region [1, 2]

Figure 2 b displays the mean skin-friction coefficient $\bar{c}_f$, the reverse-flow parameter, and the pressure coefficient c_p in the streamwise direction in a strong reverse-flow region (the origin of the streamwise coordinate x is at separation). In contrast to the WRF we notice a large pressure difference, pushing the near-wall flow upstream which, in combination with the curved shear layer, generates a "pressure and shear driven" reverse flow. This causes high negative mean skin friction and a very high level of $\tau'_{wr.m.s}$ as compared with that in the WRF (see [2]). The maximum level of χ_w is above 90% but never 100%. This may appear to be unexpected at first but will be explained later by flow-visualization. It should be mentioned here that the distributions of the pressure coefficient in several SRF configurations fall on one curve (see Fig. 5 in [4]) if normalized according to [7].

3 Instantaneous values of the skin friction and the higher moments in weak and strong reserve-flow regions

Skewness S_w and flatness F_w distributions were measured by means of a wall pulsed-wire probe in a WRF [1] and in a SRF ([8] unpublished data), and are plotted in Figs. 3a and b. At the wall the instantaneous shear stress is linearly related to the local velocity field, so S_w and F_w give information about the near-wall fluctuations.

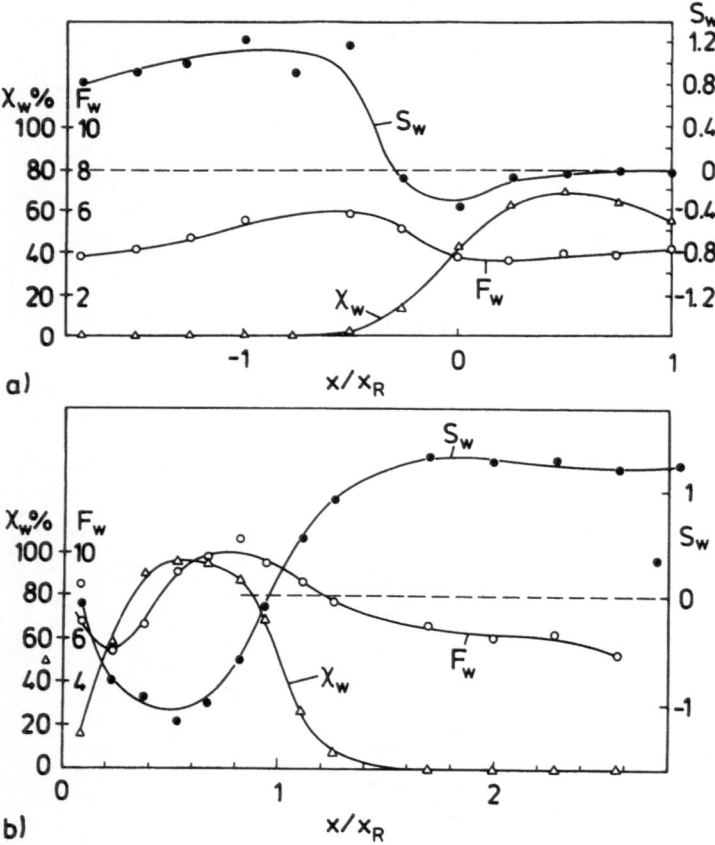

Fig. 3. Skewness S_w and flatness F_w of the wall shear stress and the reverse-flow factor χ_w in a closed WRF region **(a)** and in a SRF region **(b)** ([1] and [8], unpublished data)

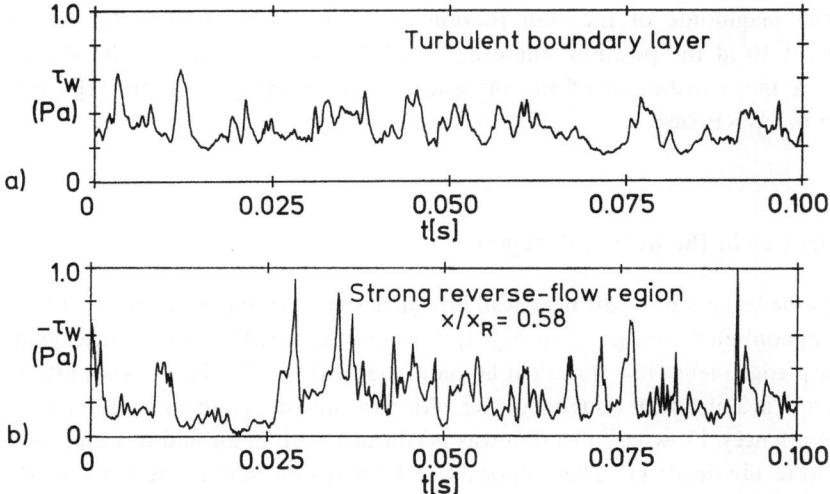

Fig. 4. Typical time record of the wall shear stress **a** in a zero pressure gradient turbulent boundary layer and **b** in a SRF region downstream of a fence. From Wagner (1991), private communication

S_w and F_w show qualitatively similar patterns in the attached forward-flow region, upstream of the bubble in the WRF and in the attached flow downstream of the bubble in the SRF case, with large positive deviations from Gaussian behaviour ($S_w = 0$ and $F_w = 3$) in the adverse pressure gradient and in the recovery regions, respectively (deviations similar to those in zero pressure gradient flow). This behaviour is consistent with what [9] found in channel flow. They showed that near the wall in channel flow a positive skewness reflects the presence of high shear events imposed on a background of lower turbulence, also indicated by the high values of the flatness factor. This flow pattern appears to be characteristic of fully attached wall bounded turbulent flows. A typical time record of τ_w was measured by means of a wall hot-wire probe (with an active wire length $l^+ = \dfrac{l u_\tau}{\nu} = 15$) in a zero pressure gradient turbulent boundary layer and is shown in Fig. 4a.

Towards separation and into the separation region of the WRF case (Fig. 3a) one notices the onset of instantaneous reverse flow, $\chi_w > 0$, and first a decrease of S_w and F_w and then an increase to almost Gaussian values which could indicate a reduction in the frequency of high-amplitude bursts and a disappearance of this dual nature. This interpretation is consistent with the observation that production of turbulent energy $\overline{u'v'}\dfrac{\partial \bar{u}}{\partial y}$ moves away from the wall in this region [1]. This is obvious since the peak of the $\overline{u'v'}$-distribution moves away from the wall and $\partial \bar{u}/\partial y$ decreases near the wall in an adverse pressure gradient boundary layer.

In the case of SRF (Fig. 3b) the onset of instantaneous reverse flow again coincides with a sharp decrease of the skewness S_w but now with a change of sign from $+1$ to -1. This reflects the change from high positive shear events in the recovery region to high negative ones in the SRF region. Production of turbulent kinetic energy near the wall is high as compared with the WRF case (deduced from measured values of $\partial \bar{u}/\partial y$ and $\overline{u'^2}$; $\overline{v'^2}$ and $\overline{u'v'}$ could not be measured). High negative instantaneous values of τ_w can be seen in the time record (Fig. 4b), measured in the reverse-flow region downstream of a fence and this points also to high values of the Reynolds shear stress near the wall. In both cases

(Fig. 4a and b) the magnitude of the skin friction was equal. The flatness F_w reaches a maximum of about 10 at the position where $S_w = -1$. Such high values of the flatness usually indicate that the distribution of the intensity of the quantity is spotty [1], again a feature absent in the WRF case.

4 Coherent structures in the near-wall region

Further insight into the behaviour of the near-wall flow in a SRF-region can be obtained from flow visualization in configurations given in Fig. 1, a normal plate with a long splitter plate (Fig. 1a) and a sharp-edged fence in a turbulent boundary layer (Fig. 1b). These flow patterns are compared with that in a turbulent boundary layer with zero-pressure gradient (configuration Fig. 1c i.e. without a fence). Flow visualization was performed both in air and in water. Two important results were obtained: (1) There appears to be a strong interaction between the separated shear layer and the near-wall flow by transport of material in the direction normal to the wall (black-hole effect). (2) The coherent structures in SRF in the vicinity of the wall are distinctly different from those in an attached turbulent boundary layer.

One explanation for the spottiness of the shear stress in the vicinity of the wall is the impingement of lumps of energetic turbulent flow which are transported from the separated shear layer almost vertically down into the near-wall flow and spread in all directions. Flow

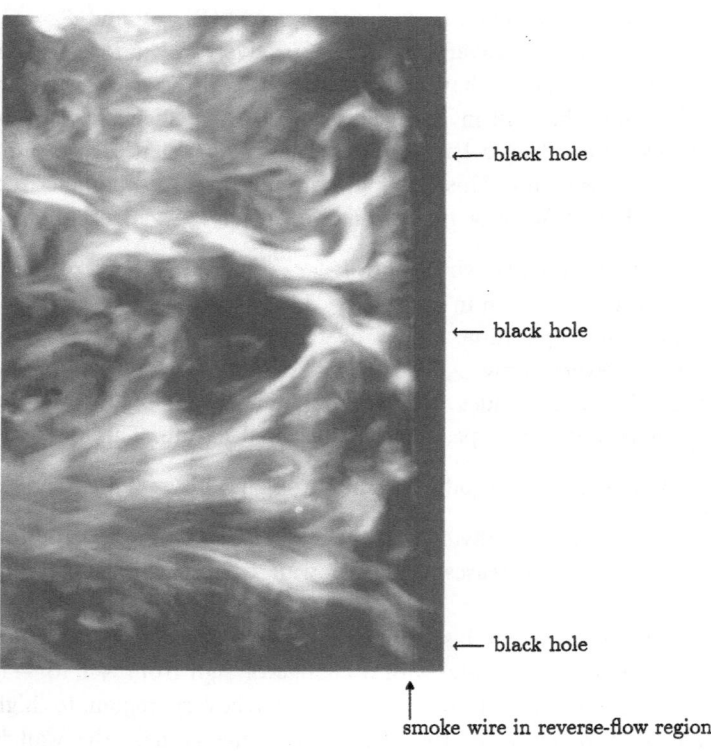

←— black hole

←— black hole

←— black hole

↑
smoke wire in reverse-flow region

⟹   main flow direction

Fig. 5. Visualization of the reverse flow near the wall: smoke from a wire (1 mm above the wall), illuminated by green and blue laser sheets parallel to the wall (green $0 \leq y \leq 4$ mm, blue $0 \leq y \leq 8$ mm). From Wagner and Gründel, private communication, 1991)

visualization (smoke illuminated by green and blue laser sheets) in a SRF-region (Fig. 1b, Wagner/Gründel, private communication) showed that the smokeless flow down from the separated shear layer displaces the smoke layer parallel to the wall revealing the "black colour" of the wall. This displacement causes an additional movement of the near-wall flow in all directions, i.e. also downstream. Such a characteristic flow situations in the reverse-flow region downstream of a fence is visualized in Fig. 5 with the main flow from left to right. Figure 5 displays several "black holes" upstream of the smoke wire and rather wide near-wall streaks, when compared with the slim flow structures in a zero pressure gradient turbulent boundary layer (not shown here, see [11]). Similar information was obtained when these flow visualization experiments (video film and still photographs) were repeated in water ($Re_H \approx 1200$) with hydrogen bubbles generated at a platinum wire and at a wire rake situated parallel to and 1 mm above the wall (Wagner/Imbert, private communication (1990)). The visualization of the flow normal to the wall showed that the impingement of the flow down from the shear layer occurred randomly over a large part of the SRF-region ($0.3 < x/x_R < 1$) and not only for $x/x_R > 0.7$ as reported by [12]. Furthermore an upward flow away from the wall was observed transporting low energy fluid away from the near-wall region. This "lifted backflow" was also reported by [12] with the plausible explanation that a vortical eddy approaching the wall would tend to lift the fluid near the wall.

The irregular occurrence of turbulent lumps in the near-wall region — which has also been observed in LES calculations on backward-facing step flow by [13] — may serve to explain three phenomena:

(a) The upstream mass-flow contribution prevents the reverse-flow parameter χ_w from approaching one;

(b) The mass flow in the reverse-flow region is not only supplied from the reattachment region but also directly down from the shear layer (see also [14] in the case of an open separation region);

(c) Low-energy fluid from the near-wall region is transported towards the separated shear layer.

The flow visualizations presented above show also that the width of the coherent streamwise structures in the near-wall region is distinctly different in a SRF from that in an attached wall bounded flow — e.g. the recovery region downstream from reattachment or a turbulent boundary layer with zero pressure gradient (see for example [15]). Preliminary correlation measurements by wall hot-wire probes — though at a higher Reynolds number ($Re_H = 14000$) — show a width at least ten times larger for the structures in the SRF-region than for the structures in a turbulent boundary layer with the same skin friction.

5 Considerations on the modeling of the flow in closed reverse-flow regions

Based on the near-wall measurements and the interpretation of the physical properties of the flow, it is appropriate to discuss the possibility of devising a model for the flow in closed reverse-flow regions. Such a zonal model may be useful until calculations by DNS or LES methods can be extended to higher Reynolds number ranges and to the near-wall region or more appropriate turbulence models (e.g. [16]) are available than at present. It will turn out that models for SRF must, of course, be supplemented by properties of the outer shear layer.

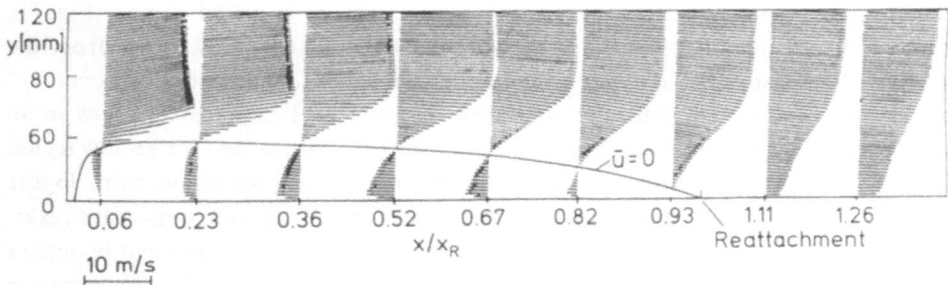

Fig. 6. Mean velocity $\underline{c} = \underline{c}(\underline{u}, \underline{v})$ distribution in a strong reverse-flow region (configuration a in Fig. 1, $Re_{hF} = 14 \cdot 10^3$) [8]

The distributions of the mean velocity profiles (Fig. 6) in both WRF-regions (e.g. [6]) and SRF-regions (e.g. [8]) resemble those of a wall jet in a counterflow with the corollaries (a) that the maximum velocity first increases and then decreases and (b) that the static pressure first decreases and then increases (seen from reattachment). These conditions are difficult to meet in a flow model but the outlook for finding a solution to the modeling problem may become slightly brighter if weak and strong reverse flows are treated separately.

5.1 The weak reverse-flow region

In WRF the mean reverse-flow velocity is low, the skin friction and with it the Reynolds shear stresses are small, and the turbulence-energy production near the wall is negligible, since both $\overline{u'v'}$ and $\partial \bar{u}/\partial y$ are small. Turbulence is sustained in this region by turbulent diffusion towards the wall [14] but this transport process is usually weak. So there is only a weak interaction between the near-wall flow in a shallow bubble and the separated boundary layer above. This means also that the influence of the wall is negligible except that the wall has the important function of constraining the normal component of the velocity v at the wall and the spreading of the separated shear layer on one side. One may further assume that the interaction of the reverse flow with the separated boundary layer is small. Under these conditions it appears to be justified to extend the boundary-layer model and not to incorporate the more complicated wall-jet model to flows which include separation regions if the artificial singularities at separation and reattachment can be overcome. This may be achieved by using a triple-deck method or an inverse boundary-layer calculation method and a boundary condition other than the external pressure. One suitable boundary condition is a length scale which takes into account the spreading of the separated shear layer normal to the wall or the bubble height. The displacement thickness δ_1 has been used in several calculation methods (e.g. [17]) as such a characteristic length and has been shown to be, indeed, more sensitive to small changes in the boundary conditions, such as the pressure distribution in streamwise direction, than other boundary-layer integral length scales [1]. If the distribution of the displacement thickness is known, then the classical boundary-layer model can be used. The only remaining problem is the lack of a satisfactory turbulence model, as shown for example by [17]. As a preliminary limit for the validity of this WRF-model one might use the asymptotic mean velocity profile for separated boundary layers, which does not hold, however, in a SRF [1].

5.2. The strong reverse-flow region

In the SRF the mean reverse-flow velocity U_N is comparatively high ($U_N/U_{max} \lessgtr 0.30$); so is the skin friction[2] ($|c_{f max}|/c_{f zPG}$) $\lessgtr 0.7$) and with it the mean-velocity gradients and the total shear stress.[3] The intensity of the velocity fluctuations parallel to the wall in the near-wall region is of the order of magnitude of the local mean velocity [2]. In contrast to the WRF case, the near-wall flow is therefore highly energetic and should interact strongly with the separated shear layer. These phenomenological facts indicate that a boundary-layer model, as applicable for a WRF, would no longer be appropriate, and it may be useful to look for similarities between the highly energetic near-wall flow of a SRF-region and a wall jet. Beginning with the mean velocity distribution $\bar{u}/\bar{u}_\tau$ normal to the wall, there is experimental evidence that the velocity distribution is linear instantaneously and in the mean (linear law of the wall). In a SRF all the available, though scarce, velocity measurements in the sublayer (e.g. [18, 2, 19]) confirm the validity of the

linear law up to a dimensionless wall distance $y^+ = \dfrac{u_\tau y}{v} \leqq 8$, where u_τ is the skin-friction velocity and v the kinematic viscosity.

There is *less* general consensus about the validity of the logarithmic part of the law of the wall although the above mentioned measurements of skin friction and mean velocity show that there is *no* logarithmic law in a SRF-region, as already indicated by [5] and shown experimentally by [2], for example, where skin friction and mean velocities were measured by pulsed-wire probes and the near wall velocities by a hot wire.

Returning to the wall jet model, it is suggested that the mean velocity distributions in a SRF in the near-wall region show a pattern which is probably similar to that of a plane wall jet in a counterflow. Although no measurements of this specific case which could be used for this discussion are known,[4] one can show that such a similarity is very probable by looking first at the behaviour of the velocity profiles in a wall jet underneath an external stream with a flow direction parallel to the wall jet and then at the velocity profiles in a wall jet without an external flow (e.g. [20]). In the former case the velocity profiles follow the log-law in a range of the wall coordinate $50 \leqq y^+ \leqq 600$. In the latter case this range is considerably shorter ($30 \leqq y^+ \leqq 60$) and it is to be expected that the log-law region will disappear completely if there is a counterflow. This is exactly what is noticed in the SRF-region. There are two physical reasons for the disappearance of the logarithmic law of the wall: (a) The strong influence of the outer flow on the inner region, mainly by the shear stress-distribution and the production mechanism of turbulent energy; (b) the existence of a log-law region is generally considered as the result of a balance between production and dissipation in this flow region. Such a balance may not exist here and might have to be replaced by a balance between production, turbulent diffusion and dissipation. This change of the importance of terms in the transport equations must be taken into account by turbulence models for the near-wall region of a SRF. Unfortunately it is not possible at present to show whether the same balance or the classical balance between production and dissipation holds for a wall jet in a counterflow since there are no measurements. Furthermore the wall-jet analogy is valid only for the flow region where the maximum velocity U_N of the wall jet decreases. This is in the downstream part of the reverse flow. In the upstream region the maximum velocity

---

[2] $c_{f zPG}$ denotes the skin friction where the pressure gradient dp/dx is practically zero downstream.

[3] [18] claim that the Reynolds shear stress is small in the near-wall region based on data obtained by [5], [12], and [21] but more data is needed in my opinion to confirm this view (see also [22]). Their point is strengthened, however, by a reduced bursting rate in a wall-bounded shear flow with a favourable pressure gradient [15] but this has not been shown yet for SRF-regions.

[4] There are two investigations of wall jets in a counterflow but they do not contain the information necessary for the present consideration [23] and [24]).

of the model wall jet would have to increase, a condition which cannot be fulfilled by a normal wall jet unless it is made up by a number of successive jets with increasing velocity. So this model is rather complicated in parts of the flow region but it is at least physically more realistic than a boundary-layer model with the boundary layer starting at the reattachment line and continuing upstream into the SRF-region (e.g. [25]). Many good arguments against such a boundary-layer model have been put forward already by [26] and by [18]. In addition to these reasons it should be remembered that the boundary-layer concept as such does not apply to such a flow and that the boundary conditions are also different from those for wall boundary layers. So there are important similarities between a wall jet in a counter-flow and the reverse-flow in a SRF region but there remains the problem of the increasing maximum velocity which cannot be reconciled easily with a simple wall-jet model.

6 Conclusion

Near-wall measurements and flow visualization in closed reverse-flow regions provide enough information to distinguish between two types of flow, named here weak and strong reverse-flow regions. Although both flows show similar mean velocity distributions, their physical properties, such as mean and fluctuating kinetic, wall shear stress, and production of turbulent energy are widely different. There is little interaction between the WRF and the separating boundary layer on the one hand and strong interaction between the SRF and the separated shear layer on the other. It appears that WRF-regions can still be incorporated with some modifications into the classical boundary-layer concept, whereas SRF can only be dealt with if a new zonal flow model can be devised. First attempts are made to suggest such a SRF-model by using the analogy of a wall jet in a counterflow. In order to connect the model flow to the main flow, that is to the separated shear layer, matching conditions must be provided which take care of the outer boundary conditions of the reverse flow and the inner boundary conditions of the shear layer. Since more experimental information is available in the outer region of such flows than in the near-wall region, such an evaluation should be possible.

References

[1] Dengel, P., Fernholz, H. H.: An experimental investigation of an incompressible turbulent boundary layer in the vicinity of separation. J. Fluid Mech. **212**, 615−636 (1990).
[2] Ruderich, R., Fernholz, H. H.: An experimental investigation of a turbulent shear flow with separation, reverse flow, and reattachment. J. Fluid Mech. **163**, 283−322 (1986).
[3] Castro, I. P., Haque, A.: The structure of a turbulent shear layer bounding a separated region. J. Fluid Mech. **179**, 439−468 (1987).
[4] Jaroch, M. P., Fernholz, H. H.: The three-dimensional character of a nominally two-dimensional separated turbulent shear flow. J. Fluid Mech. **205**, 523−552 (1989).
[5] Simpson, R. L., Chew, Y.-T., Shivaprasad, B. G.: The structure of a separating turbulent boundary layer. Part 1. Mean flow and Reynolds stresses. J. Fluid Mech. **113**, 23−51 (1981).
[6] Dengel, P.: Über die Struktur und die Sensibilität einer inkompressiblen turbulenten Grenzschicht am Rande der Ablösung. Dissertation, TU Berlin 1992.
[7] Roshko, A., Lau, J. C.: Some observations on transition and reattachment of a free shear layer in incompressible flow. In: Proc. 1965 Heat Transfer and Fluid Mech. Inst., (vol. 18, 1965, pp. 157−167).
[8] Ruderich, R.: Entwicklung des Nachlaufs einer senkrechten Platte längs einer zur Anströmung parallelen ebenen Wand. Dissertation TU Berlin D83 1985.

[9] Johansson, A. V., Alfredsson, P. H.: On the structure of turbulent channel flow. J. Fluid Mech. **122**, 295–314 (1982).

[10] Townsend, A. A.: The structure of turbulent shear flow. Cambridge: Cambridge University Press 1976.

[11] Fernholz, H. H.: Near-wall phenomena in turbulent separated flows. Institutsbericht 01/92, Hermann-Föttinger-Institut, TU Berlin 1992.

[12] Pronchick, S. W.: An experimental investigation of the structure of turbulent reattaching flow behind a backward-facing step. Ph. D. Thesis, Stanford Univ. Dept. Mech. Eng. Univ. Microfilms 8 320 765 1983.

[13] Friedrich, R., Arnal, M.: Analysing turbulent backward-facing step flow with the lowpass-filtered Navier-Stokes equations. J. Wind Eng. Industr. Aerodyn. **35**, 101–128 (1990).

[14] Simpson, R. L., Chew, Y.-T., Shivaprasad, B. G.: The structure of a separating turbulent boundary layer. Part 2. Higher-order turbulence results. J. Fluid Mech. **113**, 53–73 (1981).

[15] Kline, S. J., Reynolds, W. C., Schraub, F. A., Runstadler, P. W.: The structure of turbulent boundary layers. J. Fluid Mech. **30**, 741–773 (1967).

[16] Thangam, S., Speziale, C. G.: Turbulent separated flow past a backward facing step: A critical evaluation of two-equation turbulence models. ICASE Rep. No 91-23 (1991).

[17] Schalau, B., Dengel, P., Thiele, F.: Computation of turbulent boundary layer flow with separation – a critical evaluation of parameters influencing the numerical results. In: Advances in turbulence 2. (Fernholz, H. H., Fiedler, H. E. eds.), pp. 377–382. Berlin Heidelberg New York Tokyo: Springer 1989.

[18] Devenport, W. J., Sutton, E. P.: Near-wall behaviour of separated and reattaching flows. AIAA J. **29**, 25–31 (1991).

[19] Dianat, M., Castro, I. P.: Measurements in separating boundary layers. AIAA J. **27**, 719–724 (1991).

[20] Bradshaw, P., Gee, M. T.: Turbulent wall jets with and without an external stream. ARC, R + M 3252 (1960).

[21] Stevenson, W. H., Thompson, H. D., Craig, R. R.: Laser velocimeter measurements in highly turbulent recirculating flows. Trans ASME. J. Fluid Eng. **106**, 173–180 (1984).

[22] Launder, B. E., Rodi, W.: The turbulent wall jet. Progr. Aerospace Sci. **19**, 81–128 (1981).

[23] Volchkow, E. P., Lebedev, V. P., Nizovtev, M. I.k, Terekhov, V. I.: The flow separation from a duct wall caused by the near-wall counterjet. IUTAM Symp. Novosibirsk 1990 (Kozlov, V. V., Dovgal, A. V. eds.), pp. 493–502. Berlin Heidelberg New York Tokyo: Springer 1991.

[24] Hewedy, N. I. I.: Untersuchung eines ebenen, tangential ausgeblasenen Gegenstroms. Dissertation, RWTH Aachen 1980.

[25] Westphal, R. V., Johnston, J. P., Eaton, J. K.: Experimental study of flow reattachment in a single-sided sudden expansion. NASA CR-3765 and Stanford Univ., Dept. Mech. Eng. Rep. MD-41 (1984).

[26] Adams, E. W., Johnston, J. P.: Flow structure in the near-wall zone of a turbulent separated flow. AIAA J. **26**, 932–939 (1988).

Author's address: Professor Dr.-Ing. H. H. Fernholz, Hermann-Föttinger-Institut für Thermo- und Fluiddynamik, Technische Universität Berlin, Sekr. HF 1, Strasse des 17. Juni 135, D-10623 Berlin, Federal Republic of Germany

Acta Mechanica (1994) [Suppl] 4: 69−77
© Springer-Verlag 1994

On the interaction of wave-like disturbances with shocks — two idealizations of the shock/turbulence interaction problem

R. Friedrich and **R. Hannappel,** München, Federal Republic of Germany

Summary. Numerical simulations are presented of the interaction between a Mach-8-shock and vorticity 'waves' in one case and dilatation 'waves' in the other using the two-dimensional Euler equations. Due to compressibility effects high amplification rates of the fluctuations and strong shock deformation are observed in the second case. An explanation of some of these effects is given on the basis of rapid distortion arguments. They provide insight into more complicated cases of shock/turbulence interaction.

1 Introduction

Studies of compressible turbulence have only recently, with the help of direct numerical simulations (DNS), revealed a clearer picture of the complicated energy transfer and dissipation mechanisms typical of such flows. While vorticity fluctuations play the predominant role in incompressible (isothermal) turbulence, there are, according to [1], three modes which control compressible turbulence, namely the vorticity, acoustic and entropy modes. The latter two lead to additional correlations involving both the fluctuating thermodynamic quantities and the fluctuating dilatation (divergence of the velocity vector). Such correlations have to be taken into account in turbulence models, whenever the r.m.s. density fluctuations, ϱ_{rms}, are not small with respect to the mean density $\bar{\varrho}$ [2]. Apart from these fluctuations, there is another indicator of compressibility effects within the turbulence structure, the turbulent Mach number $M_t = k^{1/2}/\bar{a}$, where k is twice the turbulent kinetic energy and $\bar{a}$ the local mean speed of sound. When M_t exceeds values of 0.4 shocklets (small embedded shock waves) form in $3D$ homogeneous shear turbulence as demonstrated in the DNS of [3]. In [4] the importance of a third parameter is stressed, the acoustic energy partition parameter, $F = \varkappa \bar{\varrho} \bar{p} \, \overline{u_i^{c'} u_i^{c'}}/\overline{p''^2}$ [1] which describes the partition of energy in the acoustic mode between internal energy (pressure fluctuations) and compressible kinetic energy. $u_i^{c'}$ is the irrotational part of the fluctuating velocity vector in a classical Helmholtz decomposition. $\varkappa$ represents the ratio of the specific heats, $\bar{p}$ the mean pressure, $\overline{u_i^{c'} u_i^{c'}}$ twice the compressible part of the turbulence kinetic energy and $\overline{p''^2}$ the mean square pressure fluctuations. In [3] it is shown that for low M_t (e.g. $M_t < 0.1$) and a parameter F far from equilibrium (i.e. $F \gg 1$ or $F \ll 1$) an energy exchange happens on a fast (acoustic) time scale between internal and compressible kinetic energy until an acoustic equilibrium has established. F can also be expressed in terms of M_t and $\chi = 0.5 \, \overline{u_i^{c'} u_i^{c'}}/k$ as $F = 2\varkappa^2 \bar{p}^2 \chi M_t^2/\overline{p''^2}$.

---

[1] An overbar denotes Reynolds averaging and a tilda mass-weighted averaging.

In 2D direct simulations of shock/isotropic turbulence interactions we found the parameters ϱ_{rms}, M_t and χ to control the amplification of the turbulent fluctuations, the decrease in length scales and the anisotropy of the turbulence intensities, see [5]. Furthermore, an increase in ϱ_{rms} and/or χ (while M_t was unchanged) produced locally stronger shock deformations and oscillations. These phenomena had not been observed before. The first 3D DNS of shock isotropic turbulence interaction, [6], was designed in such a way that only the influence of M_t and of the mean shock strength on the distortion of the shock wave could be studied.

Trying to explain some of these effects we have performed two numerical experiments with the two-dimensional Euler equations. In the first experiment a Mach-8-shock interacts with a plane vorticity 'wave'. This experiment had been proposed in [7] in order to study effects pertinent to the amplification and generation of turbulence in shock-wave/turbulent boundary-layer interactions and to place limits on the range of validity of existing linear theories like that of [8]. We use the experiment of [7] for comparison with a new one in which the (transversal) vorticity 'wave' is replaced by a (longitudinal) dilatation 'wave' striking the mean front of the Mach-8-shock at the same angle of incidence. The influence of viscosity can be neglected in these experiments, because the interaction between shock and 'waves' and the generation of new 'waves' primarily depend on rapid mechanisms. Molecular transport phenomena are, however, slow processes even slower, in general, than nonlinear interactions between disturbances. The irreversible entropy production within the shock due to viscous dissipation and heat conduction by the mean fields is on the other hand properly taken care of in the present finite volume Euler solution.

2 Numerical model and results

The Euler equations are integrated using a code developed by R. Hannappel and Th. Hauser. Features of this code are an explicit time integration scheme of the three-step compact-storage third-order Runge-Kutta type and a finite-volume shock-capturing technique. The solution vector is stored in the cell center and extrapolated to the cell-faces using ENO-interpolations up to fourth order following [9]. This leads to one-dimensional Riemann problems at each cell-face which are solved approximately according to [10] along with the entropy correction of [11]. Other shock-capturing schemes e.g. of the MUSCL-type have been implemented for comparison. We have demonstrated in [12] that ENO-interpolations of equal order of accuracy are better suited to resolve fluctuating fields.

The following numerical results were obtained on an equidistant cartesian grid of 240×160 cells in the intervall $[0, 2 \cdot 2^{1/2}] \times [0, 2 \cdot 5]$ of the (x, y)-plane. Periodic boundary conditions were applied in the y-direction parallel to the mean shock front. In the stationary frame of reference used the shock approaches the 'waves' with eight times the mean speed of sound. Constant extrapolations of the flow variables along lines parallel to the 'wave' front provide suitable inflow conditions at each time step. On the subsonic side (downstream) nonreflecting boundary conditions are implemented. A computation is stopped before the shock leaves the computational domain. A perfect gas is considered with a constant ratio of specific heats of 1.4.

2.1 Zang et al.'s experiment

While these authors have generated vorticity 'waves' at a 30 deg angle of incidence, we have chosen a 45 deg angle so that the fluctuating velocity vectors meet the mean Mach 8 shock front in both experiments under the same angle of incidence. The wave-like velocity 'fluctuations' on the upstream side (index 1) are given by:

$$u_1' = -v_1' = -\alpha \bar{a}_1 \cos\left((0.5)^{1/2} k(x + y)\right),$$ (1)

with $\alpha = 0.5$, $k = 2\pi$, the wavenumber and $\bar{a}_1 = (\varkappa \bar{p}_1/\bar{\varrho}_1)^{1/2}$.

(1) defines pure vorticity 'waves'

$$(0, 0, \omega') = -2^{1/2} \alpha k \bar{a}_1 \sin\left((0.5)^{1/2} k(x + y)\right).$$ (2)

and there are no density or pressure fluctuations, in accordance with the incompressibility condition satisfied by Eq. (1):

$$(\partial u_k'/\partial x_K)_1 = 0.$$ (3)

Figures 1 and 2 show contour lines of vorticity (ω'), entropy (s''), pressure (p'') and density (ϱ'') 'waves'.

After $t = 0.2$ non-dimensional problem times the vorticity 'waves' have penetrated the downstream side of the shock with the mean convection speed $\tilde{u}_{\text{rel2}}$ (relative to the shock) and undergone a one-dimensional compression normal to the shock front which results in a typical deflection. At the same time entropy, pressure and density 'waves' have been generated through the coupling between the primary vorticity 'waves' and the gradients of mean flow variables within the shock. The generation of 'waves' downstream of the shock and their coupling are explained in [13] on the basis of a rapid distortion theory. While vorticity and entropy 'waves' are simply convected with the same speed the pressure and density disturbances downstream of the shock reflect acoustic waves travelling with the speed of sound with respect to the mean flow. In a reference frame moving with u_2 and within the framework of a rapid distortion theory ϱ'' satisfies the wave equation:

$$\frac{\partial^2 \varrho''}{\partial t^2} - \bar{a}^2 \frac{\partial^2 \varrho''}{\partial x_K^2} = \frac{\bar{p}}{c_v} \frac{\partial^2 s''}{\partial x_K^2}.$$ (4)

A comparison of the computed amplitudes with the predictions of the above-mentioned linear theory shows excellent agreement. Equation (4) provides an explanation for the coup-

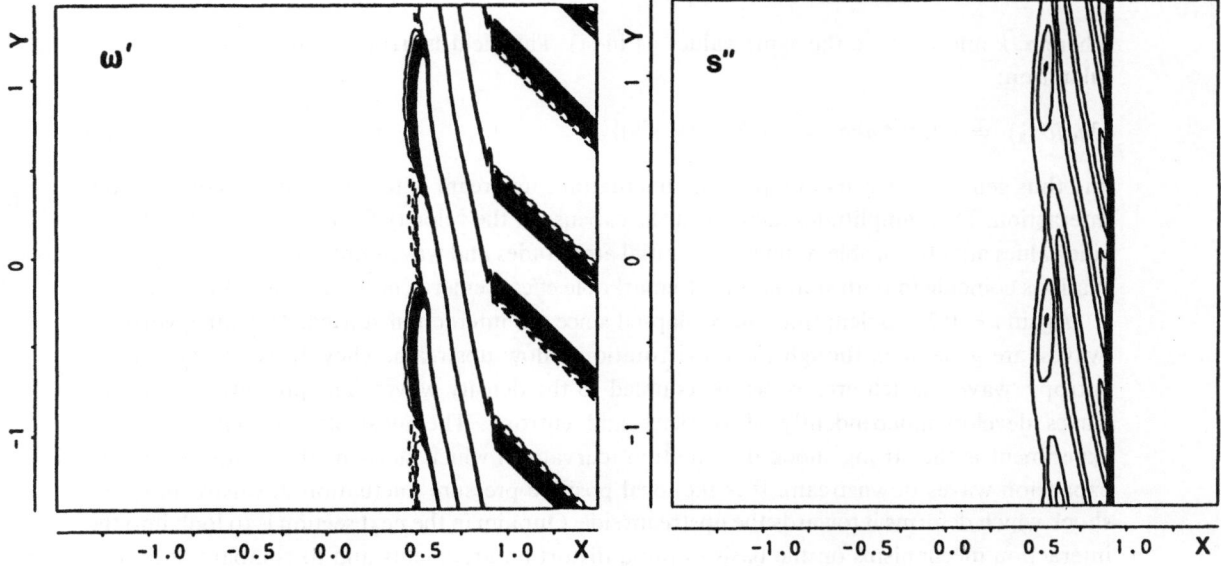

Fig. 1. Contour lines of vorticity and entropy 'waves' (transversal vorticity 'waves' upstream, $\chi = 0$)

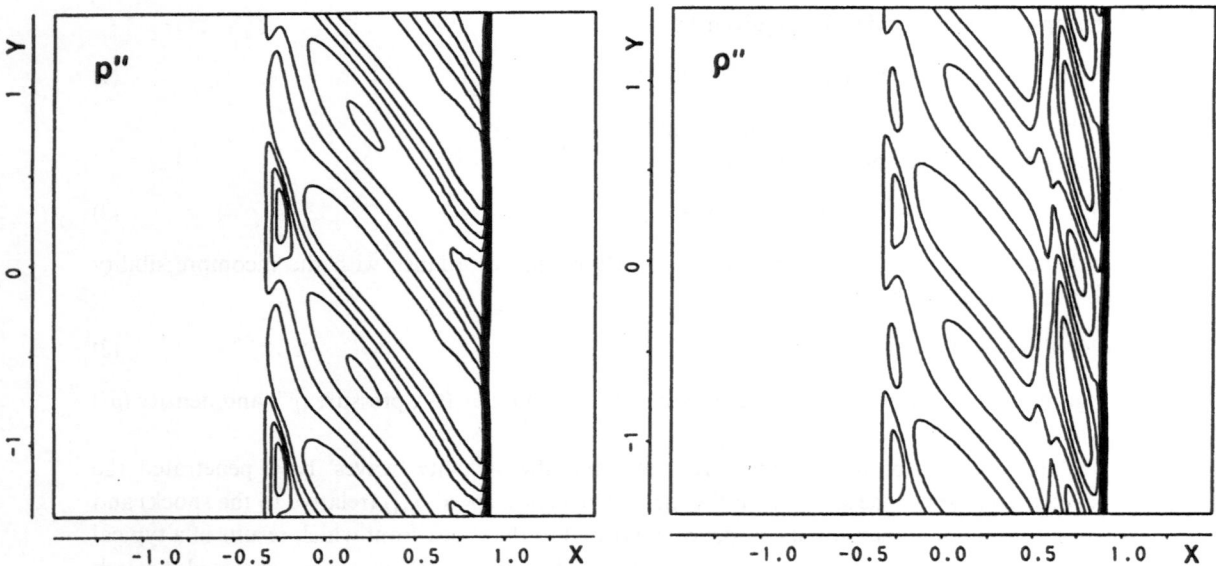

Fig. 2. Contour lines of pressure and density waves

ling between ϱ'' and s''. The entropy 'waves' have left their footprints in the density waves in Fig. 2. The pressure waves on the other hand are decoupled from the vorticity and entropy 'waves'.

2.2   Interaction between a Mach-8-shock and dilatation 'waves'

Longitudinal dilatation 'waves' hit the mean shock front at a 45 deg angle of incidence. At time $t = 0$ the interaction starts with upstream velocity fluctuations of the form:

$$u_1' = v_1' = -\alpha \bar{a}_1 \cos \left((0.5)^{1/2} k(x + y) \right), \tag{5}$$

where α, k and $\bar{a}_1$ have the same values as in (1). This field is irrotational, but has non-zero dilatation:

$$(\partial u_k' / \partial x_K)_1 = -(2)^{1/2} \alpha k \bar{a}_1 \sin \left((0.5)^{1/2} (x + y) \right) \tag{6}$$

and thus generates density and pressure fluctuations upstream of the shock in the course of the interaction. Their amplitudes increase at the expense the velocity fluctuations and reach specific values after 0.2 problem times. The initial amplitudes and wavenumbers of the velocity fluctuations coincide in both simulations. Remarkable effects emerge now, as seen in Figs. 3 and 4.

Again $t = 0.2$ problem times have elapsed since the interaction started. This time, vorticity 'waves' are generated, though there is irrotational flow upstream. They 'travel' as fast as the entropy 'waves' which are, as before, coupled to the density waves. The pressure or acoustic waves develop independently of vorticity and entropy. The most striking feature of this experiment is the strong shock deformation (curvature) which steepens the compression and expansion waves downstream. It is the local positive pressure fluctuation downstream of the shock which deforms it towards the upstream side. Our aim in the next section is to look into the interaction mechanisms on the basis of rapid distortion arguments and to compare both flow cases.

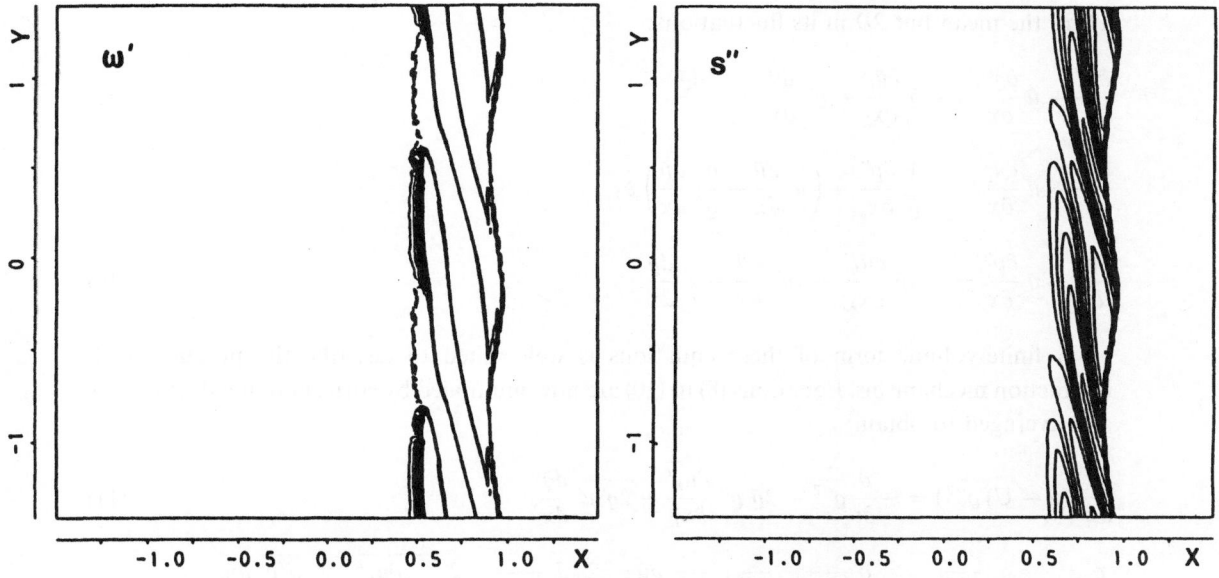

Fig. 3. Contour lines of vorticity and entropy 'waves' (longitudinal dilatation 'waves' upstream, $\chi = 1$)

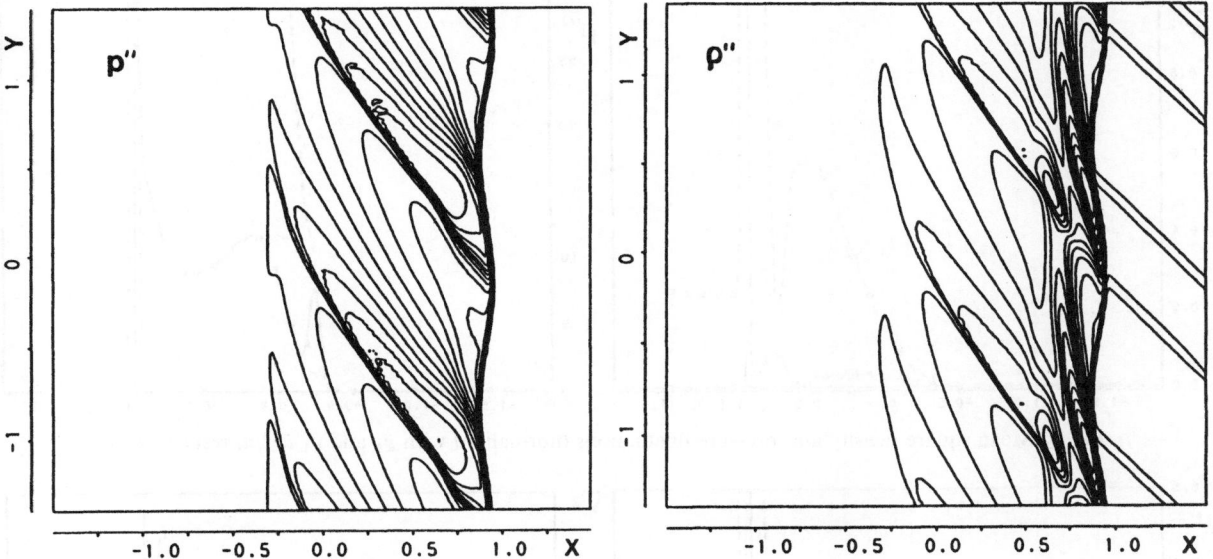

Fig. 4. Contour lines of pressure and density waves

3 Transport equations for mean square fluctuations

We start from the transport equations for the density, velocity and pressure fluctuations in which we use the following decomposition:

$$\varrho = \bar{\varrho} + \varrho'', \quad p = \bar{p} + p'', \quad u_i = \tilde{u}_i + u_i', \quad \tilde{u}_i = \overline{\varrho u_i}/\bar{\varrho} \tag{7}$$

Restricting ourselves to the study of rapid effects, i.e. neglecting molecular transport phenomena and nonlinearities in the fluctuations, these equations have the following form for a flow which is

1*D* in the mean but 2*D* in its fluctuations:

$$\frac{\partial \varrho''}{\partial t} + \tilde{u}\,\frac{\partial \varrho''}{\partial x} = -\bar{\varrho}\,\frac{\partial \bar{u}_k'}{\partial x_k} - \varrho''\,\frac{d\tilde{u}}{dx} - u'\,\frac{d\bar{\varrho}}{dx}, \tag{8}$$

$$\frac{\partial u_i'}{\partial t} + \tilde{u}\,\frac{\partial u_i'}{\partial x} = -\frac{1}{\bar{\varrho}}\,\frac{\partial p''}{\partial x_i} - \left(u'\,\frac{d\tilde{u}}{dx} - \frac{\varrho''}{\bar{\varrho}^2}\,\frac{d\bar{p}}{dx}\right)\delta_{1i}, \tag{9}$$

$$\frac{\partial p''}{\partial t} + \tilde{u}\,\frac{\partial p''}{\partial x} = -\varkappa\bar{p}\,\frac{\partial \bar{u}_K'}{\partial x_K} - \varkappa p''\,\frac{d\tilde{u}}{dx} - u'\,\frac{d\bar{p}}{dx}. \tag{10}$$

A finite-volume form of these equations is well suited to describe the present shock interaction mechanisms. Equations (8) to (10) are now multiplied by corresponding fluctuations and averaged to obtain:

$$\frac{\partial}{\partial x}\left((\tilde{u} - U)\,\overline{\varrho''^2}\right) = -\frac{\partial}{\partial t}\,\overline{\varrho''^2} - 2\bar{\varrho}\,\overline{\varrho''\,\frac{\partial u_K'}{\partial x_K}} - 2\overline{\varrho''u'}\,\frac{d\bar{\varrho}}{dx}, \tag{11}$$

$$\frac{\partial}{\partial x}\left((\tilde{u} - U)\,\overline{u_i'^2}\right) = -\frac{\partial}{\partial t}\,\overline{u_i'^2} + (\overline{v'^2} - \overline{u'^2})\,\frac{d\tilde{u}}{dx} - \frac{2}{\bar{\varrho}}\,\frac{d}{dx}\,\overline{p''u'} + \frac{2}{\bar{\varrho}}\,\overline{p''\,\frac{\partial u_K'}{\partial x_K}} + 2\,\overline{\frac{\varrho''u'}{\bar{\varrho}^2}}\,\frac{d\bar{p}}{dx}, \tag{12}$$

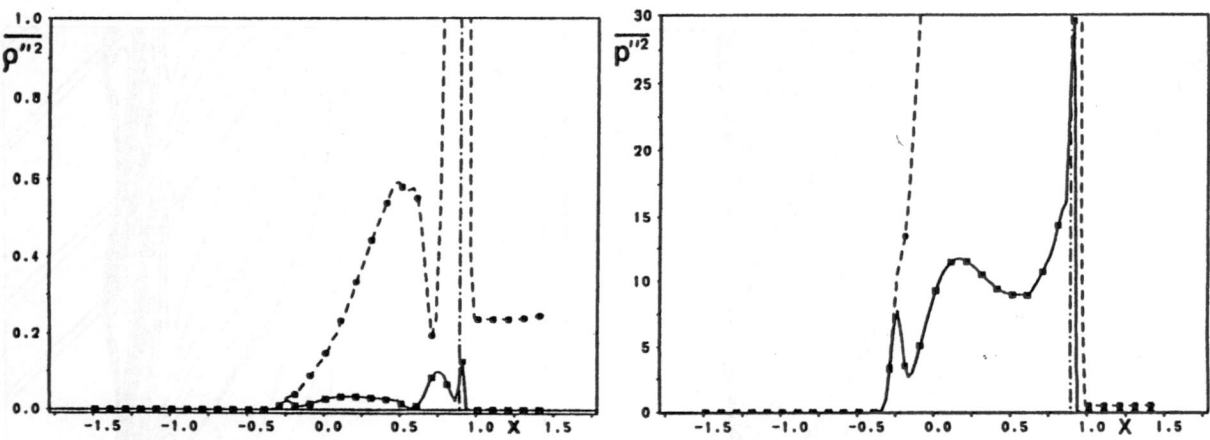

Fig. 5. Mean square density and pressure fluctuations (normalized with $\bar{\varrho}_1$ and $\bar{\varrho}_1\,\bar{a}_1^2/\varkappa$, respectively)

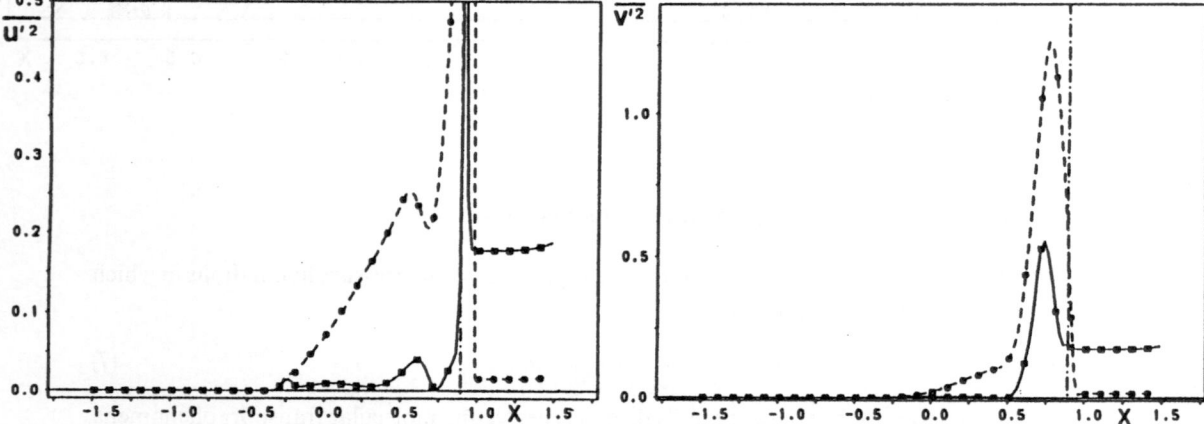

Fig. 6. Mean square velocity fluctuations in *x*- and *y*-directions (normalized with $\bar{a}_1/\varkappa^{1/2}$)

$$\frac{\partial}{\partial x}\left((\tilde{u} - U)\,\overline{p''^2}\right) = -\frac{\partial}{\partial t}\,\overline{p''^2} - 2\overline{p''u'}\,\frac{d\bar{p}}{dx} + (1 - 2\varkappa)\,\overline{p''^2}\,\frac{d\tilde{u}}{dx} - 2\varkappa\bar{p}\,\overline{p''\,\frac{\partial u_K'}{\partial x_K}}. \tag{13}$$

Note that Eqs. (11)−(13) are valid in a coordinate system moving with the shock speed U. They allow us to interpret the differences in the two numerical experiments. We first keep in mind that in the case of upstream vorticity 'waves', density and pressure waves downstream appear due to terms like $u'd\bar{\varrho}/dx$, $u'd\bar{p}/dx$ etc. which express the coupling between fluctuations and mean flow quantities within the shock.

Likewise, taking the curl of (9) it can be shown that in the case of dilatation 'waves' upstream, the baroclinic torque generates vorticity downstream of the shock [13].

We have computed profiles of mean square density, velocity and pressure fluctuations interpreting mean quantities as averages over the transverse direction and observe much higher amplification rates when the incoming fluctuating velocity field is compressible. In Figs. 5 and 6 the dashed and solid curves refer to results of incoming dilatation and vorticity 'waves', respectively.

An explanation of these phenomena starts from terms on the r.h.s. of (11)−(13). Those on the l.h.s. can be directly integrated across the shock. The unsteady terms provide contributions only

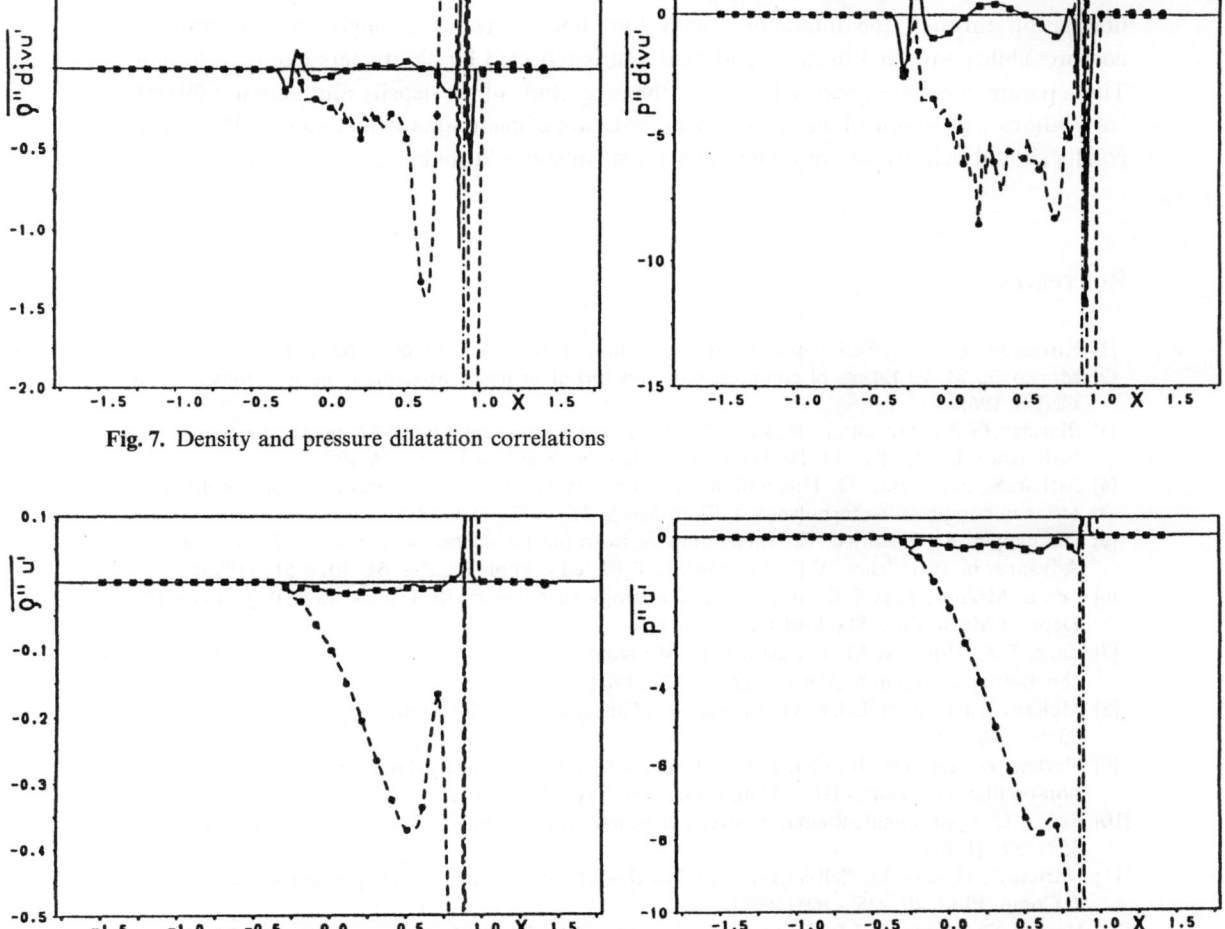

Fig. 7. Density and pressure dilatation correlations

Fig. 8. Density and pressure velocity correlations

for incoming dilatation 'waves'. In this case the density and pressure dilatation terms, $\overline{\varrho''\partial u_K'/\partial x_K}$, $\overline{p''\partial u_K'/\partial x_K}$ as well as the density velocity and the pressure velocity correlations are strongly negative behind the shock. The dilatation terms oscillate across the shock which was also observed in [6], but do contribute to the level of the fluctuations like the production terms containing mean gradients. The pressure dilatation correlation plays a double role appearing in (12) and (13). It exchanges energy between the compressible kinetic and the internal energies until an acoustic equilibrium state is reached. Figures 7 and 8 show the effect of compressibility on the dilatation correlations and the density resp. pressure velocity correlations. The fact that $\overline{\varrho''u'}$ is negative downstream of the shock (l.h.s. of the figure) in both cases indicates that the so-called Strong Reynolds Analogy which would yield $\overline{\varrho''u'} = \bar{\varrho}\,\overline{u'^2}/\tilde{u}$, see [14], leads to gross errors.

4 Conclusions

The interaction of wave-like fluctuations with strong shocks investigated in this paper with the help of the two-dimensional Euler equations is governed by rapid production and energy exchange mechanisms. They are analyzed on the basis of linearized transport equations which reveal the acoustic nature of density and pressure fluctuations. The amplification of any fluctuation through the interaction with the shock depends strongly on the amount of compressibility involved in the experiment and is reflected by parameters like ϱ_{rms}, M_t and χ. These parameters have great influence on the magnitude of the density and pressure dilatation correlations which control the downstream evolution of mean square fluctuations. The obtained results should help to interpret effects pertinent to shock/turbulence interaction.

References

[1] Kovaznay, L. S. G.: Turbulence in supersonic flow. J. Aero. Sci. **20**, 657−682 (1953).
[2] Morkovin, M. V.: Effects of compressibility on turbulent flows. Mécanique de la turbulence. Paris: CNRS, 1962.
[3] Blaisdell, G. A., Mansour, N. N., Reynolds, W. C.: Numerical simulations of compressible homogeneous turbulence. Report No. TF-50, Dept. of Mech. Eng., Stanford Univ., CA 1991.
[4] Sarkar, S., Erlebacher, G., Hussaini, M. Y., Kreiss, H. O.: The analysis and modelling of dilatational terms in compressible turbulence. J. Fluid Mech. **227**, 473−493 (1991).
[5] Hannappel, R., Friedrich, R.: Interaction of isotropic turbulence with a normal shock wave. In: Advances in Turbulence IV (Nieuwstadt, F. T. M., ed.). Appl. Sci. Res. **51**, 507−512 (1993).
[6] Lee, S., Moin, P., Lele, S. K.: Interaction of isotropic turbulence with a shock wave. Report No. TF-52, Dept. of Mech. Eng., Stanford Univ., CA 1992.
[7] Zang, T. A., Hussaini, M. Y., Bushnell, D. M.: Numerical computations of turbulence amplification in shock-wave interaction. AIAA J. **22**, 13−21 (1984).
[8] McKenzie, J. F., Westphal, K. O.: Interaction of linear waves with oblique shock waves. Phys. Fluids **11**, 2350−2362 (1968).
[9] Harten, A., Enquist, B., Osher, S., Chakravarthy, S.: Uniformly high order accurate essentially non-oscillatory schemes III. J. Comp. Phys. **71**, 231−303 (1987).
[10] Roe, P. L.: Approximate Riemann solvers, parameter vectors and difference schemes. J. Comp. Phys. **43**, 357−372 (1981).
[11] Harten, A., Hyman, M.: Self-adjusting grid methods for one-dimensional hyperbolic conservation laws. J. Comp. Phys. **49**, 235−269 (1983).
[12] Hauser, Th., Hannappel, R., Friedrich, R.: Testing high-order shock-capturing schemes in 2D super- and hypersonic flows. − In: Proceedings of the 1st European Symposium on Aerothermodynamics for

Space Vehicles, ESTEC, Noordwijk, May 28 – 30, 1991, (Berry, W., et al., eds.), pp. 393 – 399. Noordwijk: ESA Publ. 1991.

[13] Friedrich, R., Hannappel, R., Hauser, Th.: Stoß-Wellen- und Stoß-Turbulenz-Wechselwirkung. In: Turbulente Strömungen in Forschung und Praxis (Leder, A., ed.), pp. 131 – 144. Aachen: Shaker 1993.

[14] Zeman, O., Coleman, G. N.: Compressible turbulence subjected to shear and rapid compression. In: Turbulent shear flows Vol. 8 (Durst, W., et al., eds.). Berlin Heidelberg New York Tokyo: Springer 1993.

Authors' address: Professor Dr.-Ing. R. Friedrich and Dr. rer. nat. R. Hannappel, Lehrstuhl für Fluidmechanik, Technische Universität München, Arcisstrasse 21, D - 80333 München, Federal Republic of Germany

Acta Mechanica (1994) [Suppl] 4: 79—86

Some aspects of a model for coherent structures in turbulent boundary layers

M. Hackeschmidt and **M. Buschmann**, Dresden, Federal Republic of Germany

Summary. First considerations of setting up a mathematical model for coherent structures in turbulent boundary layers without pressure gradient are presented. Using the method of W. Albring the vortex transport equation is solved step by step. Starting points of the considerations are the definition of coherent structures according to F. Hussian and experimental discoveries.

1 Introduction

Leonardo da Vinci (1452—1519) probably was the first who found the vortex pattern in the wake of stakes in water so characteristic, that he portrayed them. What Leonardo da Vinci as an attentive observer saw were the organized coherent motions of thousands of millions of water molecules. However, it took about 500 years before these vortex motions were classed with the common expression *coherent structures*. Many experimental examinations showed, that the coherent structures are a substantial component of turbulent shear flows in general and of turbulent boundary layers in particular. Only the knowledge of evolution and behaviour of these structures permits the active control of fluid flow for the purpose of turbulence management. That is why we will demonstrate with the present paper one possible way to develop a mathematical model of coherent structures of turbulent boundary layers without pressure gradient.

In spite of the complexity that ensues from instationarity and three-dimensionality of the structures F. Hussain [1] was successful in setting up a definition of coherent structures. According to this definition such structures are a connected (coherent) fluid mass with strong statistical dependence of the motion in the interior of the structure. The instantaneously phase-correlated vorticity – the *coherent vorticity* – is the deciding feature of the structures. Figure 1 shows a cross-section of a coherent structure according to F. Hussain [1]. Two typical points of the structure are evident. The saddle A with negligible coherent vorticity and the centre point B with an extremum of vorticity.

The works of M. R. Head and P. Bandyopadhyay [2] provided far-reaching new aspects of the structuring of turbulent boundary layers. They could prove by experiments, that the turbulent boundary layer without pressure gradient is dominated by horseshoelike vortex structures (Fig. 2). Basically these structures may be subdivided into the foot area, the vortex legs and the head area. While the vorticity vector varies its direction in the foot and head area along the axis of the structure strongly, this vector has a nearly constant angle of inclination of about $45°$ towards the wall (x, z-plane) in the area of the vortex legs. This fact allows to take apart the horseshoe vortex in three model respective

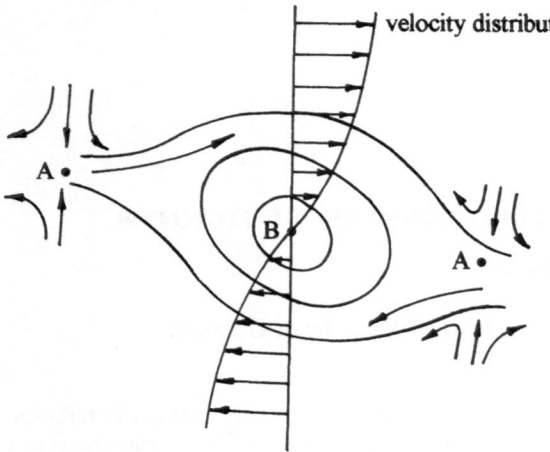

Fig. 1. Cross-section of a coherent structure according to [1]. *A* Saddle, *B* centre of the structure

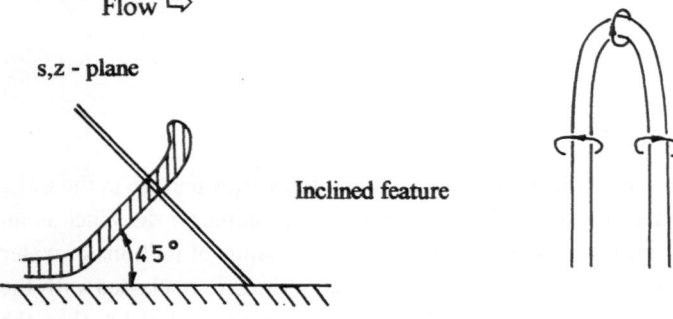

Fig. 2. Horseshoelike vortex structure according to [2]

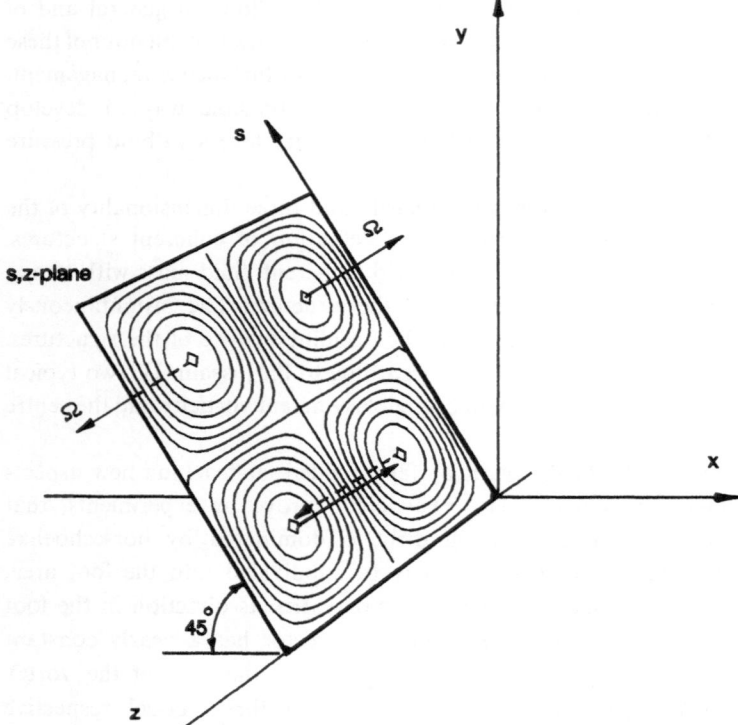

Fig. 3. Orientation of vortex structures in turbulent boundary layers according to [5]

calculation domains. While in the foot and the head area pure numerical models seem to be useful, for the vortex legs also analytical solutions appear possible. In the following the paper will turn to the mathematical description of this area.

The flow visualisations of the structures by M. R. Head and P. Bandyopadhyay permit to draw Fig. 3. The s, z-plane is identical with a spanwise cross-section of the discussed horseshoe vortices. Each pair of neighbouring counterrotating vortices is equivalent to one horseshoe vortex. If the velocity normal to the s, z-plane is negligible, these vortex pairs may be described with the step by step solution of the vorticity transport equation developed by W. Albring.

The aim of the presented research is to create a model for coherent structures in the logarithmic region of a turbulent boundary layer. Taking the vorticity transport equation as the starting point the authors use an inductive analytical method for the development of the model. This method was choosen because at the moment direct numerical simulation of turbulent boundary layers is only possible up to about $\mathrm{Re}_\theta = 1410$ (e. g. P. R. Spalart [6]).

According to the measurements of M. R. Head and P. Bandyopadhyay [2] the model is developed for $\mathrm{Re}_\theta > 2000$. It is assumed that for such high Reynolds-numbers horseshoelike vortices are more probably than streamwise vortices. With the model the authors do not believe to include all details of turbulent boundary layers, but the most important aspects of the kinematical phenomena.

2 Fundamental equations of forces and vorticity

By setting up the balance of forces for an infinitesimally small volume of fluid the Navier-Stokes equation follows in the form of Eq. (2.1). Here C is the velocity vector, p the pressure and ϱK an external force. It is assumed that the viscosity v and the density ϱ are constants. By taking the rotation of Eq. (2.1) the pressure term will be eliminated. Then the vorticity transport equation (2.2) – also known as Helmholtz's equation – follows. The vector Ω represents the rotation of the velocity field (2.3).

$$\varrho \frac{DC}{Dt} = -\nabla p - \eta \operatorname{curl} \operatorname{curl} C + \varrho K \tag{2.1}$$

$$\frac{\partial \Omega}{\partial t} - \operatorname{curl} [C \times \Omega] + v \operatorname{curl} \operatorname{curl} \Omega = \operatorname{curl} K \tag{2.2}$$

$$\Omega = i \left(\frac{\partial c_z}{\partial y} - \frac{\partial c_y}{\partial z} \right) + j \left(\frac{\partial c_x}{\partial z} - \frac{\partial c_z}{\partial x} \right) + k \left(\frac{\partial c_y}{\partial x} - \frac{\partial c_x}{\partial y} \right) = i\omega_x + j\omega_y + k\omega_z$$

i, j, k = unit vectors. $\tag{2.3}$

Essential simplifications of Eq. (2.2) result in the case of two-dimensionality. Only the ω_z – component of the vorticity vector and the velocity components c_x and c_y then exist. If ω_z is set equal at $-0.5 \Delta \Psi$ – with Ψ as a scalar stream function – then Rayleigh's differential equation follows.

$$\frac{\partial \Delta \Psi}{\partial t} + c\nabla \Delta \Psi - v\Delta \Delta \Psi = G. \tag{2.4}$$

3 Solutions of the vorticity transport equation

Thirty years ago W. Albring developed a method for solving Rayleigh's differential equation by using series (see for example [3]). The expression *coherent structure* was unknown at this time. Considerations concerning this didn't become established in those formulations. However, with the help of the knowledge about the orientation of the vorticity vector it should now be possible to use the method by W. Albring to construct mathematical models for these three-dimensional structures [4]. Therefore it is especially indispensable to analyse experimental results to get informations of coherent structures like dimensions, geometry and frequencies which will be used in the model. Figure 4 shows the way of model development.

3.1 Solutions of Rayleigh's differential equation

In the light of a homogeneous parallel mean flow that is overlayed by a vortex motion the method by W. Albring will be presented for two-dimensional flows. The mean flow has only one velocity component $c_x = c_\infty$. Rayleigh's differential equation will be split into a linear (indicated with L) and a nonlinear (indicated with Q) term (3.1). The solution is found in form of the series (3.2). The variable η is a positive real number. After putting Eq. (3.2) into (3.1), Rayleigh's differential equation may be taken apart into a system of nonlinear inhomogeneous differential equations. These are easier to solve than the original differential equation. Rayleigh's differential equation is fulfilled, when each of the equations of the system (3.3) is set at zero. The linear term $L_{(r+1)}$ for the equation of η^{r+1} is used to compensate the nonlinear term Q_r of the disturbance that is caused by the solution of the equation of η^r. The various terms of the series (3.2) have the form (3.5). The coupling with the outer flow is possible by using an appropriate function for Ψ_0.

$$\underbrace{\frac{\partial \Delta\Psi}{\partial t} + c_\infty \frac{\partial \Delta\Psi}{\partial x} - \nu\Delta\Delta\Psi}_{L} + \underbrace{\frac{\partial \Psi}{\partial y}\frac{\partial \Delta\Psi}{\partial x} - \frac{\partial \Psi}{\partial x}\frac{\partial \Delta\Psi}{\partial y}}_{Q} = g \tag{3.1}$$

$$\frac{\Psi}{c_\infty\delta} - \frac{y}{\delta} = K_0\Psi_0 + \eta K_1\Psi_1 + \eta^2 K_2\Psi_2 + \ldots + \eta^r K_r\Psi_r \tag{3.2}$$

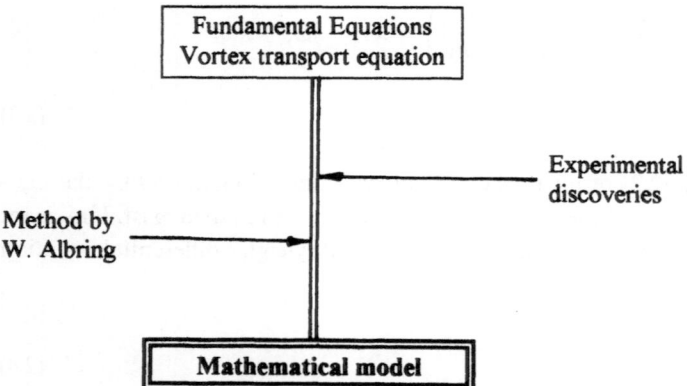

Fundamental Equations
Vortex transport equation

Experimental discoveries

Method by W. Albring

Mathematical model

Fig. 4. Model development

$$\left(\left(\frac{\partial \Delta \Psi_0}{\partial t} + c_\infty \frac{\partial \Delta \Psi_0}{\partial x} - \nu \Delta \Delta \Psi_0\right) K_0\right) \eta^0 = 0$$

$$\left(\left(\frac{\partial \Delta \Psi_1}{\partial t} + c_\infty \frac{\partial \Delta \Psi_1}{\partial x} - \nu \Delta \Delta \Psi_1\right) K_1 - \frac{g}{\eta}\right) \eta^1 = 0$$

$$\left(\left(\frac{\partial \Delta \Psi_2}{\partial t} + c_\infty \frac{\partial \Delta \Psi_2}{\partial x} - \nu \Delta \Delta \Psi_2\right) K_2 - Q_1\right) \eta^2 = 0 \tag{3.3}$$

$$\left(\left(\frac{\partial \Delta \Psi_{r+1}}{\partial t} + c_\infty \frac{\partial \Delta \Psi_{r+1}}{\partial x} - \nu \Delta \Delta \Psi_{r+1}\right) K_{r+1} - Q_r\right) \eta^{r+1} = 0$$

$$L_{r+1} = Q_r \tag{3.4}$$

$$\Psi_r = k_r \exp(\alpha) \cos(\beta) \tag{3.5}$$

3.2 Solutions in view of coherent structures

By using the relation $C = \text{curl}(\Psi)$ Eq. (2.2) may be written for three-dimensional fluid flow in terms of a vector stream function. The form of this equation would be now adequate to Rayleigh's differential equation but there would be also fourth derivations of the stream function. That is why it is more favourable to write the vortex transport equation only in terms of the velocity and vorticity vector fields (3.6).

Similar to the solution by W. Albring now a series is set up for solving the differential equation [4]. We have to insert the series into the vortex transport equation and to sort out according to powers of η. A set of nonlinear differential equations is found (3.8). The vortex transport equation for any η is fulfilled only if each equation of the system is set at zero.

$$\frac{\partial \Omega}{\partial t} - [C \nabla]\, \Omega + [\Omega \nabla]\, c + \nu \,\text{curl curl}\, \Omega = G \tag{3.6}$$

$$C = (ic_{x,0} + jc_{y,0} + kc_{z,0}) + \eta(ic_{x,1} + jc_{y,1} + kc_{z,1}) + \dots + \eta^r(ic_{x,r} + jc_{y,r} + kc_{z,r}) \tag{3.7}$$

s-component:

$$\left[\frac{\partial \omega_{s,0}}{\partial t} - \nu \nabla \omega_{s,0} - [\Omega_0 \nabla]\, c_{s,0} + [C_0 \nabla]\, \omega_{s,0}\right] = 0$$

$$\left[\frac{\partial \omega_{s,1}}{\partial t} - \nu \Delta \omega_{s,1} - [\Omega_0 \nabla]\, c_{s,1} - [\Omega_1 \nabla]\, c_{s,0} + [C_0 \nabla]\, \omega_{s,1} + [C_1 \nabla]\, \omega_{s,0} - \frac{g_s}{\eta}\right] \eta = 0$$

$$\left[\frac{\partial \omega_{s,2}}{\partial t} - \nu \Delta \omega_{s,2} - [\Omega_0 \nabla]\, c_{s,2} - [\Omega_1 \nabla]\, c_{s,1} - [\Omega_2 \nabla]\, c_{s,0}\right. \tag{3.8}$$

$$\left. + [C_0 \nabla]\, \omega_{s,2} + [C_1 \nabla]\, \omega_{s,1} + [C_2 \nabla]\, \omega_{s,0}\right] \eta^2 = 0$$

$$\left[\frac{\partial \omega_{s,r+1}}{\partial t} - \nu \Delta \omega_{s,r+1} - [\Omega_0 \nabla]\, c_{s,r} - \Sigma[\Omega_n \nabla]\, c_{s,r-n} - [\Omega_0 \nabla]\, c_{s,0}\right.$$

$$\left. + [C_0 \nabla]\, \omega_{s,r} + \Sigma[C_n \nabla]\, \omega_{s,r-n} + [C_r \nabla]\, \omega_{s,0}\right] \eta^{r+1} = 0$$

$$s = x, y, z.$$

When $c_{x,0}$, $c_{y,0}$ and $c_{z,0}$ (the first terms of Eq. (3.7)) are set at zero only the terms in bold print remain in the system (3.8). Similar to the two-dimensional case the linear term of the equation of η^{r+1} may be used to compensate the nonlinear term of this equation. A solution of the system of differential equations derived from (3.8) will be found recursively. Beginning with the disturbance G the function C_1 will be determined from the equation for η. The following summands of the series (3.7) result from gradually solving the next equations of the system. Taking into account the discussion of Section 1 regarding of the coherent structures, the differential equations may be further simplified. As Fig. 3 shows the vorticity vector is only inclined against the x, z-plane. That is why the z-component of this vector may be set at zero. Furthermore for a first mathematical model the z-component of the velocity vector is neglected.

In the most simple case the components of the disturbance vector are introduced as exponential functions. Then the solutions for the various summands of the series are also exponential functions (3.10).

$$G = i(f_x/a_y) \exp(\alpha) + j(f_y/a_x) \exp(\alpha) + k(f_y - f_x) \exp(\alpha)$$

$$\alpha = a_x x + a_y y + a_z z \tag{3.9}$$

$$c_{s,r} = \sum_{n=1}^{r} f_r \exp(r\alpha) \tag{3.10}$$

A trigonometric disturbance function (3.11) is choosen to get a periodical velocity vector. If this disturbance is inserted into the equation of η of the system (3.8) the function C_1 will also be trigonometrical. Provided v is very small and the friction term is neglected, C_1 has the form of Eq. (3.12.1). Further summands of the series of the velocity vector are in the shape of sums of trigonometrical terms. The arguments of these functions are multiples of the argument of the disturbance vector (3.12.2 to 3.12.5).

The series was developed up to $r = 5$. It was discovered that all coeffizients $q_{r,n}$ are equal or smaller than one. The series (3.12) converges if w is equal or smaller than two.

$$G = if_g a_z a_t \sin(\alpha) \sin(\beta) - jf_g a_z a_t \sin(\alpha) \sin(\beta); \quad \alpha = a_{xy}(x + y); \quad \beta = a_z z + a_t t \tag{3.11}$$

$$w = a_{xy}/a_t/2; \quad s \ldots x, y$$

$$c_{s,1} = f_g q_{11} \sin(\alpha) \sin(\beta) \tag{3.12.1}$$

$$c_{s,2} = f_g w q_{21} \sin(2\alpha) \sin(2\beta) \tag{3.12.2}$$

$$c_{s,3} = f_g w^2 [q_{31} \sin(\alpha) \sin(\beta) + q_{32} \sin(\alpha) \sin(3\beta) + q_{33} \sin(3\alpha) \sin(\beta) + q_{34} \sin(3\alpha) \sin(3\beta)] \tag{3.12.3}$$

$$c_{s,r} = f_g w^{(r-1)} \left[\sum_{n=1}^{r/2} q_{r,n} \sin(2n\alpha) \sum_{n=1}^{r/2} \sin(2n\beta) \right] \qquad r \text{ is even} \tag{3.12.4}$$

$$c_{s,r+1} = f_g w^r \left[\sum_{n=1}^{(r-1)/2} q_{r,n} \sin((2n-1)\alpha) \sum_{n=1}^{(r-1)/2} \sin((2n-1)\beta) \right] \qquad r \text{ is uneven} \tag{3.12.5}$$

Figure 5 shows sequences of contour plots of velocity and vorticity in the z, s-plane after one, two and four steps of the solution. Considering the definition of coherent structures given by F. Hussain, it is possible to interpret these fields as descriptions of such structures. In Fig. 6 the field of isolines of the vorticity and the velocity distribution are drawn after four steps. The saddles and centres which characterising the structures are also marked.

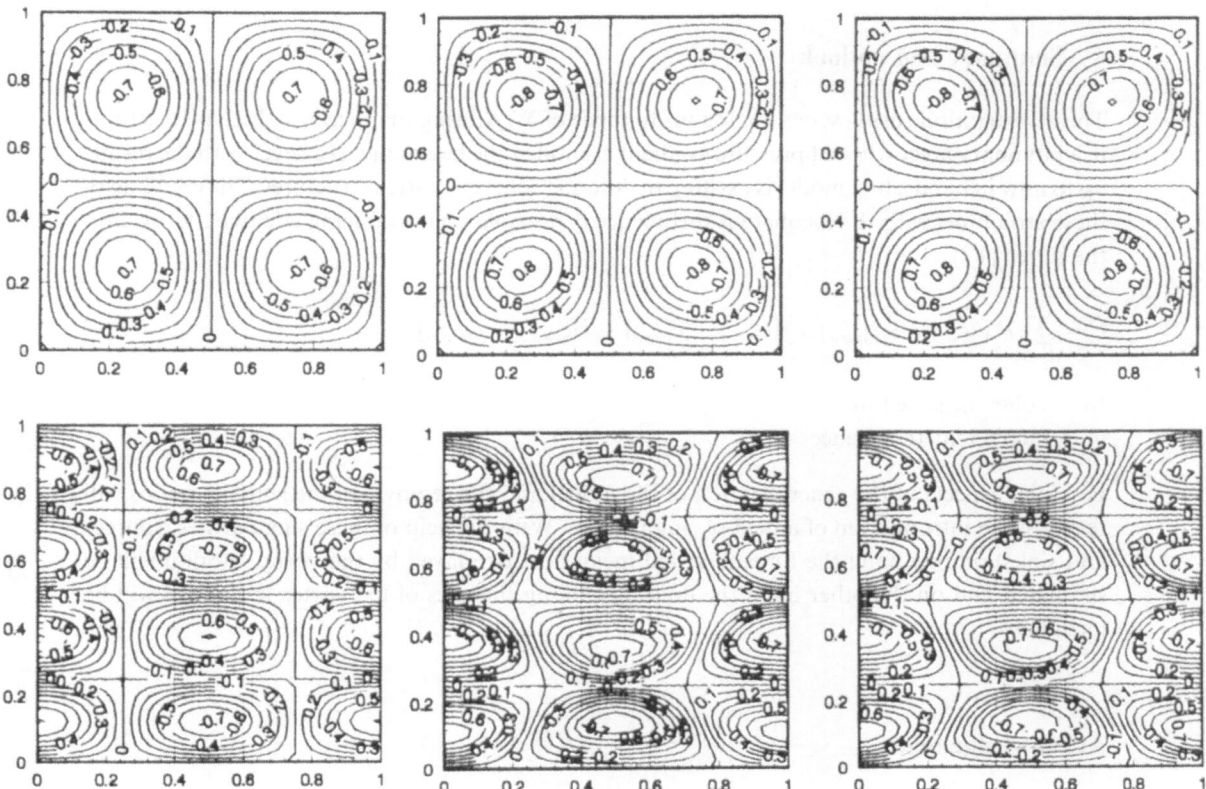

Fig. 5. Contour plots of velocity (above) and vorticity (below) with $a_{xy} = 1$; $a_z = 1$; $a_t = 1$; $f_g = 0.5$ and $\eta = 1$; abscissas: $z/a_{xy}/\pi$; ordinates: $s/a_{xy}/\pi$

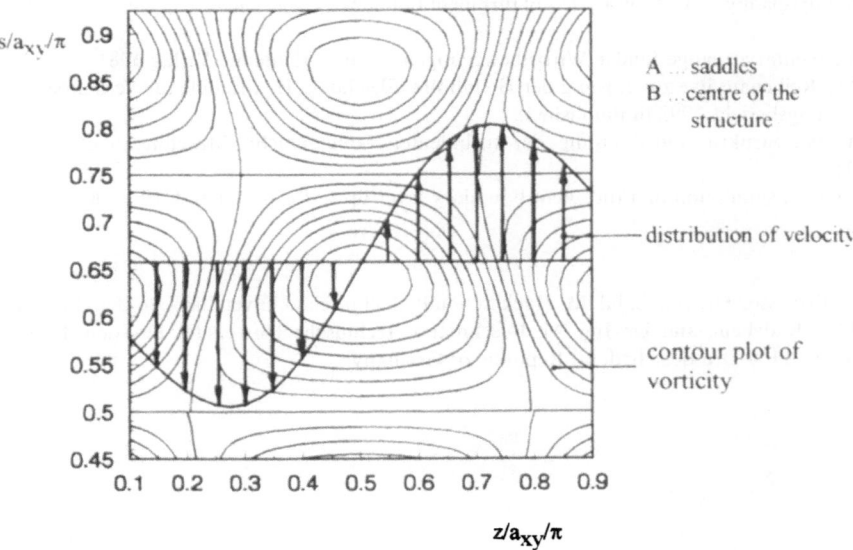

Fig. 6. Interpretation of the isolines of vorticity in the s, z-plane

4   Summary and outlook

The investigations using series analogous to those of W. Albring to describe coherent structures in a boundary layer without pressure gradient are still at the beginning. However with the results up to now in hand a first model is existing to describe coherent vortex structures. The coupling of these structures with the incoherent turbulence is possible by introducing adequate terms into the series (4.1).

$$C = \sum_{i=0}^{r} \eta^i \left[i(c_{co,x,r} + c_{in,x,r}) + j(c_{co,y,r} + c_{in,y,r}) + k(c_{co,z,r} + c_{in,z,r}) \right] \tag{4.1}$$

co ... coherent structure
in ... incoherent turbulence

Essential aspects of the generalization of the model and the removal of the simplifications will result in the introduction of a field of singularities. With the help of these singularities (sources and sinks) on one hand the time averaged mean flow overlayed by the vortex motion will be simulated and on the other hand the fluid flow along the axes of the vortex structures will be included.

Acknowledgement

We are grateful to the Deutsche Forschungsgemeinschaft which supports this research through the grant Ha 1788/2-1.

References

[1] Hussain, F.: Coherent structures and turbulence. J. Fluid Mech. **173**, 303–356 (1986).
[2] Head, M. R., Bandyopadhyay, P.: New aspects of turbulent boundary-layer structure. J. Fluid Mech. **107**, 297–338 (1981).
[3] Albring, M.: Elementarvorgänge fluider Wirbelbewegungen. Berlin: Akademie-Verlag 1981.
[4] Buschmann, M.: Reihenansätze zur Lösung der Helmholtz-Gleichung. Hochschule für Verkehrswesen Dresden, Forschungsbericht 1992 (unpublished).
[5] Hackeschmidt, M.: Strukturmodellbildung der turbulenten Grenzschicht. Maschinenbautech., **40**, 104–108 (1991).
[6] Spalart, P. R.: Direct simulation of a turbulent boundary layer up to $Re_{\theta} = 1410$. J. Fluid Mech. **187**, 61–98 (1988).

Authors' addresses: Professor Dr.-Ing. habil. M. Hackeschmidt, Technische Universität Dresden, Einsteinstrasse 10a, D-01445 Radebeul, and Dr.-Ing. M. Buschmann, Technische Universität Dresden, Rudolf-Renner-Strasse 26, D-01159 Dresden, Federal Republic of Germany

Acta Mechanica (1994) [Suppl] 4: 87−94
© Springer-Verlag 1994

On the critical condition for wall turbulence generation

M. Nishioka, Osaka, **M. Asai,** Tokyo, and **S. Furumoto,** Osaka, Japan

Summary. The critical flow condition for self-sustaining wall turbulence and the related minimum Reynolds number are examined on the basis of our previous and new experiments on the subcritical transition in channel and flat-plate flows to obtain a better understanding of the mechanism for wall turbulence generation. It is stressed that energetic hairpin eddies coming from upstream can trigger the subcritical disturbance growth, or their regeneration, in Blasius flow beyond $R_\theta = 110 \sim 130$. Downstream, the mean velocity can follow the logarithmic distribution beyond $R_\theta = 210$. We also find that the minimum Reynolds number R_θ is not much different for the pipe, channel and flat-plate wall flows.

1 Introduction

It is our great pleasure to contribute to this volume dedicated to Professor Jürgen Zierep on the occasion of his 65th birthday.

The present study is concerned with the subcritical transition in Blasius flow. When considering Transition Control and/or Turbulence Manipulation of boundary layers or even when simply trying to estimate the friction drag, one often needs to know the critical flow condition for self-sustaining wall turbulence, in particular the minimum Reynolds number. To establish such important knowledge and to improve our understanding of the mechanism of wall turbulence, we have been working on the subcritical transition in Blasius flow. In the present paper, the process of wall turbulence generation and the related critical flow condition are discussed on the basis of our previous and new experiments. We would like to begin with the important concept of the minimum Reynolds number for the self-sustaining wall turbulence.

2 Concept of minimum Reynolds number for wall turbulence

In the literature, to the authors' knowledge, Preston [1] first discussed the concept of the minimum Reynolds number in connection with the estimation of the friction drag and the choice of transition devices. For Blasius flow he proposed that the minimum Reynolds number R_θ (based on momentum thickness θ) is 320.

Time and again, Morkovin [2] emphasizes the importance of the concept as follows. There is ample evidence that boundary layers below a Reynolds number R_{mint} cannot sustain intrinsic self-energizing turbulent motions, without the occurrence of wall bursts which produce high Reynolds stresses and recoup energy from the mean flow to overcome the viscous dissipation. The bursts are really crucial in sustaining the wall turbulence. For instance, in the course of relaminarization caused by the favourable pressure gradient, the bursts first diminish in frequency and then stop altogether. This is because the pressure gradient generates new vorticity and a thin viscous layer and thus increases the dissipation to stabilize the inner wall layer. Likewise, below R_{mint}, even when vigorous free-stream turbulence is convected toward the boundary layer, the energizing wall bursts are not excited and the locally forced turbulence decays downstream. The earliest possible transition, important in design, occurs at R_{mint}.

As regards the value of R_{mint} for Blasius flow, through focusing on the Reynolds number beyond which a turbulent wedge develops downstream of a disturbance generator such as a large isolated roughness, Morkovin suggests that it nearly coincides with the critical Reynolds number R_{cr} for the growth of small-amplitude Tollmien-Schlichting waves, which is given from the stability theory to be about 200 and 520 based on the momentum thickness θ and the displacement thickness δ^* respectively (note that $\delta^*/\theta = 2.59$ for Blasius profile). Although the value $R_\theta = 200$ for R_{mint} is by no means strict as its determination cannot be free from subjective judgment, the present authors think, it is quite important that the approximate coincidence of R_{mint} and R_{cr} suggests high possibilities that the subcritical growth occurs in Blasius flow even below $R_\theta = 200$ and that the instability mechanism for the growth is closely related to that of bursts in the self-sustaining wall turbulence.

3 Wall bursts caused by hairpin eddies in transition to wall turbulence

It is reasonable to expect that the energizing bursts occur in the final stage of transition to wall turbulence. Indeed, in the literature, Nishioka et al. [3] first revealed the physical process up to the occurrence of the wall bursts which lead to the appearance of the logarithmic mean velocity distribution under still periodic flow conditions in the later stages of ribbon-induced subcritical transition in plane Poiseuille flow. They identified the disturbances triggering the bursts to be hairpin eddies (of Klebanoff type) evolving in succession from the high-shear layer away from the wall [3 − 6]. The high-shear layer is the product of the peak-valley wave development, namely the secondary instability of Herbert's fundamental mode [7 − 9]. The important findings on the physical process [3, 6] are described below.

As a result of being disturbed by the passage of Klebanoff hairpin eddies, the near-wall flow develops streaky structures, a pair of high-speed lumps of fluid with a low-speed lump in between. It should be emphasized that the spanwise distance between the high-speed streaks is about twice that between the legs of above-passing hairpin eddy. In further detail, as the high-speed streaks develop, a low-speed lump lifts up from the low-speed streak to form a shear layer, which we call the wall shear layer. And it soon breaks down into a hairpin-shaped eddy, which we call the wall hairpin. The process from the lift-up to the breakdown into the wall hairpin is strongly affected by the over-riding hairpin eddies in the next cycle. All these wall activities are called the wall burst. Initially, the lift-up starts at the downstream end of the low-speed streak, and the wall burst of the same fluid lump occurs near its upstream end since the lifted-up fluid is left behind by the high-speed lumps convecting faster. Although the flow remains almost periodic as far as the primary and secondary high-frequency (namely spike) fluctuations are concerned, the mean velocity over the streaky region (especially over the high-speed streaks) exhibits the wall-law characteristics approaching the log-law distribution for the developed wall turbulence.

Moreover, the spanwise scale of the wall streak (or the distance between the neighbouring high-speed streaks) is about 120 in wall units almost independently of the Reynolds number R_h in the range studied from 3 500 to 6 200: R_h is based on the channel half-depth h and the centerline velocity U_c (of undisturbed parabolic flow). The spanwise scale is of the same order as that in the developed wall turbulence. Comparisons [3] of these wall activities in the later stages of the transition with the near-wall coherent motions in turbulent pipe flow (namely, the repeated occurrence of strong acceleration from large negative to positive values of u-fluctuation associated with the energetic bursts) indicate no essential differences between them. Indeed, the wave forms of u-fluctuations at various corresponding positions also show surprisingly close similarities. Here, it should be noted that for the ribbon-induced transition Sandham and Kleiser [10] recently obtained numerical results which show that near-wall activities similar to the wall phenomena identified in [3, 6] occur in a cyclic sequence to sustain the wall turbulence.

4 Subcritical transition triggered by hairpin eddies

Recognizing the important role of hairpin eddies in generating and maintaining the wall turbulence as revealed in the ribbon-induced transition described above, Asai and Nishioka [11] tried to generate hairpin eddies near the leading edge of a boundary layer plate to disturb the downstream subcritical Blasius flow. The purpose was to determine the critical flow condition for the turbulence generation in Blasius flow through observing its response to nonlinear disturbances due to the high-intensity hairpin eddies acoustically excited at the leading edge. The leading-edge-generated hairpins are called LG hairpins for short. When disturbed by the LG hairpin eddies, the near-wall flow was found to develop local wall shear layers with streamwise vortices. This is seen from Figs. 1 a, b, which show smoke visualizations of the critical flow at freestream velocity $U_\infty = 4$ m/s, illustrating the side and cross-section views respectively. Note that the distance x is measured from the leading edge and y from the wall, both in mm. In the side view, the smoke released from $(x, y) = (100, 3.5)$ indicates signatures of the LG hairpins, while the smoke released from $(90, 0.5)$ visualizes the near wall activities caused by the LG hairpins. The cross-section view at $x = 150$ indicates the appearance of streamwise vortices and the related lift-up of low-speed fluid forming the wall shear layers and/or the wall hairpins. As visualized in [11] the wall shear layers develop into wall hairpins in succession, just as observed in the ribbon-induced transition, to lead to the subcritical transition. The minimum R_θ for the subcritical disturbance growth is found to be 127, less than Morkovin's 200. It should be noted that in terms of u- and v-fluctuations, the intensity of the near-wall activity at the critical Reynolds number was almost the same as (or slightly less than) that of the developed wall turbulence.

Here, in Fig. 2, we would like to present our new results, a sequence of side view photographs taken with a high-speed camera that illustrate the process of wall bursts, i.e. the wall turbulence generation. The time interval is 1/400 seconds, one eighth of the period of the acoustic forcing, and 9 photographs from top to bottom in Fig. 2 complete one whole cycle of the forcing at 50 Hz. The smoke sheets are released from almost the same positions as in Fig. 1. It is also noted that the smoke sheets are initialy about 10 mm in the spanwise width. We see that the LG hairpins are still active beyond $x = 200$, growing in size and disturbing the near wall flow. Ahead of them, high-speed fluid marked by the upper smoke sheet convects toward the wall and catches up and then interacts with the lifted-up low-speed fluid marked by the lower smoke sheet. These phenomena are quite similar to the wall bursts observed in the final stage of the ribbon-induced transition described in Section 3. In particular the process of the lift-up of low-speed fluid is clearly demonstrated behind the LG hairpins. This observation is consistent with the cross-section view in Fig. 1 b. In this connection, it should be noted that through beautiful flow visualizations, Haidari and Smith [12] revealed that secondary hairpin eddies (wall hairpin in our case) are generated not only between the legs of the primary (corresponding to LG hairpin) but

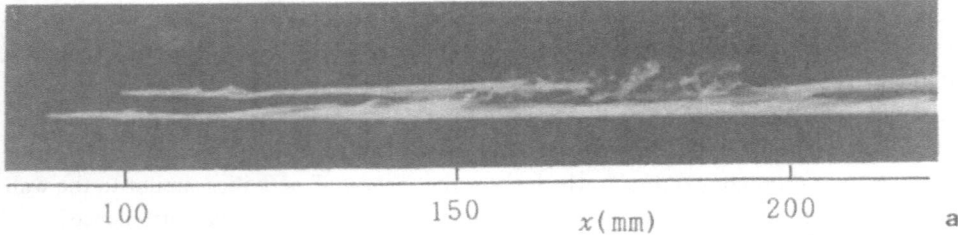

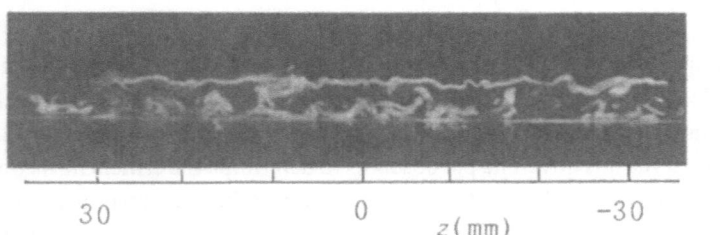

Fig. 1. Smoke-wire visualization of the subcritical disturbance growth in Blasius flow. Side view in **a** and cross-section view (at $x = 150$) in **b**

also directly behind each leg. Moreover, it is proposed that unsteady separation of the wall layer due to the pressure gradient imposed by the primary hairpin is the mechanism for lift-up (or eruption) of low-speed fluid leading to the formation of secondary hairpins; see Smith et al. [13]. Our transition observations described thus far and those of Haidari and Smith strongly suggest that hairpin eddies and the associated wall phenomena leading to their regeneration are the essential elements for the wall bursts or the mechanism for wall turbulence generation. Our continued efforts are to reveal the critical flow condition for the regeneration. So it should be stated again that in terms of u- and v-fluctuations, the intensity of the near-wall activity required for the regeneration at the minimum $R_\theta = 127$ is almost the same as (or slightly less than) that of developed wall turbulence.

In order to clarify the critical flow condition in further detail, we are currently examining the subcritical transition in Blasius flow caused by a group of circular cylinders (3 mm in dia. and 3.8 mm high) placed near the leading edge of the boundary layer plate. We would like to describe the main results of this experiment in the next Section.

5 Transition caused by cylinder-generated disturbances

The transition device is described first. The cylinders are glued on a 1.3 mm thick aluminum plate (197 mm in spanwise width and 40 mm in streamwise length), in a two-row configuration. In each spanwise row the spacing of the cylinders is 20 mm, and the two rows are separated by 15 mm in x-direction and staggered by 10 mm in z-direction. The plate with staggered cylinders are glued on another aluminum plate of 1.3 mm thick as shown in Fig. 3 to form a boundary layer trip. This trip is placed on a boundary layer plate, with the centers of the upstream cylinders at 45 mm from the leading edge. We measure x- and z-axes from the center of the cylinder standing at the

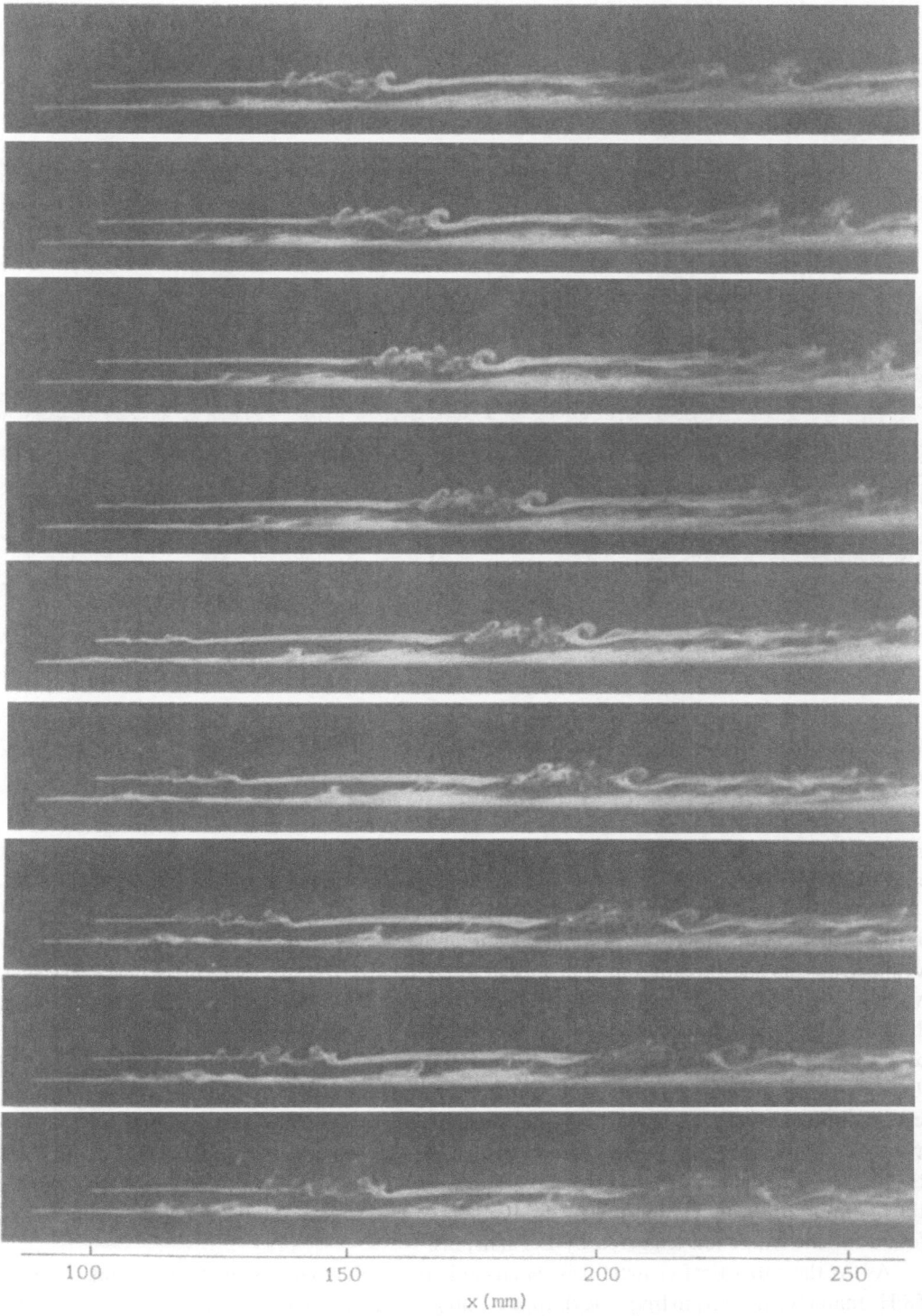

Fig. 2. Process of wall turbulence generation illustrated by a sequence of side-view photographs taken with a high-speed camera. Time interval is 2.5 ms between frames

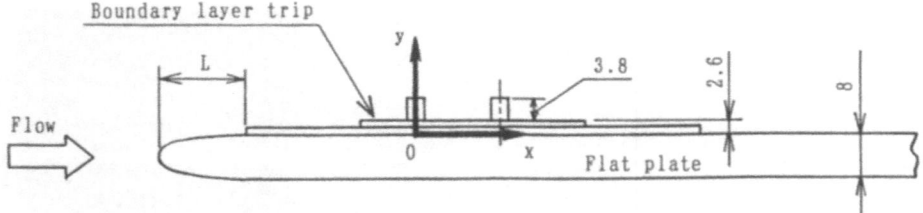

Fig. 3. Boundary layer plate with staggered cylinders placed near leading edge, with $L = 15$ mm. Dimensions in mm

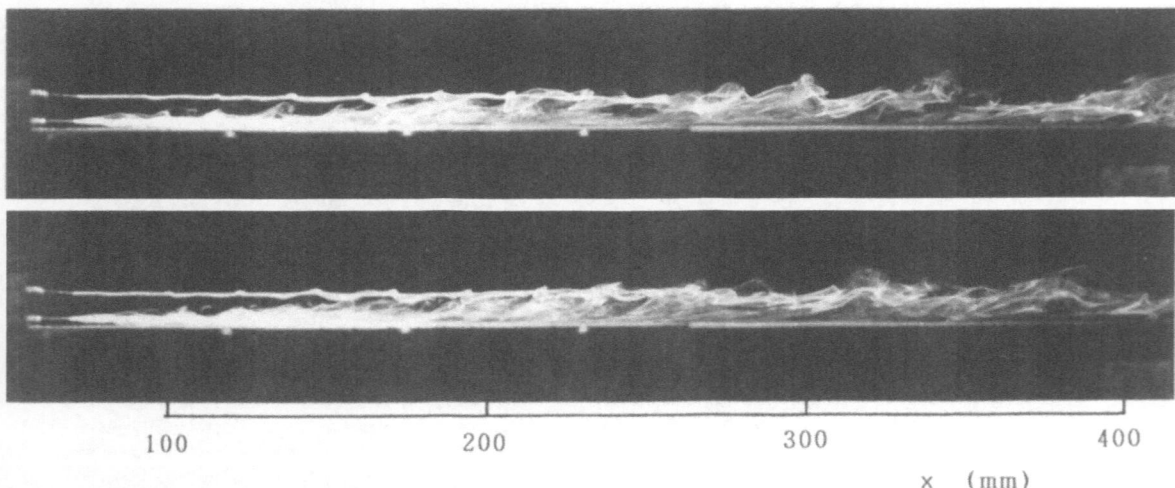

Fig. 4. Smoke-wire visualization of subcritical disturbance growth caused by cylinder-generated hairpin eddies

spanwise centre of the upstream row. The cross section of the wind tunnel used is 300 mm × 300 mm and the tunnel is capable of maintaining steady uniform flow even below 1 m/s.

We carefully examined the minimum freestream velocity U_∞ for subcritical disturbance growth. It is found to occur at about 1.42 m/s and beyond about $x = 200$. At this velocity and behind each cylinder, hairpin-shaped eddies are generated at 58 Hz, owing to the instability of the near-wake flow, with the Reynolds number being 380 based on the cylinder height. It is important to note that the hairpin-shaped eddies do decay downstream. The maximum rms intensity u_m'/U_∞, being 13% at $x = 17.2$, decays down to about 5% at $x = 160$. It decays further to $2 \sim 4\%$ at $x = 265$ for $U_\infty = 1.42$ m/s. However it grows again to 10% there for $U_\infty = 1.45$ m/s. These results show that the critical flow condition occurs near $x = 265$ at about 1.42 m/s. For the critical flow at 1.42 m/s the y-distributions of the mean velocity were measured at several spanwise z-positions so as to calculate the momentum thickness θ. Depending on the z-positions it varied from 1.10 to 1.30 mm. Hence, the minimum R_θ is in the range between 110 and 130, an average of 120, almost the same as that found for the transition triggered by LG hairpins.

As for the subcritical growth, it is remarkable that u-fluctuations of 30 Hz, about a half of 58 Hz (namely, the hairpin frequency), appear and grow. This suggests two possible causes for the subcritical growth. One is vortex amalgamation. The other is the inherent instability of the near-wall flow. To examine these, we made the flow visible with smoke wires for $U_\infty = 1.48$ m/s, slightly above the critical velocity. Figure 4 shows the side-view photographs. The upper smoke sheet from a wire at $y = 10$ indicates signatures caused by hairpins from upstream, for x up to

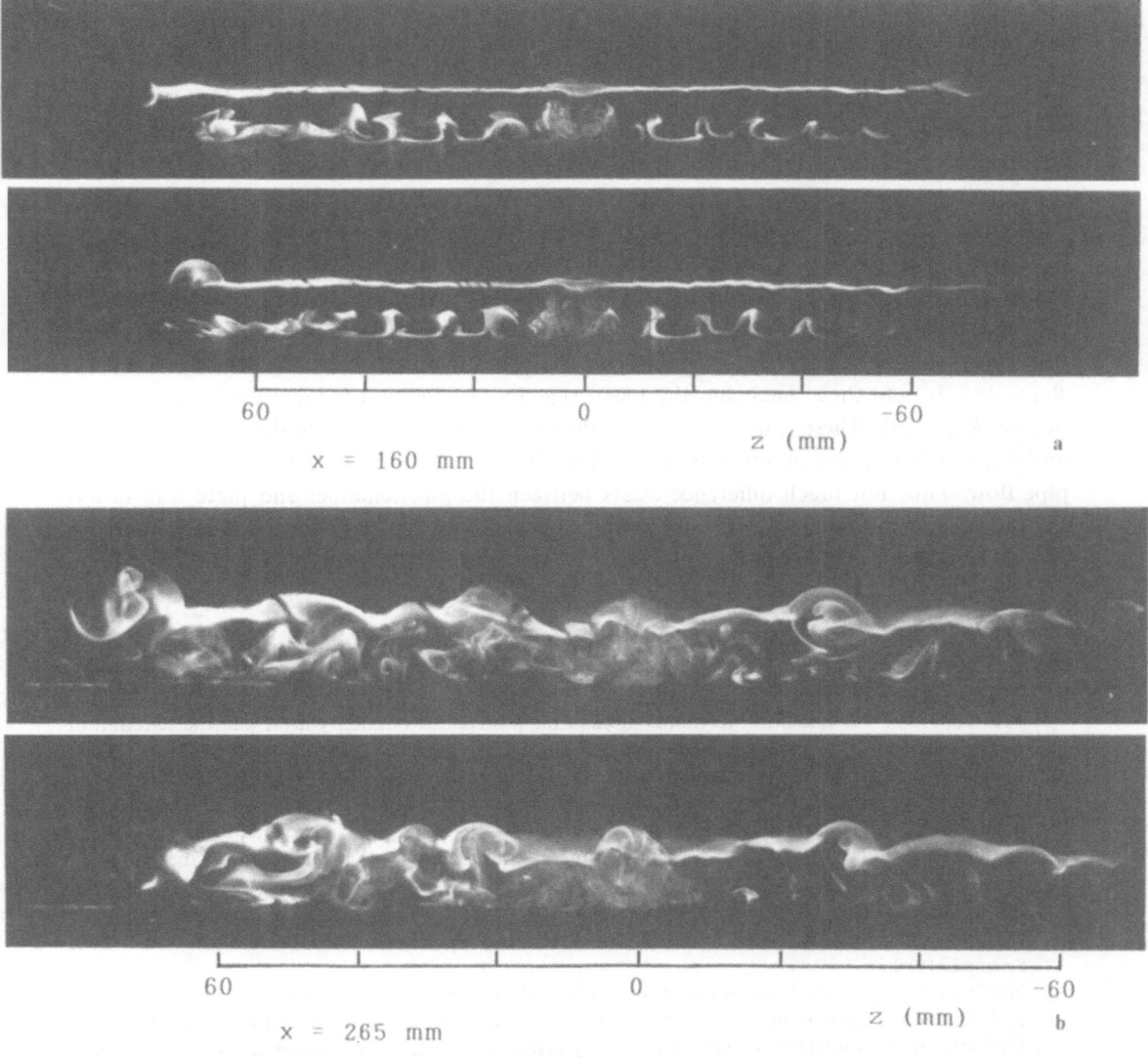

60   0   -60

x = 160 mm   z (mm)   a

60   0   -60

x = 265 mm   z (mm)   b

Fig. 5. Appearance of large scale structures illustrated by cross-section views

200. The lower smoke from another wire at $y = 2$ shows the near-wall activities and the growth of large scale structures beyond $x = 200$. (The initial spanwise width of each smoke sheet is about 10 mm.) We notice that the upper sheet is entrained into the large scales. This suggests that the large scale structures are generated through the near-wall activities. Figures 5a, b demonstrate the cross-section view at $x = 160$ and 265 respectively, and show a rather abrupt appearance of large scale mushrooms. This is consistent with the side views in Fig. 4. Examination of a number of similar photographs does not enable us to obtain firm evidence for the possible vortex amalgamation. However, we infer that together with the inherent instability of near-wall flow, the amalgamation is responsible for generating the large scale structures. Moreover, it should be stressed that Figs. 4 and 5 do indicate that when disturbed by hairpin eddies coming from upstream, the near-wall flow can regenerate vortical structures of hairpin shape, and these soon grow in scale to become much larger than the primary structures. This important finding has been confirmed for $U_\infty = 1.9$ and 2.9 m/s too.

6 Concluding remarks

As described thus far, the critical flow condition and the minimum R_θ for the wall turbulence generation have been clarified for the case of a Blasius flow that is highly disturbed by hairpin eddies generated near the leading edge. The hairpin eddies may be excited acoustically or roughness-induced. Although they quickly decay downstream, they can trigger the subcritical transition if their intensity is above $0.05 \sim 0.1$ of U_∞ (in terms of rms u-fluctuations) beyond the x-station corresponding to the minimum $R_{\theta 1}$. Independently of whether they are acoustically excited or roughness-induced, we found the minimum $R_{\theta 1}$ (for the occurrence of wall bursts or for the regeneration of hairpin-shaped vortical structures) to be $110 \sim 130$. At this minimum, the flow is transitional and intermittent. The wall flow can be continuously turbulent beyond $R_{\theta 2} = 185$. It is further noted that the mean velocity can follow the logarithmic distribution beyond $R_{\theta 3} = 210$. These values obtained in the present study are close to the values $R_{\theta 1} = 133$ and $R_{\theta 3} = 266$ observed in our channel flow [14]. Preston [1] determined $R_{\theta 3} = 193$ for circular pipe flow. Thus not much difference exists between the pipe, channel and plate wall flows! Finally, it is noted that an important finding, the growth of large scales during the regeneration process, remains to be clarified.

References

[1] Preston, J. H.: The minimum Reynolds number for a turbulent boundary layer and the selection of a transition device. J. Fluid Mech. **3**, 373 – 384 (1958).

[2] Morkovin, M. V.: Recent insights into instability and transition. AIAA-88-3675.

[3] Nishioka, M., Asai, M., Iida, S.: Wall phenomena in the final stage of transition to turbulence. In: Transition and turbulence (Meyer, R. E., ed.), pp. 113 – 126. New York: Academic Press 1981.

[4] Nishioka, M., Iida, S., Ichikawa, Y.: An experimental investigation of the stability of plane Poiseuille flow. J. Fluid Mech. **72**, 731 – 751 (1975).

[5] Nishioka, M., Asai, M., Iida, S.: An experimental investigation of the secondary instability. In: Laminar-turbulent transition (Eppler, R., Fasel, H., eds.), pp. 37 – 46. Berlin Heidelberg New York: Springer 1980.

[6] Nishioka, M., Asai, M.: Evolution of Tollmien-Schlichting waves into wall turbulence. In: Turbulence and chaotic phenomena in fluids (Tatsumi, T., ed.), pp. 87 – 92. Amsterdam: North-Holland 1984.

[7] Herbert, Th.: Secondary instability of boundary layers. Annu. Rev. Fluid Mech. **20**, 487 – 526 (1988).

[8] Nishioka, M., Asai, M.: Three-dimensional wave-disturbances in plane Poiseuille flow. In: Laminar-turbulent transition (Kozlov, V. V., ed.), pp. 173 – 182. Berlin Heidelberg New York: Springer 1985.

[9] Asai, M., Nishioka, M.: Origin of the peak-valley wave structure leading to wall turbulence. J. Fluid Mech. **208**, 1 – 23 (1989).

[10] Sandham, N. D., Kleiser, L.: The late stages of transition to turbulence in channel. J. Fluid Mech. **245**, 319 – 348 (1992).

[11] Asai, M., Nishioka, M.: Development of wall turbulence in Blasius flow. In: Laminar-turbulent transition (Arnal, D., Michel, R., eds.), pp. 215 – 224. Berlin Heidelberg New York Tokyo: Springer 1990.

[12] Haridari, A. H., Smith, C. R.: see [13].

[13] Smith, C. R., Walker, J. D. A., Haridari, A. H., Sobrun, U.: On the dynamics of near-wall turbulence. Phil. Trans. R. Soc. London Ser. A **336**, 131 – 175 (1991).

[14] Nishioka, M., Asai, M.: Some observations of the subcritical transition in plane Poiseuille flow. J. Fluid Mech. **150**, 441 – 450 (1985).

Authors' addresses: Professor M. Nishioka, Department of Aerospace Engineering, University of Osaka Prefecture, 1-1 Gakuen-Cho, Sakai, Osaka 593, Professor M. Asai, Department of Aerospace Engineering, Tokyo Metropolitan Institute of Technology, Hino, Tokyo 191, and Grad. Stud. S. Furumoto, Graduate School, University of Osaka Prefecture, Sakai, Osaka 593, Japan

Acta Mechanica (1994) [Suppl] 4: 95–104

Comparison of flamelet models for premixed turbulent combustion

N. Peters, Aachen, Federal Republic of Germany

Summary. For premixed turbulent combustion in the flamelet regime two different formulations are compared. One is the well-known Bray-Moss-Libby-model and the other is a recently developed model based on a field equation. By deriving a pdf transport equation for the latter, the correspondence between the two models is established. Finally, the balance equations for the mean conditioned velocity field that appear in the field equation formulation are presented.

1 Introduction

The flamelet concept views a turbulent flame as an ensemble of thin laminar locally one-dimensional structures embedded within the turbulent flow field. The criterion of fast chemistry may then be specified in terms of a comparison of the relevant time and length scales of chemistry and turbulence. If the characteristic chemical time scale of the oxidation reactions is short compared to the Kolmogorov time scale, combustion is completed within the turn-over time of the smallest eddies [1].

The problem to be considered is a premixed turbulent flame in the regime where the laminar flame thickness l_F is much smaller than the smallest scale of turbulence, the Kolmogorov scale η. From a macroscopic level, the flame front within the turbulent flow may then be assumed to be infinitely thin, representing a jump in temperature, density and velocity. The flame front advances with the laminar burning velocity s_L with respect to the flow field. The regime in which this description is valid is the flamelet regime.

2 The Bray-Moss-Libby-model

A well known model describing premixed turbulent combustion in the flamelet regime is the Bray-Moss-Libby (BML) model originally proposed in [2], [3]. This model is built on the progress variable c and its probability function (pdf). The progress variable may be interpreted as a non-dimensional temperature or a normalized product mass fraction

$$c = \frac{T - T_u}{T_b - T_u} = \frac{Y_P}{Y_{P,b}}. \tag{1}$$

Here T is the temperature and Y_P the mass fraction of the product, the subscripts u and b denote unburnt and burnt gas conditions, respectively. The pdf of c is assumed to be bimodal with large entries at $c = 0$ and $c = 1$, and with a negligible part in the range $0 < c < 1$. The pre-assumed pdf of c is written as

$$P(c, \boldsymbol{x}, t) = \alpha(\boldsymbol{x}, t)\,\delta(c) + \beta(\boldsymbol{x}, t)\,\delta(1 - c) + \gamma f(c) \tag{2}$$

where $\gamma \ll 1$. In the limit $\gamma \to 0$ the integration over P yields

$$\int_0^1 P(c, \boldsymbol{x}, t)\,dc = 1 = \alpha(\boldsymbol{x}, t) + \beta(\boldsymbol{x}, t). \tag{3}$$

The bimodal pdf models the physical phenomena of a thin flame front passing across the location $\boldsymbol{x}$ at time t in the flow field as shown in Fig. 1: Since $c = 0$ in the unburnt mixture and $c = 1$ in the burnt gas, $\alpha(\boldsymbol{x}, t)$ and $\beta(\boldsymbol{x}, t)$ represent the fractions of time when unburnt or burnt gas is present at the location $\boldsymbol{x}$. Then mean progress variable is by definition

$$\bar{c}(\boldsymbol{x}, t) = \int_0^1 c P(c, \boldsymbol{x}, t)\,dc = \beta(\boldsymbol{x}, t). \tag{4}$$

For steady processes the overbar defines a time average whereas ensemble averaging is required for unsteady processes.

In order to close the formulation, the BML-model uses a differential equation for the mean progress variable $\bar{c}$. The corresponding equation for the instantaneous value of c has a similar form as the temperature equation for a laminar flame with one-step chemistry. The flame propagation problem is then formulated by the two equations:

Continuity

$$\frac{\partial p}{\partial t} + \nabla(\varrho \boldsymbol{v}) = 0. \tag{5}$$

Progress variable

$$\frac{\partial}{\partial t}(\varrho c) + \nabla(\varrho \boldsymbol{v} c) = \nabla(\varrho D \nabla c) + \dot{\omega}(c). \tag{6}$$

Here ϱ is the density, $\boldsymbol{v}$ the velocity vector, D the molecular diffusivity, assumed equal for temperature and concentrations, and $\dot{\omega}$ the chemical source term, assumed to be a function of c only. When the continuity equation is subtracted from the transport equation for the progress variable and ensemble averaging is performed the equation becomes

$$\frac{\partial}{\partial t}\overline{\varrho(c - 1)} + \nabla\overline{\varrho \boldsymbol{v}(c - 1)} = \nabla\overline{(\varrho D \nabla c)} + \bar{\dot{\omega}}. \tag{7}$$

Since only events from the unburnt gas enter into the averaging, one obtains for the mean values appearing on the l.h.s. of this equation

$$\overline{\varrho(c - 1)} = \varrho_u \alpha(\boldsymbol{x}, t) = \varrho_u(\bar{c} - 1), \qquad \overline{\varrho \boldsymbol{v}(c - 1)} = \overline{\varrho_u \boldsymbol{v}_u}\alpha(\boldsymbol{x}, t) = \varrho_u \overline{\boldsymbol{v}_u}(\bar{c} - 1). \tag{8}$$

Here $\overline{\boldsymbol{v}_u}$ is the conditioned mean velocity in the unburnt gas, where the density $\varrho = \varrho_u$ is constant. No correlations between velocity and scalar fluctuations appear here because c does not fluctuate in the unburnt gas. Furthermore, it is consistent with the assumption of a thin flame of thickness l_F to neglect the diffusion term in (7), since $l_F = D/s_L$ is a diffusive thickness. When the continuity equation, conditioned for the unburnt gas, is used again, one obtains (7) in the form

$$\varrho_u \left(\frac{\partial \bar{c}}{\partial t} + \overline{v_u} \nabla \bar{c} \right) = \bar{\dot{\omega}}. \tag{9}$$

Apart from the conditioned mean velocity $\overline{v_u}$ which may, in principle, be obtained from the corresponding momentum equations, the balance equation for $\bar{c}$ contains the unknown mean chemical source term $\bar{\dot{\omega}}$. Closure of this term has remained an essentially unresolved problem. Various approaches based on the flamelet concept where recently discussed by Bray [4]. A model based on flamelet arguments is written as

$$\bar{\dot{\omega}} = \varrho_u s_L I_0 \Sigma \tag{10}$$

where Σ is the flamelet surface to volume ratio and I_0 is a correction factor representing the effect of flame stretch on the propagation of the instantaneous flame front.

3 A field equation describing the propagation of a premixed flame

An alternative approach to describe turbulent flame propagation in the flamelet regime is a formulation based on a field equation. A fundamental property of a premixed flame is its ability to propagate normal to the flame front with the laminar burning velocity s_L. Here we define s_L with respect to the unburnt gas. If n denotes the vector normal to the front and

$$v_f = \frac{dx}{dt}\bigg|_f \tag{11}$$

the velocity by which the flame propagates relative to the observer, the local kinematic balance between the propagation velocity, the flow velocity v^c at the flame front, and the burning velocity in normal directions is

$$n \frac{dx}{dt}\bigg|_f = v^c n + s_L. \tag{12}$$

The superscript c is added here to denote that v^c is a conditioned velocity, where only the value on the unburnt gas side immediately ahead of the flame front is to be used. Using differential geometry the normal vector n towards the unburnt gas may be defined as

$$n = -\frac{\nabla G}{|\nabla G|}. \tag{13}$$

Here $G(x, t)$ is a scalar field whose level surfaces

$$G(x, t) = G_0 \tag{14}$$

represent the flame surface. The choice of G_0 is arbitrary. As shown in Fig. 1, the flame surface divides the physical field into two regions where $G < G_0$ corresponds to the unburnt and $G > G_0$ to the burnt gas. Differentiating (14) with respect to t at $G = G_0$ gives

$$\frac{\partial G}{\partial t} + \nabla G \frac{\partial x}{\partial t}\bigg|_{G=G_0} = 0. \tag{15}$$

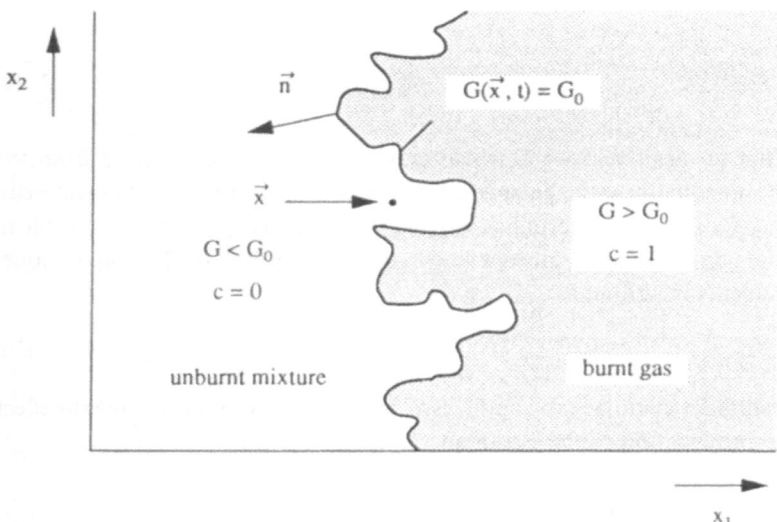

Fig. 1. Burnt and unburnt gas regions separated by a thin flame surface at $G(x, t) = G_0$

Combining this with the kinematic balance (12) and (13) one obtains the field equation

$$\frac{\partial G}{\partial t} + v^c \nabla G = s_L |\nabla G|. \tag{16}$$

It will be called the G-equation in the following. This equation has originally been derived by Markstein [5]. It was investigated numerically for constant density flow fields by Kerstein, Ashurst and Williams [6]. Recently it was used [7] to derive a consistent flamelet formulation for premixed turbulent combustion. If the conditioned flow field v^c is known, the G-equation can be solved numerically. For a constant value of s_L the solution is not unique, but cunps will be formed where different parts of the flame front intersect. This can be cured by accounting for the dependence of the burning velocity on the curvature [7].

4 Comparison of the two models

Here we want to compare the BML model with the formulation based on the G-equation. Since G is continuous at the flame front but c jumps from zero to unity the progress variable c may be defined in terms of the scalar G as

$$c = H(G - G_0) = \begin{cases} 0 & \text{for} \quad G < G_0 \\ 1 & \text{for} \quad G > G_0 \end{cases} \tag{17}$$

where H is the Heaviside function. Since the scalar G fluctuates in a turbulent flow field, we want to define the probability density function $P(G, x, t)$. For a locally planar turbulent flame we may consider a fluctuating wrinkled flame front whose mean position lies in the plane $x_1 = 0$, where x_1 is the coordinate normal to the flame towards the burnt gas. Flame front position at different times 1, 2 and 3 are schematically shown in Fig. 2.

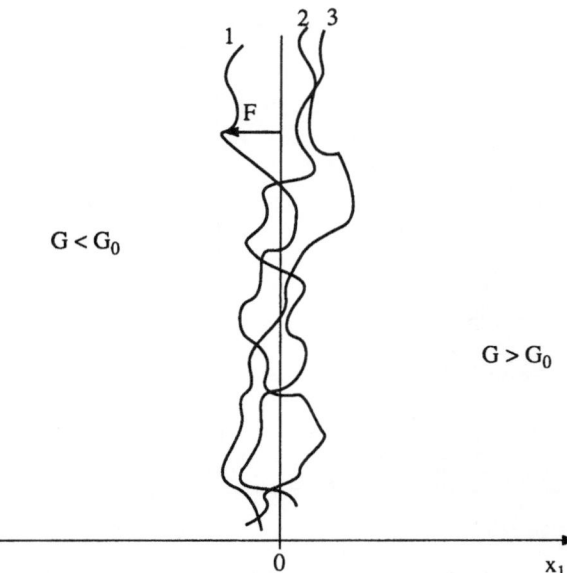

Fig. 2. Schematic illustration of several flame fronts in a turbulent flow field fluctuating around a mean position at $x_1 = 0$

We will assume that the displacement $F(x_2, x_3, t)$ from its mean position is a unique function of a x_2 and x_3, such that the flame front does not turn around itself. Then (13) is equally valid if one relates $G(x, t)$ to F as

$$G(x, t) - G_0 = x_1 + F(x_2, x_3, t). \tag{18}$$

Since the mean displacement is $\bar{F}(x_2, x_3, t) = 0$, the mean value $\bar{G}(x, t)$ is independent of x_2, x_3 and t and increases linearly with x_1

$$\bar{G}(x, t) = G_0 + x_1. \tag{19}$$

Combining this with (18) leads to

$$G'(x, t) \equiv G(x, t) - \bar{G}(x_1) = F(x_2, x_3, t). \tag{20}$$

The displacement F of individual flame fronts is a local property of the turbulent flame brush and does not depend on x_1. Since $G = G_0$ at the flame front it follows from (18) that its statistics can be obtained by measuring the statistics of the flame displacement from $x_1 = 0$ towards the negative x_1-direction as shown in Fig. 2. As an example the probability density function $P(F)$ measured inside a spark ignition engine [8] is shown in Fig. 3. According to (20) the fluctuation G' is also a local property of the flame brush only. By assuming G to fluctuate at a fixed location one transforms spatial displacements normal to the mean location of the turbulent flame into scalar level fluctuations.

When the pdf $P(F)$ is known, the pdf $P(G, x, t)$ can be written as

$$P(G, x, t) = P\big((G - \bar{G}(x_1)\big) = P(G - G_0 - x_1) = P(F). \tag{21}$$

Therefore the dependence of $P(G, x, t)$ on the spatial location x enters through $\bar{G}$.

In order to proceed further we will derive a transport equation for $P(G, x, t)$ following the work of O'Brien [9] by first defining a "fine grained density"

$$\mathscr{P}(\hat{G}, x, t) = \delta[G(x, t) - \hat{G}] \tag{22}$$

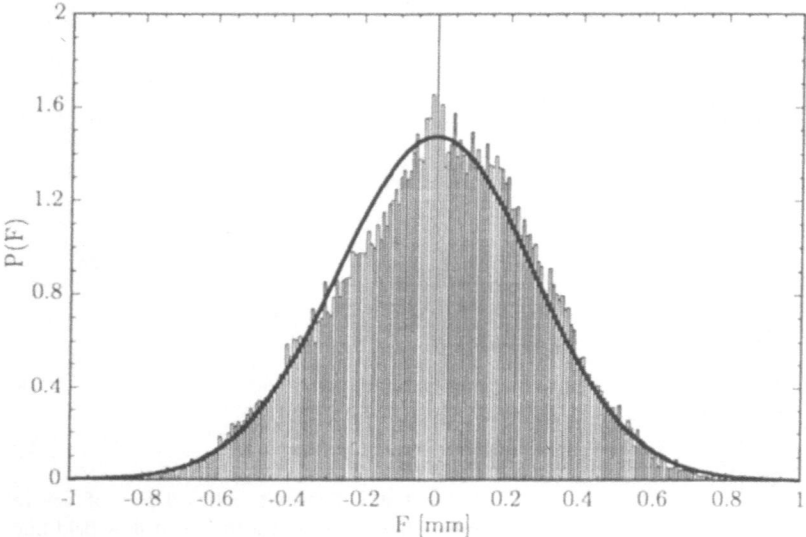

Fig. 3. Probability density function of flame surface displacement measured by Laser-Mie scattering in a transparent engine

such that $\mathscr{P}(\hat{G}, x, t)\, d\hat{G}$ is the probability that at x and t, G will be in the range $\hat{G} < G(x, t)$ $< \hat{G} + d\hat{G}$. In the definition of the fine grained density the variable $\hat{G}$ is a nonrandom quantity whereas $G(x, t)$ is the random field whose pdf is sought. The ensemble average of $\mathscr{P}(\hat{G}, x, t)$ is

$$P(\hat{G}, x, t) = \overline{\mathscr{P}(\hat{G}, x, t)}. \tag{23}$$

This is the quantity for which a transport equation shall be derived. From the definition of $\mathscr{P}$ it follows that its derivatives with respect to space and time are

$$\frac{\partial \mathscr{P}}{\partial t} = \frac{\partial \delta}{\partial G}\frac{\partial G}{\partial t} = -\frac{\partial \delta}{\partial \hat{G}}\frac{\partial G}{\partial t} = -\frac{\partial \mathscr{P}}{\partial \hat{G}}\frac{\partial G}{\partial t}$$

$$\nabla \mathscr{P} = \frac{\partial \delta}{\partial G}\nabla G = -\frac{\partial \delta}{\partial \hat{G}}\nabla G = -\frac{\partial \mathscr{P}}{\partial \hat{G}}\nabla G. \tag{24}$$

When the gradient is multiplied with the velocity v^c and both terms are added, one obtains with (16)

$$\frac{\partial \mathscr{P}}{\partial t} + v^c \cdot \nabla \mathscr{P} = -\frac{\partial}{\partial \hat{G}}\, s_L\, |\nabla G|. \tag{25}$$

Taking the ensemble average of this equation one obtains

$$\frac{\partial P}{\partial t} + \overline{v^c \cdot \nabla \mathscr{P}} = -s_L \frac{\partial}{\partial \hat{G}}\, \overline{|\nabla G|\, \mathscr{P}}. \tag{26}$$

Denoting the gradient by $\sigma = |\nabla G|$ the ensemble average appearing in the last term may be expressed as

$$\overline{|\nabla G|\, \mathscr{P}(\hat{G}, x, t)} = \int \hat{\sigma} P(\hat{G}, \hat{\sigma}, x, t)\, d\hat{\sigma}, \tag{27}$$

where $P(\hat{G}, \hat{\sigma}, x, t)$ is the joint pdf of G and σ. Using Bayes' theorem

$$P(\hat{G}, \hat{\sigma}, x, t) = P(\hat{\sigma}|\hat{G}, x, t)\, P(\hat{G}, x, t) \tag{28}$$

where $P(\hat{\sigma}|\hat{G}, x, t)$ is the dpf of $\hat{\sigma}$, conditioned on a fixed value G at x, t. The subsequent discussion would be much simplified if one could assume statistical independence of G and σ.

Conditioning on G amounts to conditioning on a fixed value of F. The gradient $|\nabla G|$ measures the inclination of the flame front with respect to the x_1-coordinate. In a high Reynolds number turbulent flow one may in fact assume that the inclination of the flame front is independent of its displacement (except for large displacements where the front will have zero inclinations towards the x_1 axis more frequently). Assuming statistical independence

$$P(\hat{G}, \hat{\sigma}, x, t) = P(\hat{\sigma}\ x, t)\, P(\hat{G}, x, t) \tag{29}$$

one obtains from (27)

$$\overline{|\nabla G|\, \mathscr{P}} = \overline{|\nabla G|}\; \mathscr{P}(\hat{G}, x, t). \tag{30}$$

Similar arguments may be used to express the ensemble average of the convective term in (26). It should be noted that the conditioned velocity v^c in the unburnt gas immediately ahead of the flame is influenced by gas expansion within the flame front. Nevertheless, this velocity should to first approximation be statistically independent of the displacement F within the flame brush. If the turbulent is curved as at the tip of a turbulent Bunsen flame, additional considerations will be necessary [7].

With these assumptions one may write the transport equation for $P(\hat{G}, x, t)$ as

$$\frac{\partial P}{\partial t} + \overline{v^c} \cdot \nabla P = -s_L \frac{\partial}{\partial \hat{G}}\left[\overline{|\nabla G|}\, P\right]. \tag{31}$$

Equation (31) is now multiplied with $\hat{c}$, which is defined in terms of $\hat{G}$ as in (17). Since it is a fixed variable, it is independent of the time and the spatial coordinate and may be written within the derivatives,

$$\frac{\partial(\hat{c}P)}{\partial t} + \overline{v^c} \cdot \nabla(\hat{c}P) = -s_L \hat{c}\, \frac{\partial}{\partial \hat{G}}\left[\overline{|\nabla G|}\, P\right]. \tag{32}$$

Integrating over $\hat{G}$ from $\hat{G} = -\infty$ to $\hat{G} = \infty$ and using partial integration for the term on the r.h.s. one obtains

$$\frac{\partial \bar{c}}{\partial t} + \overline{v^c} \cdot \nabla \bar{c} = s_L \int\limits_{-\infty}^{\infty} \frac{\partial \hat{c}}{\partial \hat{G}}\, \overline{|\nabla G|}\, P(\hat{G})\, d\hat{G}. \tag{33}$$

Noting that $\partial \hat{c}/\partial \hat{G} = \delta(\hat{G} - G_0)$ and that $\overline{|\nabla G|}$ is assumed independent of G, we obtain

$$\frac{\partial \bar{c}}{\partial t} + \overline{v^c} \nabla \bar{c} = s_L \overline{|\nabla G|}\, P(G_0; x). \tag{34}$$

Here $P(G_0, x)$ is the conditioned pdf where G is set equal to G_0. In view of (21) it is

$$P(G_0, x) = P(-x_1). \tag{35}$$

If (34) is compared with (7) and (8) it is seen that both formulations agree if $\overline{v_u}$ is assumed equal to $\overline{v^c}$ and if we set

$$I_0 \Sigma = \overline{|\nabla G|}\, P(G_0; x). \tag{36}$$

In the context of the present comparison we may set the correction factor I_0 equal to unity, since flame stretch effects were not included in the derivation of the G-equation. Then the flame surface to volume ratio Σ is identified as the mean gradient of G times the pdf of the negative distances to the flame front from the mean position $x_1 = 0$.

5   Balance equations for the mean conditioned velocity $\overline{v^c}$

Turbulent modeling of the equations for the mean scalar $\bar{G}$, its variance $\overline{G^{12}}$ and the conditioned mean velocity $\overline{v^c}$ is based on the assumption that scales smaller than the integral length scale are not resolved. It was shown in [7] that the flame brush thickness defined as

$$l_{F,t} = \frac{\overline{(G'^2)^{1/2}}}{|\nabla \bar{G}|} \tag{37}$$

is proportional to and of the order of the integral length scale. It was noted above that scalar fluctuations and therefore the scalar variance $\overline{G'^2}$ are constant across the flame brush thickness. Outside of it $\overline{G'^2}$ is not defined. The same is true for the mean scalar gradient $\overline{|\nabla G|}$ and the conditioned mean velocity $\overline{v^c}$. Since scales of the order of the flame brush thickness are not resolved, we may define the conditioned mean velocity $\overline{v^c}$ as the one immediately ahead of the mean turbulent flame front defined by $\bar{G} = G_0$. As the instantaneous velocity jumps in density, the conditioned mean velocity must jump by the same amount at $\bar{G} = G_0$.

Modeling for the equation for the mean scalar $\bar{G}$ has been proposed in [7] as

$$\frac{\partial \bar{G}}{\partial t} + \overline{v^c}\nabla\bar{G} = s_T |\nabla\bar{G}| - D_T\bar{\varkappa}\,|\nabla\bar{G}| \tag{38}$$

where D_T is a turbulent diffusivity and s_T is the turbulent velocity containing flame stretch effects. The curvature $\bar{\varkappa}$ of the mean turbulent flame brush is defined as

$$\bar{\varkappa} = \nabla \cdot \boldsymbol{n} = -\frac{\nabla^2\bar{G}}{|\nabla\bar{G}|} + \frac{\nabla|\nabla\bar{G}| \cdot \nabla\bar{G}}{|\nabla\bar{G}|^2}. \tag{39}$$

The turbulent burning velocity s_T has been obtained from empirical correlations and direct simulation as

$$s_T = (s_L + 0.8(s_L v')^{1/2} + 1.5v') \cdot \left(1 - 1.4\,\frac{\mathscr{L}}{\ell_t}\,\frac{v'}{s_L}\right). \tag{40}$$

Here v' is the turbulence intensity immediately ahead of the flame front. The first term in brackets accounts for the increase of s_T with increasing v'. It approaches Damköhler's limit $s_T \sim v'$ for large values of v'/s_L. The second term in brackets accounts for the influence of flame stretch and reduces the burning velocity. Here the ratio of the Markstein-Length $\mathscr{L}$ to the turbulent length scale ℓ_t appears [7]. This approximation is compared with other recent correlations in Fig. 4.

The continuity and momentum equations determining the mean velocity $\overline{v_u}$ on the unburnt and $\overline{v_b}$ burnt gas side are those for constant density flows

$$\nabla\overline{v_u} = 0, \qquad \nabla\overline{v_b} = 0, \tag{41}$$

$$\left(\frac{\partial\overline{v_u}}{\partial t} + \overline{v_u}\nabla\overline{v_u}\right) = -\frac{\nabla\overline{p_u}}{\varrho_u} + \nabla\big((v + v_t)_u\,\nabla\overline{v_u}\big) \tag{42}$$

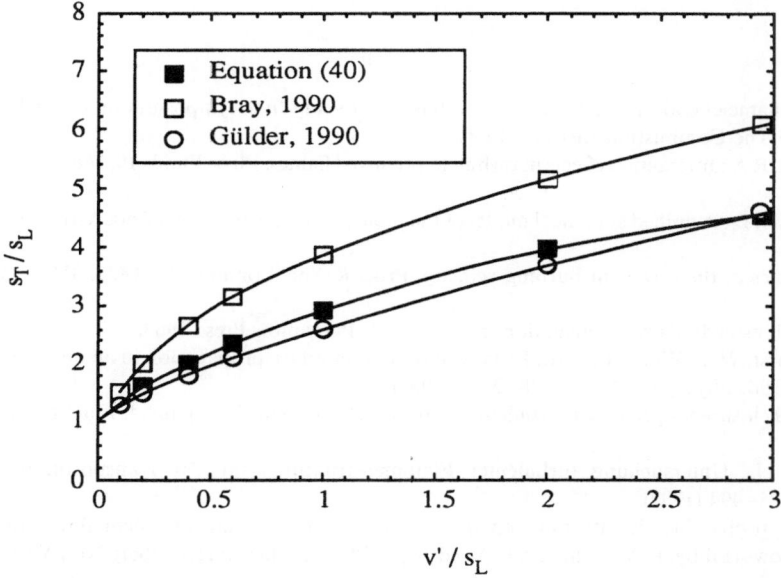

Fig. 4. Comparison of different correlations for the ratio of the turbulent and the laminar burning velocity as a function of v'/s_L for ℓ_F/ℓ_t and $\mathscr{L}/\ell_F = 4.0$

$$\left(\frac{\partial \overline{v_b}}{\partial t} + \overline{v_b}\nabla\overline{v_b}\right) = -\frac{\nabla\overline{p_b}}{\varrho_b} + \nabla[(\nu + \nu_t)_b \nabla\overline{v_b}]\tag{43}$$

The eddy viscosity ν_t must be obtained from a suitable turbulence model such as the k-ε model. At the position the mean flame front $\bar{G} = G_0$ the following jump conditions are applied (cf. Williams [11], p. 15):

Continuity

$$\varrho_u\left(\frac{\partial \bar{G}}{\partial t} + \overline{v_u}\nabla\bar{G}\right) = \varrho_b\left(\frac{\partial \bar{G}}{\partial t} + \overline{v_b}\nabla\bar{G}\right)\tag{44}$$

Momentum

$$\varrho_u\overline{v_u}\left(\frac{\partial \bar{G}}{\partial t} + \overline{v_u}\nabla\bar{G}\right) + \overline{p_u}\nabla\bar{G} = \varrho_b\overline{v_b}\left(\frac{\partial \bar{G}}{\partial t} + \overline{v_b}\nabla\bar{G}\right) + \overline{p_b}\nabla\bar{G}\tag{45}$$

In (44) and (45) $\overline{v_u}$ corresponds to $\overline{v^c}$ while (41)–(43) apply to the entire velocity field.

6 Conclusions

The flamelet formulation based on the mean $\bar{G}$ equation has the advantage over the BML model that $\bar{G}$ various smoothly in the vicinity of the turbulent flame front while $\bar{c}$ changes from 0 to 1 over scales of the order of the integral length scale. The conditioned velocity field, however, jumps at the mean flame front in the same way as in a laminar flame. Numerical techniques must be designed to take this discontinuity into account.

References

[1] Peters, N.: Laminar flamelet concepts in turbulent combustion. Twenty-first Symposium on Combustion, pp. 1231–1256. The Combustion Institute 1986.

[2] Bray, K. N. C., Libby, P. A.: Interaction effects in turbulent premixed flames. Phys. Fluids **19**, 1687–1701 1976.

[3] Bray, K. N. C., Moss, J. B.: A unified statistical model of the premixed turbulent flame. Acta Astronaut. **4**, 291–319 (1977).

[4] Bray, K. N. C.: Studies of the turbulent burning velocity. Proc. R. Soc. London Ser. **A431**, 315–335 (1990).

[5] Markstein, G. H.: Nonsteady flame propagation, p. 8. Oxford: Pergamon Press 1964.

[6] Kerstein, A. R., Ashurst, W. T., Williams, F. A.: Field equation for interface propagation in an unsteady homogeneous flow field. Phys. Rev. A **37**, 2728–2731 (1988).

[7] Peters, N.: A spectral closure for premixed turbulent combustion in the flamelet regime. J. Fluid Mech. **242**, 611–629 (1992).

[8] Wirth, M., Peters, N.: Untersuchung turbulenter Flammenstrukturen im VW-Transportmotor. VDI-Berichte **922**, 485–494 (1992).

[9] O'Brien, E. E.: The probability density function (pfd). Approach to reacting turbulent flows. In: Turbulent reacting flows (Libby, P. A., Williams, F. A., eds.), pp. 185–218. Berlin Heidelberg New York: Springer 1980.

[10] Gülder, Ö. L.: Turbulent premixed flame propagation models for different combustion regimes. 23rd Symposium (International) on Combustion, pp. 743–750. The Combustion Institute 1990.

[11] Williams, F. A.: Combustion theory, 2nd edn. Menlo Park: The Benjamin Cummings Publishing Company 1985.

Author's address: Professor Dr. N. Peters, Institut für Technische Mechanik, RWTH Aachen, Templergraben 64, D-52062 Aachen, Federal Republic of Germany

Acta Mechanica (1994) [Suppl] 4: 105–111
© Springer-Verlag 1994

Correlations in homogeneous stratified shear turbulence

U. Schumann, Oberpfaffenhofen, Federal Republic of Germany

Summary. Based on the budget of kinetic energy and simple estimates to relate dissipation and temperature or concentration fluctuations to shear, stratification, and the vertical velocity fluctuations, a consistent set of equations is deduced to estimate vertical fluxes of momentum and heat or mass. The estimates are designed for strongly sheared, neutral and stratified flows at high Reynolds numbers under approximately homogeneous conditions. The set is closed by using basically two empirical coefficients together with the turbulent Prandtl number and the growth rate of kinetic energy as a function of the gradient Richardson number. The correlations are tested using data from previous laboratory experiments and numerical simulations.

1 Introduction

For many applications, one needs estimates of the rate of turbulent mixing in neutrally and stably stratified shear flows. This is a particularly difficult topic because turbulence tends to degenerate to wavy motions under strongly stable stratification, and many models have been proposed in the past, some based on extensive second- or higher-order closure models with many model parameters. In view of the difficulties to determine the turbulence scales and even the mean profiles in such flows, simple relationships are required to estimate the magnitude of the mixing properties. Such relationships have been deduced, mainly for strongly stratified atmospheric and oceanic flows, on the basis of the energy budgets using simple closure assumptions for stationary flows [1–3].

Here, a new set of equations is deduced which takes into account the deviation from stationarity and applies to both pure shear flows and to moderately stratified shear flows. The paper is formulated for thermal stratification but the results can also be applied to density variations due to variable salt concentration. Hence, we consider the turbulence properties of a flow with given vertical velocity shear S and positive vertical potential temperature gradient s,

$$S = dU/dz, \quad s = d\Theta/dz, \tag{1}$$

which define the Brunt-Väisälä frequency N and the gradient Richardson number Ri,

$$N = (\beta g s)^{1/2}, \quad \mathrm{Ri} = N^2/S^2. \tag{2}$$

Here, β is the thermal volumetric expansion coefficient, and g is the acceleration of gravity. We consider flows in between Ri = 0, and Ri of order one.

It is assumed that the density variations affect the buoyancy only, i.e. we employ the Boussinesq approximation. The analysis is restricted to flows at high Reynolds numbers, so that the molecular diffusivities are small in comparison to turbulent diffusivities. The discussion concentrates on approximately homogeneous but time-dependent flows in which the divergence of energy fluxes is small in comparison to its local dissipation rate. Moreover, we assume approximately uniform vertical shear and stratification, remote from boundaries, so that the turbulence lengthscales are smaller than the scales of any variations in the mean profiles. The turbulence is assumed to be strongly sheared so that the timescale S^{-1} of shear is smaller than the time scale N^{-1} of stratification and both should be smaller than the turbulence timescales. Finally we assume that the exchange of energy between its kinetic and potential form has approached a local equilibrium so that the ensemble averaged fields are free of gravity wave oscillations.

Reliable data for homogeneous stratified shear flow have been measured by Rohr et al. [4], who used salt to produce the density stratification. The data are taken from appendix 2 of Rohr's thesis, as cited in [4], at shear times $tdU/dz > 6$ when the flow has approached structural equilibrium. Otherwise, data for homogeneous shear flows are available only for neutral stratification in wind tunnels [5–7]. These data will be used to calibrate and verify the model equations. Note the rather large molecular Prandtl or Schmidt number in salt (about 500) while that of air is about 0.7.

2   General consequences of the energy budget

In homogeneous turbulences the ensemble averaged kinetic energy $E_{kin} = (\overline{u^2} + \overline{v^2} + \overline{w^2})/2$ of the turbulent velocity fluctuations is a pure function of time t, and satisfies the budget

$$\frac{dE_{kin}}{dt} = P - B - \varepsilon. \tag{3}$$

It states that the local rate of change in kinetic energy equals the sum of shear production P, bouyancy destruction B, and viscous dissipation ε. If vertical shear and stratification dominate, the production terms are functions of the vertical turbulent fluxes of momentum and heat and of the related turbulent diffusivities,

$$P = -\overline{uw}S = K_m S^2, \quad B = -\overline{w\theta}N^2/s = K_h N^2. \tag{4}$$

Their ratio defines the flux Richardson number Ri_f and the turbulent Prandtl number Pr_t,

$$Ri_f = \frac{B}{P} = \frac{Ri}{Pr_t}, \quad Pr_t = \frac{K_m}{K_h}. \tag{5}$$

Now the first essential assumption is introduced, namely that the parameter G,

$$G = P/(\varepsilon + B), \tag{6}$$

is of similar universal importance for viscous flows as is the flux Richardson number for inviscid flows and controls the growth rate of kinetic energy,

$$\frac{dE}{dt} = (G - 1)(\varepsilon + B), \tag{7}$$

i.e. $G = 1$ for stationary flows, $G = 0$ for decaying flows without shear production, and $G = P/\varepsilon$ in neutral shear flows. As a consequence of the budget of kinetic energy and the above definitions of Ri_f and G, the rates of shear forcing and bouyancy destriction are related to the rate of dissipation by

$$B = \frac{\mathrm{Ri}_f G}{1 - \mathrm{Ri}_f G} \varepsilon, \quad P = \frac{G}{1 - \mathrm{Ri}_f G} \varepsilon. \tag{8}$$

Together with Eq. (4), these relationships determine the turbulent diffusivities,

$$K_m = \frac{G}{1 - \mathrm{Ri}_f G} \frac{\varepsilon}{S^2}, \tag{9}$$

$$K_h = \frac{\mathrm{Ri}_f G}{1 - \mathrm{Ri}_f G} \frac{\varepsilon}{N^2}. \tag{10}$$

We also obtain estimates for the "structure parameter" of the momentum flux and for the correlation coefficient of the heat flux,

$$\alpha_{uw} \equiv -\frac{\overline{uw}}{w'^2} = \frac{G}{1 - \mathrm{Ri}_f G} \frac{\varepsilon}{w'^2 S}, \tag{11}$$

$$\alpha_{w\theta} \equiv -\frac{\overline{w\theta}}{w'\theta'} = \frac{\mathrm{Ri}_f G}{1 - \mathrm{Ri}_f G} \frac{\varepsilon s}{N^2 w'\theta'} = \frac{G}{\mathrm{Pr}_t(1 - \mathrm{Ri}_f G)} \frac{\varepsilon s}{S^2 w'\theta'}, \tag{12}$$

where $w' = (\overline{w^2})^{1/2}$ and $\theta' = (\overline{\theta^2})^{1/2}$ are the root-mean square values of the turbulent fluctuations of vertical and temperature.

3 Approximations for strongly sheared turbulence

According to Hunt et al. [8], for strong shear but for moderate stratification, i.e. for $\mathrm{Ri} \leqq 1$, the dissipation due to small-scale mixing in turbulent flows (remote from boundaries) is controlled by shear S and the induced vertical motion velocity w'. Dimensional analysis and Prandtl's classical eddy mixing concept suggest

$$\varepsilon = A_S w'^2 S. \tag{13}$$

The temperature fluctuations are controlled by turbulent motions at the larger scales and are more sensitive to buoyancy. Therefore, the impact of buoyancy gets important at values of Ri considerably less than one. Hence, the mixing concept and dimensional analysis give

$$\theta' = \zeta_S w' s / S \quad \text{for} \quad \mathrm{Ri} \leqq 0.25, \quad \theta' = \zeta_N w' s / N \quad \text{for} \quad \mathrm{Ri} > 0.25. \tag{14}$$

The limit $\mathrm{Ri} = 0.25$ is certainly only approximately valid and is taken in correspondance with the linear stability criterion of inviscid flows. Here, A_S, ζ_S, and ζ_N are the yet undetermined model coefficients.

The two versions for the temperature fluctuations given in Eq. (14) are consistent with each other if

$$\zeta_S = \text{const}, \quad \zeta_N = \zeta_S \mathrm{Ri}^{1/2}, \tag{15}$$

for $\mathrm{Ri} \leqq 0.25$, and

$$\zeta_S = \zeta_N \mathrm{Ri}^{-1/2}, \quad \zeta_N = \text{const}, \tag{16}$$

for $Ri > 0.25$, with $\zeta_N = 0.5\zeta_S$ at the limit between the two ranges. Hence, only one of these coefficients is an independent model parameter.

Without any further assumption, these relations can be used to estimate

$$\alpha_{uw} = \frac{A_S G}{1 - Ri_f G},$$
(17)

$$\alpha_{w\theta} = \frac{\alpha_{uw}}{\zeta_S Pr_t} = \frac{A_S G}{\zeta_S Pr_t(1 - Ri_f G)},$$
(18)

$$K_h = c_S w'^2/S = c_N w'^2/N, \qquad K_m = K_h Pr_t,$$
(19)

with

$$c_S = \alpha_{uw}/Pr_t, \qquad c_N = \alpha_{uw} Ri^{1/2}/Pr_t.$$
(20)

4 Closure assumptions

In order to close the set of equations, one needs to specify A_S and ζ_S (for $Ri \leq 0.25$) as well as the growth-rate parameter $G(Ri)$, and the turbulent Prandtl number $Pr_t(Ri)$, which we assume to be pure functions of the Richardson number Ri. One expects that G and Pr_t are not unique functions of Ri, in particular for strong stratification, but we take this approach as a pragmatic procedure.

The function $G(Ri)$ is set up such that it equals unity at the stationary Richardson number Ri_S, for which the forcing by shear just balances dissipation and buoyancy destruction, and decays exponentially with Richardson number,

$$G = G_0^{(1 - Ri/Ri_S)}.$$
(21)

The value of Ri_S is less than the inviscid stability limit 0.25 because of finite dissipation in real flows.

The Prandtl number is specified to vary as

$$Pr_t = Pr_{t0} \exp\left\{-Ri/(Pr_{t0} Ri_{f\infty})\right\} + Ri/Ri_{f\infty},$$
(22)

where Pr_{t0} is the Prandtl number for neutral stratification. The model is specified such that $Pr_t \geq Pr_{t0}$ with zero gradient at $Ri = 0$, and $Pr_t \to Ri/Ri_{f\infty}$ for $Ri \gg 1$. Also, it is assumed that $Ri_{f\infty} = 0.25$. The data do not allow to determine this parameter very precisely.

5 Determination of the model coefficients

The coefficients can only be fixed when suitable measurements are given. Table 1 collects the best available data. We found that the coefficients differ depending on the molecular Prandtl number (or Schmidt number). In salt-water, the damping of concentration fluctuations is much smaller than that of temperature fluctuations in air. Therefore, we have to give two sets of coefficients. For salt-water, based on the measurements of Rohr [4], one obtains for $Ri = 0$: $G_0 = P/\varepsilon = 1.8 \pm 0.36$, and $\zeta_S = 2.88 \pm 0.15$, so that $\zeta_N = 1.44$. The structure parameter was measured to be $\alpha_{uw} = 0.87 \pm 0.08$, and the scalar flux correlation coefficient as $\alpha_{w\theta} = 0.42 \pm 0.03$. From Eq. (17) one obtains $A_S = \alpha_{uw}/G_0 = 0.5$, and from Eq. (18) $\mathrm{Pr}_{t0} = \alpha_{uw}/(\alpha_{w\theta}\zeta_S) = 0.72$. For $\mathrm{Ri} > 0$, the data suggest $G(0.36) = 0.5 \pm 0.3$, which defines $\mathrm{Ri}_S \cong 0.16 \pm 0.06$.

From the wind-tunnel data given in [5–7] for $\mathrm{Ri} = 0$, one obtains $G_0 = 1.47 \pm 0.13$, $\alpha_{uw} = 0.73 \pm 0.05$, $\alpha_{w\theta} = 0.45 \pm 0.03$, and $\zeta_S = 1.65 \pm 0.1$. Hence, $A_S = 0.48$, $\zeta_N = 0.825$, and $\mathrm{Pr}_{t0} = 0.98$. In principle, there is no reason why G_0 should be different in air and salt-water flows. However, the differences are within the scatter of the data. No measurements exist for $\mathrm{Ri} > 0$ in air, but from large-eddy simulations [9], which were performed for a Prandtl-number of one with respect to the subgrid-scale motions, we determine $\mathrm{Ri}_S \cong 0.13$. The value is close to results from direct numerical simulations [10]. Compared to salt-water, a smaller value of Ri_S has to be expected in air because of the enhanced dissipation of total energy (kinetic and potential) by the larger thermal diffusion at the smaller Prandtl number.

6 Comparison to measurements and simulation results

The value of A_S is close to the value 0.45 deduced in [8] from the logarithmic law of the wall in the boundary layer. The value $\zeta_N = 0.825$ for air is very close to the values $\zeta_N = 0.8 \pm 0.25$ and $\zeta_N \cong 0.96$ found for the stable atmospheric boundary layer in [1] and [11].

Table 1 lists the data as obtained from the experiments together with the values which result from the above equations. We see that the model approximates the measurements mostly within the standard deviation of the measured data. Data for $\mathrm{Ri} > 0$ are available only for the salt-water experiment. However, the comparison shows that A_S and ζ_S are indeed close to a constant, at least for $\mathrm{Ri} \leq 0.25$. The growth-rate parameter G decreases with Ri, and the turbulent Prandtl number increases with Ri, as assumed above.

Table 1. Data and model results

Quantity	Air: Ri = 0	Salt: Ri = 0	0.018 ± 0.004	0.062 ± 0.008	0.186 ± 0.01	0.356 ± 0.005
G	1.47 ± 0.13	1.81 ± 0.36	1.47 ± 0.34	1.44 ± 0.13	0.84 ± 0.25	0.43 ± 0.38
model:	1.47	1.80	1.68	1.43	0.91	0.49
α_{uw}	0.73 ± 0.05	0.87 ± 0.07	0.70 ± 0.13	0.65 ± 0.04	0.43 ± 0.09	0.23 ± 0.18
	0.73	0.86	0.84	0.78	0.53	0.26
$\alpha_{w\theta}$	0.45 ± 0.03	–	0.38 ± 0.03	0.36 ± 0.03	0.156 ± 0.034	0.086 ± 0.026
	0.45	0.42	0.41	0.36	0.18	0.072
ζ_S	1.65 ± 0.1	–	2.63 ± 0.25	3.04 ± 0.09	3.29 ± 0.14	2.55 ± 0.14
	1.65	2.88	2.88	2.88	2.88	2.41
A_S	0.50 ± 0.08	0.50 ± 0.07	0.48 ± 0.07	0.46 ± 0.04	0.53 ± 0.07	0.53 ± 0.09
	0.50	0.48	0.48	0.48	0.48	0.48
Pr_t	1.1 ± 0.1	–	0.71 ± 0.14	0.60 ± 0.06	0.85 ± 0.18	1.16 ± 1.34
	0.98	0.72	0.72	0.76	1.00	1.52
c_S	0.75 ± 0.08	–	0.99 ± 0.06	1.09 ± 0.09	0.51 ± 0.11	0.22 ± 0.07
	0.75	1.20	1.17	1.03	0.52	0.17
c_N	0	–	0.13 ± 0.02	0.27 ± 0.04	0.22 ± 0.05	0.13 ± 0.04
	0	0	0.16	0.26	0.23	0.10

For each quantity, the first line gives the experimental mean value and its standard deviation, and the second line gives the model result. The first column of data refers to turbulence in air at neutral stratification [5–7], the following columns apply to neutrally and stably stratified shear flows in salt-water [4]

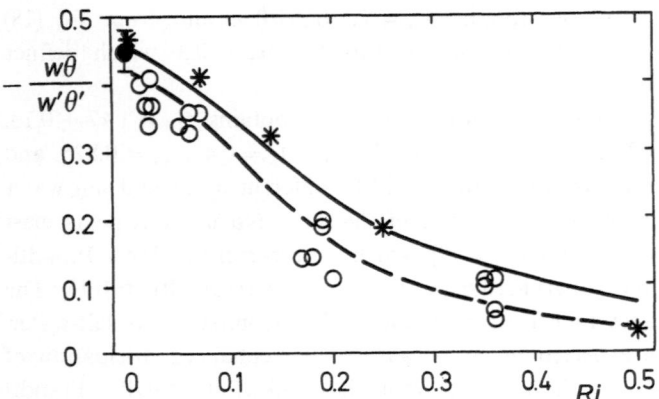

Fig. 1. Vertical scalar flux correlation coefficient $\alpha_{w\theta} = -\overline{w\theta}/(w'\theta')$ versus gradient Richardson number Ri, based on the data of Rohr et al. [4] in salt-water (circles), the measurements in neutrally stratified wind tunnel shear flows of Tavoularis and Corrsin [6] (full circle with error bar), and the LES results of Kaltenbach [9] with a subgrid-Prandtl number of one (stars). The full curve corresponds to the present model, Eq. (18), for air, the dashed curve is the result for salt-water

In Fig. 1, the approximations are compared with the data for air and salt-water in terms of the correlation coefficient for vertical scalar fluxes. The curves are the consequences of the assumptions, and the data for Ri > 0 have not been used to calibrate the model parameters. Therefore, the comparison provides a check for the internal consistency of the present model. We see that the agreement is generally within the scatter of the data. The heat flux decays more quickly than the momentum flux, which is consistent with an increasing turbulent Prandtl number because of more efficient momentum than heat transport in wavy flows.

7 Conclusions

A consistent set of equations has been deduced for strongly sheared and stratified flow, based mainly on the local energy budget, and the basic assumptions of $\varepsilon = A_S w'^2 S$, and $\theta' = \zeta_S w's/S$. If either ε or w' are given, the vertical diffusivities can be estimated from Eqs. (19) and (13). The coefficient c_N, see Eq. (20) and Table 1, is obviously a strong function of Ri. In [1], c_N was assumed to be a constant, but the data reported there show large scatter. It turns out that the results depend heavily on the variation of the growth rate G and the turbulent Prandtl number Pr_t with the gradient Richardson number Ri. Also, the coefficients are different for salt-water and air, in particular for the temperature fluctuations and the related coefficient ζ_S, because of different molecular mixing properties.

The given relationships let one conclude that A_S = const implies a constant shear number Sw'^2/ε, and that the ratio of Ellison scale $L_\theta \equiv \theta'/s$ to the Ozmidov scale $L_0 \equiv (\varepsilon/N^3)^{1/2}$ equals $L_\theta/L_0 = \zeta_S A_S^{-1/2} Ri^{3/4}$ for Ri ≤ 0.25, and the linear dependence of this scale ratio on $Ri^{3/4}$ is strongly supported by the data shown in Fig. 15 of Rohr et al. [4]. Hence, the present analysis supports in understanding and in predicting transport and dissipation properties of stratified and neutral shear flows. It shows, moreover, that a simple analysis using similarity properties and dimensional considerations, as I learned from Zierep [12], may sometimes be better suited to explain observations than complex closure models.

References

[1] Hunt, J. C. R., Kaimal, J. C., Gaynor, J. E.: Some observations of turbulence structure in stable layers. Q. J. R. Meteorol. Soc. **111**, 793–815 (1985).

[2] Lilly, D. K., Waco, D. E., Adelfang, S. I.: Stratospheric mixing estimated from high-altitude turbulence measurements. J. Appl. Meteorol. **13**, 488–493 (1974).

[3] Osborn, T. R.: Estimates of the rate of vertical diffusion from dissipation measurements. J. Phys. Oceanogr. **10**, 83–89 (1980).

[4] Rohr, J. J., Itsweire, E. C., Helland, K. N., Van Atta, C. W.: Growth and decay of turbulence in a stably stratified shear flow. J. Fluid Mech. **195**, 77–111 (1988).

[5] Tavoularis, S. Corrsin, S.: Experiments in nearly homogeneous turbulent shear flow with a uniform temperature gradient. Part 1. J. Fluid Mech. **104**, 311–347 (1981).

[6] Tavoularis, S., Corrsin, S.: Effects of shear on the turbulent diffusivity tensor. Int. J. Heat Mass Transfer **28**, 256–276 (1985).

[7] Tavoularis, S., Karnik, U.: Further experiments on the evolution of turbulent stresses and scales in uniformly sheared turbulence. J. Fluid Mech. **204**, 457–478 (1989).

[8] Hunt, J. C. R., Stretch, D. D., Britter, R. E.: Length scales in stably stratified turbulent flows and their use in turbulence models. In: Stably stratified flows and dense gas dispersion (Puttock, J. S., ed.), pp. 285–321. Oxford: Clarendon Press 1988.

[9] Kaltenbach, H.-J.: Turbulente Diffusion in einer homogenen Scherströmung mit stabiler Dichterschichtung. Diss. TU München, Report DLR-FB 92–26, p. 142 (1992).

[10] Gerz, T., Schumann, U., Elghobashi, S. E.: Direct numerical simulation of stratified homogeneous turbulent shear flow. J. Fluid Mech. **200**, 563–584 (1989).

[11] Nieuwstadt, F. T. M.: The turbulent structure of the stable, nocturnal boundary layer. J. Atmos. Sci. **41**, 2202–2216 (1984).

[12] Zierep, J.: Ähnlichkeitsgesetze und Modellregeln der Strömungslehre, p. 138. Karlsruhe: G. Braun, 1972.

Author's address: Professor Dr.-Ing. habil. U. Schumann, Deutsche Forschungsanstalt für Luft- und Raumfahrt, Institut für Physik der Atmosphäre, D-82234 Oberpfaffenhofen, Post Weßling, Federal Republic of Germany

Acta Mechanica (1994) [Suppl] 4: 113—123

Multiplicative models for turbulent energy dissipation*

K. R. Sreenivasan and **G. Stolovitzky**, New Haven, Connecticut

Summary. We consider models for describing the intermittent distribution of the energy dissipation rate per unit mass, ε, in high-Reynolds-number turbulent flows. These models are based on a physical picture in which (in one-dimensional space) an eddy of scale r breaks into b smaller eddies of scale r/b. The energy flux across scales of size r is $r\varepsilon_r$, where ε_r is the average of ε over a linear interval of size r. This energy flux can be written as the product of factors called multipliers. We discuss some properties of the distribution of multipliers. Using measured multiplier distributions obtained from atmospheric surface layer data on ε, we show that quasi-deterministic models (multiplicative models) can be developed on a rational basis for multipliers with bases $b = 2$ and 3 (that is, binary and tertiary breakdown processes). This formalism allows a unified understanding of some apparently unrelated previous work, and its simplicity permits the derivation of explicit analytic expressions for quantities such as the probability density function of $r\varepsilon_r$, which agree very well with measurements. Other related applications of multiplier distributions are presented. The limitations of this approach are discussed when bases larger than three are invoked.

1 Introduction

A Gaussian process is completely described in a statistical sense by its mean and standard deviation. It is conceivable that a nearly Gaussian process can be described well by its first few moments — at least well enough for most purposes. This is the situation with respect to velocity or temperature traces obtained in high-Reynolds-number fully turbulent flows not too close to physical boundaries. On the other hand, the situation is quite different for quantities such as the velocity derivatives and the energy dissipation rate. In high-Reynolds-number turbulence where the small-scale motion is isotropic to a good approximation [1], the average energy dissipation rate per unit mass can be written as $\langle \varepsilon \rangle = 15\nu \langle (\partial u/\partial x)^2 \rangle$, where the angular brackets indicate averages, and u is the x-component of the velocity and ν is the fluid viscosity. ∂t is often assumed, as we shall do, that $(\partial u/\partial x)^2$ is a representative component of ε instantaneously. Figure 1 is a one-dimensional section through the field of ε in the atmospheric surface layer a few meters over land. In contrast to Gaussian or nearly Gaussian processes, information about the first few low-order moments does not describe the signal in any detail. Peaks which are hundreds of times the mean are not uncommon, and the signal is at other times of very low amplitude; this strongly intermittent character is a typical property of ε in high-Reynolds-number turbulence. It is becoming increasingly clear that intermittency plays a fundamental role in the understanding and modeling of turbulent flows. The intermittency has important implications also in contexts such as the structure of turbulent flames.

In the last few years, much work based on multifractals has occurred on the description and modeling of the intermittent character (and other similar characteristics) of the energy

---

* This is the enlarged version of the text to appear in the Proceedings of the International Congress of Theoretical and Applied Mechanics, Haifa, 1992.

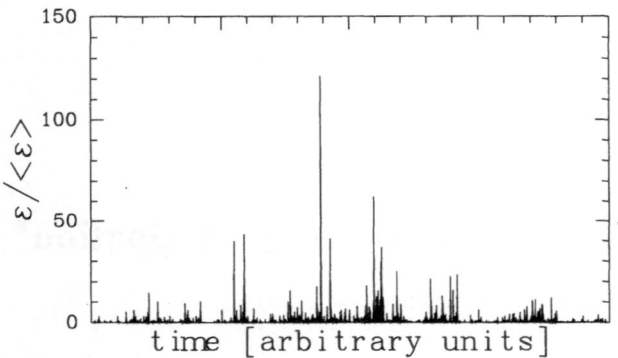

Fig. 1. A typical signal of a representative component of ε, namely $\varepsilon' \sim (du/dx)^2 \sim (du/dt)^2$, normalized by its mean. Here, u is the velocity fluctuation in the direction x of the mean velocity U. In writing the last step of the above approximation, it has been assumed that Taylor's frozen flow hypothesis, namely that the spatial derivative can be approximated by the temporal derivative, holds. The velocity fluctuation u was obtained by a hot-wire mounted on a pole 6 m above the ground level over a wheat canopy. The microscale Reynolds number is of the order 2 500. The resolution is of the order of the Kolmogorov scale

dissipation rate. For a summary, see [2]. In [3 − 5], several simple quasi-deterministic multifractal models were shown to describe the statistical properties of the energy dissipation rate quite accurately. Here, we provide a basis for developing such simplified models and show that it brings unity to some of the existing and seemingly unrelated work.

The energy dissipation rate is a positive definite quantity which is additive (in the sense that ε over two non-overlapping intervals equals the sum of ε values distributed over the sum of the two intervals). In this sense, it is convenient to think of ε as a measure distributed on an interval.

2 Multiplicative processes and multiplier distributions

The thought behind the models to be described is that the nonlinear processes occurring in the so-called inertial range (i.e., the scale range much larger than dissipative scales and much smaller than an external length scale such as the integral scale) may be abstracted by a breakdown process in which each eddy of size r subdivides into b pieces of size r/b. In this process, the energy flux per unit mass is redistributed in some unequal fashion without loss among b sub-eddies. This unequal distribution among sub-eddies is the heart of the observed intermittency. While, in reality, turbulence dynamics involves vortex stretching at many scales, the eddy breakdown processes do contain some essential physics of the inertial range.

In this model, the k-th sub-eddy ($1 \leq k \leq b$) will receive a fraction $M_k^{(r)}$ of the energy flux carried by its parent eddy; the conservation of the energy flux implies that $M_1^{(r)} + \cdots + M_b^{(r)} = 1$. In this manner, if we traced the history of all forefathers (up to the external scale L) of a given eddy at scale r, we would find that the energy flux $r\varepsilon_r$ being transferred to this eddy is related to the energy flux at the scale L as

$$r\varepsilon_r = L\varepsilon_L M^{(L)} M^{(L/b)} \dots M^{(L/b^{n-1})} \tag{1}$$

where $r/L = b^{-n}$, and $\varepsilon_L = \langle \varepsilon \rangle$ (since the average over the scale L is not distinguishable from the global average indicated by $\langle \ \rangle$).

At different scales, the multipliers are assumed to be independent stochastic variables. Since the energy flux, as it cascades down to smaller scales, is ultimately converted into energy dissipation, the two quantities have equal averages and, for present purposes, will be considered identical. The dynamics described above is, of course, much simpler than the reality. For instance, it is not obvious what the appropriate value of b (that is, the base for the cascade process) must be, or whether it remains the same from one step of the cascade to another. In spite of this ambiguity, it is clear that if the breakdown process has statistical scale-similarity, the histograms of multipliers $M^{(r)}$ should be identical at each step of the subdivision or cascade, i.e., independent of r.

How can we obtain the distribution of multipliers? Consider a long data string of ε distributed over an interval which is N integral scales in extent, N being some large integer. Divide each interval of size L into b equal-sized sub-intervals, and obtain the ratios of the measures (i.e. the integrals of ε) in each of the sub-intervals to that in the entire interval. These ratios, which are what we called multipliers, are clearly positive (since ε is positive definite) and lie between zero and unity. Subdivide each sub-interval into b pieces as before, and repeat the procedure. Proceeding with further subdivisions, there will be $N \cdot b^n$ multipliers $M_i^{(r)}$ at the n-th subsequent level, where each sub-interval is of size $r/L = b^{-n}$. Construct the histogram of the multipliers at each level, and repeat the procedure until the smallest sub-interval reached is of the order of the Kolmogorov scale $\eta = L(v^3/\langle \varepsilon \rangle)^{1/4}$.

The probability density $P(M)$ of the multipliers M — here and subsequently, we omit the indices on the $M_i^{(r)}$ and denote them simply by M — has been obtained from measurement for different stages of subdivision of the interval. Since the value of the base b is not known *a priori*, Chhabra and Sreenivasan [5] obtained $P(M)$ for various bases. Figure 2 shows the results for $b = 2$, 3 and 5. The shape of each of the distributions is invariant over a certain range of scales, suggesting that some type of self-similarity occurs in this scale range, whatever the assumed base. This range over which $P(M)$ is self-similar agrees quite well with the inertial range determined by

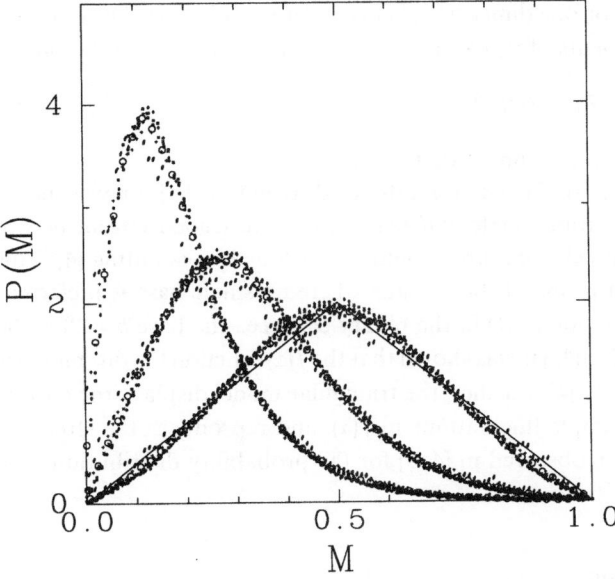

Fig. 2. Multiplier distributions $P(M)$ for bases (from right to left) $b = 2$, 3 and 5. The larger symbols show averaged multiplier distribution, which are the mean of multiplier distributions obtained by comparing measure in boxes of size m to those of $m*b$, where m ranged from 50 to 1 000 in units corresponding roughly to the Kolmogorov scale. The smaller symbols show the distributions obtained for $m = 50, 80, 150, 200, 400$ and 1 000. The solid line is the triangular approximation to the binary case. The figure is adapted from [5]

the scaling of the energy spectrum and structure functions, and covers about a decade and a half. The larger symbols show an average over steps involving comparisons between boxes of size m and those of size $m*b$, where m ranged from 50 to 1 000 in units of sampling intervals. (For the very smallest scales, the distributions have a concave shape. This concavity is related to the divergence of moments [7] and will be discussed elsewhere. For very large scales, multiplier distributions approach a delta function centered around 0.5, as would be the case for random measures.)

A disadvantage of $P(M)$ is that it is base-dependent. However, if the cascades giving rise to the observed intermittency are randomly multiplicative, then the multiplier distributions corresponding to different bases are related by convolution, and one can scale out this base-dependency [5]. In particular, for any two bases 'a' and 'b', we have

$$\log \langle (M_a)^q \rangle / \log (a) = \log \langle (M_b)^q \rangle / \log (b). \tag{2}$$

The scale-invariant multiplier distributions obtained in Fig. 2 are fundamental to the understanding of the observed scaling in turbulence. One can compute from them not only the asymptotic scaling properties of ε_r, but also finite-size fluctuations of scaling properties [5]. In order to understand this, it is helpful to note the relation between the multiplier distributions of Fig. 2 and the multifractal description of turbulence. In the latter, $r\varepsilon_r$ is assumed to scale as

$$r\varepsilon_r = L\varepsilon_L (r/L)^\alpha, \tag{3}$$

where the exponent α (the so-called Hölder exponent) depends on the spatial coordinate position.

The probability of finding a scaling index α, $P(\alpha)$, also scales as

$$P(\alpha) \sim (r/L)^{d - f_d(\alpha)}, \tag{4}$$

where d is the dimension of the space under consideration, and $f_d(\alpha)$ is called the multifractal spectrum [8], and is dependent on d. For one dimensional cuts (as those of Fig. 1), $d = 1$ and we drop the suffix on $f(\alpha)$. Then the moments of $r\varepsilon_r$ can be computed from the above relations as

$$\langle (r\varepsilon_r)^q \rangle = (L\varepsilon_L)^q \int d\alpha (r/L)^{q\alpha + (1 - f(\alpha))} \sim (L\varepsilon_L)^q (r/L)^{\tau(q) + 1} \tag{5}$$

where $\tau(q) = \min_\alpha \{q\alpha - f(\alpha)\}$, and it was assumed that $r \ll L$.

The $f(\alpha)$ function can also be easily derived from the multiplier distribution. In [5], it was shown that the $f(\alpha)$ curves computed from the distributions for different bases b were in good agreement with each other as well as with those obtained from direct methods such as box-counting [4]. This agreement indicates the existence of a probabilistic cascade where no single base is preferred.

Incidentally, a good approximation for $P(M)$ in the binary case (i.e., the base $b = 2$) is the triangular distribution shown in Fig. 2. In [5] it was shown that the $f(\alpha)$ function for this model is in agreement with that obtained directly [4]. Further, the triangular model displays the correct behavior with respect to sample-to-sample fluctuations in $f(\alpha)$, and reproduces the stretched exponential tails, $P(\varepsilon) \sim \exp\left(-\beta(\varepsilon)^{1/2}\right)$, observed in [4, 9] for the probability distribution of ε.

3 Simple quasi-deterministic models

The multiplier distributions shown in Fig. 2 are extracted directly from the experiment and their analytical forms are yet to be found from the theory. The question meanwhile is a simple representation of these distributions in a way that permits one to evaluate most of the measured properties accurately. The goal is to seek models that are simple enough to be tractable mathematically and realistic enough to represent the spirit of the underlying physics. A simple possibility is the p-model [3], which is a model for a binary cascade ($b = 2$). We first discuss the p-model and show how it can be obtained as a rational approximation to the measured multiplier distribution for the binary case. We will then discuss how models in the same spirit can be obtained for the tertiary case ($b = 3$). The limitations of the procedure for high order subdivisions ($b > 3$) will be highlighted.

From a physical point of view, the cascade process with $b = 2$ can be thought of as the break-up of a structure (the parent eddy, scale or structure) into two sub-structures. For the one-dimensional case corresponding to Fig. 1, a pertinent question is the following: is there any difference between the left and right offsprings in terms of the energy flux they receive from the parent structure? One can determine experimentally that left and right are statistically indistinguishable. (This is not true for the velocity signal itself, as can be concluded from Kolmogorov's 4/5 law [10].) Now, for the sake of simplicity and modeling, let us assume that one of the two sub-eddies always receives a fixed fraction p of the energy contained in the parent eddy; naturally, the other will receive $1 - p$. In this sense, this model is deterministic. However, it is only quasi-deterministic in the sense that either one of the two eddies could receive the fraction p; because of the left-right symmetry mentioned above, a given piece will receive p as often as $1 - p$. Then, the multiplier distribution for the p-model can be written in terms of delta functions as

$$P_{b=2}(M) = \frac{\delta(M - p) + \delta(M - (1 - p))}{2}. \tag{6}$$

If $p = 1/2$, there is no intermittency and the situation corresponds to Kolmogorov's 1941 theory [1]. To obtain intermittency, we should have a value of p different from $1/2$.

How can we determine the value of p? A natural way is to match the moments of $P_{b=2}(M)$ from Eq. (6) with those of the real $P(M)$. For both distributions, the zero-order moment (normalization) and the first-order moment (mean value) coincide trivially, and are 1 and $1/2$, respectively. The first non-trivial condition is obtained by matching the second order moment. By this matching condition, we obtain the value $p = 0.697$, or $1 - p = 0.303$.

In a completely different way, it was shown in [3] that a good fit to the $f(\alpha)$ curve can be achieved if a binary cascade with parameters $p = 0.7$ (and $1 - p = 0.3$) is assumed as the energy transfer mechanism. In [3], the numbers 0.7 and 0.3 appeared a little mysteriously as the effective fractions of energy split in the breakdown process. Even though it was recognized that a multiplicity of multipliers may occur (the multiplier distribution had not been measured at that time), it was surprising that such a simple model could do so well. We now have an explanation, which is that the binary p-model is an approximation that fits the zero-th, first and the second moments of the measured multiplier distributions. It fortunately turns out that high-order moments computed for the p-model also agree with those computed for the real data (see Fig. 3). The binary p-model can now be considered as derived from the real multiplier distribution.

We now discuss a general scheme for developing for all $b \neq 2$ quasi-deterministic models of the type developed for the binary case. Again, we attempt to do this by matching moments. The general multiplier distribution for any b in the p-model scheme is

$$P_b(M) = \frac{1}{b} \sum_{i=1}^{b} \delta(M - p_i) \tag{7}$$

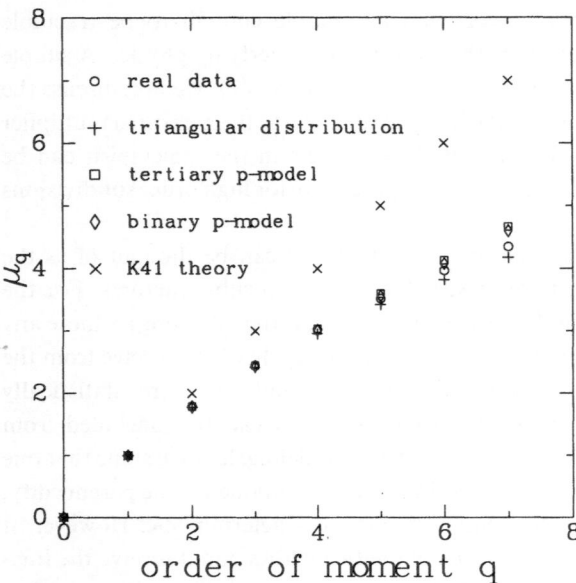

Fig. 3. A comparison between the scaling exponents μ_q (Eq. (16)) computed from the measured multiplier distributions and those computed for the different models considered in the text. Experimental data were obtained from a record length of 810 000 data points. The convergence of moments was reasonable; for example, in the last half decade of the record length, the variations observed were smaller (in the log scale) than the symbol size

where

$$\sum_{i=1}^{b} p_i = 1, \quad (0 \leqq pi \leqq 1). \tag{8}$$

We may now equate the moments of $P_b(M)$ to the moments of the real multiplier distributions. Since multiplier distributions for any base yield the same results, we may take the distribution corresponding to the binary cascade of Fig. 2. Computing the moments $\langle M^q \rangle$ from it, we are led to the equations

$$\sum_{i=1}^{b} p_i{}^q = b \langle M^q \rangle^{\log(b)/\log(2)}, \quad q = 1, 2, ..., b. \tag{9}$$

This is a system of b equations with b number of p_i's to be determined. Using Girard's rule [11], it is easy to find a polynomial of degree b whose roots are the desired p_i's. The problem thus reduces to the determination of the roots of the polynomial. It turns out *a posteriori* that this problem has physical solutions only for $b = 2$ and $b = 3$: for larger values of b, some of the roots turn out to be complex, and have no physical meaning. The values of p_i for the tertiary cascade ($b = 3$) are $p_1 = 0.155$, $p_2 = 0.283$ and $p_3 = 0.562$. We designate this as the tertiary p-model. The classical Kolmogorov theory for a tertiary breakup scheme yields $p_1 = p_2 = p_3 = 1/3$. Figure 3 shows that the comparison with the experimental values is good.

It is interesting to note that Viczek and Barabasi [12] empirically devised the tertiary p-model without any reference to the multiplier distribution. Briefly, they discussed an iterative scheme for generating a velocity signal that possessed many features of the actual turbulent velocity time trace. The starting step picks at random one of the two generators $z_c(x)$ (Fig. 4a) and $z_d(x)$ (Fig. 4b) shown in Fig. 4. The magnitudes c and d will be specified shortly. Each of the 3 linear segments resulting after the first step is randomly replaced with a rescaled version of the generators, so that at each step the function is continuous. If the slope of the segment to be replaced is positive, we use with equal probability $z_c(x)$ or $z_d(x)$. If it is negative, we use with equal probability $1 - z_c(x)$ or $1 - z_d(x)$. This is illustrated in Fig. 4c.

This scheme represents, by construction, a tertiary process. Denote by $\Delta u(r)$ the velocity increment $u(x + r) - u(x)$ across a distance $r = L3^{-k}$. At the step $k + 1$, the velocity difference across an interval $r/3$ (included in the interval $[x, x + r]$), can be one of the following three possibilities:

$$|\Delta u(r/3)| = |\Delta u(r)| \begin{cases} c \\ c - d \\ 1 - d \end{cases} \tag{10}$$

where the parameters c and d are defined in Fig. 4. Knowing that $\langle |\Delta u(r)|^q \rangle \sim (r/L)^{\zeta_q}$, Viczek and Barabasi computed ζ_q for their model, and found that in order to match experiments, the values of c and d had to be $c = 0.67$ and $d = 0.13$.

Interestingly, our tertiary p-model yields a very similar result. In effect, according with Kolmogorov's refined similarity hypotheses [13] — see also Section 4 — $\Delta u(r) \sim (r\varepsilon_r)^{1/3}$. For our

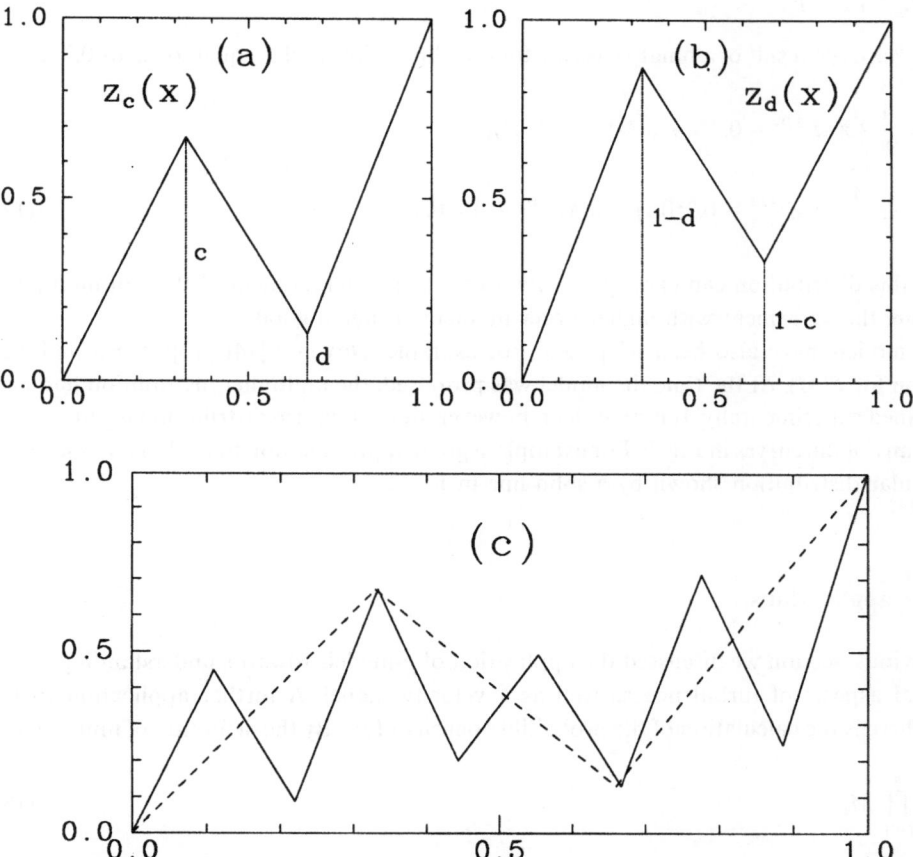

Fig. 4. Generators $z_c(x)$ and $z_d(x)$ used for the iterative scheme discussed in [12] to model a velocity trace (see **a** and **b**). At each step, a linear segment is replaced, with equal probability, by a rescaled version of $z_c(x)$ or $z_d(x)$ if the segment derivative is positive, and by $1 - z_c(x)$ or $1 - z_d(x)$ if the segment derivative is negative. **c** Example of the scheme at the first step (dashed line) and the second step (solid line)

multiplicative model, $r\varepsilon_r = L\varepsilon_L M_1 \ldots M_n$, and we conclude from Eq. (10) that $c, c - d$ and $1 - d$ have to be related with the cubic root of our multipliers $p_1^{1/3}$, $p_2^{1/3}$ and $p_3^{1/3}$ for the tertiary p-model. These values are 0.66, 0.54 and 0.82, while Viczek and Barabasi's values for $c, c - d$ and $1 - d$ are 0.67, 0.5 and 0.83. It is important to realize that the two methods of finding c and d, and the p's, are completely different. Hence, as in the case of the binary p-model, we note that the parameters c and d can be derived from the multiplier distribution, although such was not its original derivation.

Returning to the generation of quasi-deterministic models, it is clear that increasing the number of p-parameters would enable us to fit the measurements better and better. While a model with too many parameters is not very useful, it would be desirable to be able to fit moments up to order 8 — the reason being that the three-dimensional analogue of the one-dimensional binary breakdown $b = 2$ is the case of a cube breaking into 8 pieces ($b = 8$). We mentioned previously that quasi-deterministic p-models can be constructed only up to $b = 3$. While this tells us about the impossibility for quasi-deterministic models to fit experiments with any *predetermined accuracy*, it does not mean that working quasi-deterministic models cannot be constructed for $b > 3$. For example, for the three dimensional binary cascade, we can write a typical multiplier as

$$M^{(3D)} = p_1^{(1D)} p_2^{(1D)} p_3^{(1D)} \tag{11}$$

where $p_j^{(1D)}$ are the result of a binary cascade for one dimension, and is equal to 0.3 or 0.7. Then

$$P(M^{(3D)}) = \frac{1}{8} \left\{ \delta(M^{(3D)} - 0.7^3) + \delta(M^{(3D)} - 0.3^3) \right\}$$

$$+ \frac{3}{8} \left\{ \delta(M^{(3D)} - 0.3^2 0.7) + \delta(M^{(3D)} - 0.7^2 0.3) \right\}. \tag{12}$$

Although this distribution can exactly fit only up to the second moment of the real multiplier distribution, the agreement with higher order moments is again good.

Other models have also been proposed. For example, Novikov [14] proposed a uniform distribution for $P(M)$. At the time the model was proposed, the multiplier distribution had not been obtained experimentally. It is now clear, however, that a uniform distribution is not a good model for any of the curves in Fig. 2. For example, a good approximation to the binary cascade is the triangular distribution shown by a solid line in Fig. 2.

4 Further applications

In the previous Section we discussed the application of p-models towards understanding some multifractal aspects of turbulence as well as a velocity model. A further application to be discussed here is the calculation of the probability density of $r\varepsilon_r$. By the definition of multipliers,

$$r\varepsilon_r = L\varepsilon_L \prod_{i=1}^{n} M_i, \tag{13}$$

where $n(r) = \log_b (r/L)$, and L (as before) is the large-eddy size or the integral scale. Within the binary p-model scheme ($b = 2$), $x = r\varepsilon_r/L\varepsilon_L$ can take the discrete values $p^k(1 - p)^{n-k}$, where $p = 0.3$ and $0 \leq k \leq n$. The probability of occurrence of a given k is simply $2^{-n} n!/(k!(n - k)!)$. Replacing factorials by gamma functions (to allow for continuous k), the probability density function P for the ratio x is

$$P(x) = \frac{A}{x} \frac{r}{L} \frac{\Gamma\left(\log_2\left(r/L\right) + 1\right)}{\Gamma\left(k(x) + 1\right) \Gamma\left(\log_2\left(r/L\right) - k(x) + 1\right)}, \tag{14}$$

where A is a normalization constant of order unity (and would be equal to unity in the discrete case), Γ is the gamma function and

$$k(x) = \frac{\log\left(x(r/L)^{\log_2(1-p)}\right)}{\log\left(p/(1-p)\right)}. \tag{15}$$

Figure 5 shows a comparison between experimentally measured distribution of $r\varepsilon_r$ (normalized with its standard deviation) and the prediction of Eq. (14). The experimental data used are the same as those used to compute the multiplier distributions of Fig. 2. We found that to fit the data at $r/\eta = 75$ we had to use $n(r = 75\eta) = 14$ (Fig. 4a). For $r/\eta = 1\,200$ (Fig. 4b), we computed $n(r = 1\,200\eta) = n(r = 75) - \log_2\left(1\,200/75\right) = 10$. The agreement with experiment is good.

Another application is in computing the exponents μ_q, defined as

$$\langle (r\varepsilon_r)^q \rangle \sim (r/L)^{\mu_q}. \tag{16}$$

The general result is that $\mu_q = -\log_b \langle M^q \rangle$. The numerical value of μ_q depends upon the distribution used to compute $\langle M^q \rangle$. The results for the different models considered in this

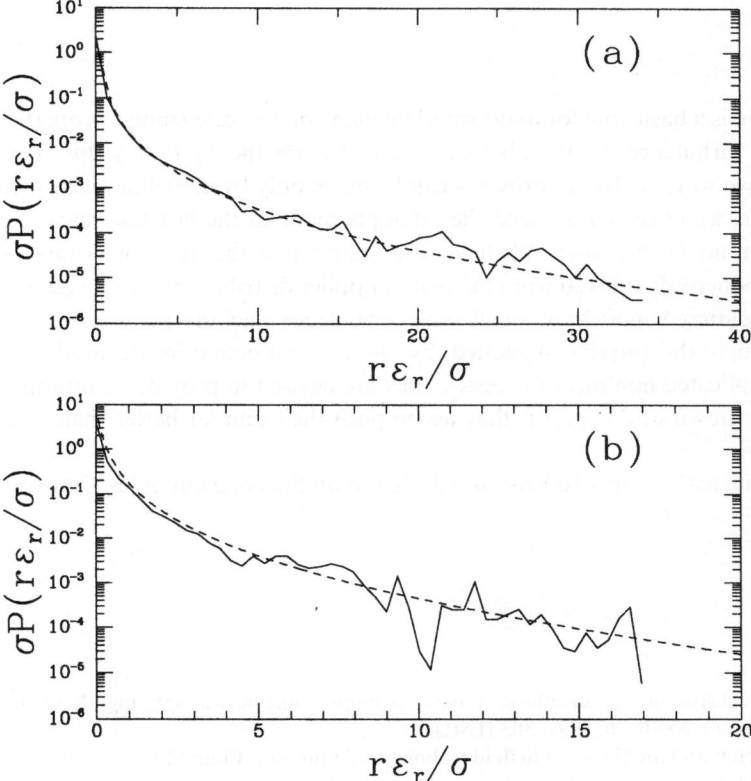

Fig. 5. Experimental (solid line) and binary p-model prediction (Eq. (14), dashed line) for the probability density function of $r\varepsilon_r$ for two values of the separation distance r, normalized by their respective standard deviations. **a** Separation distance $r = 75\eta$, number of steps in the cascade $= 14$. **b** Separation distance $r = 1\,200\eta$, number of steps in the cascade $= 10$

paper are

$$\mu_q = \begin{cases} -\log_2\left\{4\left(\dfrac{1}{2^{q+1}}\right)\dfrac{1}{(q+1)(q+2)}\right\} & \text{triangular distribution} \\ -\log_2\left[(0.3^q + 0.7^q)/2\right] & \text{binary } p\text{-model} \\ -\log_3\left[(0.155^q + 0.283^q + 0.562^q)/3\right] & \text{tertiary } p\text{-model.} \end{cases} \tag{17}$$

These different results have already been plotted along with the experimental values in Fig. 3. It is worth remembering that, although the binary and tertiary p-models, respectively, are generated to possess the first three and four moments correctly, they agree well with the measured moments up to, say, order 7.

The probability density functions of the velocity increments can also be computed. On using the second refined similarity hypothesis [13], the velocity increments can be written as

$$\Delta u(r) = V(r\varepsilon_r)^{1/3}, \tag{18}$$

where V is a universal stochastic variable independent of r and $r\varepsilon_r$. In [15], we obtained the probability density of V. The probability density of Δu can be computed if one of the previously described models is assumed for $r\varepsilon_r$. In [9], those probability density functions were computed using the binary p-model, and were found to be in good agreement with the data.

5 Conclusions

The multiplier distribution is a basic tool for understanding many of the scale-similar properties of energy dissipation in turbulence. In the absence of an *ab initio* theory that yields these distributions in a deductive way, analytical progress can be made only by modeling them with reasonable schemes. Here, we have summarized the attempts made in the last few years and mentioned a few applications of this work. We have also shown that the quasi-deterministic models described here, deductively derived from the real multiplier distribution, can be used to bring unity to various scattered models of small scale turbulence and the phenomenon of intermittency. Even though the physics suggested by these quasi-deterministic models is a caricature of more complicated nonlinear processes, they are devised to provide quantitative information. As we have shown in this paper, they accomplish their aim far better than their simplicity might suggest.

It is a pleasure to dedicate this paper to Professor J. Zierep on the occasion of his sixty-fifth birthday.

References

[1] Kolmogorov, A. N.: Local structure of turbulence in in an incompressible fluid at very high Reynolds numbers. Dokl. Akad. Nauk SSSR **30**, 299−303 (1941).

[2] Sreenivasan, K. R.: Fractals and multifractals in fluid turbulence. Annu. Rev. Fluid Mech. **23**, 539−600 (1992).

[3] Meneveau, C., Sreenivasan, K. R.: Simple multifractal cascade model for fully developed turbulence. Phys. Rev. Lett. **59**, 1424−1427 (1987).

[4] Meneveau, C., Sreenivasan, K. R.: The multifractal nature of energy dissipation. J. Fluid Mech. **224**, 429−484 (1991).

[5] Chhabra, A. B., Sreenivasan, K. R.: Scale-invariant multiplier distributions in turbulence. Phys. Rev. Lett. **62**, 2762 – 2765 (1992).

[6] Mandelbrot, B. B.: Intermittent turbulence in self-similar cascades: Divergence of high moments and dimensions of the carrier. J. Fluid Mech. **62**, 331 – 358 (1974).

[7] Schertzer, D., Lovejoy, S.: Nonlinear variability in geophysics. Dordrecht: Kluwer 1991.

[8] Frisch, U., Parisi, G.: On the singularity structure of fully developed turbulence. In: Turbulence and predictability in geophysical fluid dynamics and climate dynamics (Ghil, M., Benzi, R., Parisi, G.), pp. 84 – 88. Amsterdam: North-Holland 1985; Halsey, T. C., Jensen, M. H., Kadanoff, L. P., Procaccia, I., Shraiman, B. I.: Fractal measures and their singularities: The characterization of strange sets. Phys. Rev. A **33**, 1141 – 1151 (1986).

[9] Kailasnath, P., Sreenivasan, K. R., Stolovitzky, G.: Probability density of velocity increments in turbulent flows. Phys. Rev. Lett. **68**, 2766 – 2769 (1992).

[10] Kolmogorov, A. N.: Energy dissipation in locally isotropic turbulence. Dokl. Akad. Nauk SSSR **32**, 19 – 21 (1941).

[11] Rey Pastor, J., Pi Calleja, P., Trejo, C.: Analisis matematico, vol I. Kapelusz: Buenos Aires, 1957.

[12] Vicsek, T., Barabási, A.-L.: Multi-affine model for the velocity distribution in fully turbulent flows. J. Phys. A, **24**, L 845 – L 851 (1991).

[13] Kolmogorov, A. N.: A refinement of previous hypotheses concerning the local structure of turbulence in a viscous incompressible fluid at high Reynolds number. J. Fluid Mech. **13**, 82 – 85 (1962).

[14] Novikov, E. A.: Intermittency and scale similarity in the structure of a turbulent flow. Prikl. Mat. Mekh. **35**, 266 – 277 (1971).

[15] Stolovitzky, G., Kailasnath, P., Sreenivasan, K. R.: Kolmogorov's refined similarity hypotheses. Phys. Rev. Lett. **69**, 1178 – 1181 (1992).

Authors' address: Professor Dr. K. R. Sreenivasan and Dr. G. Stolovitzky, Department of Engineering and Applied Sciences, Mason Laboratory, Yale University, P.O. Box 2159, New Haven, CT 06520, U.S.A.

Acta Mechanica (1994) [Suppl] 4: 125–131
© Springer-Verlag 1994

Part 3: Transonic Flow

Near critical transonic nozzle flows with homogeneous condensation

C. F. Delale, Ankara, Turkey, and **G. H. Schnerr**, Karlsruhe, Federal Republic of Germany

Summary. The 1-*D* asymptotic theory of transonic nozzle flows with homogeneous condensation is applied to both subcritical and supercritical moist air expansions under atmospheric supply conditions through relatively slender nozzles. Good agreement with experiments is achieved for sufficiently low initial relative humidity in subcritical and high initial relative humidity in supercritical flows when the condensed phase is taken to consist purely of water drops. In near critical flows where the initial relative humidity ranges in between and the flow borders on supercritical heat addition, regardless of nozzle geometry the 1-*D* asymptotic theory predicts deviations from the onset of condensation and, in case of flows with normal shocks, from the visualized shock locations due to the 2-*D* nature of the flow field.

1 Subcritical and supercritical transonic nozzle flows

We herein give a brief discussion of the recent 1-*D* asymptotic solution of transonic nozzle flows of a mixture of a condensible vapor and a carrier gas with initial specific humidity ω_o, reservoir temperature T_o and initial relative humidity φ_o [1, 2]. With the normalization carried out in [3] the flow equations together with the thermal equation of state yield the functional relations

$$u(g, x) = \frac{[1 + u_s{}^2 + R(g, x)]/(2u_s) \pm \sqrt{\Delta(g, x)}}{[\gamma + 1 + (\gamma - 1)\,H^{-1}g]/(2\gamma)},\tag{1}$$

$$\varrho(g, x) = \frac{u_s}{u(g, x)\ A(x)},\tag{2}$$

$$p(g, x) = \frac{1 + u_s{}^2 + R(g, x) - u_s u(g, x)}{A(x)},\tag{3}$$

$$T(g, x) = T_o + \frac{L(T)\,g}{c_{po}} - \frac{[u(g, x)]^2}{2c_{po}}\tag{4}$$

where

$$R(g, x) \equiv \int_{x_s}^{x} p[g(\xi), \xi]\,\frac{dA}{d\xi}\,d\xi\tag{5}$$

and

$$\Delta(g, x) \equiv T_o\Theta(g)\left[\frac{q^*(g, x)}{c_{po}T_o} - \frac{q(g, x)}{c_{po}T_o}\right]\tag{6}$$

with

$$\frac{q(g, x)}{c_{po}T_o} \equiv \frac{gL(T)}{c_{po}T_o}, \tag{7}$$

$$\frac{q^*(g, x)}{c_{po}T_o} \equiv \frac{([1 + u_s{}^2 + R(g, x)]/[2u_s])^2}{T_o\Theta(g)} - 1 \tag{8}$$

and

$$\Theta(g) \equiv \frac{[1 + \gamma + (\gamma - 1)H^{-1}g]\ [1 - H^{-1}g]}{2\gamma} \tag{9}$$

and where in Eq. (1) it is understood that the $(+)$ sign be chosen in the supersonic and the $(-)$ sign in the subsonic regions. In relations (1)–(9) ϱ, p and T are respectively the density, the pressure and the temperature of the mixture, $c_{po} \equiv \gamma/(\gamma - 1)$ is the dimensionless specific heat of the mixture at constant pressure with γ denoting the adiabatic exponent of the mixture, $L(T)$ is the latent heat of condensation at the local temperature T, u is the flow speed, A is the local cross-sectional area of the nozzle, g is the normalized condensate mass fraction, $q(g, x)$ and $q^*(g, x)$ are respectively the normalized amount of heat released by condensation and the normalized critical amount of heat not to be exceeded in shock-free condensing flows at any location x along the nozzle. All of the flow variables, except u and g, are normalized with respect to saturation conditions with subscript s so that $\varrho_s = p_s = T_s = L_s = 1$. The cross-sectional area is also normalized with respect to its saturation value and the axial coordinate x with origin at the throat is normalized with respect to the throat height $2y^*$. The flow speed u and the condensate mass fraction g are normalized somewhat differently by $u \equiv u'/\sqrt{\Re T_s'/\mu_o}$ and $g \equiv \mu_o Hg'/\mu_v$ where μ_0 and μ_v are respectively the mixture molecular weight in the reservoir and the vapor molecular weight, $\Re$ is the universal gas constant, g' is the actual condensate mass fraction, u' is the actual flow speed, T_s' is the actual saturation temperature and the constant H is defined by $H \equiv \mu_v L_s'/(\Re T_s')$ with L_s' denoting the actual latent heat of condensation at saturation. In this work all of the primed variables denote the actual flow variables, subscript v is reserved for the vapor phase, subscript o for the reservoir state and subscript s for the saturation state.

For a complete description of 1-D nozzle flows with nonequilibrium condensation the functional relations (1)–(6) should be supplemented by the rate equation for g. The nonequilibrium condensation rate equation for g is constructed from a nuclei production rate J conveniently normalized as

$$J = \Sigma(p, T, g) \exp\left[-K^{-1}B(p, T, g)\right] \tag{10}$$

together with a radius independent droplet growth rate

$$\frac{dr}{dx} = \lambda\Omega(p, T, g) \tag{11}$$

as

$$g(x) = \int_{x_s}^{x} \left[r^*(\xi) + \lambda \int_{\xi}^{x} \Omega(\eta)\, d\eta\right]^3 \Sigma(\xi)\, A(\xi) \exp\left[-K^{-1}B(\xi)\right] d\xi. \tag{12}$$

Here $\Sigma(p, T, g)$ is a normalized multiplicative factor in the nucleation rate equation, $B(p, T, g)$ is the normalized activation function, K is the nucleation parameter (assumed small $K \ll 1$), $\Omega(p, T, g)$ is a conveniently normalized droplet growth function, λ is the droplet growth parameter (assumed large $\lambda \gg 1$) and $r^*(p, T, g)$ is the normalized critical radius for nuclei production (for details see [1, 3]). The functional relations (1)−(6) yield a local implicit solution if the function $g(x)$ can be solved from Eq. (12) coupled to the equations of flow and state. This is achieved by asymptotics in the double limit as $K \to 0$ and $\lambda \to \infty$. In this case the behavior of the activation function B distinguishes the distinct condensation zones (Fig. 1).

In subcritical flows B is a smooth function of x over its domain. In this case the double limit as $K \to 0$ and $\lambda \to \infty$, ordered in a natural fashion as in [3], distinguishes the condensation zones shown in Fig. 1a. The initial growth zone IGZ is defined as the zone where B almost equals B_f (subscript f denotes condensate-free frozen flow). In the further growth zone FGZ, B is distinct from B_f with small deviations and $dB/dx = O(1)$. The rapid growth zone RGZ begins as dB/dx diminishes to $O(K^{1/2})$ as $K \to 0$ and ends at x_l, the turning point of B where the nucleation rate is maximum. The onset zone OZ containing the onset point is embedded in this zone. The nucleation zone with growth (NZ) extends downstream of x_l where both nucleation and droplet growth are important. Finally the droplet growth zone DGZ is reached as the nucleation rate completely diminishes. The approach to saturated thermodynamic equilibrium is reached in this zone. The asymptotic expressions for $g(x)$ in each zone (exhibited in [1, 3]) together with the functional relations (1)−(9) yield a complete implicit solution for the local flow field.

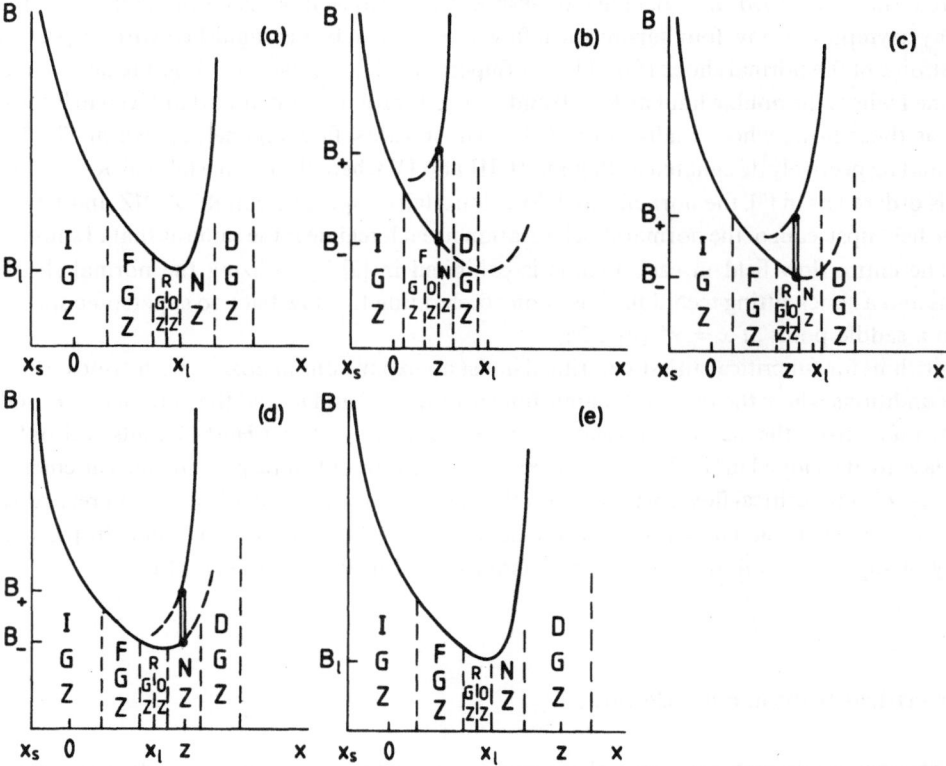

Fig. 1. Variation of the activation function along the nozzle axis in the classification scheme for transonic nozzle flows with nonequilibrium condensation. **a** Subcritical flow, **b** Regime I supercritical flow, **c** Regime II supercritical flow, **d** Regime III supercritical flow and **e** Regime IV supercritical flow (x_s is the saturation point, $x = 0$ is the throat location, z is the normal shock location and x_l is the turning point of B)

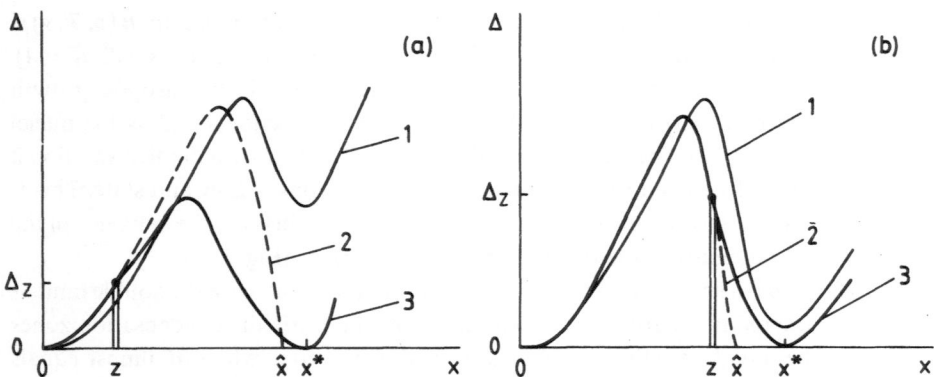

Fig. 2. Typical behavior of the functional Δ along the nozzle axis for constant latent heat in **a** subcritical flow with relatively low and supercritical flow with relatively high initial relative humidities and **b** near critical flow (curve 1: subcritical flow, curve 2: thermally choked flow, curve 3: supercritical flow)

When the amount of heat released by condensation given by Eq. (7) exceeds the critical amount given by Eq. (8), the functional relations (1)−(6) together with the solution of Eq. (12) can not yield a continuous solution for the flow field since Δ vanishes at some $x = \hat{x}$ and $\Delta < 0$ for $x > \hat{x}$ (Fig. 2a). In this case the flow is *thermally choked* and the inclusion of a stationary normal shock wave becomes necessary (details of the theory of thermal choking can be found in [4]). Such flows are termed *supercritical*. A detailed analysis of supercritical flows is given in [2] where it is shown by asymptotics how four supercritical flow regimes can be distinguished with respect to the location z of the normal shock (Fig. 1 b−e). Supercritical flows shown in Fig. 1 b, designated as Regime I where the double limit as $K \to 0$ and $\lambda \to \infty$ is ordered similar to that given in [5], are defined as those flows where z falls in FGZ. For supercritical flow regimes, shown in Fig. 1 c, d and e and respectively designated as Regime II, III and IV where the double limit as $K \to 0$ and $\lambda \to \infty$ is ordered as in [3], the normal shock location z falls respectively in RGZ, NZ and DGZ. The flow field upstream of the normal shock location in each regime is the same as that of smooth flows. The entire flow field in each regime is exhibited in [2] by utilizing the normal shock relations and a shock fitting technique based on accelerating the flow back to supersonic speeds through a saddle point at $x = x^*$ (Fig. 2a).

Algorithms for subcritical and supercritical moist air expansions in nozzles with atmospheric supply conditions where the thermodynamic functions B, Σ, Ω and r^* and the parameters K and λ are identified from the classical nucleation theory together with the Hertz-Knudsen droplet growth law are developed in [1, 2]. Good agreement with the recent static pressure measurements and, in case of supercritical flows, with the visualized shock locations [6] is achieved in relatively slender nozzles when the condensed phase is assumed to consist purely of water drops. The majority of supercritical flows under these conditions fall in Regime I (Fig. 1 b).

2 Near critical transonic nozzle flows

In this section we discuss near critical transonic nozzle flows utilizing the 1-D asymptotic solution summarized in Section 1. For these flows the heat released by condensation borders on the critical amount; therefore, they can be realized with or without a normal shock depending on whether or not the condition $\Delta = 0$ is reached at some $x = \hat{x}$ downstream of the onset zone.

Utilizing Eq. (6) we define stationary *near critical flows* either as shock-free flows satisfying

$$\frac{q^*(g, x) - q(g, x)}{q(g, x)} = o(1) \qquad\qquad (13)$$

in some interval which assumes x_l as a lower bound, or as flows with weak normal shock waves whose locations proceed the point x_l along the nozzle axis. It is now clear that stationary near critical flows are either subcritical flows satisfying (13) or supercritical flows falling in Regime III or Regime IV (Fig. 1 d — e). Typical behavior of the functional $\Delta(g, x)$ with constant latent heat in near critical flows is exhibited in Fig. 2 b in contrast to that given in Fig. 2 a for subcritical or supercritical flows where the above near critical conditions are not realized. It follows directly by definition that downstream of the onset zone the axial flow gradients become very steep in shock-free near critical flows whereas a precompression region of relatively small thickness appears just upstream of the normal shock location along the nozzle axis in near critical flow with shocks.

As mentioned earlier in Section 1, the 1-*D* asymptotic solution of transonic moist air expansions under atmospheric supply conditions [1, 2] yields satisfactory results in relatively slender nozzles (e.g. the circular arc nozzle of Fig. 3 a with throat height $2y^* = 30$ mm and arc radius $R^* = 400$ mm) when the condensed phase is assumed to consist purely of water drops. In this case the initial relative humidities range from 30 to 40 % in subcritical and from 60 to 80 % in supercritical flows. It would be interesting to find out the predictions of the 1-*D* asymptotic theory for flows through the same nozzle with the same reservoir temperature and initial specific humidity, but with the initial relative humidity ranging from 40 to 60 %, i.e. flows which presumably are near critical. Unfortunately no static pressure measurements and schlieren pictures typical of stationary near critical flows in relatively slender nozzles are available (the observed flow patterns by schlieren photographs in this range of relative humidity are nonsymmetric with respect to the nozzle axis and show weak nonstationary disturbances); therefore, we choose to consider the nearest supercritical flow example resembling near critical flows for which such data are available.

Figure 3 a shows the schlieren picture of such a supercritical example with initial relative humidity $\varphi_o = 59{,}4$ %, initial specific humidity $\omega_o = 7.6$ g/kg and reservoir temperature $T_o' = 290.8$ °K in the relatively slender circular arc nozzle with throat height $2y^* = 30$ mm and arc radius $R^* = 400$ mm. A precompression region of thickness $4 - 5$ mm is observed just upstream of the normal shock located at $z_{ob}' = 21$ mm along the nozzle axis. Near the walls weakening in the axial flow gradients of compression followed by a weak oblique shock is observed. This emphasizes the importance of 2-*D* effects in supercritical flows through relatively slender nozzles when the initial relative humidity is not sufficiently high. Figure 3 b shows the 1-*D* asymptotic predictions for the pressure distribution, the nucleation rate and the condensate mass fraction through the same nozzle with the same conditions on the assumption that the condensed phase consists purely of water drops. In this case a supercritical flow pattern which falls in Regime II with a normal shock located at $z' = 17$ mm is observed. Thus, contrary to supercritical flows in Regime I, no good agreement with the normal shock location is achieved. The upstream shift in the predicted shock location follows from the neglected 2-*D* structure of the flow field as is exhibited in Fig. 3 a.

As the initial relative humidity φ_o is gradually decreased with the rest of the conditions kept fixed, the normal shock of Fig. 3 b weakens and shifts downstream with simultaneous transition from Regime II to Regime III and ultimately to Regime IV. During this transition thickening of the precompression region along the nozzle axis is observed and the flow remains two-

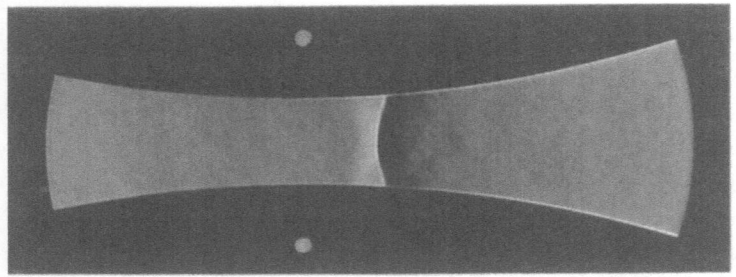

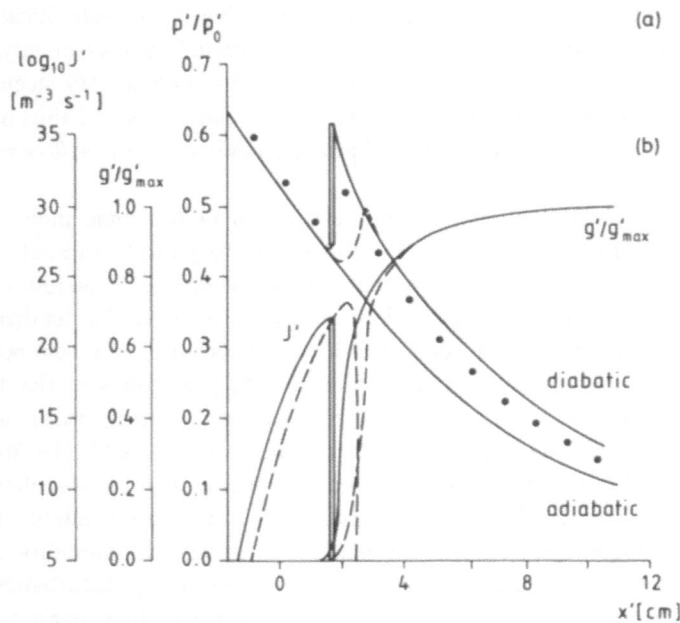

Fig. 3. a Schlieren picture of moist air expansion in the circular arc nozzle with throat height $2\,y^* = 30$ mm and arc radius $R^* = 400$ mm under the atmospheric supply conditions $T_o' = 290.8°$ K, $\omega_o = 7.6$ g/kg and $\varphi_o = 59.4\,\%$. **b** Distribution of the pressure, the nucleation rate and the condensate mass fraction along the above circular arc nozzle axis in the 1-D asymptotic theory ($T_o' = 290.8°$K, $\omega_o = 7.6$ g/kg). ———— supercritical flow with $\varphi_o = 59.4\,\%$, – – – shock-free near critical flow with $\varphi_o = 49.0\,\%$

dimensional. When the lowered initial relative humidity reaches a value of $\varphi_o = 49\,\%$, the normal shock disappears and a shock-free near critical flow pattern with very steep axial flow gradients in the condensation zones following onset sets in (Fig. 3 b). Although no experimental data are available for this case, it is believed that the compression effects arising from heat released by condensation are strong enough that variations in the axial flow gradients over the cross-sectional area remain important (the axial flow gradients are somewhat less steeper near the walls than along the axis in the condensation zones). This means that near critical flows are two-dimensional regardless of nozzle geometry. On the other hand if the initial relative humidity is further decreased keeping the rest of the conditions fixed, the compression along the nozzle axis weakens and the variations of the axial flow gradients over the cross-sectional area become less and less significant. This is why the 1-D asymptotic predictions yield satisfactory results in subcritical flows of relatively low initial relative humidities through slender nozzles (in supercritical flows not bordering on critical heat addition the 1-D asymptotic predictions are also satisfactory in relatively slender nozzles since in this case the condensation zones occur closer to the throat, where variations of the axial flow gradients over the cross-sectional area are even smaller, with much stronger compression effects resulting in a normal shock extending almost from wall to wall).

In conclusion for the expansion of moist air under atmospheric supply conditions in relatively slender nozzles:

(i) the predictions of the 1-D asymptotic theory are satisfactory in subcritical flows where the initial relative humidity is sufficiently low (30 to 40 %) and in supercritical flows where the initial relative humidity is sufficiently high (60 to 80 %),

(ii) in near critical flows where the initial relative humidity ranges from 40 to 60 %, the flow field becomes two-dimensional and the predictions of the 1-D asymptotic theory do not satisfactorily agree with experiments.

These conclusions are in agreement with the early observations of Pouring [7].

Acknowledgements

One of the authors (C.F.D.) acknowledges the support by the Alexander von Humboldt Foundation during the performance of this work in the Institut für Strömungslehre und Strömungsmaschinen at the University of Karlsruhe (TH).

References

[1] Delale, C. F., Schnerr, G. H., Zierep, J.: Asymptotic solution of transonic nozzle flows with homogeneous condensation. I. Subcritical flows. Phys. Fluids A **5,** 2969−2981 (1993).

[2] Delale, C. F., Schnerr, G. H., Zierep, J.: Asymptotic solution of transonic nozzle flows with homogeneous condensation. II. Supercritical flows. Phys. Fluids A **5,** 2982−2995 (1993).

[3] Clarke, J. H., Delale, C. F.: Nozzle flows with nonequilibrium condensation. Phys. Fluids **29,** 1398−1413 (1986).

[4] Delale, C. F., Schnerr, G. H., Zierep, J.: The mathematical theory of thermal choking in nozzle flows. ZAMP **44,** 943−976 (1993).

[5] Blythe, P. A., Shih, C. J.: Condensation shocks in nozzle flows. J. Fluid Mech. **76,** 593−621 (1976).

[6] Schnerr, G. H.: 2-D Transonic flow with energy supply by homogeneous condensation. Exp. Fluids **7,** 145−156 (1989).

[7] Pouring, A. A.: Thermal choking and condensation in nozzles. Phys. Fluids **8,** 1802−1810 (1965).

Authors' addresses: Professor C. F. Delale Ph. O., Department of Mathematics, Bilkent University, 06533 Bilkent, Ankara, Turkey; and Professor Dr.-Ing. habil. G. H. Schnerr, Institut für Strömungslehre und Strömungsmaschinen, Universität Karlsruhe (TH), D-76128 Karlsruhe, Federal Republic of Germany

Acta Mechanica (1994) [Suppl] 4: 133—140

Normal shock λ-foot topography at turbulent boundary layer

P. Doerffer, Gdansk, Poland

Summary. The normal shock λ-foot configuration at turbulent boundary layer has been investigated for $M = 1.35 \div 1.47$ and for $Re_\delta = 5.5 \div 15.3 \times 10^4$. The flow structure in respect to the shocks' inclination is discussed. It is shown that λ-foot topography within the parameters range contradicts the generally accepted assumption of equal static pressure and flow direction on both sides of a slip line downstream the triple point. A new approach has been proposed which displays a very good agreement with the experimental results.

1 Introduction

The experimental investigation of a normal shock wave-boundary layer interaction has been started by the fundamental work of Ackeret et al. [1] published in 1946. This research has been carried out for $M < 1.32$ at a laminar as well as a turbulent boundary layer. It has been proven that at the same Mach number value λ-foot appears by a laminar boundary layer and is absent by a turbulent one. However, Seddon [2] in 1967 has shown that at $M = 1.47$ by turbulent boundary layer a large λ-foot is formed. It is recognised at present that at the Mach numbers above $M = 1.3$ a large compression zone is induced upstream the shock at the boundary layer. For higher Mach numbers, already below $M = 1.4$, the compression becomes an oblique shock wave and a distinct λ-foot appears. The Mach number range of interest is about $M\ 1.3 \div 1.5$ because below the interaction effect is not significant and above the flow separation is already very strong.

The topography of the interaction flow field is presented in Fig. 1. The pressure jump induced by the interacting normal shock is stretched up- und downstream along the wall within the boundary layer. Therefore the beginning of the compression at the wall precedes the shock location. Existing compression causes a stream deflection from the wall. This generates

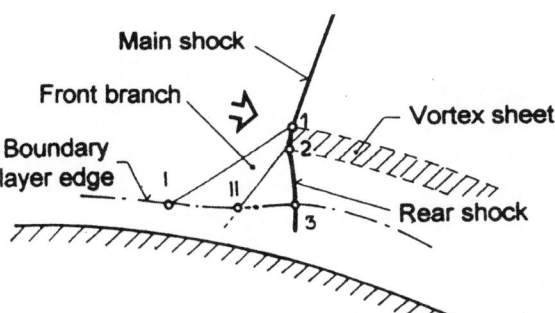

Fig. 1. Sketch of a λ-foot topography

a compression fan in the supersonic outer flow part, between points I and II. At low Mach numbers ($<$ 1.35) the compression fan is located very close to the shock wave but it shifts upstream with increasing Mach number and it finds enough space to develop an oblique shock wave outside the boundary layer. This compression fan coincides with the main shock between points "1" and "2". The distance between these points depends on the Mach number value. At low Mach numbers it is large and at high Mach numbers it shrinks becoming a shock wave. Between these points the main shock wave curves and becomes a rear branch of a λ-foot at point "2". Between points "2" and "3" the rear shock is apparently straight confirming the uniformity of the flow field between λ-foot branches in area "II-2-3". Within the boundary layer the rear shock is weakened to disappear at the sonic line close to the wall.

Downstream the λ-foot junction zone "1-2" (called further a triple point) a strip of transverse variation of entropy is to be expected due to a difference of the shock losses at the main shock and a sequence of front and rear ones. It is smeared at low Mach numbers but when front compression becomes a shock wave, at higher Mach numbers, it becomes narrow and well visible in Schlieren pictures. This vortex sheet appears to be very stable showing hardly any change in the accessible flow area. It is on the contrary to strong tangential discontinuities for which the mixing zone or vortex sheet increases in thickness so quickly that the mixing is often characterised by the angle of its spreading.

2   Experiment

The experimental work has been partly described in [3, 4] and in the references given there. The desired flow configuration is obtained in a curved channel where at the convex wall a local supersonic area develops. It is terminated by a normal shock wave which interacts with a turbulent boundary layer at the wall. This configuration preserves the characteristics features of the natural flow field and corresponds to the test section of Ackeret et al. [1]. The range of flow parameters covered by the carried out experimental work and the parameters of the actually measured cases is presented in Fig. 2.

The carried out measurements concerned:

— static pressure along the test wall
— boundary layer profile development through the interaction zone
— flow field parameters distribution in the interaction area
— optical visualisation of the flow structure
— oil visualisation of the separation flow structure at the test wall.

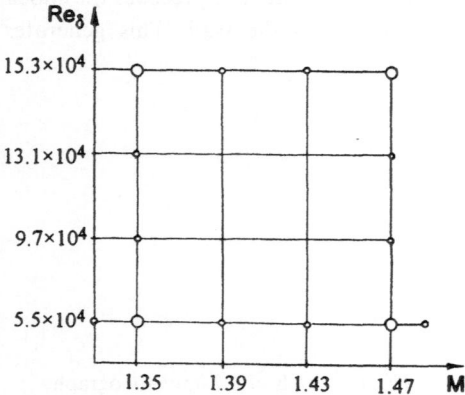

Fig. 2. Flow parameters of measured cases

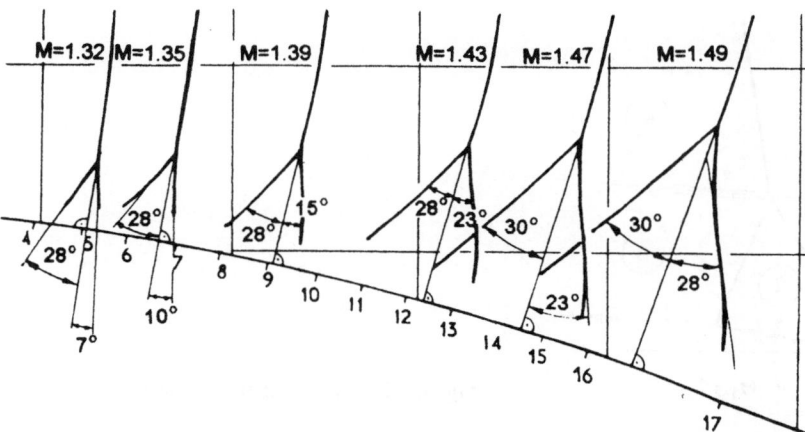

Fig. 3. λ-foot topography at various Mach numbers

In this paper we shall concentrate on the results concerning λ-foot details only. Flow visualisation by means of Schlieren pictures allows to determine the details of the λ-foot topography. Figure 3 presents the λ-feet for several Mach numbers and provides the shock inclination angles. In order to complete the details of the λ-foot geometry a flow deflection angle value at the front shock of the λ-foot is necessary to know. The direct measurement was not possible and therefore two indirect methods have been used. One is based on the boundary layer displacement thickness distribution along the interaction area. The other one uses the Mach number reduction in the front branch of the λ-foot. The deflection angles obtained are enclosed between $3.5 \div 6°$ for all measured flow cases.

3 λ-Foot shocks inclination analysis

The carried out measurements have provided exceptionally large amount of measurement cases what allows verification of statements made previously mainly on qualitative basis. There are very few papers dealing with the problem of shocks inclination at a triple point. The works of McGregor [5] and of Henderson [6] should be mentioned here which treat generally a coincidence point of three oblique shock waves under the condition of equal static pressure and flow direction downstream the triple point on both sides of a vortex sheet. There are also papers concerning a triple point formed at the Mach-type shock reflection [7−9].

Determination of shocks inclination forming a λ-foot concerns the analysis of the flow around a triple point area. The front compression may be substituted by an oblique shock of a weak type, for the model simplicity, because at the low Mach numbers used and the small deflection angles the flow remains nearly isentropic. In such a case the distance between points $1−2$ (Fig. 1) is reduced to one point where there oblique shocks interfere and a slip line originates. The model used to determine the inclination angle of the shocks is presented in Fig. 4. The considered flow area is located away from the boundary layer and the three shocks are treated as straight. Therefore the standard oblique shock theory is used for further analysis, like in other above-mentioned papers.

As the arrows show the flow is from left to right. Given are the parameters of the oncoming stream area-(1) and the deflection angle at the front shock of the λ-foot Θ_{1-2}. These data allow

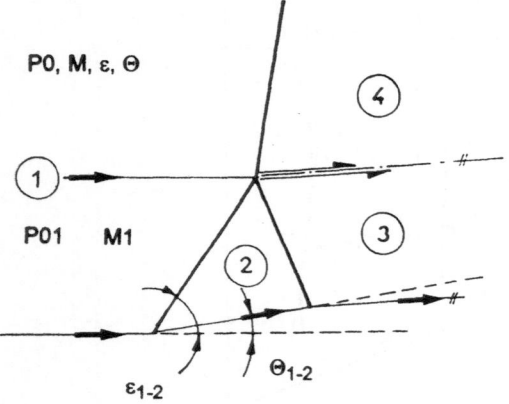

Fig. 4. Flow model at triple point

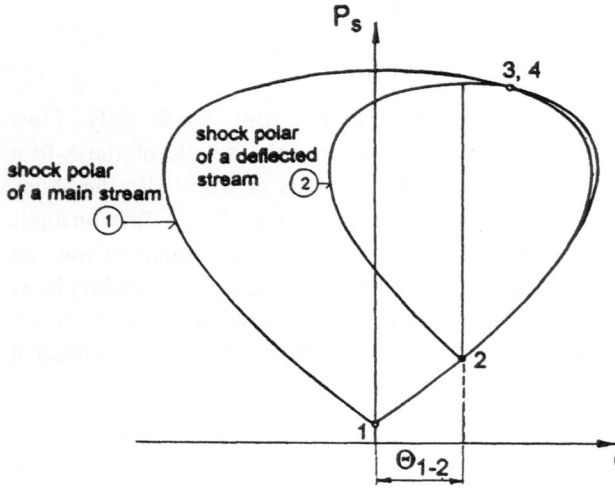

Fig. 5. Solution in the shock polar diagram

a unique determination of the front shock inclination and the flow parameters downstream of this shock in (2) for a weak type shock. In order to find a solution of the main shock (between (1) and (4)) and the rear shock (between (2) and (3)) it is necessary to make an assumption concerning conditions at the slip line (between (4) and (3)). The condition of equal static pressure and flow direction in (3) and (4) seems to be generally accepted. This, so called von Neumann solution, is confirmed by measurements of Abbiss et al. [10] for a single λ-foot case at M = 1.5.

The solution discussed is conveniently illustrated in the static pressure P_S and deflection angle Θ diagram (Fig. 5). The shock polars provide the possible solutions of the stream parameters downstream of an oblique shock. The bottom of the polar marked by the point "1" and "2" provide flow parameters upstream the shocks. So for flow parameters in area-(1) a unique polar is determined. On this polar, conditions in area-(2) as weak solution and in area-(4) as strong solution are to be found. Θ_{1-2} allows to determine the parameters of area-(2) and unique determination of the polar for area-(2). In the coordinate system used the solution becomes a crossing point of both polars. A simple analysis indicates that the proportions between the presented two polars change considerably for various Mach numbers of oncoming stream. A very important conclusion is that below M = 1.35 the polars for deflected stream (2) lie entirely inside the polar (1). In consequence no solution within the mentioned model could be found. It has turned out that this problem has been already faced in the analysis of Mach-type reflection at low Mach numbers [7, 8].

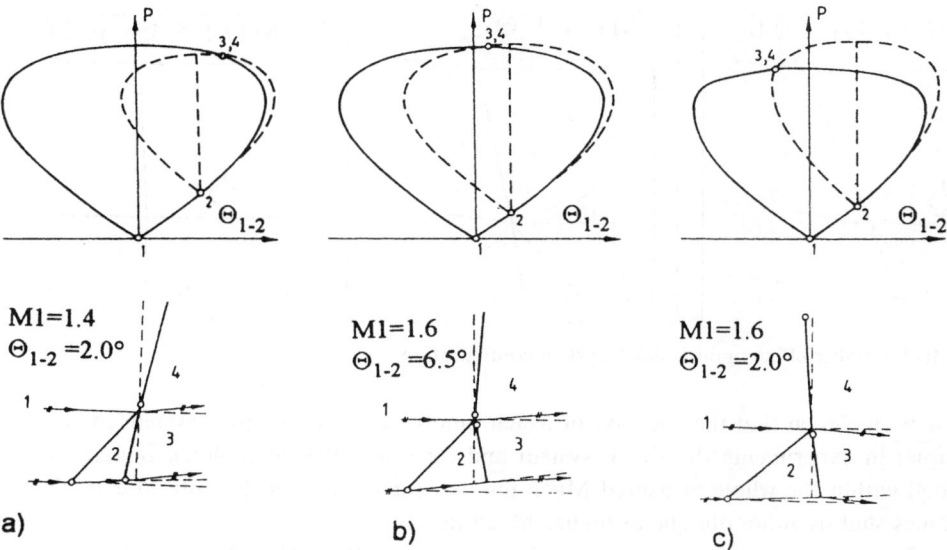

Fig. 6. Three types of solution

A careful analysis of the obtained solutions revealed an existence of three types of solution which correspond to different physical shock configurations. The differences concern the location of the solution point in respect to the symmetric lines of polars and in consequence the location of origin points of shock waves. These are presented in Fig. 6 where for each type a polar diagram and a λ-foot sketch are presented and the shock origin points are marked with circles. The case in Fig. 6a is typical of low Mach numbers. The flow deflection at the rear λ-foot shock is in the same direction as at the front shock and the general deflection in the sketch is upwards. The solution point is located in polar diagrams to the right of both symmetric lines of the polars. It means that the origin of the front and rear shock is away from the triple point. Only the main shock originates at the triple point. It therefore corresponds to the case of two oblique shocks merging into one. The case in Fig. 6b corresponds to the higher Mach numbers, starting with $M = 1.5$ and small deflection angles Θ. In this case the solution point is located between the symmetric lines of two polars. The deflection at the rear shock (downwards) is in other direction than at the front shock. In consequence the rear and main shocks originate at the triple point. This corresponds to the λ-foot structure.

The case of Fig. 6c appears at even higher Mach numbers and low deflection angles. The solution point in this case is found at the left side of both symmetric lines of the polars. This time the λ-foot shock system deflects the stream downwards. The front shock and the main shock take their origin away from the triple point and only the rear shock originates at the triple point. This case concerns the crossing of a weak and a strong type shock what may happen at the strong oblique shock interference with a wall.

A sequence of λ-foot configurations for interesting Mach number values in Fig. 7 indicate that at $M = 1.5$ the λ-foot seems to be correct but the Mach number decrease causes the whole shock system to become oblique.

The above presented analysis of the model inspires following conclusions:

— the solution does not exist at Mach numbers below $M = 1.35$; in case it is true one could expect a qualitative change of λ-foot behaviour below this Mach numbers; however, nothing of the sort has been observed by the experiment

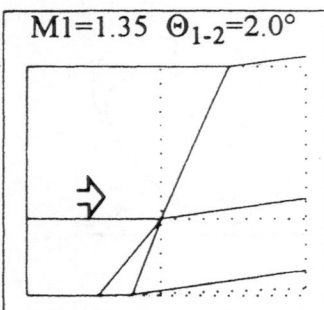 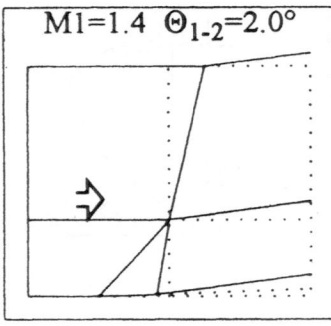 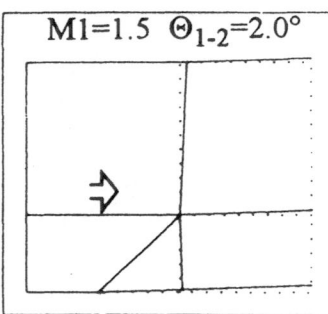

Fig. 7. Mach number effect upon a shock system configuration

— it has been shown that the decrease of Mach number causes the shock system to become oblique; in experiments the shock system and especially the main shock remain nearly normal within the whole measured Mach number range; it could be noticed even that it becomes slightly more oblique at higher Mach numbers

— at the flow parameters of interest, namely $M = 1.35 \div 1.47$ and $\Theta = 3.5° \div 6°$, the solution describes a merging of two oblique shocks instead of a λ-foot formation.

4 Discussion of an alternative model

The large experimental material provides a reliable basis for verification of various models that would better describe the physics of the λ-foot formation. The flow parameters of interest are such that polars cross at the right side of both symmetric lines as in Fig. 6a and it does not provide correct inclination of shock waves. The experimental observation that between areas (1) and (4) as well as between (2) and (3) oblique shocks are present and originate at the triple point imposes that the effective flow direction downstream the triple point must be contained within the angle between flow direction of area (1) and (2). Therefore the solution has to be looked for between both symmetric lines of polars as marked in Fig. 8 (thick pieces of polars). This however means that the condition of equal pressure may not be fulfilled. It should be taken into account that in a matter of fact the triple point and a slip line are simplifications introduced for the model

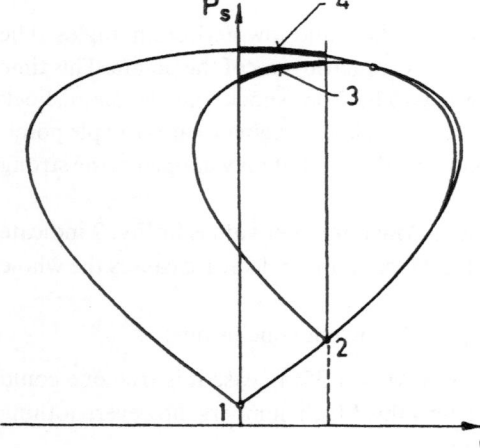

Fig. 8. Solution zone

purposes. In real flow these mean some areas which are much thicker than the shock wave thickness and within which gradients of pressure or flow direction may take place [7, 8]. Measured static pressure distribution in a number of traverses normal to the flow direction upstream and downstream the λ-foot have indicated that the triple point induces a static pressure gradient which decays quickly downstream.

In general it is possible that also the deflection angle changes across the vortex sheet. However looking only for a more appropriate model let us assume that the flow direction downstream the triple point is equal on both sides. The triple point may be considered as an apex where two streams of different (converging) flow directions collide. In case the both streams have equal momentum an outflow angle should coincide with the bisectrix of the angle formed by colliding streams. The momentum of stream (2) should be only slightly smaller from stream (1) because the decrease of velocity is accompanied by an increase of density at the front branch. The calculations carried out for the flow parameters in our experimental cases proved that the difference between so calculated outflow angle and the bisectrix is below twenty minutes. Therefore it has been decided that in a first approximation a simple assumption of outflow angle equal half of the deflection at the front branch of the λ-foot is appropriate. The new model first of all does not induce any problems of solution existence at low Mach numbers. Moreover the λ-foot configuration seems to be similar to the one observed in experiments and its behaviour at Mach number variations is in agreement with experiment, too.

A comparison of the λ-foot configuration obtained by means of the present model with experimental results for two extreme Mach number cases is presented in Fig. 9. The dashed lines concern calculated configuration and the continuous lines present λ-feet from Schlieren pictures. The comparison task was set up in the following manner. For given Mach number value upstream of the triple point (from experiment) the deflection angle should be found for which the shocks inclination coincide the best. By this means the verification of the experimentally defined deflection angle could be performed. The coincidence of the results appeared to be very good. Deflection angels agree with an accuracy below 1° with the values obtained from the Mach number change at the front branch of the λ-foot.

The very good coincidence of the shocks' inclination angles together with the deflection angles at the front shock waves provides a convincing verification of the proposed new model.

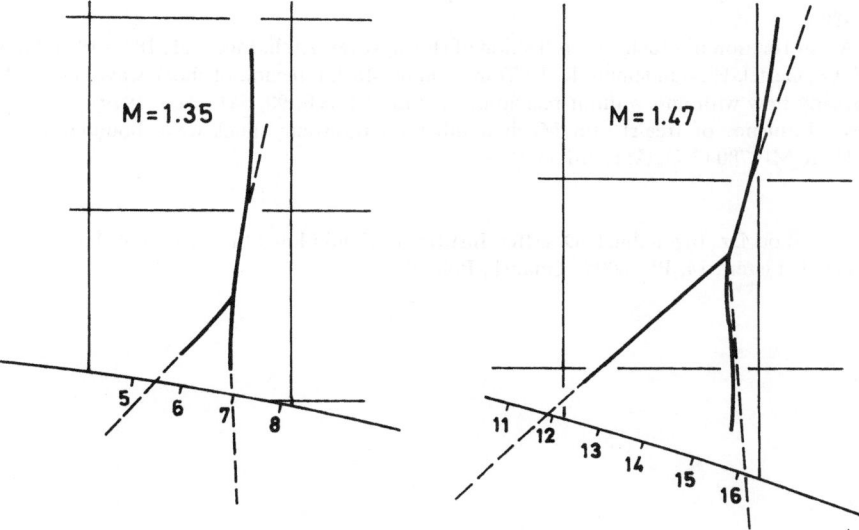

Fig. 9. Comparison of calculated and measured λ-foot configuration

Remarkable is the simplicity of proposed new approach. It could be still supplemented by the condition allowing for the outflow angle difference, however, the obtained quality of results dismisses such necessity.

5   Conclusions

1. An assumption of equal static pressure and flow direction on both sides of a slip line just downstream the triple point provides shock inclinations that are in evident disagreement with experiment at lower Mach numbers. Additionally at Mach number below about $M = 1.35$ such solution do not exist.

2. If the shock configuration is to be of a λ-foot type the static pressure downstream the main shock and the rear shock can not be equal. However the pressure difference is small.

3. The assumption of equal flow direction securing the momentum balance provides results which are qualitatively as well as quantitatively in very good agreement with available experimental results.

References

[1] Ackeret, J., Feldmann, F., Rott, H.: Untersuchungen an Verdichtungsstößen und Grenzschichten in schnell bewegten Gasen. Bericht No. 10 des Institutes für Aerodynamik, ETH Zürich 1946.
[2] Seddon, J.: The flow produced by interaction of a turbulent boundary layer with a normal shock wave of strenght sufficient to cause separation. ARC R&M **3502** (1967).
[3] Doerffer, P., Zierep, J.: An experimental investigation of the Reynolds number effect on a normal shock wave — turbulent boundary layer interactions on a curved wall. Acta Mech. **73**, 77—93 (1988).
[4] Doerffer, P.: An experimental investigation of the Mach number effect upon a normal shock wave — turbulent boundary layer interaction on a curved wall. Acta Mech. **76**, 35—51 (1989).
[5] McGregor, I.: Some calculations of conditions at the intersection of a week shock wave with strong shock. ARC R&M **3728** (1973).
[6] Henderson, L. F.: On the confluence of three shock waves in a perfect gas. Aeronaut. Q. **15**, 181—197 (1964).
[7] Walenta, Z. A.: Microscopic structure of the Mach-type reflexion of the shock wave. Arch. Mech. **32**, 819—825 (1980).
[8] Walenta, Z. A.: Formation of Mach-type reflection of shock waves. Arch. Mech. **35**, 187—196 (1983).
[9] Hornung, H. G., Oertel, H., Sandeman, R. J.: Transition to Mach reflexion of shock waves in steady and pseudosteady flow with and without relaxation. J. Fluids Mech. **90**, 541—560 (1979).
[10] Abbiss, J. W.: Influence of free-stream Mach number on transonic shock-wave boundary-layer interaction. NLR MP 78013 U, Amsterdam 1978.

Author's address: Priv. d.oc. Dr.-Ing. habil. P. Doerffer, Institute of Fluid Flow Machinery, Polish Academy of Sciences, ul. Gen. J. Fiszera 14, PL-80952 Gdansk, Poland

Acta Mechanica (1994) [Suppl] 4: 141–146

On the stand-off distance of detached shock waves in internal transonic flows

R. Dvořák, Praha, Czech Republic

Summary. The stand-off distance of detached shock waves in internal transonic flows depends not only on the shape of the profile (body) in front of which the shock occurs: it is strongly affected by the channel geometry. The front shock waves can easily be misinterpreted with shock waves terminating the region of supersonic flow in the channel.

1 Introduction

There are many excellent studies on detached front shock waves on airfoils and axially symmetric bodies, and Professor Zierep and his school have substantially contributed to this problem (see, e.g., [1]–[4]). In what follows are only three minor additions to this subject, two of them being inspired by internal aerodynamics.

2 The stand-off distance at low supersonic velocities

There exist extensive analytical studies (namely by Shifrin and Belocerkovskij), (for a brief survey of these results see [5]) analyzing the shape of the detached front shock wave close to the profile (body), as well as at large distances from the profile (body), in supersonic flows with $M_\infty \to 1$. These analyses, however, do not tell anything of the stand-off distance of the shock waves although this is a value which can play a decisive role, especially in internal flows.

At higher supersonic velocities the stand-off distance is given almost entirely by the leading edge shape. At lower supersonic velocities (in the range of transonic velocities) this is not the case. An additional parameter to determine the stand-off distance of the front shock wave is the mass flow density in the neighbourhood of the sonic line. If we choose two arbitrary streamlines having the airfoil (body) in between (see Fig. 1), then the Mach number M_∞ is determined by the "critical mass flow", corresponding to the "critical cross section" ($S_1 S_A + S_2 S_B$). In unbounded flows extending to infinity this cross section is given by the length and shape of the sonic line. Thus the airfoil thickness at the sonic points determine M_2, and assuming a normal shock wave even $M_1 = M_\infty$.

The stand-off distance d (i.e., the distance $O_R O$ as measured on the stagnation point streamline) depends on M_1, as well as on the distance of the shock wave from the sonic line (or sonic points on the airfoil) and on $\partial M/\partial x$ (or $\partial A/\partial x$) between the boundary streamlines ψ_A and ψ_B. It is this parameter through which the effect of the flow field boundaries or, in internal

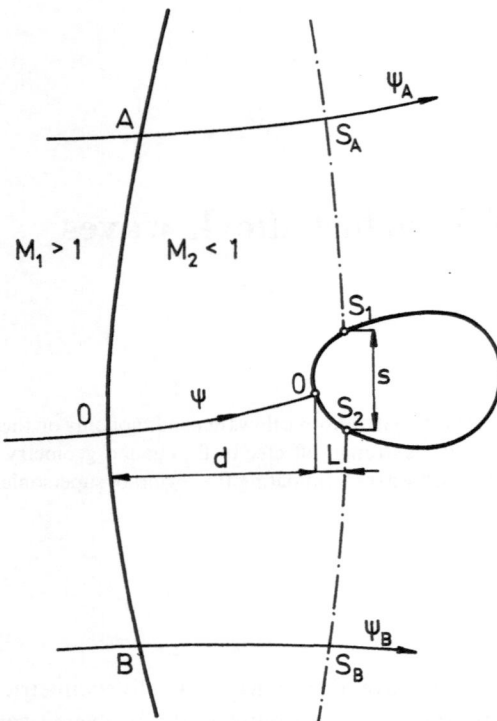

Fig. 1. The detached front shock wave

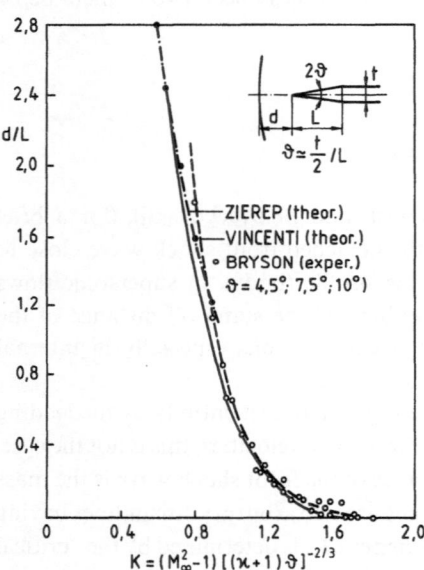

Fig. 2. The stand-off distance d, nondimensionalized by L as a function of the transonic similarity parameter K

aerodynamics of the channel width, enters the problem. If the shock wave is sufficiently weak and the stand-off distance great, the $\partial A/\partial x$ factor does not depend so much on the leading edge shape. Thus the stand-off distance as measured to the sonic points and normalized by L (the distance of the sonic points to the leading edge) should have the same value for any profile (body), even when the actual shock wave distance from the leading edge varies substantially. A small rounding of the leading edge on transonic airfoils cannot have any effect on the detached front shock wave, it can only improve the airfoil aerodynamics when the incidence is changed.

In Fig. 2 the nondimensionalized stand-off distance d/L is plotted against the transonic similarity parameter $K = (M_\infty^2 - 1) [(\varkappa + 1) \vartheta]^{2/3}$, where $\vartheta = \dfrac{t}{2}\Big/L$. Let us assume $M_\infty = $ const, $t = $ const and change the L from 1 to 2 (changing thus also ϑ). Then, starting, for example, with $L = 1, K = 0.64$ we have $d/L = 2.4$, i.e., $d = 2.4, d + L = 3.4$. For $L = 2$ we have $K = 1.0112$, i.e. $d/L = 0.6, d = 1.2$ and $d + L = 3.2$. Apparently, while the stand-off distance $d + L$ as measured from the sonic point has not changed considerably, the same is not true for the stand-off distance d from the leading edge.

3 The stand-off distance of the system of front shock waves in transonic cascades

In internal aerodynamics the stand-off distance at transonic velocities is controlled by different rules.

It has been shown by the present author (see, e.g., [5]) that detached front shock waves in two-dimensional cascades appear already at subsonic inlet Mach numbers, having their origin on the neighbouring profile where they develop from the shock waves terminating the local supersonic regions or their front parts (see Fig. 3). Therefore, the stand-off distance of these shock waves on a particular profile depends primarily on the development of the supersonic flow on the neighbouring profiles. The stand-off distance thus cannot exceed the leading edge of the profile on which it is being built-up and with increasing Mach numbers it approaches asymptotically the value pertinent to an isolated airfoil of the same leading edge shape.

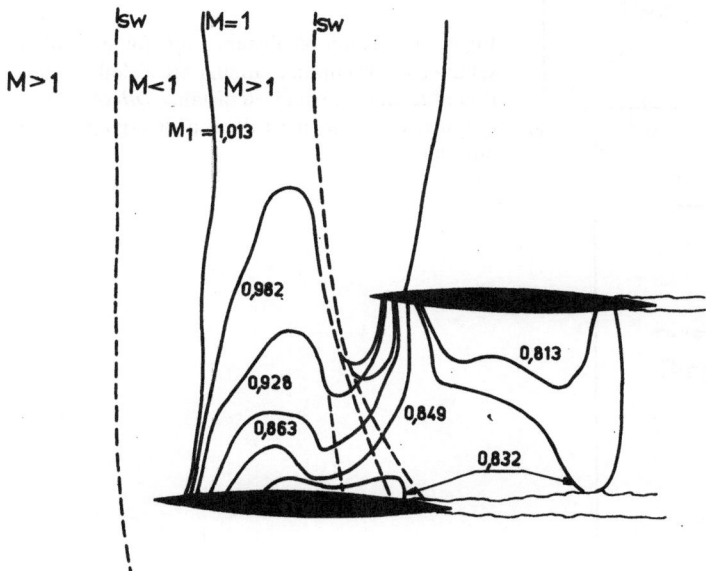

Fig. 3. Development of the system of front shock waves (dotted lines) in a cascade of double-circular-arc symmetric profiles with increasing Mach number. The solid lines are the sonic lines. The numbers attached to them indicate the inlet Mach number

4 The stand-off distance on profiles placed in closed channels (nozzles)

If an isolated profile is placed in a closed channel at very low supersonic velocities, its stand-off distance is often controled by parameters which are not at all related to the shape of the profile, like, e.g., the back pressure, channel geometry and the proximity of the walls with all the accompanying viscous effects. We shall not consider here the last one which is a problem of its own.

As a rule, in the case considered, the detached front shock wave in front of the profile coincides with the shock wave in which transition from supersonic to subsonic velocities for the given channel geometry takes place. The effect of the back pressure on the position of this shock wave can be assessed by using the theory of one-dimensional gasdynamics.

Consider a diverging channel with a single symmetric double circular arc 10% airfoil inserted in the exit cross section so that only its front shock wave is inside the nozzle and contributes to the pressure loss in the nozzle. Due to the fact that in the range $1.0 < M_\infty < 1.32$ the wave drag

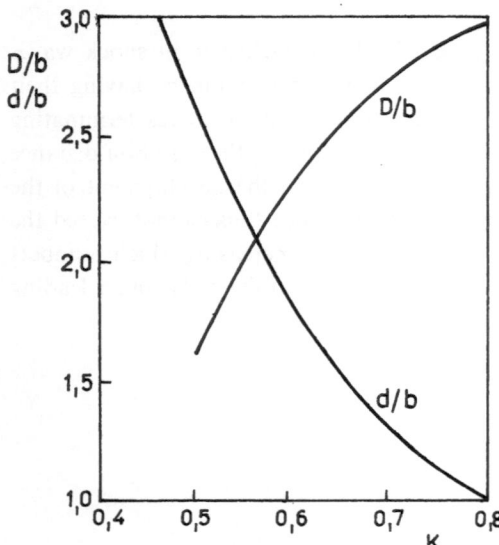

Fig. 4. The stand-off distance d/b for an isolated symmetric 10% double circular arc airfoil as a function of K and the stand-off distance D/b of the same airfoil placed into the exit of a diverging channel (nozzle)

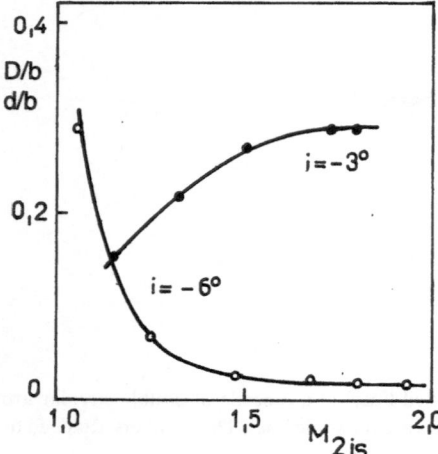

Fig. 5. Anomalous stand-off distance for a turbine cascade of flat profiles at incidence $i = -3°$

increases with M_∞ (drag of the forward portion only) even the pressure loss increases and the distance D of the nozzle terminal shock wave from the profile is increasing. The shock is then moving upstream into the nozzle. On the other hand, the same profile placed in an unbounded space, exhibits in the same range of M_∞ a decreasing stand-off distance d of its front shock wave with increasing M_∞.

Figure 4 shows the nondimensional stand-off distance d/b of a single symmetric double circular arc 10% airfoil plotted versus the similarity parameter K, and, at the same time, a similar plot of the nondimensionalized distance D/b, representing the terminal shock wave distance from the nozzle exit, as explained above and calculated using simple one-dimensional analysis.

The most interesting result of this figure is the qualitatively different shape of both curves. While with increasing K (or increasing M_∞) d/b decreases, the distance of the terminal shock wave from the nozzle exit D/b (i.e., also from the profile) increases due to the profile drag increase. For $M_\infty > 1.32$, however, even this dependence will change (i.e., D/d will decrease with increasing M_∞). For back pressure smaller than that corresponding to the shock wave position in the nozzle exit, the $D/b = f(K)$ curve is shifted to higher values of K and the range of Mach numbers in which it is only the profile that determines the stand-off distance is greater.

In [8] the author presented experimental results concerning a turbine cascade at negative incidences. The stand-off distance d (normalized by the blade chord b) dropped down monotonously with increasing M_1 (or M_{2is}) for most incidences i, except for $i = -3°$ (see Fig. 5). This behaviour could be explained only by the above suggested analysis.

5 Conclusions

The stand-off distance of a front shock wave at very low supersonic Mach numbers is only exceptionally given by the shape of the leading edge of the airfoil (body).

At isolated airfoils it is the distance between the sonic points on either sides of the leading edge determining the stand-off distance.

In internal flows it is necessary to analyze how the front shock wave is being built up. In cascades it is almost always the neighbouring profile that determines the stand-off distance.

On airfoil (bodies) placed into closed channels (profiles) the front shock wave often coincides with the wave terminating the region of supersonic flow in the channel (nozzle) and it is then the nozzle shape or the back pressure that determine the stand-off distance.

References

[1] Zierep, J.: Der Kopfwellenabstand bei einem spitzen schlanken Körper in schallnaher Überschallströmung. Acta Mech. **5**, 204−208 (1968).
[2] Zierep, J.: Ähnlichkeitsgesetze und Modellregeln der Strömungslehre, 3rd ed. Karlsruhe: G. Braun 1991.
[3] Frank, W.: Die Untersuchung der schallnahen Überschallströmung um schlanke Profile. Diss. TH Karlsruhe 1972.
[4] Schnerr, G. H., Zierep, J.: Airfoils in supersonic source and sink flows. Z. Flugwiss. Weltraumforsch. **13**, 281−290 (1989).

[5] Dvořák, R.: Transonic flows (in Czech). Prague: Academia 1986.

[6] Vincenti, W. G., Wagoner, C. B.: Transonic flow past a wedge profile with a detached bow wave — general analytical method and final calculated results. NACA TN 2339; TN 2588 (1951).

[7] Bryson, A. E.: An experimental investigation of transonic flow past two-dimensional wedge and circular arc sections using a Mach-Zehnder interferometer. NACA TN 2560 (1951).

[8] Dvořák, R., Šafařík, P.: Aerodynamic research of tip sections of last stage rotor blades for steam turbines of large output. Paper C 185/79, Inst. Mech. Eng., London 1979.

Author's address: Dr.-Ing. R. Dvořák DrSc., Institute of Thermomechanics, Czech Academy of Sciences, Dolejškova 5, CS-18200 Praha 8, Czech Republic

Acta Mechanica (1994) [Suppl] 4: 147–154

Internal flows with multiple sonic points

G. H. Schnerr and **P. Leidner**, Karlsruhe, Federal Republic of Germany

Summary. In real gas flow several effects are inverted if the fundamental gasdynamic derivative Γ becomes negative. Here we investigate stationary flows with multiple sonic points. In a nozzle with two throats three sonic points occur where the first or the last is related with the absolute maximum of the mass flux density; the location of this absolute maximum depends on the reservoir state. Then we calculate 2-D flows in a circular arc nozzle by solving the Euler equation with a time dependent finite volume method (FVM) of Jameson. For a high exit pressure ($p_e/p_{01} = 0.94$) two sonic shocks occur whereas the flow remains entirely subsonic in between. In order to demonstrate nonclassical effects in strongly bended channels we present results of potential vortex flow of dense gases. Here we observe the formation of separated circular ring shaped supersonic and subsonic regions in the interior of the vortex.

1 Introduction

In classical gasdynamics the fluid is usually treated as perfect gas. In a region near the saturated vapor pressure curve this model is no longer valid. Duhem [1] introduced the quantity

$$\Gamma = 1 + \frac{\varrho}{a}\left(\frac{\partial a}{\partial \varrho}\right)_s = \frac{v^3}{2a^2}\left(\frac{\partial^2 p}{\partial v^2}\right)_s \tag{1}$$

to describe real gas effects. There ϱ, a, p, v, s denote respectively the density, sound speed, pressure, specific volume, and the entropy. Thompson [2] referred to this parameter as fundamental gasdynamic derivative to emphasize its importance in a wide range of dense gas flows and Cramer and Kluwick designated the locus of $\Gamma = 0$ as transition line [3]. The perfect gas model yields $\Gamma = (\gamma + 1)/2$ with γ as the specific heat ratio ($\Gamma = 1.2$ if $\gamma = 1.4$). In the dense gas region Γ is no longer constant and in a wide range of pressure and density $\Gamma < 1$ is possible. In an area partially bounded by the saturated vapor pressure curve close to the thermodynamical critical point Γ even may become negative. Following Cramer [4] we refer to fluids which exhibit these behavior as BZT fluids, due to the fundamental work of Bethe [5], Zel'dovich [6], and Thompson [2]. Thompson investigated the influence of Γ on isentropic flow and reported the existence of expansion shocks in the negative Γ region. Thompson and Lambrakis [7] presented a detailed description of this phenomenon. In the negative Γ region only expansion shocks are possible and compression shocks are prohibited by the second law of thermodynamics. As a rule the Rayleigh line connecting the upstream and the downstream state of a shock must be entirely "right" of the shock adiabat. Sonic shocks, i.e. shocks where either the upstream or the downstream Mach number is unity and double sonic shocks where both Mach numbers are unity are also possible if Γ changes sign across the shock. Then the Rayleigh line is tangential to the shock adiabat.

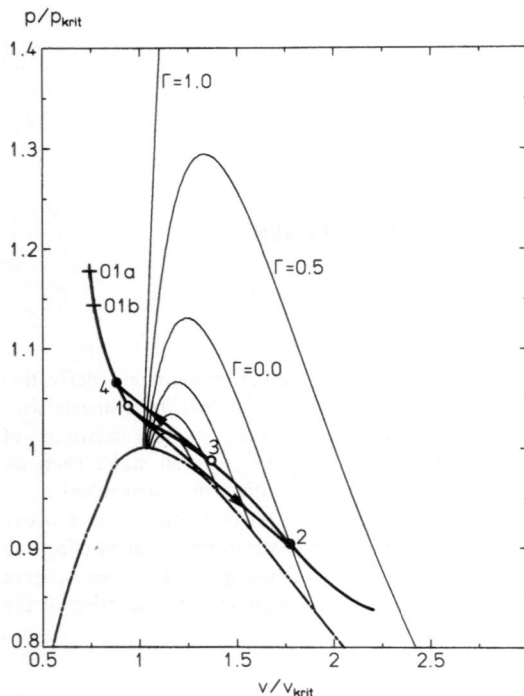

Fig. 1. p, v-Diagramm of PP11 $(C_{14}F_{24})$: $p_{crit} = 14.6$ bar, $v_{crit} = 1.6 \cdot 10^{-3}$ m³/kg, $T_{crit} = 650.1$ K. Reservoir values 01a and 01b correspond to Figs. 2−5; isentrope and shock locations correspond to Fig. 4, 1 → 2: sonic expansion shock 3 → 4: sonic compression shock. − Γ-contours, $\Delta\Gamma = 0.5$, - · · -two phase boundary

Stationary isentropic flow of dense gases and the related phenomena have been investigated by Cramer [4] and Cramer and Best [8]. Applying the Navier-Stokes equation, Kluwick [9] derived the transonic small disturbance equation for 1-D viscous flow of dense gases. In the inviscid limit he presented a classification of possible shock combinations in nonclassical nozzle flows. Cramer and Fry [10] recently derived exact results for 1-D nozzle flows applying the exact shock equation. Schnerr and Leidner [11] employed a FVM of Jameson to derive solutions of 2-D stationary inviscid nozzle flow. In the simple wave flow of a Prandtl-Meyer expansion the characteristics steepen and intersect if Γ becomes negative [2]. As a result expansion shocks develop in flows around convex corners. Cramer and Crickenberger [12] recently investigated the Prandtl-Meyer function for various dense gases.

Considering Eq. (1) the sound speed increases during isentropic expansion if $\Gamma < 1$. Γ also indicates the pattern of the isentropes in the p, v-plane, i.e. both Γ and the curvature of isentropes have the same sign. These effects arise in flows of hydrocarbons and fluorocarbons of high molecular weight which are frequently used as working media in organic Rankine cycles (ORC). Figure 1 exhibits $\Gamma = $ const. contours of PP11 $(C_{14}F_{24})$ in the p, v-plane bounded by the two-phase boundary which is calculated by the formula of Gomez and Thodos [13].

In this paper we present examples of real gas flow of PP11 with multiple sonic points. In all cases under consideration the entropy of the reservoir values is kept constant, i.e. the expansion follows the same isentrope, only the geometry is varied. Special attention is drawn on the mass flux density. First stationary flow in a nozzle with two throats is considered. The reservoir values are varied to achieve the maximum mass flux density at the first or at the second throat. Then we investigate the 2-D inviscid flow with double shocks in a circular arc nozzle. In the last chapter the flow of a potential vortex with maximum mass flux is considered.

2 Mass flux density

To describe real gas effects we employ the equation of van der Waals

$$p = \frac{T}{Z_c(v - b)} - \frac{27}{64Z_c^2} \frac{1}{v^2}. \tag{2}$$

Here the static pressure p, the temperature T, and the specific volume v are normalized with respect to their critical values ($p_{crit} = 14.6$ bar, $v_{crit} = 1.6 \cdot 10^{-3}$ m³/kg, $T_{crit} = 650.1$ K). To ensure that the critical isotherm has a horizontal tangent at the thermodynamical critical point the constants are taken to be $b = 1/3$, $Z_c = 3/8$. As a further condition the specific heat at constant volume c_v is assumed to be constant. For isentropic flow the relationship between Mach number and density has been given by Cramer [4] as

$$\frac{dM}{d\varrho} = \frac{M}{\varrho} \left(\frac{M^2 - 1}{M^2} - \Gamma \right). \tag{3}$$

Thus the Mach number decreases with decreasing density (expansion) if the right hand side of Eq. (3) becomes positive (Fig. 2). In transonic flow this leads to three sonic points and particularly to three critical sound speeds with the related critical Mach numbers. Figure 2 shows the M, ϱ-distribution for two different reservoir states (points 01a and 01b in Fig. 1 with $p_{01a} = 18.2$ bar, $\varrho_{01a} = 828$ kg/m³, $T_{01a} = 669.5$ K; $p_{01b} = 17.9$ bar, $\varrho_{01b} = 815$ kg/m³, $T_{01b} = 669.3$ K).

The related shape of the nozzle can be derived from [2]

$$\frac{1}{M} \frac{dM}{dx} = \frac{1 + (\Gamma - 1) M^2}{M^2 - 1} \frac{1}{A} \frac{dA}{dx}. \tag{4}$$

As in classical gasdynamics the stationary expansion to supersonic flow requires as a necessary condition $dA/dx = 0$ at Mach number unity, i.e. in the region of $\Gamma > 0$ a throat and in negative Γ flow an anti-throat. This yields a nozzle geometry with two throats and an anti-throat in between. For a more detailed analysis we investigate the mass flux density. The conservation of mass yields simply

$$\frac{\varrho w}{\varrho^* w^*} = \frac{y^*}{y} \tag{5}$$

where w, y and the superscript ($*$) denote the flow velocity, the height of the nozzle and the reference point at the absolute maximum of the mass flux density at the minimum cross sectional

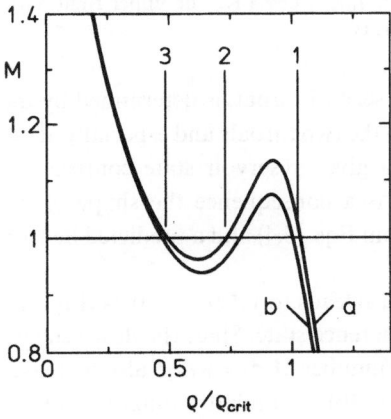

Fig. 2. M, ϱ-Diagramm of PP11 ($C_{14}F_{24}$): Reservoir conditions: $p_{01a} = 18.2$ bar, $\varrho_{01a} = 828$ kg/m³, $T_{01a} = 669.5$ K; $p_{01b} = 17.9$ bar, $\varrho_{01b} = 815$ kg/m³, $T_{01b} = 669.3$ K

150   G. H. Schnerr and P. Leidner

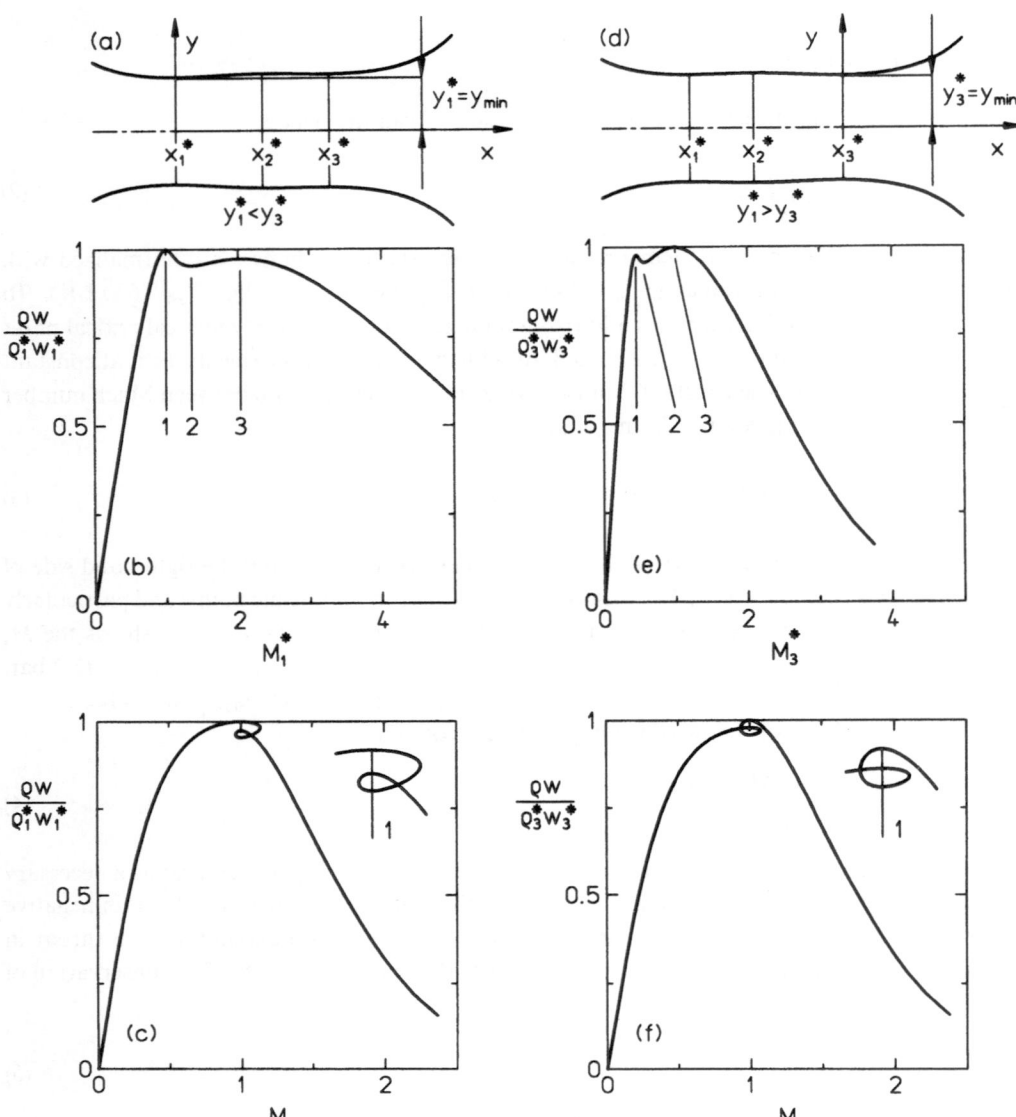

Fig. 3. Nozzle geometry and mass flux density of PP11 ($C_{14}F_{24}$). First case (**a** — **c**, left) reference point $x_1{}^*$, reservoir conditions: $p_{01a} = 18.2$ bar, $\varrho_{01a} = 828$ kg/m^3, $T_{01a} = 669.5$ K; Second case (**d** — **f**, right) reference point $x_3{}^*$, reservoir conditions: $p_{01b} = 17.9$ bar, $\varrho_{01b} = 815$ kg/m^3, $T_{01b} = 669.3$ K. The upper right hand parts of **c, f** shows enlargement near the maximum mass flux density

area. The location of this point, either at the first or at the second throat, is determined by the reservoir values. As a further result the cross section between the two throats and especially at the anti-throat is controlled by the mass flux density. Thus a given reservoir state controls the area variation between the first and the second throat. As a consequence the shape of the nozzle and the Mach number distribution are calculated from Eqs. (3,4) and a predicted density distribution [8].

The expansion starting from the reservoir conditions (01a) leads to $y_1{}^*/y_3{}^* = 0.94$ (Fig. 3a), therefore the location of the first sonic point is chosen as reference state. Since the flow velocity w always increases with decreasing density the critical Mach number $M_1{}^* = w/a_1{}^*$ also increases monotonically and $M_1{}^* = 1$ can only be attained once (Fig. 3b). The relative minimum of the mass flux density corresponds with y_2^* and the second maximum with y_3^*. Regarding the mass flux density as a function of the local Mach number M three sonic points occur, where the first is the point of absolute maximum of the mass flux density (Fig. 3c).

In the second case the reservoir values are slightly decreased isentropically (point 01b in Fig. 1). Now we get $y_1^*/y_3^* = 1.06$. The total minimum of the nozzle and consequently of the maximum mass flux density are located at x_3^*, and the third sonic point is chosen as reference point (Figs. 3d – f).

3 Two-dimensional nozzle flow

We continue with the same reservoir values (01b) and investigate the 2-D stationary inviscid flow in a plane circular arc nozzle (radius $R^* = 200$ mm, total throat height $2y^* = 30$ mm). The 2-D Euler equation is solved on a 512×64 H-grid by an explicit time dependent FVM of Jameson [14]. An exit pressure of 16.85 bar ($p_e/p_{01} = 0.94$) leads to a phenomenon which is unknown in perfect gas flow (Fig. 4). In the converging nozzle part Mach number unity is attained, there a sonic expansion shock appears with downstream state close to the transition line. The shock is curved downstream and at the wall followed by a post-shock compression. This effect has been reported by Schnerr and Leidner [15, 16]. They applied the theory of Oswatitsch and Zierep for normal shocks near curved walls [17] to real gases. Due to the numerical spread of the shock thickness there are overshoots in Γ, the sound speed, and the Mach number. Behind this shock the pressure and density decrease and the Mach number increases. At the throat the flow remains

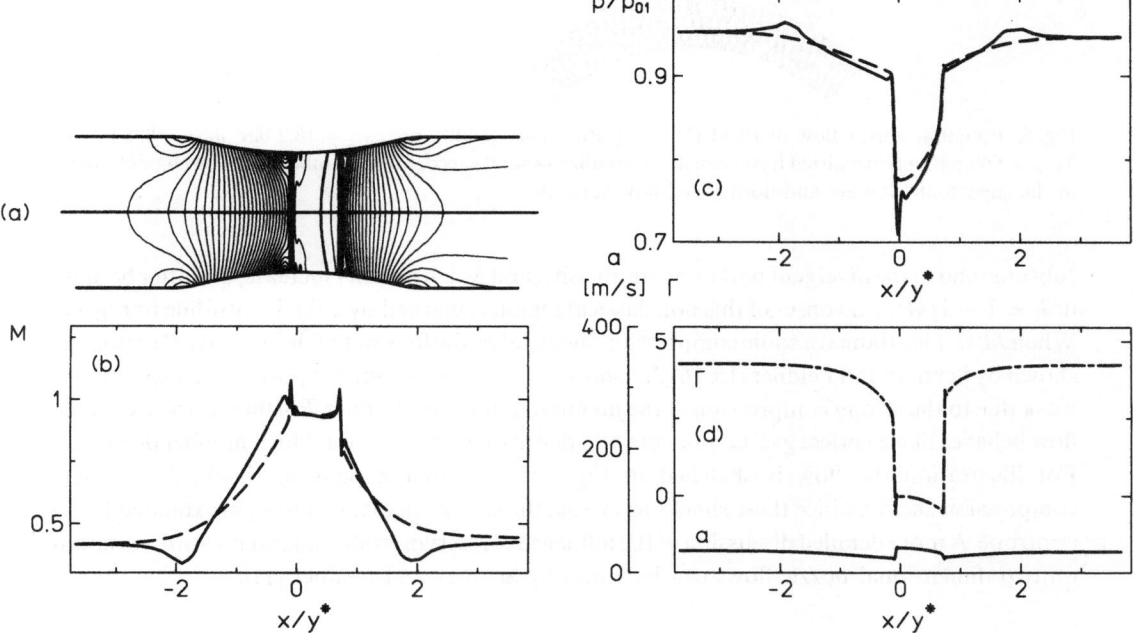

Fig. 4. Laval nozzle flow of PP11 ($C_{14}F_{24}$), flow from left. Sonic expansion shock in the converging part, sonic compression shock in the diverging part, subsonic outflow, exit pressure $p_e = 16.85$ bar. Circular arc nozzle $y^* = 30$ mm, $R^* = 200$ mm; reservoir conditions: $p_{01b} = 17.9$ bar, $\varrho_{01b} = 815$ kg/m^3, $T_{01b} = 669.3$ K. **a** Mach number contours, $\Delta M = 0.02$; **b** Mach number (axis: - - -, wall: —). The overshoots are due to the numerical spread of the shock thickness; **c** static pressure (axis: - - -, wall: —); **d** sound speed (—) and Γ (- · · -) along the axis

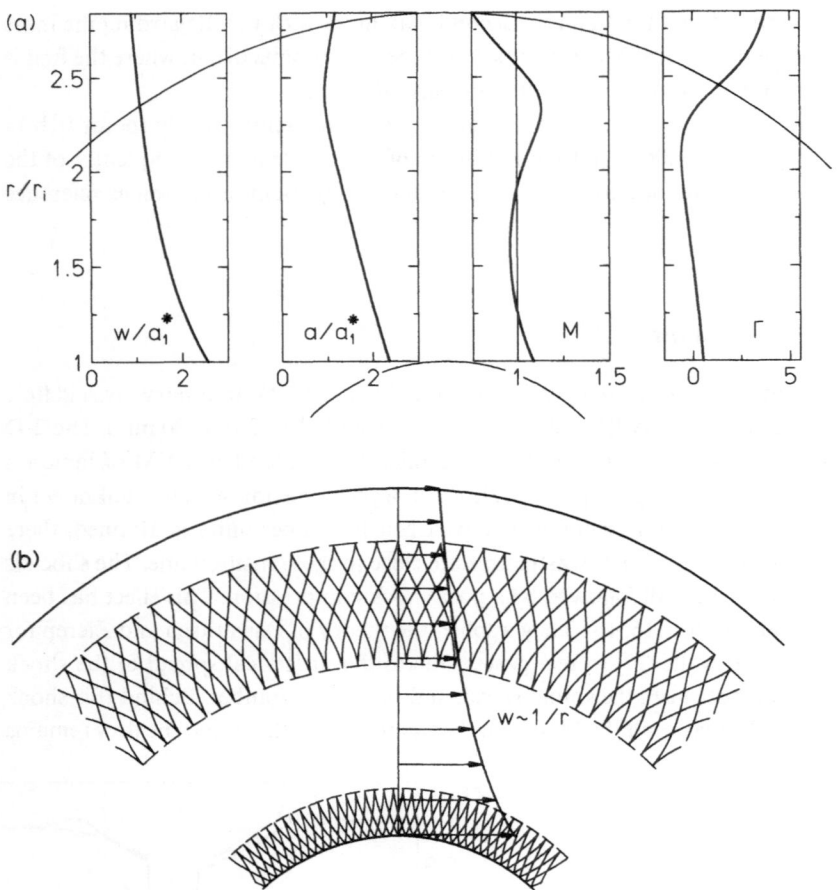

Fig. 5. Potential vortex flow of PP11 ($C_{14}F_{24}$). Reservoir conditions: $p_{01a} = 18.2$ bar, $\varrho_{01a} = 828$ kg/m^3, $T_{01a} = 669.5$ K. **a** Normalized flow velocity, normalized sound speed, Mach number, and Γ; **b** characteristics in the supersonic domain and normalized flow velocity

subsonic and in the divergent part pressure, density, and Mach number increase, the latter because of $\Gamma < 1 - 1/M^2$. The onset of this nonclassical region is marked by a thicker iso-line in Fig. 4a. When $M = 1$ is attained a sonic compression shock takes the flow out of the negative Γ region. As shown by Schnerr and Leidner [15, 16] this shock is curved upstream, the post-shock expansion is weak due to the strong compression at the junction to the parallel duct. Further downstream the flow behaves like a perfect gas, i.e. pressure and density increase and the Mach number decreases. For illustration the flow is sketched in Fig. 1 ($1 \rightarrow 2$ sonic expansion shock, $3 \rightarrow 4$ sonic compression shock). Since these shocks are weak, the shock adiabat can be approximated by an isentrope. A more detailed discussion of the influence of reservoir state and exit pressure variation on two-dimensional nozzle flows can be found by Schnerr and Leidner [11].

4 Potential vortex flow

In this Section the flow starts from the reservoir values (01a), therefore the M, ϱ-distribution (Fig. 2) including three sonic points remains unchanged. We consider stationary inviscid flow in a bended duct bounded by concentric circles. Assuming irrotational flow yields with $w_r = 0$

independent from the gas model [18]

$$w_\varphi = \frac{\text{const.}}{r}, \tag{6}$$

here w_r and w_φ denote the velocity components in polar coordinates. In a flow with reservoir state 01a the maximum mass flux density is achieved at the first sonic point (Fig. 3a − c), thus this point is chosen as reference state (superscript "*"). This leads to $w/a_1^* = r_1^*/r$. Figure 5a depicts the normalized velocity, normalized sound speed, Mach number, and the Γ distribution of the potential vortex. The flow velocity decreases monotonically with increasing r, whereas the sound speed distribution exhibits a local minimum which causes three sonic points. As a consequence separated circular rings of supersonic and of subsonic regions develop. These rings are embedded between a subsonic region at the upper wall and a supersonic region at the lower wall. The flow velocity and the characteristics are sketched in Fig. 5b.

5 Conclusions

We examined internal flows of a BZT fluid (PP11) with multiple sonic points in different geometries. In a nozzle with two throats the mass flux density is maximum at the location of the throats and has a local minimum at the location of the anti-throat, whereas the location of the absolute maximum depends on the reservoir state. Due to the different behavior of the Mach number and the critical Mach number the mass flux density is a non-unique function of M but not of M^*.

We then changed the geometry in a circular arc nozzle and kept the reservoir values constant. For a given exit pressure ratio $p_e/p_{01} = 0.94$ a sonic expansion shock develops in the converging part, at the throat the flow remains subsonic and in the diverging part a sonic compression shock takes the flow out of the nonclassical region. Thus the flow remains entirely subsonic.

As a further result in potential vortex flow additional supersonic and subsonic regions develop in the interior part of the flow field.

Further investigations will be done with different nozzle geometries, i.e. varying the wall curvature and the throat height of straight and bended nozzles. More accurate solutions can be obtained introducing other equations of state like the Redlich-Kwong equation or the Martin-Hou equation.

These results may be regarded as a first step towards the numerical investigation of axial turbines for ORC processes. The aim of this further research is to improve the efficiency of these machines and to reduce or even avoid shock losses.

Acknowledgement

The authors would like to express their thanks to Professor M. S. Cramer from Virginia Polytechnic Institute and State University for providing the fluid data for the calculation of the equation of state.

References

[1] Duhem, P.: Sur la propagation des ondes de choc au sein des fluides. Z. Phys. Chem. **69**, 169 − 186 (1909).
[2] Thompson, P. A.: A fundamental derivative in gasdynamics. Phys. Fluids **14**, 1843 − 1849 (1971).
[3] Cramer, M. S., Kluwick, A.: On the propagation of waves exhibiting both positive and negative nonlinearity. J. Fluid Mech. **142**, 9 − 37 (1984).

[4] Cramer, M. S.: Nonclassical dynamics of classical gases. In: Nonlinear waves in real fluids (Kluwick, A., ed.), pp. 91–145. Wien New York: Springer 1991.

[5] Bethe, H.: The theory of shock waves for an arbitrary equation of state. Off. Sci. Res. Dev. Rep. **545** (1942).

[6] Zel'dovich, Y.: On the possibility of rarefaction shock waves. Zh. Eksp. Teor. Fiz. **4**, 363–364 (1946).

[7] Thompson, P., Lambrakis, K.: Negative shock waves. J. Fluid Mech. **60**, 187–208 (1973).

[8] Cramer, M. S., Best, L.: Steady, isentropic flows of dense gases. Phys. Fluids A **3**, 219–226 (1991).

[9] Kluwick, A.: Transonic nozzle flow of dense gases. J. Fluid Mech. **247**, 661–688 (1993).

[10] Cramer, M. S., Fry, R.: Nozzle flows of dense gases. Phys. Fluids A **5**, 1246–1259 (1993).

[11] Schnerr, G. H., Leidner, P.: Two-dimensional nozzle flow of dense gases. ASME Paper 93-FE-8, ASME Fluids Engineering Conference, Washington, DC, June 20–24, 1993.

[12] Cramer, M. S., Crickenberger, A. B.: Prandtl-Meyer function for dense gases. AIAA J. **30**, 561–564 (1992).

[13] Reid, R., Prausnitz, J., Poling, B.: The properties of gases and liquids, 4th edn. New York: Mc Graw-Hill 1987.

[14] Jameson, A., Schmidt, W., Turkel, E.: Numerical solution of the Euler equations by finite volume methods using Runge Kutta time stepping schemes. AIAA Paper 81–1259 (1981).

[15] Schnerr, G. H., Leidner, P.: Realgaseinflüsse auf einen senkrechten Stoß an einer gekrümmten Wand. Z. Angew. Math. Mech. **73**, T548–T551 (1993).

[16] Schnerr, G. H., Leidner, P.: Real gas effects on the normal shock behavior near curved walls. Phys. Fluids A **5**, 2996–3003 (1993).

[17] Zierep, J.: Der senkrechte Verdichtungsstoß am gekrümmten Profil. Z. Angew. Math. Phys. **IXb**, 764–776 (1958) (Festschrift Jakob Ackeret).

[18] Oswatitsch, K.: Grundlagen der Gasdynamik. Wien New York: Springer 1976.

Authors' address: Professor Dr.-Ing. habil. G. H. Schnerr and Dipl.-Ing. P. Leidner, Institut für Strömungslehre und Strömungsmaschinen, Universität Karlsruhe (TH), D-76128 Karlsruhe, Federal Republic of Germany

Acta Mechanica (1994) [Suppl] 4: 155–162

The boundary value problem for low aspect ratio, pointed wings at sonic speed
Recent developments in theory

S. Turbatu, Bucharest, Romania

Summary. In this paper a review of the recent developments in the theory of a slowly oscillating low aspect ratio, pointed wing at sonic speed is presented, using the concepts of the parabolic method, the equivalence rule and a iteration method to approximately incorporate the thickness effect in the analysis.

1 Introduction

Wings in unsteady transonic flow have been extensively treated in the recent literature in which the analysis is based on the linearized unsteady transonic small perturbation potential equation and the linearized tangent flow condition. The small perturbation equation, governing transonic flow, has become well established. Its use in the prediction of transonic phenomena is greatly complicated by the fact that the equation is nonlinear.

The introduction of the parabolic method [1] for steady flow over a body of revolution at sonic speed has facilitated a method to restore some of the thickness effect in the unsteady flow analysis as can be seen in [2], [3] for two-dimensional flow and in [4] for the body of revolution. In this paper, an approach is presented, using the concepts of the parabolic method [1], the equivalence rule for the parabolic method [5] and an iteration method [6] to approximately incorporate the thickness effect in the analysis of a slowly oscillating low aspect ratio, pointed wing at sonic speed.

2 Problem formulation

Consider a rigid pointed thin wing which performs harmonic pitching oscillations of small amplitude in a steady uniform transonic flow. With the flow directed along the positive x axis and the wing oriented along the x axis, we may introduce nondimensional variables x, y, z by the relations $\bar{x} = l \cdot x$, $\bar{y} = l \cdot y$, $\bar{z} = l \cdot z$ and $\bar{t} = (U_0/2) \cdot t$, where l is the body length. U_0 is the freestream speed, and the bar coordinates are the physical coordinates. The steady-state position of the wing is assumed the be symmetric. Denoting the thickness ratio by ε, the requirement that the wing has to be thin becomes $\varepsilon \ll 1$. We assume the existence of a velocity potential Φ such that the x, y and z components of the flow velocity are $1 + \Phi_x$, Φ_y and Φ_z respectively. It has been shown [7] that the potential Φ must satisfy the following equation:

$$(1 - M_\infty^2) \, \Phi_{xx} + \Phi_{yy} + \Phi_{zz} - 2M_\infty^2 \Phi_{xt} - M_\infty^2 \Phi_{tt} = (\gamma + 1) \, M_\infty^2 \Phi_x \Phi_{xx}, \tag{1}$$

where M_∞ is the freestream Mach number and γ is the ratio of specific heats. The nonlinear term in Eq. (1) yields the fundamental difficulty in solving the transonic problem.

Consistent with the requirement that the wing performs small oscillations of amplitude δ about its steady-state position, we may write the equation for points on the wing as [8]:

$$z = \varepsilon g(x, y)\, \mathrm{sgn}\, z + \delta\, \mathrm{Re}\, [f(x, y)\, \exp\, (ikt)], \tag{2}$$

where $g(x, y)$ is the steady-state wing shape of order 1 and $f(x, y)$, also of order 1, represents the change in shape due to oscillation. k is the reduced frequency of oscillation equal to $\omega l/U_0$, where ω is the physical frequency. We assume the oscillation to be a small disturbance to the steady-state solution, so $\varepsilon \gg \delta$.

If the wing is assumed to be nearly planar so that $\sigma \gg \varepsilon$ (and $\sigma \gg \delta$), where σ denotes the semispan to chord ratio, the boundary condition is [8]:

$$\Phi_z|_{\mathrm{body}} = \mathrm{Re}\, \{\varepsilon g_x\, \mathrm{sgn}\, z + \delta(f_x + ikf)\, \exp\, (ikt)\}. \tag{3}$$

Moreover, we assume that δ is sufficiently small so that Eq. (3) may be evaluated on $z = \pm 0$ instead of Eq. (2).

Following the ideas of papers [2], [3], [8], [9] and [10], we confine ourselves to flows for which $M_\infty = 1$ and

$$\Phi(x, y, z, t) = \chi(x, y, z) + \psi(x, y, z, t), \tag{4}$$

where χ is the steady and ψ the unsteady small perturbation potential respectively. Then, the unsteady part of ψ satisfies:

$$\psi_{yy} + \psi_{zz} - (\gamma + 1)\, \chi_x\psi_{xx} - (\gamma + 1)\, \chi_{xx}\psi_x - 2\psi_{xt} - \psi_{tt} = 0. \tag{5}$$

The term $(\gamma + 1)\, \chi_x\psi_{xx}$ is neglected, based on the argument that ψ is a small disturbance to Φ and the contribution of this term is less than that of the term retained [3], [8], [9] and [10]. Here we assume that the solution ψ is periodic in time, i.e.

$$\psi(x, y, z, t) = \mathrm{Re}\, [\varphi(x, y, z)\, \exp\, (ikt)],$$

where φ is the complex amplitude of the oscillatory perturbation velocity potential. Substituting this into (5) yields the time independent equation for φ. If the symmetry with respect to z is noted, then we may consider the boundary value problem:

$$\varphi_{yy} + \varphi_{zz} - [(\gamma + 1)\, \chi_{xx} + 2ik]\, \varphi_x + k^2\varphi = 0, \quad z > 0 \tag{6}$$

$$\varphi_z(x, y, + 0) = \delta(f_x + ikf) = W(x, y), \quad \text{for} \quad |y| < s(x) \tag{7}$$

$$\varphi(x, y, + 0) = 0, \quad \text{for} \quad |y| \geq s(x), \tag{8}$$

where $s(x)$ is the body planform shape.

The three-dimensional boundary value problem defined by Eqs. (6), (7) and (8) can be solved in several different ways. This paper will summarize this boundary value problem the results for a slowly oscillating low aspect ratio, pointed wing.

3 The parabolic method

For $(\gamma + 1)\chi_{xx} = 0$ (linear theory) the boundary value problem for a slowly oscillating low aspect ratio, pointed wing has been solved by Landahl [7]. In this case, the frequency of oscillation is assumed to be high and the thickness effect, a steady state property, is completely uncoupled from the unsteady equation.

The incorporation of the thickness effect in the unsteady flow analysis has been of interest to many investigators. The introduction of the parabolic method [1] for steady body of revolution at sonic speed has facilitated a method to restore some of the thickness effect in the unsteady flow analysis as can be seen in [2], [3] for two-dimensional flow and in [4], [11] for the body of revolution and for slender bodies respectively.

For the case $(\gamma + 1)\chi_{xx} = \text{const.} = \lambda > 0$ (the parabolic method) in [8], [9] and [10] an approach is presented, using the concepts of the parabolic method [1] the equivalence rule for the parabolic method [5] and an iteration method [6] to approximately incorporate the thickness effect in the analysis of a slowly oscillating low aspect ratio, pointed wing at sonic speed. It has been shown by Zierep [5] that the equivalence rule of Oswatitsch [12] between wings and bodies holds also for the parabolic method.

The numerical results in [9] show that the thickness has the same influence as the frequency on the damping-in-pitch, and that for moderately thin wings the thickness effect is not very significant. The effect of aspect ratio, however, is more pronounced than that of either thickness or frequency. In [9] only the very low frequency case has been considered for the case of a delta wing whose leading edge is expressed as $s(x) = \sigma x$. To illustrate the effect of body geometry and thickness, i.e., the parabolic constant λ influence, numerical examples were calculated for three basic wing planforms in [8].

Sapunkova [10] considers the case $(\gamma + 1)\chi_{xx} = \lambda(x)$, where $\lambda = \lambda(x)$ is a continuous differentiable function and introduced a now variable:

$$x_1 = \int_0^x \frac{2ik}{\lambda(\xi) + 2ik}\, d\xi. \tag{9}$$

Substituting this in the Eq. (6) yields that

$$\varphi_{yy} + \varphi_{zz} - 2ik\varphi_{x_1} + k^2\varphi = 0. \tag{10}$$

Applying the Fourier transform with respect to x_1 and introducing the linearized tangent flow condition on the wing, Eq. (7), after taking the inverse transformation, we obtain the solution in the physical plane:

$$W(x, y) = -\frac{1}{\pi} \int_{-s(x)}^{+s(x)} \frac{\varphi_\eta}{y - \eta}\, d\eta + P^{(1)}\{\varphi\} + P^{(2)}\{\varphi\}, \tag{11}$$

where

$$P^{(1)}\{\varphi\} = \frac{ik}{2\pi} D \left\{ \left(\ln \frac{k}{2} + C + \frac{i\pi}{2} - 1 \right) \int_{-s(x)}^{+s(x)} \varphi\, d\eta + 2 \int_{-s(x)}^{+s(x)} \ln|y - \eta|\, \varphi\, d\eta \right.$$

$$- D \int\limits_0^x \exp\left[-\frac{ik}{2}\left(x_1(x) - x_1(\alpha)\right)\right] \ln\left(x_1(x) - x_1(\alpha)\right) \cdot \int\limits_{-s(\alpha)}^{+s(\alpha)} \frac{2ik}{\lambda(\alpha) + 2ik}\, \varphi\, d\eta\, d\alpha \bigg\},$$

$$P^{(2)}\{\varphi\} = -\frac{k^2}{8\pi} D^2 \left\{ \left(\ln\frac{k}{2} + C + \frac{i\pi}{2} - \frac{5}{2}\right) \int\limits_{-s(x)}^{+s(x)} \varphi(y - \eta)^2\, d\eta + 2 \int\limits_{-s(x)}^{+s(x)} (y - \eta)^2 \ln|y - \eta|\, \varphi\, d\eta \right.$$

$$\left. - D \int\limits_0^x \exp\left[-\frac{ik}{2}\left(x_1(x) - x_1(\alpha)\right)\right] \ln\left(x_1(x) - x_1(\alpha)\right) \cdot \int\limits_{-s(\alpha)}^{+s(\alpha)} (y - \eta)^2 \frac{2ik}{\lambda(\alpha) + 2ik}\, \varphi\, d\eta\, d\alpha \right\},$$

$$D = \left(1 - \frac{i\lambda(x)}{2k}\right) \frac{\partial}{\partial x} + \frac{ik}{2} \text{ and } C = \text{Euler's constant} = 0{,}5772\ldots$$

For $\lambda(x) = 0$ and $\lambda(x) = \text{const.} > 0$ Eq. (11) reduces to Eq. (3.7) of Landahl [7] and Eq. (6) of Ruo [9], respectively.

An approximate solution of Eq. (11) can be obtained by iteration following Adams and Sears [6]. We set:

$$\varphi = \varphi^{(1)} + \varphi^{(2)} + \varphi^{(3)} + \cdots, \tag{12}$$

where $\varphi^{(1)}$ is the slender-wing theory value and $\varphi^{(2)}$, $\varphi^{(3)}$, ... the higher-order terms provided by the present theory. Substituting φ as given by Eq. (12) into Eq. (11) and equating terms of the same order gives:

$$W^{(n)} = -\frac{1}{\pi} \int\limits_{-s(x)}^{+s(x)} \frac{\varphi_\eta^{(n)}}{y - \eta}\, d\eta, \quad n = 1, 2, 3, \ldots, \tag{13}$$

where

$$W^{(1)} = W, \quad W^{(2)} = -P^{(1)}\{\varphi^{(1)}\}$$
$$W^{(3)} = -P^{(1)}\{\varphi^{(2)}\} - P^{(2)}\{\varphi^{(1)}\}, \cdots \tag{14}$$

The solution of the Söhngen integral equation (13) is well known. It is assumed that the leading edge of the wing is pointed, i.e., $\varphi^{(n)}(x, \pm s, +0) = 0$, and the solution is given by:

$$\varphi_y^{(n)} = \frac{1}{\pi \sqrt{s^2(x) - y^2}} \int\limits_{-s(x)}^{+s(x)} \frac{W^{(n)} \sqrt{s^2(x) - \eta^2}}{y - \eta}\, d\eta \tag{15}$$

With $W^{(n)}$ being given in terms of the lower order solutions as shown in Eq. (14), $\varphi^{(n)}$ can be derived from Eq. (15). The problem is thus reduced to the evaluation of a number of fairly simple integrals.

The simplest planform which fulfils the above requirements is the triangular one [7], [8] and [9]. In the subsequent analysis, only the first two terms on the right hand side of Eq. (12) will be considered for the case of a delta wing whose leading edge is expressed as $s(x) = \pm \sigma x$ and $\chi_{xx} = a + bx$, where a is the arbitrary pitching axis location measured from the nose of the delta wing.

Let the displacement of the mean surface of the wing, performing an oscillatory motion in a mode j with a zero mean angle of attack, be given by

$$z = \delta_j f_j(x, y) \exp(ikt)$$ (16)

and let the linearized tangent flow condition be written as

$$\varphi_{z_j}(x, y, +0) = \delta_j(f_{jx} + ik f_j), \quad j = 1, 2.$$ (17)

Straightforward application of the formulas in the previous Section of the vertical-translations oscillations ($f_1 = 1$) gives then for the potential distribution on the upper surface of the wing:

$$\varphi_1^{(1)} = -ik\,\delta_1(\sigma^2 x^2 - y^2)^{1/2}$$

$$\varphi_1^{(2)} = -\frac{ik\,\delta_1 \sigma^2(\gamma + 1)\,b}{8}\,(\sigma^2 x^2 - y^2)^{1/2}\left\{[2x(x + \mu) - Ax^2]\right.$$

$$\times \left[\ln\frac{ki\sigma^2 x^2}{8B} + 2C - 1\right] + \frac{2}{3}Ax^2 - \frac{2}{3}A\frac{y^2}{\sigma^2} - (x + \mu)^2(2 - A)$$ (18)

$$\times [\mathrm{li}\,(\xi^{4-2}) - \ln(2 - A)] + 2\mu(x + \mu)(1 - A)[\mathrm{li}\,(\xi^{4-1}) - \ln(1 - A)]$$

$$\left. + \mu A^2[\mathrm{li}\,(\xi^4) - \ln(-A)]\right\},$$

where $A = k^2/b(\gamma + 1)$, $B = 2ik/b(\gamma + 1)$, $\mu = B + a/b$, $\xi = (x + \mu)/\mu$ and li is the logarithm-integral.

The results for pitching oscillations ($f_2 = x - a$) are

$$\varphi_2^{(1)} = -\delta_2(1 + kx)(\sigma^2 x^2 - y^2)^{1/2}$$

$$\varphi_2^{(2)} = -\frac{\delta_2 \sigma^2(\gamma + 1)\,b}{8}\,(\sigma^2 x^2 - y^2)^{1/2}\left\{x[(x + \mu)(2 + 3ikx) - Ax(1 + ikx)]\right.$$

$$\times \left[\ln\frac{\sigma^2 x^2(\gamma + 1)\,b}{16} + 2C - 1\right] - ik(x + \mu)^3(3 - A)[\mathrm{li}\,(\xi^{4-3}) - \ln(3 - A)]$$ (19)

$$- (1 - 3ik\mu)(x + \mu)^2(2 - A)[\mathrm{li}\,(\xi^{4-2}) - \ln(2 - A)]$$

$$- \mu(3ik\mu - 2)(x + \mu)(1 - A)[\mathrm{li}\,(\xi^{4-1}) - \ln(1 - A)]$$

$$\left. + A\mu^2(1 - ik\mu)[\mathrm{li}\,(\xi^4) - \ln(-A)] + 2[ik(x + \mu) - \lambda(1 + ikx)]\left(\frac{y^2}{3\sigma^2} - \frac{x^2}{3}\right)\right\}.$$

The pressure coefficient may be obtained as for example in Landahl [7] from

$$C_p = -2\,\mathrm{Re}\,[(\varphi_x + ik\varphi)\exp(ikt)]|_{z=0}$$ (20)

and the generalized aerodynamic force coefficients by

$$L_{mj} = \frac{4}{\delta_m S}\iint_S (\varphi_{mx} + ik\varphi_m)_{z=0}\,f_j\,dx\,dy$$ (21)

where $m, j = 1, 2, f_1 = 1, f_2 = x - a$ and S is the wing planform area. φ_m is the unsteady potential due to mode m. For pitching about an axis $x = a$, $z = 0$, we have:

$$C_{M\dot{a}} = \frac{4}{k \, \delta S} \text{Im} \left[\iint_S (\varphi_x + ik\varphi)(x - a) \, dx \, dy \right] = \frac{1}{k} \text{Im} [L_{22} - a(L_{21} + L_{12}) + a^2 L_{11}], \quad (22)$$

where $\varphi = \varphi_2 - a\varphi_1$ and $L_{mj} = L_{mj}^{(1)} + L_{mj}^{(2)} + \cdots$.
The terms $L_{mj}^{(1)}$, $L_{mj}^{(2)}$, ... are given by Sapunkova [10]:

$$\frac{L_{11}^{(1)}}{2\pi\sigma} = -ik + \frac{k^2}{3}, \quad \frac{L_{21}^{(1)}}{2\pi\sigma} = -1 - \frac{4}{3}ik + \frac{k^2}{4}, \quad (23)$$

$$\frac{L_{12}^{(1)}}{2\pi\sigma} = -\frac{2}{3}ik + \frac{k^2}{4}, \quad \frac{L_{22}^{(1)}}{2\pi\sigma} = -\frac{2}{3} - ik + \frac{k^2}{5}, \quad (24)$$

$$\frac{L_{11}^{(2)}}{2\pi\sigma} = \frac{k(\gamma + 1)b\sigma^2}{4} \left\{ \left[\ln \frac{\sigma^2 b(\gamma + 1)}{16} + 2C - 1 \right] \left[k\left(\frac{1}{5} + \frac{\mu}{4} - \frac{\lambda}{10} \right) - i\left(1 + \mu - \frac{\lambda}{2} \right) \right] \right.$$

$$- \frac{2k}{25} - \frac{\mu k}{8} + \frac{\lambda k}{20} - \frac{i\lambda}{4} + \frac{\lambda - 2}{2} \left[k\left(\frac{1}{5} + \frac{\mu}{2} + \frac{\mu^2}{3} \right) - i(1 + \mu)^2 \right] [\text{li}\,(\xi_0^{\lambda - 2}) - \ln(2 - \lambda)]$$

$$+ \mu(1 - \lambda) \left[k\left(\frac{1}{4} + \frac{\mu}{3} \right) - i(1 + \mu) \right] \cdot [\text{li}\,(\xi_0^{\lambda - 1}) - \ln(1 - \lambda)] + \frac{\mu^2 \lambda}{2} \left(\frac{k}{3} - i \right)$$

$$\times [\text{li}\,(\xi_0^{\lambda}) - \ln(-\lambda)] + k\mu^5 \left[-\frac{3 + \lambda}{60} \text{li}\,(\xi_0^{\lambda + 3}) + \frac{\lambda + 2}{12} \text{li}\,(\xi_0^{\lambda + 2}) \right.$$

$$- \frac{\lambda + 1}{6} \text{li}\,(\xi_0^{\lambda + 1}) + \frac{\lambda}{6} \text{li}\,(\xi_0^{\lambda}) + \frac{1 - \lambda}{12} \text{li}\,(\xi_0^{\lambda - 1}) - \frac{2 - \lambda}{60} \text{li}\,(\xi_0^{\lambda - 2})$$

$$\left. \left. + \ln \frac{(\lambda + 3)^{(\lambda + 3)/60} (\lambda + 1)^{(\lambda + 1)/6} (\lambda - 1)^{(\lambda - 1)/12}}{(\lambda + 2)^{(\lambda + 2)/12} \lambda^{\lambda/6} (\lambda - 2)^{(\lambda - 2)/60}} \right] \right\}, \quad (25)$$

$$\frac{L_{12}^{(2)}}{2\pi\sigma} = -\frac{ik\sigma^2(\gamma + 1)b}{4} \left\{ \left(\ln \frac{\sigma^2 b(\gamma + 1)}{16} + 2C - 1 \right) \left[\frac{4}{5} + \frac{3\mu}{4} - \frac{2\lambda}{5} + ik\left(\frac{1}{6} + \frac{\mu}{5} - \frac{\lambda}{12} \right) \right] \right.$$

$$+ \frac{2}{25} + \frac{\mu}{8} + \frac{4\lambda}{25} + ik\left(\frac{5\lambda}{72} - \frac{2\mu}{25} - \frac{1}{18} \right) + (\lambda - 2) [\text{li}\,(\xi_0^{\lambda - 2}) - \ln(2 - \lambda)]$$

$$\times \left[\frac{2}{5} + \frac{3\mu}{4} + \frac{\mu^2}{3} + ik\left(\frac{1}{12} + \frac{\mu}{5} + \frac{\mu^2}{8} \right) \right] - \mu(\lambda - 1) [\text{li}\,(\xi_0^{\lambda - 1}) - \ln(1 - \lambda)]$$

$$\times \left[\frac{3}{4} + \frac{2\mu}{3} + ik\left(\frac{1}{5} + \frac{\mu}{4} \right) \right] + \mu^2 \lambda [\text{li}\,(\xi_0^{\lambda}) - \ln(-\lambda)] \left(\frac{1}{3} + \frac{ik}{8} \right) - \frac{ik(\lambda + 4)}{120} \mu^6 \text{li}\,(\xi_0^{\lambda + 4})$$

$$+ \frac{\mu^5(\lambda + 3)}{60} \text{li}\,(\xi_0^{\lambda + 3})(1 + 3\mu ik) - \mu^5(\lambda + 2) \text{li}\,(\xi_0^{\lambda + 2}) \left(\frac{1}{12} + \frac{ik\mu}{8} \right) + (1 + \lambda) \mu^5 \text{li}\,(\xi_0^{1 + \lambda})$$

$$\times \left(\frac{1}{6} + \frac{ik\mu}{6} \right) - \lambda \text{li}\,(\xi_0^{\lambda}) \mu^5 \left(\frac{1}{6} + \frac{ik\mu}{8} \right) + \mu^5(\lambda - 1) \text{li}\,(\xi_0^{\lambda - 1}) \left(\frac{1}{12} + \frac{ik\mu}{20} \right)$$

$$- \mu^5(\lambda - 2)\, \text{li}\,(\xi_0^{\lambda - 2}) \left(\frac{1}{60} + \frac{ik\mu}{120} \right) - \mu^5 \ln \frac{(\lambda + 3)^{(\lambda + 3)/60} (\lambda + 1)^{(\lambda + 1)/6} (\lambda - 1)^{(\lambda - 1)/12}}{(\lambda + 2)^{(\lambda + 2)/12} \lambda^{\lambda/6} (\lambda - 2)^{(\lambda - 2)/60}}$$

$$+ ik\mu^6 \ln \frac{(\lambda + 4)^{(\lambda + 4)/120} (\lambda + 2)^{(\lambda + 2)/2} \lambda^{\lambda/6} (\lambda - 2)^{(\lambda - 2)/120}}{(\lambda + 3)^{(\lambda + 3)/20} (\lambda + 1)^{(\lambda + 1)/6} (\lambda - 1)^{(\lambda - 1)/20}} \Bigg\}, \tag{26}$$

where $\xi_0 = \dfrac{1 + \mu}{\mu}$, $L_{mj}^{(1)}$ represents the slender-wing solution and $L_{mj}^{(2)}$ is the term of relative order $k\sigma^2$. For the profiles with the relatively thickness parameter $\tau = 0.12$, $x|_{z\max} = 0.7$ and $x|_{z\max} = 0.3$, Sapunkova [10] obtained $a = 0.6409$, $b = 0.1700$ and respectively $a = 2.1313$, $b = -3.5177$.

In Fig. 1 the results from Landahl [7], Ruo [9] and Sapunkova [10] for thin delta wings with different σ pitching at an axis $x = a$ are plotted. It shows that the Sapunkova-method can be used to provide the case for $k \to 0$ where the linearized theory of Landahl [7] fails. It also shows that the thickness effect yields the same influence trend as that of the frequency on damping-in-pitch, and the influence of σ is far more pronounced than that of either frequency or thickness. By increasing σ the influence of term χ_{xx} is intensified accordingly.

It is of interest to note from [9] that the solution for constant λ-values, as being based on the equivalent axis symmetrical body of rotation, is not fully adequate [8]. Here it is considered to depend on the total cross-sectional area distribution along the x-axis only and not on the cross-sectional shape of the profiles in x- and y-direction of the wing. This deficiency can be improved by the Sapunkova-method.

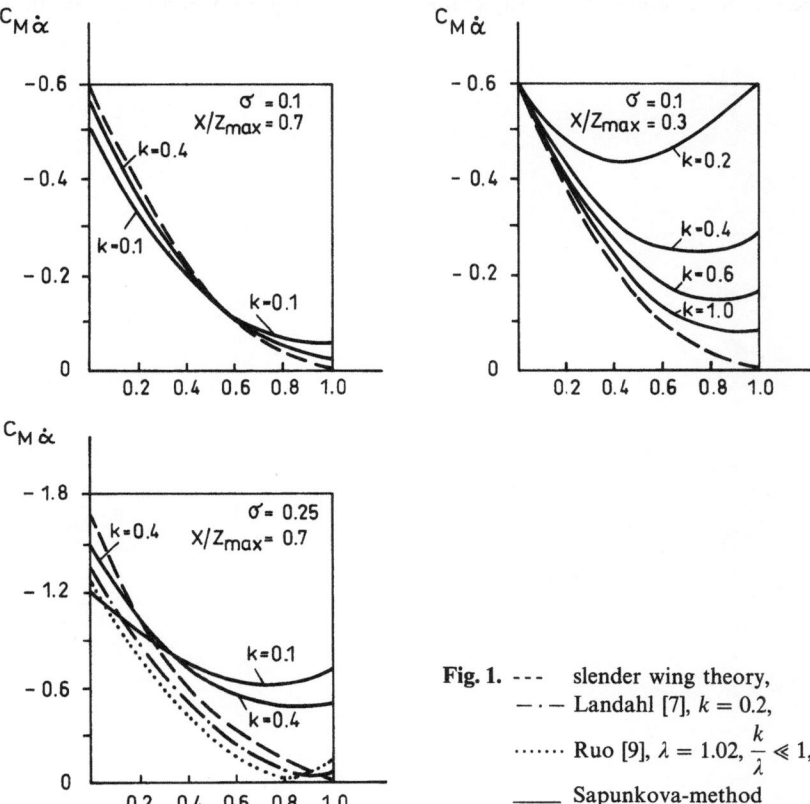

Fig. 1. --- slender wing theory,
   $- \cdot -$ Landahl [7], $k = 0.2$,
   .... Ruo [9], $\lambda = 1.02$, $\dfrac{k}{\lambda} \ll 1$,
   ____ Sapunkova-method

References

[1] Oswatitsch, K., Keune, F.: The flow around bodies of revolution at Mach number 1. Proceedings of the conference on high-speed aeronautics, pp. 113–131. Polytechnic Institute of Brooklyn, New York (1955).

[2] Hosokawa, I.: A simplified analysis for transonic flows around thin bodies. In: Symposium Transsonicum (Oswatitsch, K. ed.), pp. 184–199. IUTAM Symposium, Aachen, Germany 1962. Berlin Göttingen Heidelberg: Springer 1964.

[3] Teipel, I.: Die instationären Luftkräfte bei der Machzahl 1. Z. Flugwiss. 12, 6–14 (1964).

[4] Ruo, S. Y., Liu, D. D.: Calculation of stability derivatives for slowly oscillating bodies of revolution at Mach 1. Lockheed Missiles and Space Co. LMSC/HREC D162375 (1971).

[5] Zierep, J.: Der Äquivalenzsatz und die parabolische Methode für schallnahe Strömungen. ZAMM 45, 19–27 (1965).

[6] Adams, M. C., Sears, W. R.: Slender-body theory – review and extension. J. Aero. Sci. 20, 85–98 (1953).

[7] Landahl, M. T.: Unsteady transonic flow. New York: Pergamon Press 1961.

[8] Kimble, K. R., Liu, D. D., Ruo, S. Y., Wu, J. M.: Unsteady transonic flow analysis for low aspect ratio, pointed wings. AIAA J. 12, 516–522 (1974).

[9] Ruo, S. Y.: Slowly oscillating pointed low aspect ratio wings at sonic speed with thickness effect. ZAMM 54, 119–121 (1974).

[10] Sapunkova, O. M.: Slowly oscillating pointed low aspect ratio wings at sonic speed. Aerodynamica (Transonic Flow), University of Saratov, 52–60 (1981).

[11] Liu, D. D., Platzer, M. F., Ruo, S. Y.: Unsteady linearized transonic flow analysis for slender bodies. AIAA J. 15, 966–973 (1977).

[12] Oswatitsch, K.: Similarity and equivalence in compressible flow. Advances in applied mechanics, Vol. VI, pp. 153–271. New York: Academic Press 1960.

Author's address: Professor Dr. S. Turbatu, University of Bucharest, Str. BOCSA 3, Sector 2, Cod. 70224, Bucharest, Romania

Acta Mechanica (1994) [Suppl] 4: 163—170
© Springer-Verlag 1994

Part 4: Hypersonic Flow

Hypervelocity flow over spheres

H. Hornung, C. Wen, Pasadena, California, and **G. Candler,** Minneapolis, Minnesota

Summary. Some aspects of the principle of binary scaling of hypervelocity flows with chemical reactions are discussed and tested both numerically and experimentally. The experiments, obtained in a new free-piston shock tunnel, show the value and limitations of binary scaling in very good agreement with the numerical computations. The use of spherical models eliminates end-effect problems previously encountered with cylindrical models. Global quantities, such as the bow shock stand-off distance, follow binary scaling very well. The results include differential interferograms and surface heat transfer measurements of nitrogen, air and carbon dioxide flows.

1 Introduction

The experimental study of the field of high-enthalpy real-gas dynamics, as it applies to transport into and back from space through planetary atmospheres, requires facilities that are at the upper end of the scale of what is possible at universities. The reason for this becomes clear from the features of the problem: Take the atmosphere of earth, where the orbital speed is 8 km/s, and the critical heating rate, the most serious concern of designers, occurs at approximately 6 km/s. The thermal energy per unit mass in the stagnation region of a body traveling at this speed is 18 MJ/kg. This may be compared to the characteristic energies of vibration and dissociation of the molecular components of air:

$$D_{N_2} = 33.6 \text{ MJ/kg} \qquad E_{vN_2} = 0.992 \text{ MJ/kg}$$

$$D_{O_2} = 15.5 \text{ MJ/kg} \qquad E_{vO_2} = 0.579 \text{ MJ/kg}$$

$$D_{NO} = 20.9 \text{ MJ/kg} \qquad E_{vNO} = 0.751 \text{ MJ/kg}.$$

It is clear that very substantial chemical activity will take place and the vibrational energy may often be considered to be in equilibrium with the translational energy. Because of the complex idiosyncrasies of the substance air at such high specific energy, it is not possible to simulate them with other gases, and we are necessarily forced to reproduce the actual flow speed that the body sees.

A second feature of the problem is that in many parts of the flow field around a body, the chemical reactions occur at speeds which are comparable to the local flow speed, so that it is necessary to study not only equilibrium high-enthalpy effects but also those that arise because of the finiteness of the chemical reaction rates. These, by virtue of the flow speed, introduce characteristic lengths into the problem, which depend on the properties of the gases as well as on the flow variables. Two kinds of chemical reactions may broadly be distinguished: Binary and ternary reactions, for which the characteristic lengths vary as the inverse of the density and as the inverse of the square of the density respectively. In a simulation, where one wishes to reproduce the ratio of body size to characteristic chemical length, it is therefore necessary to reproduce

$$\varrho L, \quad \text{or} \quad \varrho^2 L$$

depending on whether binary or ternary reactions are dominant in the phenomenon of interest. It is not possible to simulate situations involving both types correctly except at full scale.

Binary reactions dominate the features of blunt body flows, so that for the simulation of the important problems that occur there, we see that the scale of a facility has a lower bound that is dictated by the pressure at which one is prepared to operate. If the scaling ratio (model scale/prototype scale) is 1:100, the density scaling ratio has to be 100:1. Since the temperature has to be simulated correctly (assured by correct speed), the pressure also has to be 100 times that of the real flow. If the pressure is limited by one or more of strength, safety, noise, environment, all of which are concerns at universities, the size of the facility has a lower bound.

The above considerations formed the core of the constraints of the design of a new facility at GALCIT which had the specific aim of studying the fundamental features of high-enthalpy reacting flows. The facility was named T5, because it built heavily on the experience gained in the Australian facilities T1 to T4, which, like T5, use the principle of free-piston adiabatic compression of the driver gas of a shock tunnel to achieve the high shock speeds and densities required to generate the high enthalpy and reaction scaling. The shock tunnel has the additional essential feature that the test duration is sufficiently short to avoid destruction of the machine by melting. A more detailed description of T5 and its performance envelope, flow quality and repeatability may be found in Hornung [1]. The largest facility of this type has been completed in 1992 at Göttingen, see Eitelberg [2].

The features of hypervelocity blunt body flows and of the limits of binary scaling have been studied extensively theoretically. However, only very few experimental results are available. The most detailed of these are the interferometric studies of cylinder flows by Hornung [3]. The reason for the choice of cylinders in those experiments was that the density achievable in the facility used (T3) was insufficient to get good fringe resolution in optical interferograms of flows over spheres. The difficulty with "two-dimensional" flows is that they are an idealization that does not exist in reality. In the blunt body problem, the end effects on cylinder flows manifest themselves in interferograms in just the same manner as the finite-rate effects one wishes to study, and may therefore obscure them seriously. The size and density achievable in T5 has now extended the available parameter range sufficiently to avoid this problem by studying the axisymmetric situation, i.e., by using spheres as models. The following pages present the first steps of an investigation in T5, and of an accompanying theoretical effort, towards gaining more insight into the blunt body problem and the limits of binary scaling.

At the same time, the development of numerical methods for the computation of blunt body flow fields has become very much quicker since the 1970's, so that even interferograms of three-dimensional chemically reacting flows can be computed with reasonable speeds, see e.g., Rock et al. [4]. For laminar flow, the viscous case is also accessible, so that heating rates too may be compared with theoretical models, see e.g. Candler [5]. The experiments with spheres to be described in the following pages therefore are equipped with heat transfer gauges and differential interferometry for the quantitative exploration of the flow field and comparison with computation. Some comparisons of theoretical and experimental results were also made by Macrossan [6] who showed that the simple model of Hornung [7] (which is based on matched asymptotic expansions of the near-equilibrium case at relatively small densities, where the recombination rate may be neglected) has restricted applicability, especially in the tunnel, where the recombination rate is too high if binary scaling is applied.

2 Design of the experiment

Because of the relative ease with which the flow may be computed, a computational exploration of the parameter space affords an excellent method of designing an experiment. With this in mind, the inviscid reacting flow over a sphere was computed with the Candler code. This program uses a finite-volume method and is described in detail in Candler [5]. The flow field is described by coupled partial differential equations for the conservation of species, mass, mass-averaged momentum, vibrational energy of each diatomic species and total energy. The steady-state solution to these fully coupled equations is obtained for different gases characterized by a two-temperature model using an implicit Gauss-Seidel line relaxation technique. The success of this code in reproducing in great detail those experimental results that are available in the form of field measurements is impressive, and extensive documentation of examples exists. The code is actually set up for viscous flow, but was used in the context of this section in the inviscid form. Figure 1 shows an example of such a computation for the case of a nitrogen flow, in the form of the density field and of the infinite-fringe interferogram that would be observed in such a flow. As may be seen, a large number of fringes would be observed, so that even a relatively insensitive differential interferogram would give good quantitative resolution of the flow field. The case shown corresponds to a representative condition achievable in T5.

Computations of this type may be used to plot the dimensionless shock stand-off distance, for example, and this is shown in a plot of this quantity against the dimensionless reaction rate at a point immediately downstream of the normal shock. This is shown in Fig. 2 for the range of values achievable in the tunnel. In the regime where binary scaling is applicable, this curve should not separately depend on ϱ and d, but only on their product. As may be seen from the scatter in the diagram, this is not quite the case, but the stand-off distance does not seem to be strongly affected by the failure of binary scaling.

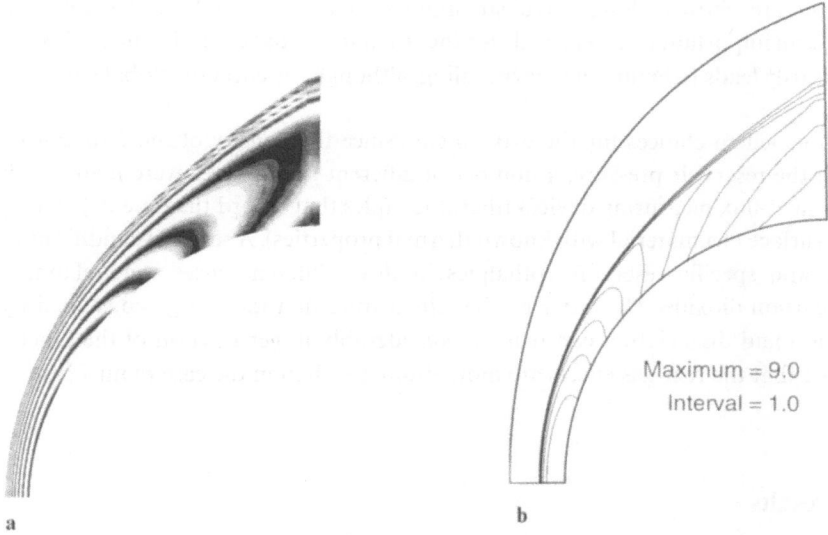

a   b

Fig. 1. Result of a computation of the flow field over a sphere. Free stream conditions: 5.08 km/s, nitrogen, 0.04 kg/m^3, 16.56 MJ/kg. **a** Infinite fringe interferogram constructed from the numerical solution. **b** Lines of constant density. The numbers on the right indicate the density values as multiples of the free-stream density. The border of the diagram on the right is the edge of the computational domain. The glitch at the right is an artefact of the interpolation routine

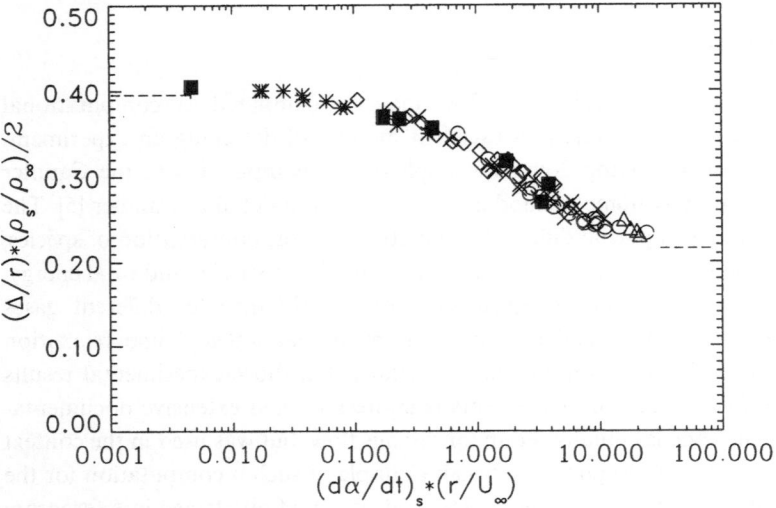

Fig. 2. Stand-off distance variation with reaction rate parameter. The abscissa is the reaction rate parameter, defined as the dissociation rate immediately after the shock scaled by sphere radius and flow speed. The ordinate is the dimensionless stand-off distance scaled by the density ratio at the shock. The square symbols in this diagram represent experimental results in T5, while the other symbols are computational results. The asymptotes shown are for equilibrium and frozen flow respectively

A second, more sensitive test is to plot the reaction rates against the distance along a streamline. If the model of Hornung [7] applies, the reaction stops while the gas is still dissociating, because of the temperature drop brought about by the curvature of the shock. As had been shown by Macrossan [6], this assumption fails in the higher density flows considered here. This may also be seen in Fig. 3, where dissociation rate, recombination rate and dissociation fraction are shown along three streamlines in a nitrogen flow. Clearly, the recombination plays an important role, especially for the streamlines that cross the shock close to the axis. This necessarily leads to failure of binary scaling, although the effect on global variables may be small.

The computations led to choices for the experiment. Since the density of the flow can be varied by changing the reservoir pressure, a number of different-size spheres were made, each equipped with four heat-flux measuring devices (thermocouples that record the time-dependent temperature of the surface of a material with known thermal properties). A series of conditions at different pressures and specific reservoir enthalpies in three different gases were planned: Nitrogen, air and carbon dioxide. The latter gas has the feature that the energy consumed by vibrational excitation and dissociation can reach a considerably larger fraction of the kinetic energy of the flow so that the real-gas effects are more dramatic, than in the case of nitrogen.

3 Experimental results

The experiments are still ongoing at the time of writing, so that the range of parameters studied is still incomplete, and the quality of the interferograms needs improvement. The first interest was the high-density range, where near-equilibrium cases could be studied in contrast to the situation in T3, so that this presentation is limited to that region.

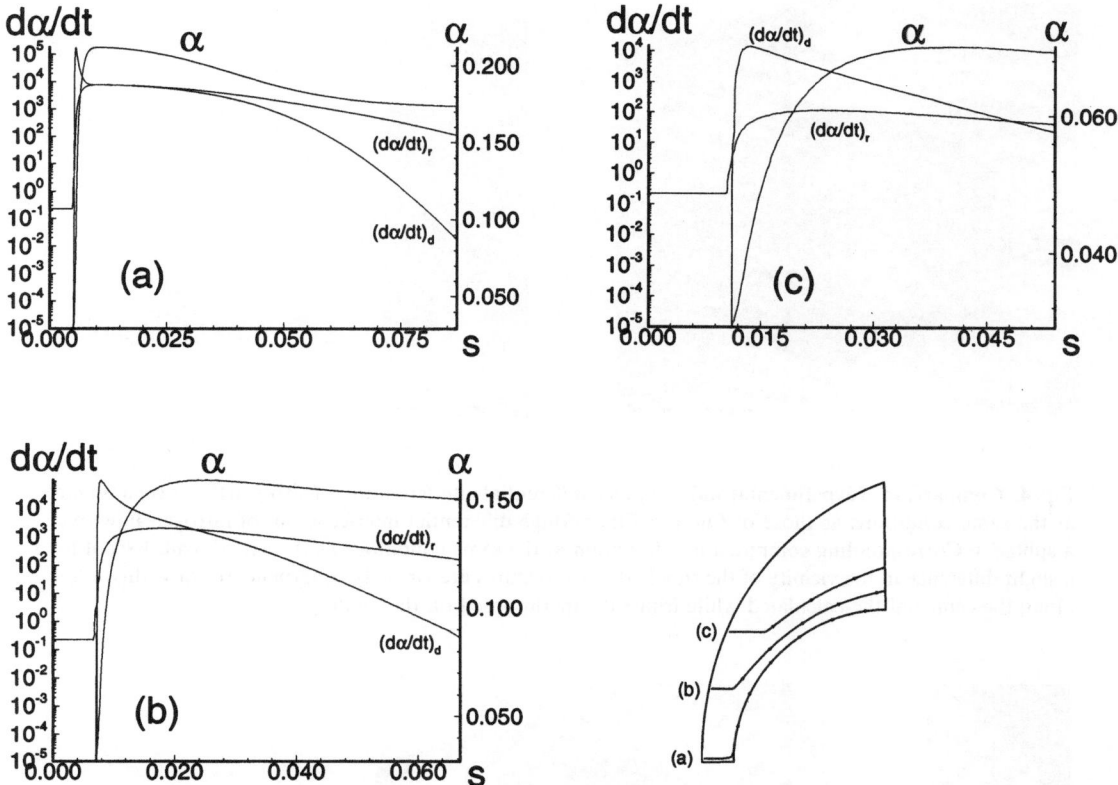

Fig. 3. Reaction rate and composition variation along three selected streamlines. The streamlines are labelled *a*, *b* and *c* in the diagram on the right. The kink in the streamline indicates the shock location. The dissociation rate and recombination rate are plotted separately, and the dissociation fraction, α reaches a maximum at the cross-over point. As may be seen, recombination may not be neglected for the inner streamlines

Figure 4 shows a differential interferogram of the flow of nitrogen over a 10 cm diameter sphere at a specific reservoir enthalpy of 16 MJ/kg and a reservoir pressure of 60 MPa. Differential interferometry measures gradient of optical path length. Specifically, it measures the component of the gradient in the image plane and in a direction that can be chosen by rotating an optical element in the setup. The parallel fringes in the undisturbed free stream of the photograph are oriented at right angles to the direction of this component. In order to translate the fringe shift observed in the flow field behind the shock into a refractive index field, from which the density might be inferred, it would be necessary to perform an inversion of the Abel integral equation using the measured fringe shift as input. While this does not present a problem, the result is still ambiguous, since the refractive index may not be interpreted directly as density in a gas whose composition is not uniform. In nitrogen flow, this problem is relatively mild, but if one deals with air, where some 8 species may be expected to be present, the problem is harder. A better comparison with computed results is to compute the differential interferogram from the calculated flow field, using the computed density and composition and the known refractive indices of the components of the gas mixture.

This was done by first calculating the infinite fringe interferogram and subtracting from it a displaced copy of itself. This gives a double image of the boundaries, such as the body, of course,

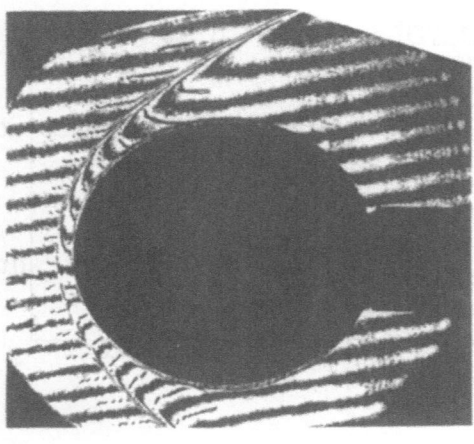

a   b

Fig. 4. Comparison of experimental and computed differential interferograms of nitrogen flow over a sphere, at the same conditions as those of Fig. 1. **a** Finite-fringe differential interferogram of nitrogen flow over a sphere. **b** Corresponding computed interferogram at the same conditions as the experiment. Except for a slight difference in the vicinity of the shock, the two pictures are virtually congruent. To show this, a line along the center of the calculated white fringe is superimposed on the photograph

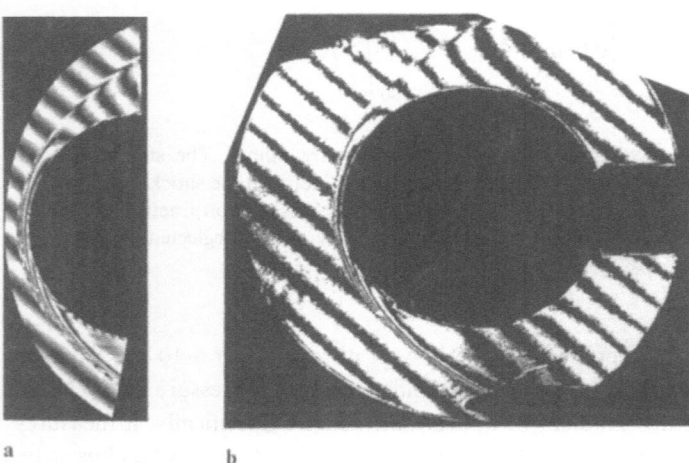

a   b

Fig. 5. Comparison of differential interferograms in the case of carbon dioxide flow over a sphere. Free stream: 3.18 km/s, 0.106 kg/m³, 8.52 MJ/kg. In this case advantage has been taken of the symmetry of the flow to get both components of the gradient measured by the differential interferogram, by setting the angle of the fringes at 45 deg. in the free stream. **a** Computation, **b** experiment

just as in the experimental differential interferogram. This computed differential interferogram is also shown in Fig. 4 for comparison with the measured one. As may be seen, the features of the photo are faithfully reproduced by the computation, both qualitatively and quantitatively. It should be noted that if the density of the computation is changed by 10%, this change can be resolved easily in the interferogram. An interesting feature of symmetrical flows is that the differential interferogram can be chosen to measure the component of gradients at 45 deg. to the axis (or plane) of symmetry, so that one half of the photo gives one component, and the other half the orthogonal component of the gradient.

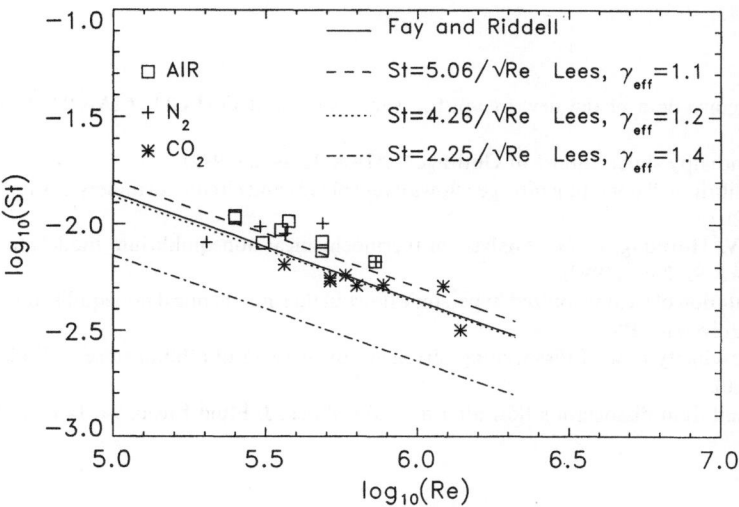

Fig. 6. Stagnation-point heat flux in dimensionless form, and comparison with Lees' theory, and Fay and Riddell's correlation

To show another example, Fig. 5 shows a similar exercise for a carbon dioxide flow. Again, the comparison shows excellent agreement. Clearly, viscous effects are not important in the determination of the major features of the flow field.

Viscous and thermal effects are, however, the main reason for studying these flows in the first place. The heat flux measurements are thus an important component of our study. With the extremely high values of heat flux encountered at the stagnation point of the sphere, the classical thin-film metal gauges are not suitable. Thermocouple gauges were found to serve our purpose very well. The maximum value of the heat flux measured was 60 MW/m². The heat transfer rate was normalized to form the Stanton number, and a plot of this against the Reynolds number based on the sphere diameter is shown in Fig. 6.

4 Conclusions

A free-piston shock tunnel was used to obtain new data on flow over spheres at high enthalpy and Reynolds number, where non-equilibrium dissociation is important. By using spheres, the end effects of previous data on flow over cylinders were eliminated and much better agreement with computations was observed, in interferometric flow-field measurements. Though computations show that recombination is important in parts of the flow field, binary scaling is a satisfactory model for global quantities such as the stand-off distance. Stagnation-point heat flux measurements broadly follow expected behavior, though systematic differences between different gases are apparent.

Acknowledgement

This work was supported by AFOSR Grant No. F49610-92-J-0110 monitored by Dr. L. Sakell.

References

[1] Hornung, H. G.: Performance data of the new free-piston shock tunnel at GALCIT. AIAA 92-3943 Nashville (1992).

[2] Eitelberg, G.: The high-enthalpy shock tunnel in Göttingen. AIAA 92-3942 (1992).

[3] Hornung, H. G.: Nonequilibrium dissociating nitrogen flows over spheres and circular cylinders. J. Fluid Mech. **53**, 149−176 (1972).

[4] Rock, S. G., Candler, G. V., Hornung, H. G.: Analysis of thermochemical nonequilibrium models for carbon dioxide flows. AIAA 92-2852 (1992).

[5] Candler, G. V.: The computation of weakly ionized hypersonic flows in thermo-chemical nonequilibrium. Ph. D. Thesis, Stanford University 1988.

[6] Macrossan, M. N.: Hypervelocity flow of dissociating nitrogen downstream of a blunt nose. J. Fluid Mech. **217**, 167−202 (1990).

[7] Hornung, H. G.: Nonequilibrium dissociating flow after a curved shock. J. Fluid Mech. **74**, 143−160 (1976).

Authors' addresses: Professor Dr. H. Hornung and Dr. C. Wen, Graduate Aeronautical Labs., California Institute of Technology, Pasadena, CA 91125, and Prof. Dr. G. Candler, Dept. of Aerospace Engineering and Mechanics, University of Minnesota, 110 Union St. S. E., Minneapolis, MN 55455, U.S.A.

Acta Mechanica (1994) [Suppl] 4: 171—182

Basics of aerothermodynamics

H. Oertel jr., Braunschweig, Federal Republic of Germany

Summary. The article provides an introduction to the basics of aerothermodynamics with the example of the reentry of an orbital space capsule into the earth's atmosphere. The gaskinetical and continuum mechanical methods and phenomena, which are required for the prediction of the heat flux into the heat shield of the reentry capsule, will be presented. The emphasis in the discussion of the results is on the chemical sensitivity studies and the mathematical formulation of the physical-chemical boundary conditions along the reentry trajectory. In this case, the chemical modelling of the interaction of the chemically reacting hot gas with the technical surface of the heat shield of the reentry capsule and its implementation into developed gaskinetical and continuum mechanical numerical simulation methods are the keys to the design of heat shields for future reusable reentry vehicles. The article attempts to provide a shortened summary of the scientific results from a text book on the same subject by H. Oertel jr. et al. [1].

1 Introduction

We follow the reentry trajectory of a ballistic reentry capsule, as sketched in Figs. 1 and 2, from the orbit until it lands on the earth. It flies through the gaskinetical regime of the upper atmosphere. With decreasing altitude, the increasing density of the atmosphere causes the reentry capsule to decelerate from the orbital speed to enter the continuum mechanical regime of the reentry trajectory, which is associated with a heating of the flowing gases up to $11\,000$ K and a thermal stress on the capsule of up to $2{,}5$ MW/m^2. The ballistic reentry of the reentry capsule in question lasts for about 100 seconds. In the upper atmosphere the gas around the reentry vehicle becomes so hot that the oxygen dissociates almost completely and the nitrogen molecules partially dissociate. At altitudes below 30 km, the decisive factor is the vibrational excitation of N_2 and O_2 molecules, which determines the aerothermodynamics of the reentry flow in addition to classical gasdynamics. The gasdynamical flow field around the reentry capsule is sketched in Fig. 3.

The oncoming hypersonic flow causes a bow shock around the vehicle. Distinguished points in the flow field are the stagnation points on the vehicle nose and in the wake, where respective maximum temperatures are reached. The curved shock in front of the vehicle causes an entropy layer, which in interaction with the vehicle boundary layer, can cause another relative temperature maximum on the vehicle. The wake of the reentry capsule is characterized by the flow acceleration in the Prandtl-Meyer expansion fan and the region of reverse flow. The vehicle boundary layer continues as a shear layer into the wake region and leads through the recompression shock at high altitudes in the atmosphere to the laminar-turbulent transition in the wake. As the altitude decreases along the reentry trajectory, the transition zone moves over the shear layer towards the vehicle, up to the stagnation point on the vehicle. Due to the laminar-turbulent transition in the boundary layer, the heat flux into the heat shield of the reentry capsule rises.

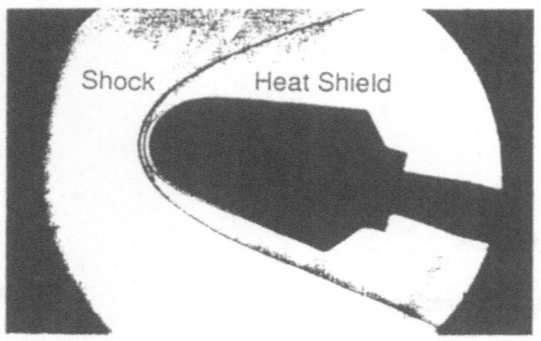

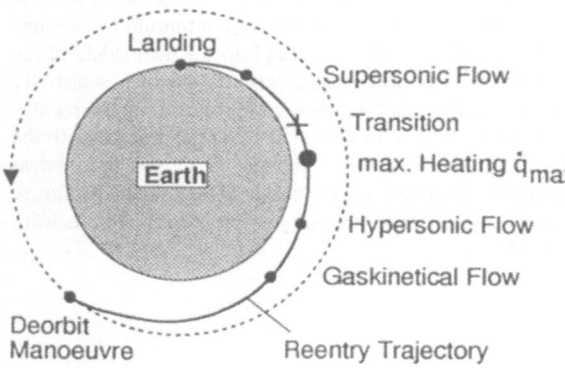

Fig. 1. Gasdynamics of a reentry capsule (Gun Tunnel Experiment $M_\infty = 12$) and principal sketch of the reentry trajectory

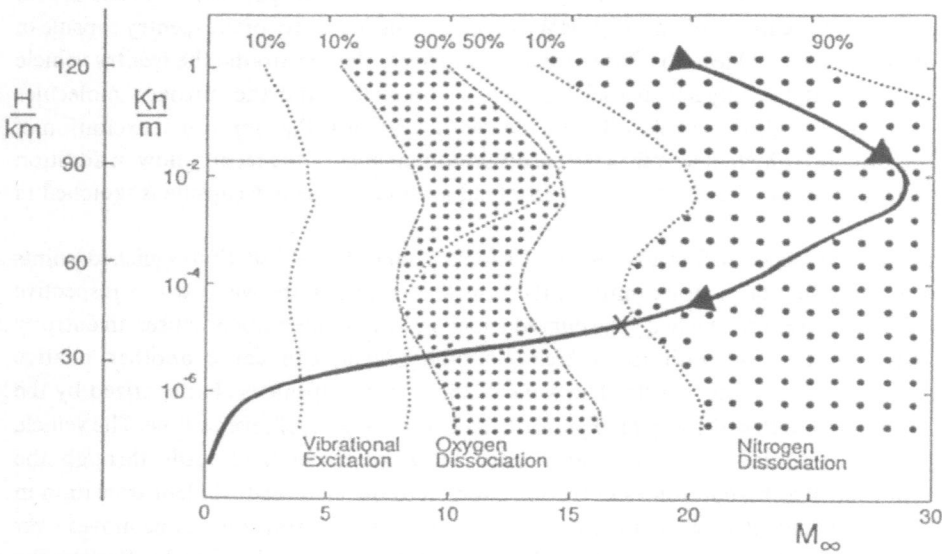

Fig. 2. Sketch of the reentry trajectory including the major chemical reactions in the earth's atmosphere. Ballistic reentry: ▲ Regimes of numerical simulations and chemical sensitivity studies, x laminar-turbulent transition regime

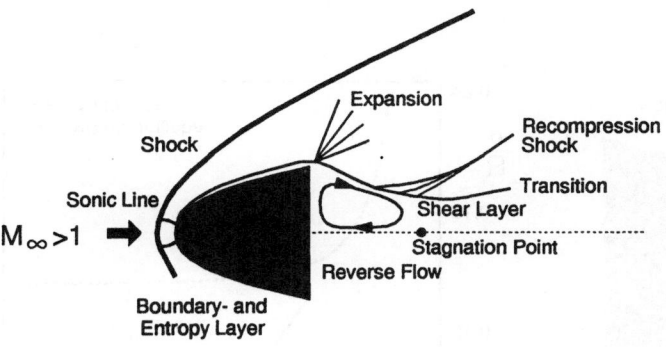

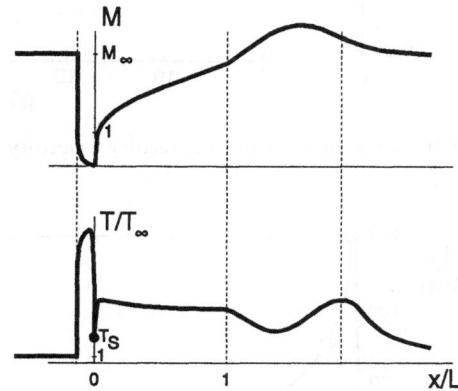

Fig. 3. Gasdynamical sketch of the reentry flow around a capsule. Qualitative Mach-number and temperature profiles along the stagnation point boundary-layer and wake-shear-layer streamline

The organization of this article is derived from the sequence of the aerothermodynamic phenomena along the reentry trajectory. First, the gaskinetical and gasdynamical fundamentals of the reentry flow problems will be described. Next, the selection of the chemical models in the flow field by means of chemical sensitivity studies follows. The stability analysis of the boundary layer flow allows for the determination of the transition regime on the reentry trajectory. The gaskinetical and continuum mechanical numerical simulation computations lead, in conjunction with reentry free flight experiments, to engineering design methods for future reusable heat shields of reentry capsules.

2 Gasdynamical basics

The gasdynamics of the reentry problem is determined by the bow shock in front of the reentry capsule as shown in Fig. 1. Figure 4 illustrates the general sketch of the shock visualized by interferometry in Fig. 1. Along the stagnation stream line, the Mach number jumps, as seen in Fig. 3, across the normal shock to subsonic values. This part is mathematically elliptical. Due to the body bluntness, the flow is accelerated across the sonic line again to supersonic speeds in the hyperbolic part of the flow field. The shock heats the gas so that the molecular vibration of the molecules is excited and the gas partially dissociates. The associated increase in densitiy causes a decrease of the shock standoff distance by a factor of up to 2. For the reentry trajectory under consideration, the heat flux on the heat shield of the reentry capsule rises, as seen in Fig. 5, to a value of $2\,460$ kW/m^2 at an altitude of 45 km. The calculated gas temperature behind the shock is $6\,500$ K. The surface temperature of the approximate radiation adiabatic surface of the heat shield is assumed to be $2\,550$ K.

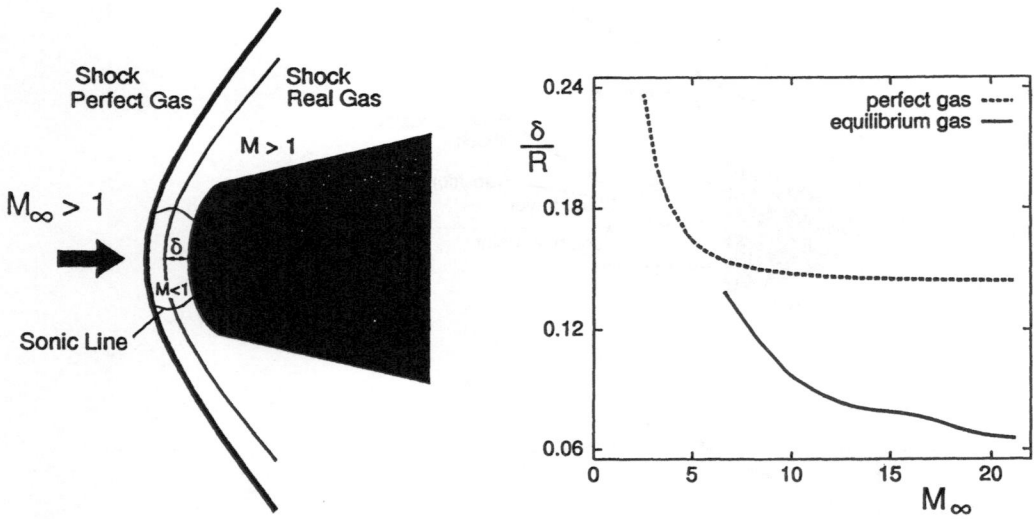

Fig. 4. Principle sketch of the shock distance δ and numerical results along the reentry trajectory of Fig. 2

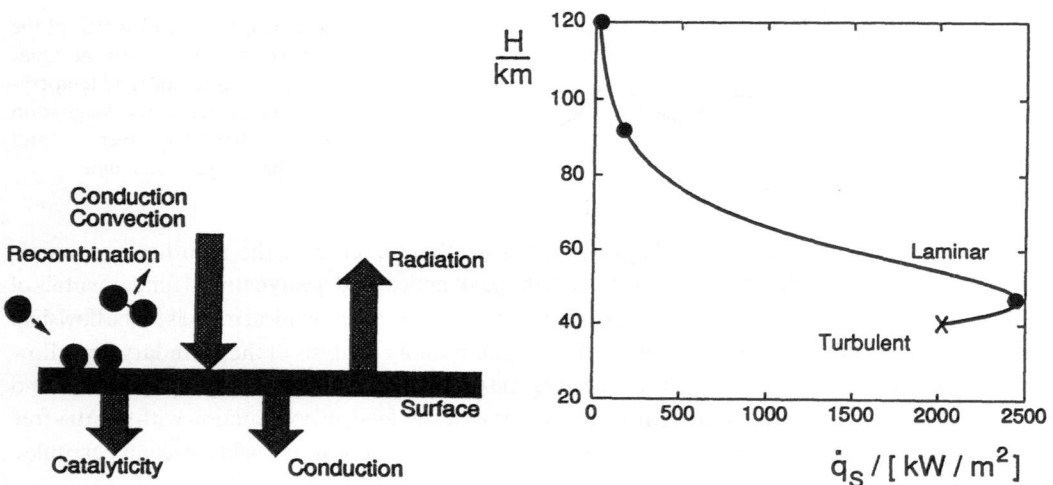

Fig. 5. Boundary conditions at the heat shield surface and calculated stagnation point heat flux

3 Aerothermodynamical basics

A requirement for the numerical simulation of the hot flow field and the prediction of the heat flux into the heat shield is the determination of a chemical model. This changes along the reentry trajectory. In the gaskinetical regime at an altitude of 120 km there is little chemical activity. The chemical composition remains essentially constant along the stagnation streamline. In the transitional regime at an altitude of 92 km there is a significant amount of dissociation of oxygen and nitrogen. At an altitude of 45 km, at the point of maximal heat transfer into the heat shield, NO and N_2 are in chemical nonequilibrium. The oxygen dissociation and the associated exchange reactions are in thermodynamical equilibrium. The oxygen-recombination becomes sensitive only in the vehicle boundary layer. The basics for the determination of the chemical models are chemical sensitivity studies, J. Warnatz [2]. Here, the Damköhler number of the respectively considered chemical reaction

$$\text{Da} = \frac{\text{Characteristic flow time}}{\text{Time required for equilibrium}} = \frac{\tau_F}{\tau_R}$$

determines which of the chemical reactions in the considered flow field are in thermodynamical equilibrium or respectively in relaxation. High values for the Damköhler number mean that thermodynamical equilibrium, compared to the characteristic time of the flow field, is rapidly reached. Respectively, the numerical simulation can be performed in thermodynamical equilibrium. As we single out the oxygen dissociation reaction, the Damköhler number for the considered points of the reentry trajectory has the following values: H = 92 km, Da = O(1); H = 75 km, Da $\approx$ 25; H = 45 km, Da $\approx$ 100. From this, we can make the following conclusion: At an altitude of 92 km, the oxygen dissociation is in nonequilibrium and at lower altitudes it is in equilibrium.

The method of the sensitivity analysis deals with the time history of concentration changes of suddenly heated gases across a shock. The considered fluid elements in front and behind the shock are sketched in Fig. 6. The chemical model from C. Park [3] with the following reactions is used:

$$O_2 + M \rightleftharpoons O + O + M, \quad N_2 + M \rightleftharpoons N + N + M, \quad NO + M \rightleftharpoons N + O + M, \quad .$$

$$N_2 + O \rightleftharpoons NO + N, \quad NO + O \rightleftharpoons O_2 + N, \quad N + O \rightleftharpoons NO^+ + e^-.$$

There, M represents the collision partners O_2, N_2, NO, O, N, NO^+ and e^-. For the reaction rate coefficients k_f, a modified Arrhenius-Ansatz is made, which, in the scope of a two-temperature model, takes into account the translation and vibration temperature. A total of 24 forward reaction coefficients and 6 equilibrium constants $K(T)$ are taken into account, which have measurement errors of up to one order of magnitude (see e.g. H. Oertel jr. [4]). The influence of the measurement uncertainties on the numerical simulation of the gas composition is determined through the sensitivity analysis.

Due to the heating across the bow shock, the molar concentrations $[x_1], ..., [x_n]$ of the individual air species change. We consider the time history of the concentration change along the stagnation streamline:

$$\frac{d[x_i]}{dt} = f_i([x_1], ..., [x_n], k_{f1}, ..., k_{fm})$$

with n = 7 species and m = 24 reaction rate coefficients k_{fm}.

For the determination of the influence of the reaction velocity on the mole concentration $[x_i](t)$, a sensitivity coefficient is defined:

$$W_{ij}(t) = \frac{1}{\sum\limits_{l=1}^{n} [x_l]} \frac{\partial [x_i]}{\partial (\ln k_{fi})}.$$

Coupling this definition with the governing equations of the chemical model yields the following differential equation:

$$\frac{\partial W_{ij}}{\partial t} = \frac{1}{\sum\limits_{l=1}^{n} [x_l]} \left(\frac{\partial f_i}{\partial (\ln k_{fj})} + \sum\limits_{k=1}^{n} \frac{\partial f_i}{\partial [x_k]} W_{kj} \right)$$

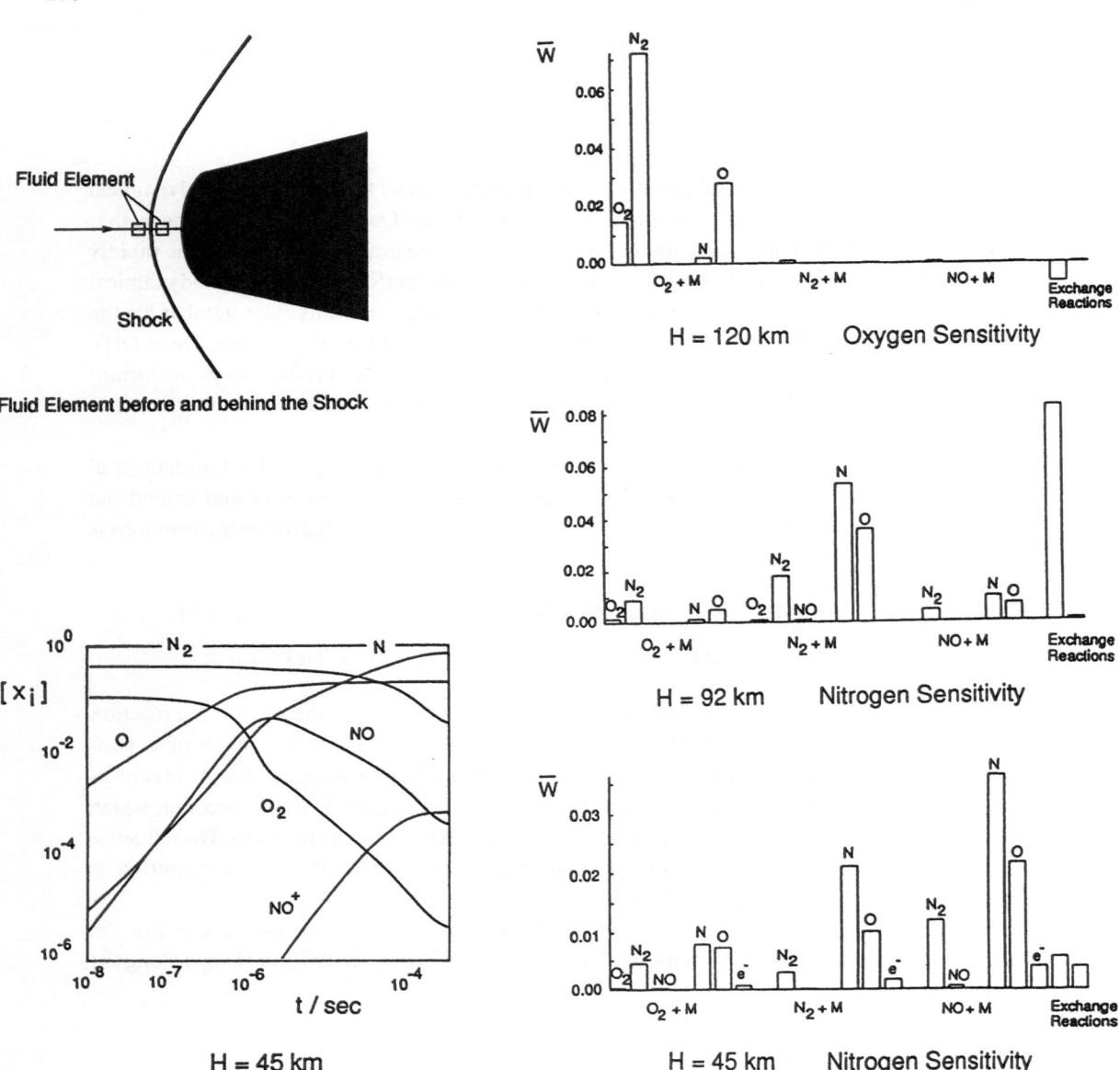

Fig. 6. Chemical sensitivity studies for the oxygen- and nitrogen-atom-concentration (M. Böhle et al. [5, 6])

for the determination of the sensitivity coefficients $W_{ij}(t)$. We define the time average of the sensitivity coefficients as

$$\overline{W_{ij}} = \frac{1}{T} \int_0^T W_{ij}(t) \, dt.$$

The chosen time interval $T = 4 \cdot 10^{-4}$ seconds (45 km altitude) results from the numerically calculated time histories of the considered volume element along the stagnation streamline (Sections 5 and 6). The significance of the time averaged sensitivity coefficients can be explained by the following consideration. A change of the reaction rate coefficient k_{fj} leads to a change of the mole concentration of a certain air component i expressed by the average value $\overline{W_{ij}}$. For example, the value $\overline{W_{ij}} = 0.03$ means a change of the mole concentration of species i, with respect to the total mole concentration of about 7%, provided that the reaction rate coefficient changes by one order of magnitude.

In Fig. 6 the results of the sensitivity analysis for the O and N concentrations along the considered reentry trajectory are illustrated. In the gaskinetical regime, at an altitude of 120 km, only the forward reaction rate coefficients of the O_2 dissociation act sensitively on the O_2 and O concentration. The nitrogen molecule is the dominant collision partner.

However, the forward reaction rate coefficient of the exchange reaction $N_2 + O \rightleftharpoons NO + N$ is sensitive to flows at altitudes of 92 km with respect to the N_2 and N concentration. This is not the case for the flow at 45 km altitude. For flows at 92 km, two coefficients of the N_2 dissociation (collision partners N and O) effect the N_2 and N concentration.

At an altitude of 45 km, the oxygen dissociation reaches equilibrium just behind the shock. For this reason, no coefficient may influence the oxygen concentration sensitively.

4 Transition

Whether or not the maximum of the heat transfer on the heat shield, with the laminar supersonic flow considered up to now, really occurs at an altitude of 45 km on the reentry trajectory is determined by a stability analysis of the laminar-turbulent transition in the boundary layer. The transition to a turbulent boundary layer may raise the nondimensional heat flux St, according to Fig. 7, by more than a factor 2.

The characteristics of the classical Orr-Sommerfeld stability analysis of harmonic disturbance waves are described in H. Oertel jr. [7]. With the use of the numerical boundary layer solutions of Section 6, the laminar-turbulent transition was examined down to an altitude of 40 km. No transition was found. At lower altitudes the boundary layer will become convectively unstable. This means that the transition process will not occur suddenly, but over several unstable individual processes (see H. Oertel jr. [7]). The question whether absolute unstable crossflow instabilities may occur in hypersonic flows is under investigation.

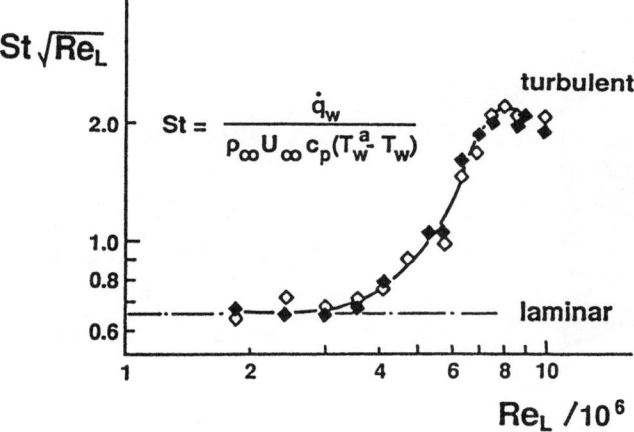

Fig. 7. Non-dimensional heat-flux on a cone (Stanton-number St), $M_\infty = 10$, cone angle $8°$, T_W-wall temperature, $T_W{}^a$-adiabatic wall temperature, $\dot{q}_W$-wall heat flux, V. di Christina [8]:

$$\left(\frac{\partial}{\partial t} + c\frac{\partial}{\partial r} + F\frac{\partial}{\partial c}\right) f(r, c, t) = \frac{1}{Kn}\left(\frac{\partial f}{\partial t}\right)_{coll} \cdot \quad \left(\frac{\partial f}{\partial t}\right)_{coll} = \iiint (f'f_1' - ff_1)\, c_{rel} b\, db\, d\varepsilon\, dc_1, \quad Kn = \bar{\lambda}/L$$

5 Gaskinetical simulation

The conditions for the chemical sensitivity studies in Section 3 and the transition prediction in Section 4 are the numerical simulation results of the aerothermodynamical flow field around the reentry capsule.

We begin with the gaskinetical simulation in the upper atmosphere, H. Oertel jr. [9]. Figure 8 shows a primary sketch in the regime of rarefied gases. In contrast to the gasdynamical modelling of a hypersonic flow around a reentry capsule, with a sharply limited shock from Fig. 1, in the gaskinetical regime, the shock is smeared over approximately 10 mean free paths. The sonic line is oriented along the vehicle surface and separates the entire flow field into subsonic and supersonic parts. According to the sensitivity studies from Section 3, at an altitude of 92 km, for example, the oxygen dissociation is in nonequilibrium and turns to frozen flow further downstream.

The gaskinetical simulation is based on the Boltzmann equation. In this equation, $f(r, c, t)$ represents the distribution function which is the basis for the microscopic description of the flow. It describes the statistical temporal and spatial distribution of the particles and their velocities in the phase space. Integration of the distribution function over the velocity space $f(r, c, t)\, dc$ results in the particle density in a volume element of the physical space. Further integration over the entire physical space $f(r, c, t)\, dcdr$ then gives the total number of particles. If the molecular velocity distribution is known, the local macroscopic parameters can be calculated by momentum formation with integration over the entire velocity space.

We employ the classical Monte-Carlo simulation method. With this method, the distribution function is represented directly by the gas particles in the phase space. The real gas flow with 10^{19} particles per cubic centimetre is simulated by several thousand model particles for this purpose. This is possible, since the Boltzmann equation can be normalized in such a way that the solution is independent of the number of particles. It must be ensured that the product $n \cdot \sigma$ is kept constant, n being the particle density and σ the cross section of the gas particles.

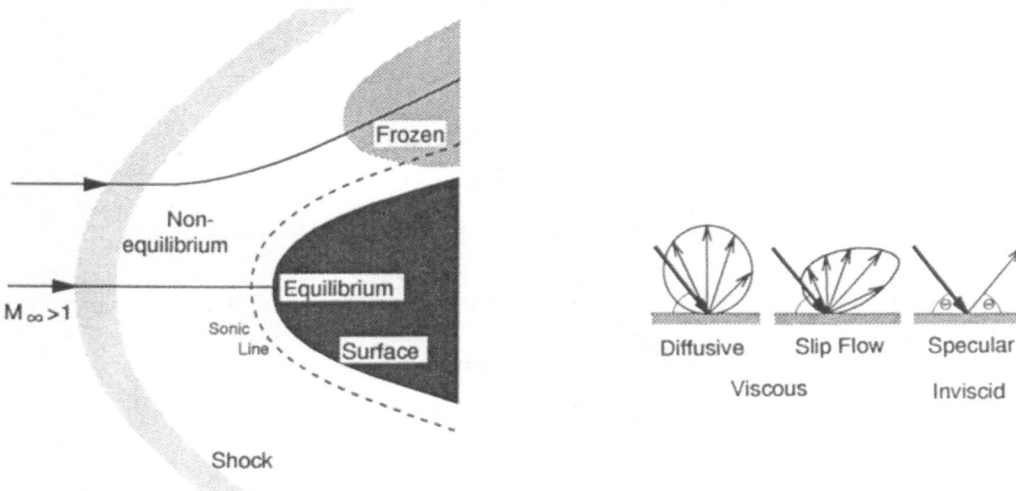

Fig. 8. Basic of the gaskinetical simulation: principle sketch of the aerothermodynamical flow field, boundary conditions, and Boltzmann equation

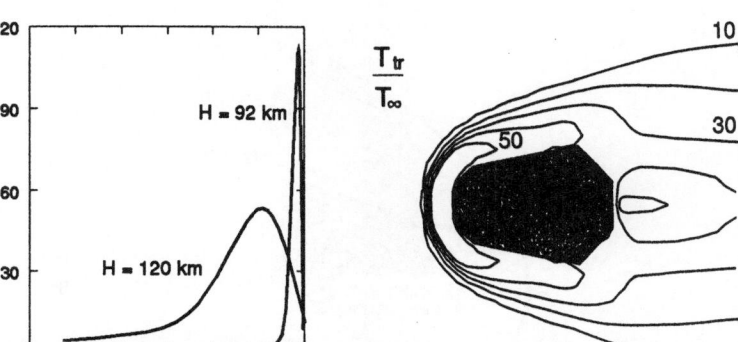

Fig. 9. DSMC Monte Carlo simulation including Park's chemistry: heat fluxes, temperature distributions and isothermal lines, D. Hafermann [10]:

$$\frac{\partial U}{\partial t} + \sum_i \left(\frac{\partial F_i}{\partial x_i} + \frac{\partial G_i}{\partial x_i} + \frac{\partial H_i}{\partial x_i} \right) = W, \quad U = (\varrho, \varrho u_1, \varrho u_2, \varrho u_3, \varrho e_{tot}, \varrho e_{vib}, \varrho_1, \varrho_2, ..., \varrho_{n-1})^T$$

The momentum wall-interaction with the heat shield is formulated as a diffuse reflection of the particle with a Maxwellian velocity distribution of the assumed isothermal wall-temperature. The hypersonic outer flow is superimposed to this thermodynamic equilibrium distribution. For the particle energy, we assume complete accommodation at the wall for the first formulation. The vibrational excitation of the molecules and the chemical reactions will be taken into account, which describes the reaction kinetics phenomenologically. According to the Larsen Borgnakke-model, with reaction rates k_f, a statistical selection of the exchange of the internal energy of the respective collision partners is made, corresponding to the reaction mechanism in Section 3.

The gaskinetical results for the reentry capsule are presented in Fig. 9 at altitudes of 120 km and 92 km of the considered reentry trajectory. Across the shock, along the stagnation streamline, the translational temperature T_{tr} rises to a factor 110 at an altitude of 92 km, compared with the free stream temperature T_∞, and then decreases to the assumed constant value of the wall temperature in the thermal boundary layer. The heat flux into the heat shield increases along the reentry trajectory. But the nondimensional heat flux decreases according to the increasing density.

At an altitude of 92 km, the influence of a fully catalytic wall is studied. Due to the condition of an isothermal wall, the temperature distribution around the capsule is not affected. The concentrations of the atomic species decrease in the vicinity of the wall. The integral heat flux increases about 16% in comparison to a noncatalytic wall.

6 Continuum mechanical simulation

The continuum mechanical simulation is based on the Navier-Stokes equations in conservation form of Fig. 10 under consideration of the chemical source term W, which accounts for the thermodynamical nonequilibrium effects. The vector of the conservative state variables consists for n-species of the partial densities $\varrho_1, ..., \varrho_{n-1}$, the total density ϱ, the velocity components u_1, u_2, u_3, the vibrational and total energies e_{vib} and e_{tot} per unit mass. F_i represents the vector of the conservative fluxes in the directions x_i, G_i, the vector of the diffusive fluxes due to friction and heat conduction, and H_i, the vector of mass and heat fluxes due to concentration changes.

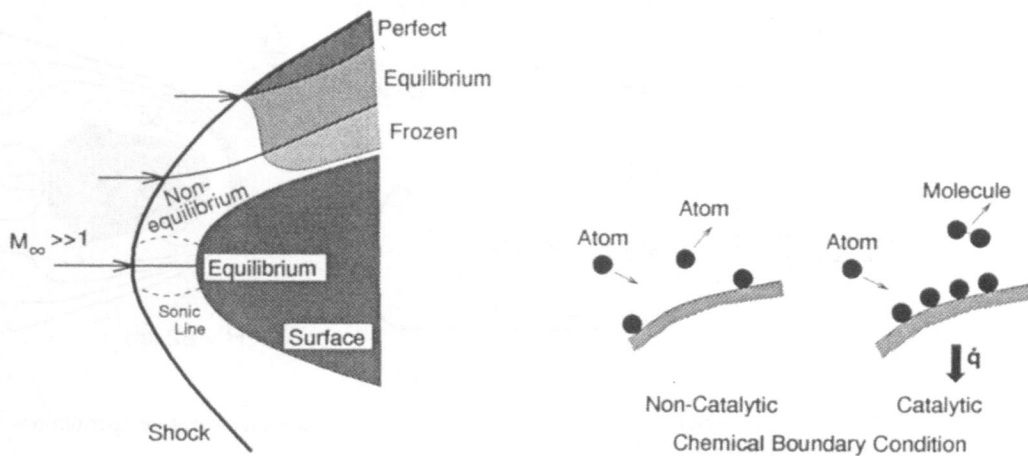

Fig. 10. Basics of continuummechanical simulation: principle sketch of the aerothermodynamical flow field, catalytic and non-catalytic boundary conditions, continuity, Navier-Stokes and energy equations

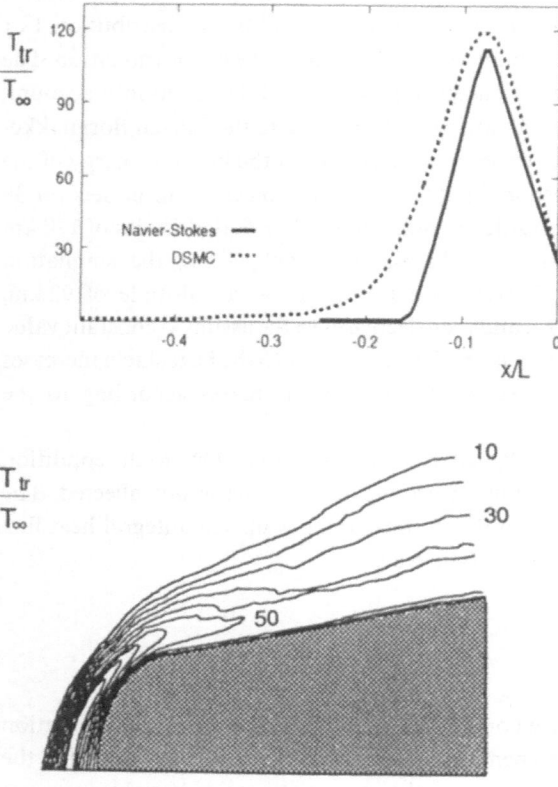

H = 92 km

Fig. 11. Navier-Stokes-simulation including PARK's chemistry in comparison with the gaskinetical results, temperature distributions and isothermal lines. See D. Hafermann [10], H. Holthoff [13]

The system of equations is solved in consideration of the boundary conditions from Fig. 5 for the limiting cases of the chemical-wall interaction of catalytic, respectively noncatalytic surfaces with an explicit Taylor-Galerkin Finite Element method (see e.g. E. Laurien et al. [11]). The catalytic gas-surface interaction of the air with the heat shield of the reentry capsule considers the recombination, especially of oxygen atoms, at the wall, which causes an additional heat flux into the heat shield. The limiting case of the noncatalytic wall does not consider the recombination of atoms at the wall and assumes that the approaching atoms leave the wall again as atoms.

Figure 11 shows the results of the numerical simulation in comparison to the gaskinetical results of Section 5 at an altitude of 92 km of the reentry trajectory for the considered limiting case of a noncatalytic wall. The peak translational temperatures differ by less than 10%. It is not expected that the Navier-Stokes Simulation resolves the rise of the translational temperature exactly because the Navier-Stokes equations are not valid within the shock structure.

7 Conclusion

We have presented the aerothermodynamic and numerical basics which are necessary for the numerical design of a reentry capsule. The chemical sensitivity studies have reduced the reentry problem from original 24 reaction coefficients and 6 equilibrium constants at the point of maximum of heat transfer to 7 relevant quantities which are sufficient for design purposes. The catalytic and noncatalytic wall-surface boundary conditions lead to heat flux variations in the heat shield of a factor 2, which has a decisive influence on the design of the heat shield.

As a complete simulation of aerothermodynamics in a wind tunnel is, in principle, not possible, the interaction parameters remain to be determined through quantitative reentry free flight experiments. The engineering sciences call this validation of numerical algorithms, boundary conditions and chemical models. It is performed by comparison to the experimental free flight data basis, with regard to the adaptation of the numerical to the measured real heat flux into the heat shield, H. Oertel jr. [12]. The free flight experiment HYPERBA consists of a phase compensated Michelson interferometer (PMI) integrated in a reusable C-SIC heat shield tile for the measurement of the density at the stagnation point, as a reference for the free flight database.

The present article gives the scientific fundamentals for the quantitative evaluation of free flight experiments. The goal here is to match the numerical simulation along the flown reentry trajectory, the gaskinetical and continuum mechanical gas surface interaction models, and the chemical parameters determined in sensitivity studies by comparison with the experimental free flight data basis, such that, for the design of a reusable heat shield, the numerical prediction of the heat flux into the heat shield tiles is possible. At present, the numerical design tools for the reusable heat shield tiles of reentry capsules are developed through the described intermediate step of aerothermodynamic validation.

References

[1] Oertel, H. jr., Böhle, M., Delfs, J., Hafermann, D., Holthoff, H.: Grundlagen der Aerothermodynamik. Berlin Heidelberg New York Tokyo: Springer 1993.

[2] Warnatz, J., Jäger, W.: Complex chemical reaction systems: mathematical modelling and simulation. Berlin Heidelberg New York Tokyo: Springer 1987.

[3] Park, C.: On convergence of computation of chemically reacting flows. AIAA 85-0247 (1985).

[4] Oertel, H. jr.: Berechnungen und Messungen der Dissoziationsrelaxation hinter schief reflektierten Stößen in Sauerstoff. Dissertation Universität Karlsruhe (TH) 1974.

[5] Böhle, M., Holthoff, H., Laurien, E.: Zur Sensitivität von Wiedereintrittsströmungen gegenüber aerothermodynamischen Modellparametern. DGLR-Jahrbuch II, 831—839 (1992).

[6] Böhle, M.: Zur Sensitivität von Wiedereintrittsströmungen gegenüber aerothermodynamischen Modellparametern. ZLR-Bericht 93-01 (1993).

[7] Oertel, H. jr.: Turbulenzentstehung (Stabilitätstheorie). In: Geschichttheorie (Gersten, K., Schlichting, H., eds.). Berlin Heidelberg New York Tokyo: Springer 1994.

[8] di Christina, V.: Three-dimensional laminar boundary layer transition on a sharp 8 cone at mach 10. AIAA J., 8, 852—856 (1970). See also: Anderson jr., J. D.: Hypersonic and high temperature gas dynamics, p. 285. New York: McGraw-Hill 1970.

[9] Oertel, H. jr.: Gaskinetical and Navier-Stokes simulation of reentry flows. In: Proceedings of INRIA and GAMNI-SMAI workshop on hypersonic flows for reentry problems I, pp. 63—84 (1990).

[10] Hafermann, D.: Gaskinetische Simulation der Wiedereintrittsaerothermodynamik um stumpfe Körper. Dissertation TU Braunschweig, ZLR-Bericht 93-02, 1993.

[11] Laurien, E., Böhle, M., Holthoff, H., Wiesbaum, J., Lieseberg, A.: Finite element algorithm for chemically reacting hypersonic flows. AIAA 92-0754 (1992).

[12] Oertel, H. jr.: The aerothermodynamic validation reentry experiment HYPERBA, Proceedings 3. In: Aerospace Symposium Braunschweig (Oertel, H. jr., Körner, H., eds.), pp. 103—128. Berlin Heidelberg New York Tokyo: Springer 1991.

[13] Holthoff, H.: Chemische Sensitivitätsanalysen zur Bestimmung eines Wand-Wechsel-Wirkungsmodells für die Berechnung des Wärmeübergangs auf stumpfe Körper in Hyperschallströmungen. Dissertation TU Braunschweig, ZLR-Bericht 93-03, 1993.

Author's address: Professor Dr.-Ing. H. Oertel jr., Institute for Fluid Mechanics, Technical University of Braunschweig, Bienroder Weg 3, D-38106 Braunschweig, Federal Republic of Germany.

Acta Mechanica (1994) [Suppl] 4: 183−188

Hypersonic flow of real gases with relaxation

W. Wuest, Göttingen, Federal Republic of Germany

Summary. A hypersonic high enthalpy wind tunnel with stagnation temperatures up to 14 000 K has come into operation at the DLR Göttingen in February 1992. It may offer new possibilities to study high temperature effects with relaxation in high velocity flows. A survey is given of the expected behavior of such flows including relaxation effects at low density and new diagnostics methods for high temperature flows.

1 Introduction

The thermodynamic behavior of gases at high temperatures is essentially determined by the number of degrees of freedom. Monoatomic molecules can only possess translational energy, i.e. they have three degrees of freedom. The average energy of each degree of freedom is equal and has the value $kT/2$, were k is a gas constant and T the absolute temperature. Bi- und multiatomic molecules may have additional rotational and vibrational energy. Biatomic gases have two rotational degrees of freedom because the rotation about the molecular axis has vanishing energy. However, for quantum mechanics reasons the rotational energy is only activated above a definite temperature level. There is a transition region from purely translational energy to additional rotational energy. E.g. for hydrogen this transition region is from $T = 70$ K with $c_p = 5/2k$ to $T = 300$ K with $c_p = 7/2k$, where c_p is the specific heat at constant pressure. The same behavior is true for vibrational energy. However, the transitional regime is at a higher temperature level, e.g. for nitrogen from $T = 400$ K to $T = 2000$ K. If the vibrational energy exceeds a limit dissociation or even ionization takes place.

2 Equilibrium flow at hypersonic velocities

A "perfect" gas with constant specific heat coefficient c_p may only approximately be verified in a lower and limited temperature range. Generally c_p is a function of temperature. With beginning dissociation the chemical composition of air is changing as shown in Fig. 1. The most important reactions are the dissociation of oxygen $O_2 = O + O$ (at 3000 K) and of nitrogen $N_2 = N + N$ (at 5000 K) and the formation of NO, NO_2, NO_2, No^+ and O_3. At temperatures exceeding 10 000 K ionization releases electrons. Dissociation requires large amounts of heat. Thus in the specific heat chart for high temperature air (Fig. 2) extreme values accompany the dissociation process. In the general gas equation the gas constant is variable because of the increase of particles $p/\varrho = (1 + \alpha)$ RT, where the magnifying factor $1 + \alpha$ is also shown in Fig. 2.

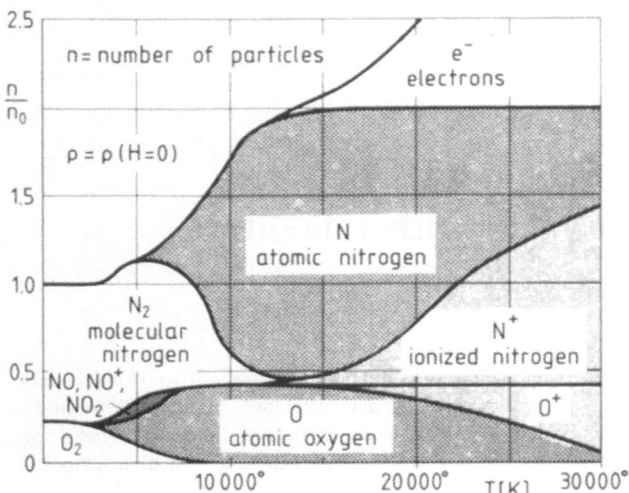

Fig. 1. Composition of air at high temperatures

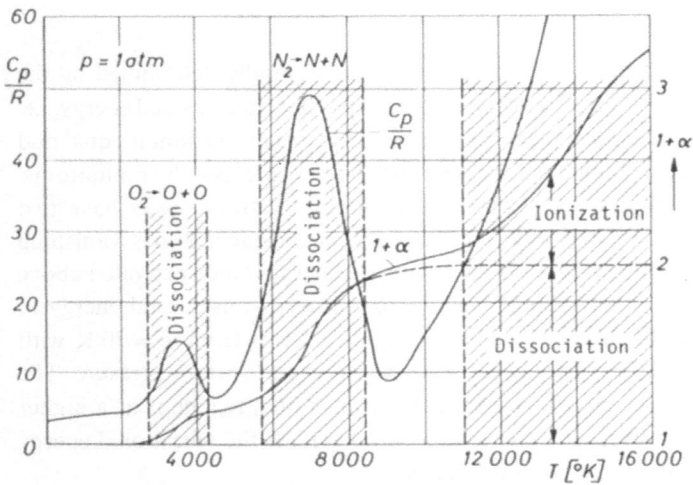

Fig. 2. Specific heat coefficient c_p and magnification factor $1 + \alpha$ of gas constant R for air at high temperature

3   Hypersonic flow with relaxation

All alterations of chemical equilibrium need a certain amount of molecular collisions. Translational equilibrium only needs a few collisions and is therefore quickly reached. Higher energy levels need more and more collisions and therefore finite time to reach equilibrium. It is convenient to attribute to each degree of freedom a temperature instead of an energy. At equilibrium this temperature T reaches the equilibrium temperature T_e and the simplest assumption to reach equilibrium is $dT/dt = (T - T_e)/\tau$, where τ is the relaxation time. In a flow of velocity u we can also define a relaxation length $L = u\tau$. If a characteristic length of the flow field l is small compared with L, the flow is "frozen", because no equilibrium can be attained within the characteristic length l of the flow field. It is evident that in relaxation gasdynamics [1] besides such ordinary variables as pressure, density, gas temperature, and flow velocity additional parameters must be introduced which characterize non-equilibrium gaseous state. In the following we prefer a phenomenological description of relaxation processes. A statistical description is more general and does not require local equilibrium.

3.1 Translational relaxation

If the mean free path of the molecules is much larger than a characteristic length of the flow field, the number of collisions is insufficient to reach equilibrium. E.g., a free jet flow merges into a free molecular flow ("Knudsen effusion"). The molecular velocity distribution is no more a Maxwellian.

3.2 Rotational relaxation

To reach rotational equilibrium some 50 collisions are necessary. For this reason relaxation may only be found at relatively strong rarefaction. Therefore the measurement of rotational temperature generally serves to determine the equilibrium temperature in a flow field.

3.3 Vibrational relaxation

As approximately 1 000 collisions are needed to reach vibrational equilibrium relaxation is nearly always present in hypersonic flow problems. Vibrational relaxation is very important because it is the first step to relaxation of dissociation. The product of relaxation time τ_v and pressure p has been found to be only dependent on the temperature T_a. A large number of experimental and theoretical results are available for N_2, O_2 and air. For nitrogen C. E. Treanor and P. V. Marrone [2] give the interpolation formula $\tau_v p = 1.1 \times 10^{-11} \sqrt{T_a} \exp(154/\sqrt{T_a})$. Using this formula W. Wuest [3] has calculated the flow through a hypersonic nozzle and has found that the vibrational energy freezes in the subsonic part of the nozzle long before the throat. Dissociation and vibrational degrees of freedom are strongly coupled and the correlation may be expressed by the CVD-scheme (coupling vibrational degrees). It assumes that the vibrational degrees of freedom relax independently from dissociation.

3.4 Relaxation of dissociation and recombination

Measurements of dissociation in high velocity flows were made especially with light gas guns and spheres from nylon or polyethylen up to velocities of 6 km/s. Only by lowering the gas pressure below 2.5 mm Hg relaxation especially of nitrogen dissociation became evident. Measurements of the inverse problem, the recombination of a dissociated gas in an expansion flow were made in a shock tube.

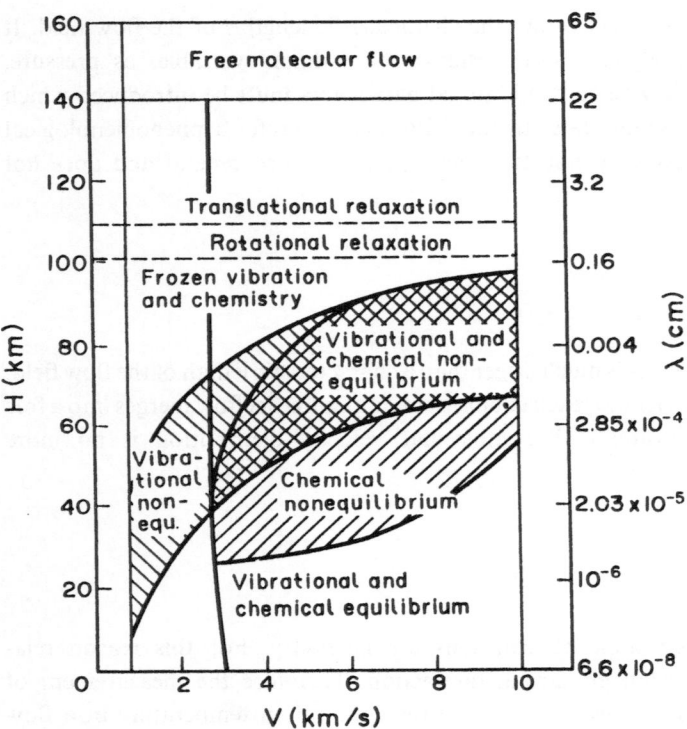

Fig. 3. Different regimes of relaxation phenomena in hypersonic rarefied flow for a blunt nosed body with curvature radius of 0.3 m

3.5 Relaxation of ionization

Different theories have been developed in order to calculate ionization time. Ionization may be introduced by electrons already present before the shock or gas pollutions may be responsible for the initiation of ionization. Measurements of ionization time were made by microwave technology.

To summarize the relaxation phenomena a chart was established which shows the different regimes in hypersonic rarefied flow for a special geometry (Fig. 3).

4 Experimental flow-diagnostics methods for high velocity flows

At the reentry of space vehicles into the atmosphere the velocity is of the order of 8 km/s. The compression behind the leading edge shock leads to extreme temperatures. A simulation of the flow conditions is possible in hypersonic low density facilities. A wind tunnel of this type with stagnation temperatures up to 2 500 K was completed at Göttingen in 1965 and allowed a study of rotational and vibrational relaxation processes. The range of stagnation temperatures was extended to 14 000 K by a new high enthalpy wind tunnel come into operation at the DLR-Göttingen in 1992 [4]. Now also relaxation processes of dissociation and ionization may be studied in the new facility. However, the high enthalpy wind tunnel has a measuring time of only 10^{-3} s, whereas the hypersonic low density tunnel could be operated continuously. Different non-intrusive methods have been developed to measure rotational and vibrational temperatures and species composition and even the absolute velocity.

4.1 Electron beam fluorescence technique

The electron beam fluorescence technique is well established for hypersonic wind tunnel testing since more than 25 years. A comprehensive survey on this technique, which has been widely used in the hypersonic low density tunnel at Göttingen was given in [5]. If an electron beam is passed through a neutral gas an analysis of the light emission allows not only to determine the local density but the relative intensities of the rotational fine structure yield information about the rotational temperature, and the population distribution of the vibrational energy levels allows to determine the vibrational temperature. By pulsing the electron beam it is even possible to measure the absolute velocity. However the density range of the electron beam fluorescence technique is 10^{13} to 10^{16} molecules/cm^3 [6]. At higher density quenching and beam scattering problems occur and make this technique difficult.

4.2 Laser-induced fluorescence diagnostics

The LIF-technique can measure translational, rotational, and vibrational temperatures by exciting selectively (in contrast to the electron beam) each molecular state even at very low densities. The LIF-technique is becoming a standard tool for non-intrusive measurements in supersonic wind tunnel flow fields in a wide density range. At higher densities it is possible to modify this technique and detect predissociative molecular states (LIPF), as stated by C. Dankert et al. [7]. A LIF apparatus for application in the new high enthalpy wind-tunnel at Göttingen is being constructed by Prof. P. Andresen from the Max-Planck-Institute for Flow Research at Göttingen together with the firm LaVision 2D-Meßtechnik at Göttingen [8]. Some other advanced optical diagnostics methods for temperature and density measurements in hypersonic research like Coherent Laser Raman (CARS), Laser Rayleigh scattering, and Laser Raman scattering have only been applied in wind tunnels at relatively high density.

4.3 Initial calibration results of the high enthalpy shock tunnel in Göttingen (HEG)

The high enthalpy shock tunnel in Göttingen is a free piston driven shock tunnel operating in the reflected mode. The first commissioning results show that conditions similar to atmospheric reentry conditions of a space flight vehicle, with specific stagnation enthalpies $h = 20$ corresponding to flight velocities $v = 6.3$ km/s (or flight Mach number 21), have been achieved with test gas nitrogen providing reservoir temperatures between 9000 and 10000 K. A holding time of approximately 2 ms has been obtained and the recovered pressure in the nozzle reservoir has been observed to be approximately equal to the diaphragm rupture pressure. The first experiments included a simple flow visualizaton method with the help of the self luminosity of the shock heated gas with a cylinder placed in the flow. From the proximity of the bow shock to the cylinder it is clear that strong dissociation of the molecular nitrogen has had to occur.

References

[1] Loser, S. A., Osipow, A. I.: Modern problems of relaxation. In: Int. Symposium on Rarefied Gas Dynamics 1982 (Belotserkovskii, O. M. , Kogan, M. N., Kutateladze, S. S., Rebrov, A. K., eds.), pp. 145−159. New York, London: Plenum Press 1985.
[2] Treanor, C. E., Marrone, P. V.: The effects of dissociation on the rate of vibrational relaxation. Cornell Aeronaut. Lab. CAL-Rep. No. QM-1626-A-4 (1962).

[3] Wuest, W.: Hypersonic nozzle flows with vibrational relaxation (in German). Int. Rep. AVA Göttingen 66 A 40 (1966).

[4] Eitelberg, G., McIntyre, T. J., Beck, W. H., Lacey, J.: The high enthalpy shock tunnel in Göttingen. AIAA 92-3942 (1992).

[5] Bütefisch, K. A., Vennemann, D.: The electron beam technique in hypersonic rarefied gas dynamics. Progr. Aerospace Sci. **15**, 217−255 (1974).

[6] Lewis, J. W.: Optical diagnostics of low-density flow fields. Progr. Astron. Aeron. **117**, 107−132 (1989).

[7] Dankert, C., Cattolica, R., Sellers, W.: Local measurement of temperature and concentrations — a review for hypersonic flows. In: Hypersonic ARW, 70th Fluid Dynamics Panel Meeting of AGARD -CP-514 B 1−10 (1993).

[8] Beck, W. H.: Simulation of Hermes-reentry into the terrestrial atmosphere (in German). DLR-Nachrichten **63**, 24−28 (1991).

Author's address: Professor Dr. rer. nat. W. Wuest, DLR-Institut für Experimentelle Strömungsmechanik, Bunsenstrasse 10, D-37073 Göttingen, Federal Republic of Germany

Acta Mechanica (1994) [Suppl] 4: 189–197

Part 5: Fluid Machinery

Effect of sickle shaped blades on sound generated by axial flow fans

K.-O. Felsch, Karlsruhe and W. Stütz, Künzelsau, Federal Republic of Germany

Summary. In this paper the point in question is whether fan noise reduction is possible by means of sickleshaped blades. Lighthill's extended acoustic analogy is used to verify this effect. For the special case of the aeroacoustic interaction between an radiator and an impeller, numerical and measured results are compared and discussed. The noise reduction using the forward sickled fan is substantial in comparison with the sound generated by a conventionally shaped fan. The numerical calculations and the experimental results show the same tendencies as well for the sound power level as for the spectrum.

1 Introduction

Fans are produced in large numbers. They are used in all domains of daily technical life. In most applications not only a good efficiency but also low noise emission are requested, because in the majority of cases secondary soundproofing methods are not possible or too expensive.

In an earlier publication of the second author [1] the effect of "sickle-shaped" blades[1] on sound generated by axial flow fans has been investigated. "Sickle-shaped" means that the line from blade root to blade tip on which the blade sections are arranged is no longer straight and radial but curved (but still in a plane normal to the fan rotation axis; Fig. 1).

The sickle-shaped blades, of course, strongly influence the aerodynamical behaviour of the fan. Due to the changed blade geometry radial components are caused in addition to tangential and axial blade forces. Therefore, the sickle-shape has to be taken into account during the design process, otherwise the efficiency will decrease. Stütz [1] and Wright [2] have consequently developed sickled axial flow fans with good efficiencies. Furthermore, Stütz [1] has shown that there is no essential noise improvement — at least in the design-point — if the inlet flow of the fan is homogeneous and undisturbed. However, the advantage of the sickle-shaped blades is significant, if there are inlet flow distortions. This is the case in most of the fan installations caused for instance by struts, guide wheels, bends, obstructions etc. In this paper the special case of the aeroacoustic interaction between a radiator and an axial flow fan is studied (Figs. 2 and 3).

2 Model to calculate acoustic radiation due to sickle-effects

Lighthill [3] developed an acoustic analogy approach from the exact equations of continuity and momentum to calculate acoustic radiation from relatively small regions of turbulent flow embedded in an infinite homogeneous fluid. This equation has been applied to high-speed jets.

---

[1] In known references one finds "skewed" oder "swept" blades or similar expressions. But these are misleading notations.

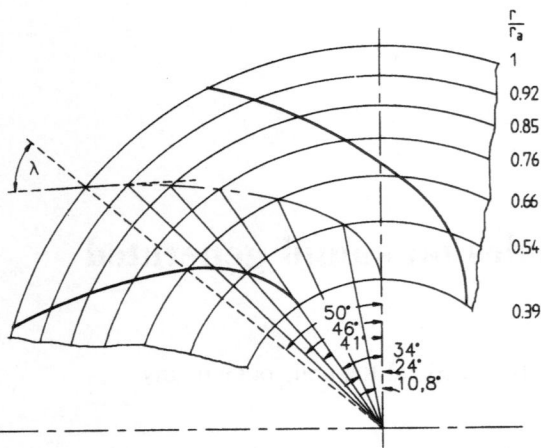

Fig. 1. Shape of the forward-sickled fan rotor blades

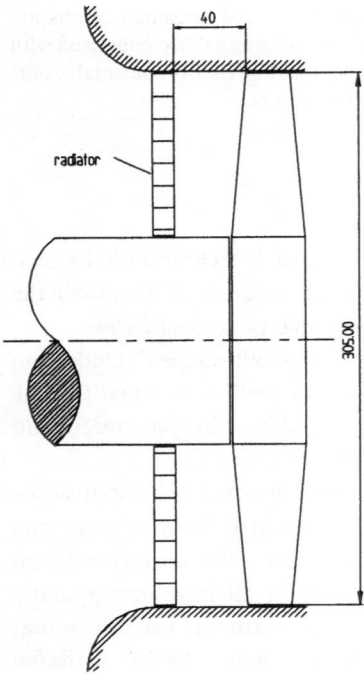

Fig. 2. View of experimental arrangement

Later, Lighthill's equation has been extended to include the effects of solid boundaries (see e.g. [4, 5]). Without going into the details of the derivation we quote the extended Lighthill equation of Heckl [4]:

$$\frac{1}{a^2}\frac{\partial^2 p}{\partial t^2} - \Delta p = \frac{\partial q_V}{\partial t} - \frac{\partial f_i}{\partial x_i} + \frac{\partial^2(\varrho w_i w_j)}{\partial x_i\,\partial x_j}. \tag{1}$$

This equation has clearly the same form as the inhomogeneous acoustic wave equation. Δ means the Laplacian operator, a the sound speed, p the pressure, q_V the mass flow, f the force per unit volume. The lefthand side describes the propagation of sound waves, the righthand side shows the acoustic source terms. The first term represents an acoustic monopole produced by mass-fluctuations; the second represents the sound generated by flow forces (dipole), and the

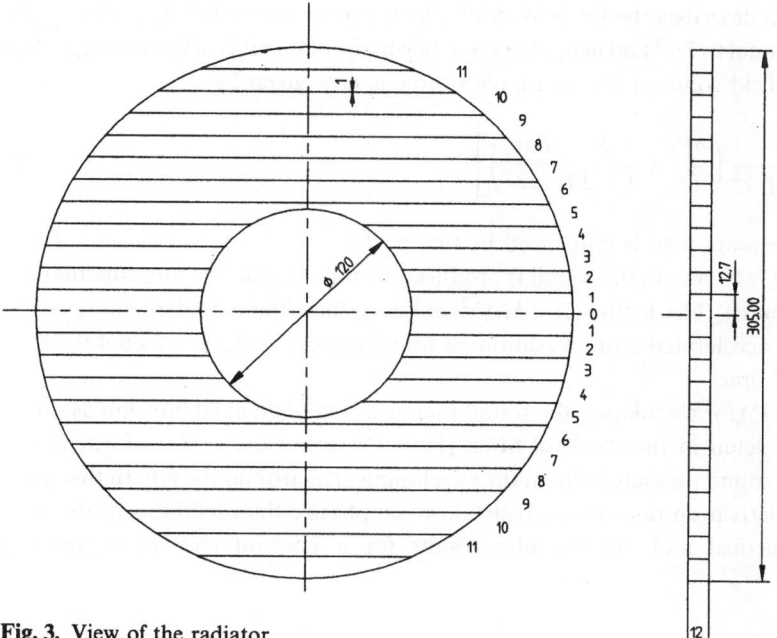

Fig. 3. View of the radiator

third represents the quadrupole source caused by turbulent stresses, which is of interest only for high speed flows. For all further steps only the dipole-term is considered.

The general solution of the inhomogeneous wave equation is [6]

$$p(t, x_b, y_b, z_b) = \frac{1}{4\pi} \iiint \frac{f\left(t - \dfrac{r}{a}, x, y, z\right)}{r} \, dx \, dy \, dz,$$ (2)

where r is the distance between the observation point and the source point and $(t - r/a, x, y, z)$ the retarded time (the time difference which the sound needs from his source to the observation point).

We now use Lowson's solution [7] and calculate the time-dependent pressure-field concentrating the source on one point with the help of the Dirac delta function. After some transformations one finds the following solution for the sound pressure [7]:

$$p - p_0 = \left[\frac{x_{b_i} - x_{q_i}}{4\pi a(1 - M_r)^2 r^2} \left(\frac{\partial F_i}{\partial t} + \frac{F_i}{1 - M_r} \frac{\partial M_r}{\partial t} \right) \right]$$
$$+ \left[\frac{1 + M_r}{4\pi(1 - M_r)^2 r^2} \left(\frac{x_{b_i} - x_{q_i}}{r} F_i \frac{1 - M^2}{1 - M_r^2} + F_i M_i \right) \right].$$ (3)

Now, the force components F_i are only time dependent. M means the Mach-number of the source motion, M_r is the Mach-number component in the direction from the source to the observation point.

$$M = (M_x, M_y, M_z), \qquad M_r = \frac{1}{r}(x_{b_i} - x_{q_i}) M_i; \quad i = x, y, z$$ (4)

The first term in Eq. (3) describes the far-field effect which is proportional to $1/r$, whereas the second term is proportional to $1/r^2$ and hence may be neglected, because it is of interest only close to the source. The far-field solution for the dipole source is now given by

$$p - p_0 = \left[\frac{x_{b_i} - x_{q_i}}{4\pi a (1 - M_r)^2 \, r^2} \left(\frac{\partial F_i}{\partial t} + \frac{F_i}{1 - M_r} \frac{\partial M_r}{\partial t} \right) \right]. \tag{5}$$

The resulting sound pressure field is influenced by two parts:

For steady motion ($\partial M_r / \partial t = 0$) the sound is produced by time-dependent force fluctuations similar to a resting source. The factor $(1 - M_r)^{-2}$ only modifies the sound pressure in the observation point. For accelerated motion additional sound is generated, even then if the force itself is independent of time.

The solution of Eq. (5) for calculating the sound pressure field of an axial flow fan as dipole source is described in detail in the thesis of Stütz [1]. A Discrete-Fourier-Transformation is used for evaluating the sound pressure field on the enveloping area around the fan. In this paper we do not repeat the derivation once more. Instead, we emphasize the usefulness of the basic concept showing numerical and experimental results for a relevant case of engineering application.

3 Comparison of experimental and theoretical investigations

3.1 Tested impeller

The influence of sickle shaped blades on the noise emission under distorted inflow conditions will be investigated for two different impellers: A strongly crooked stacking line in circumferential direction, and for comparison a traditional one with a straight stacking line. Figure 1 shows the sickle shaped blade with correct scale. To obtain the sickle shaped blades of the impeller the profile sections were moved on a defined radius in circumferential direction. The local angles of inclination are given in Table 1.

The following comparisons are always based on the operating points of best efficiency. Both impellers reach efficiencies η_{fa} of 60%. For undisturbed inflow the sound power levels (tonal, A-weighted) are nearly the same for both types. The operating point is defined as follows:

pressure coefficient	$\Psi_{fa} = 0.116$	rpm	$n = 3000 \; 1/\text{min}$
flow coefficient	$\varphi_r = 0.158$	density	$\varrho = 1.2 \; \text{kg/m}^3$
tip diameter	$D_a = 0.305$	tip clearance	$s/D_a = 1.1\text{‰}$
hub diameter	$D_i = 0.120$		

The two fans have five blades each and therefore the blade passage frequency will be 250 Hz. This corresponds to the Strouhal number $St = (f D_a / u_a)(\pi / Z) = 1$.

All experimental investigations are made using the aeroacoustic teststand [8] of our institute.

Table 1. Local angles of inclination for the forward curved impeller

r/r_a	0.39	0.54	0.66	0.76	0.85	0.92	1.00
$\lambda[°]$	13	42	52	51	47	45	43

3.2   Influence of a radiator in the impeller inflow

In [1] the influence of struts in the fan inflow on the noise emission was experimentally and theoretically investigated with success. This was done for the same fans described in Chapter 3.1. In this contribution the attempt is made to show the influence of radiators (automotive industry) in connection with fans mounted downstream. The tested fans are the same. This is a very important application, indeed, because sound absorbers cannot be used because of lack of space and other constraints.

3.2.1   Test conditions — experimental results

Concerning the investigations on the influence of a radiator a test configuration was chosen as drawn in Figs. 2 and 3. At the beginning of the cylindrical part of a standard nozzle the radiator dummy was positioned. It consists of 23 parallelly fitted sheet segments 1 mm thick.

The result of the noise measurement is given in Fig. 4 in which the sound power levels (L_{WA}) radiated from the pressure side of the two impellers are directly compared with each other. The fan with the sickle shaped blades radiates less noise in a wide range around the operating point of best efficiency. In Fig. 5 the result of a narrowband FFT-measurement ($\Delta f = 7.81$ Hz) is summarized for the blade passage frequencies at the design point. It is obvious that the sickled fan is much better only for Strouhal numbers greater than 10. But those belong to a frequency range in which the human ear is very sensitive.

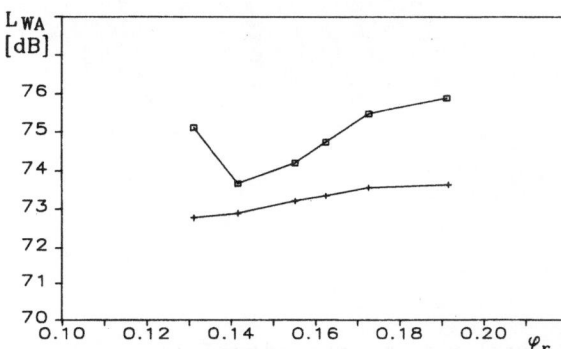

Fig. 4. Sound power level L_{WA} dependent on flow coefficient φ_r ($\square$ reference impeller, + sickled impeller)

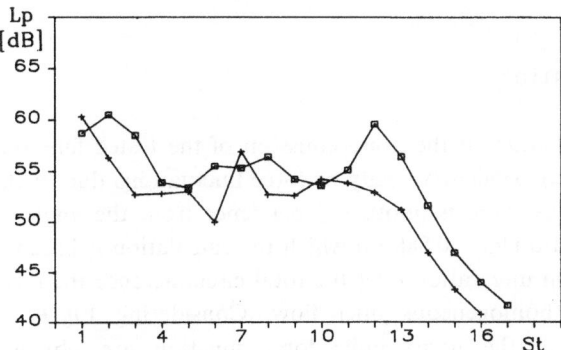

Fig. 5. Blade passage frequency analysis of sound pressure level at the fan design point ($\square$ reference impeller, + sickled impeller)

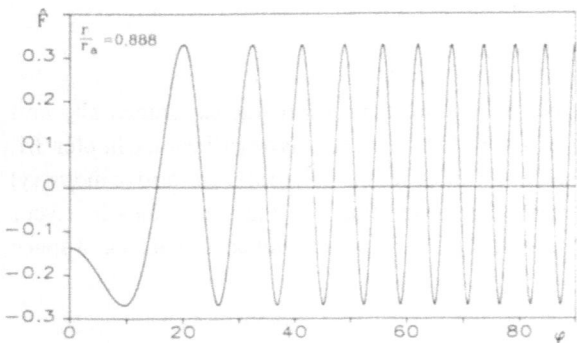

Fig. 6. Assumed pattern of force fluctuations dependent on angle φ

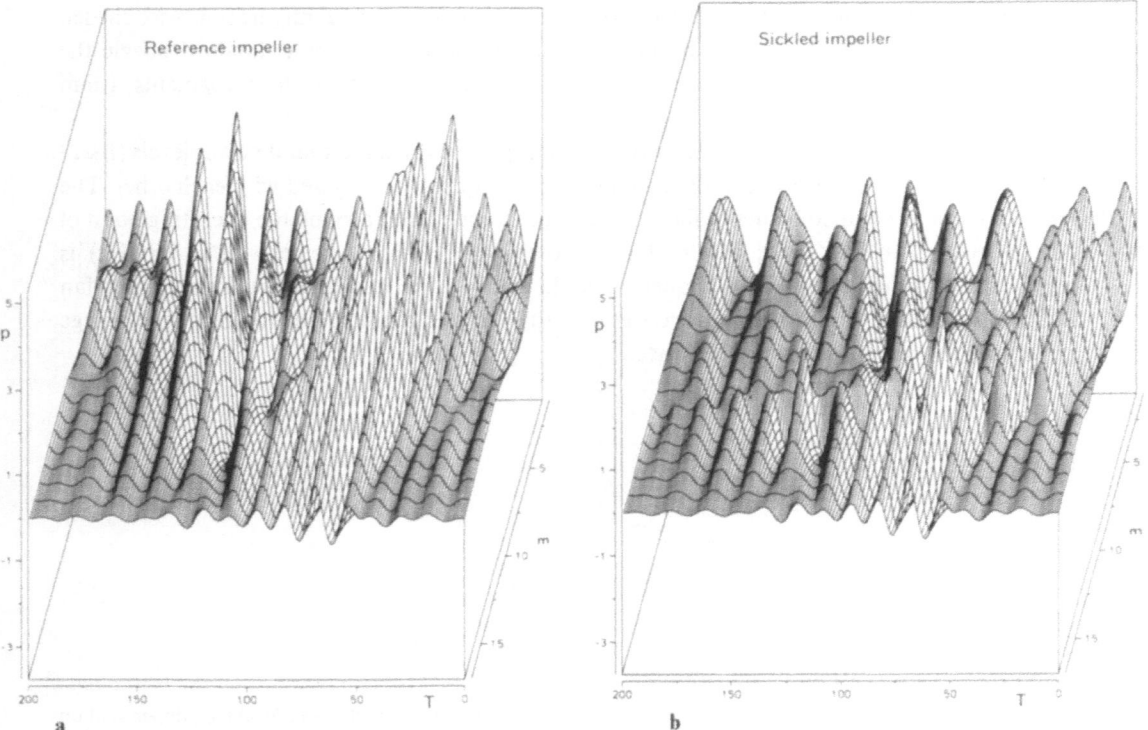

Fig. 7. Calculated sound pressure p[Pa] dependent on observation point m and time T: **a** reference impeller, **b** sickled impeller

3.2.2 Sound pressure field calculation

When simulating the influence of the radiator on the noise emission of the tested fans one meets the difficulty to describe the angle dependent velocity or force fluctuations due to the parallel radiator elements. On each radius there is another dependence from the angle φ. The function of the force at the radius ratio $r/r_a = 0.888$ on which the calculation is based is shown over a range of 90° in Fig. 6. By an integration over the total circumference the force becomes zero, this corresponds to a homogeneous inlet flow. Considering Fig. 6 all difficulties concerning the description of the nonperiodic force function are obvious.

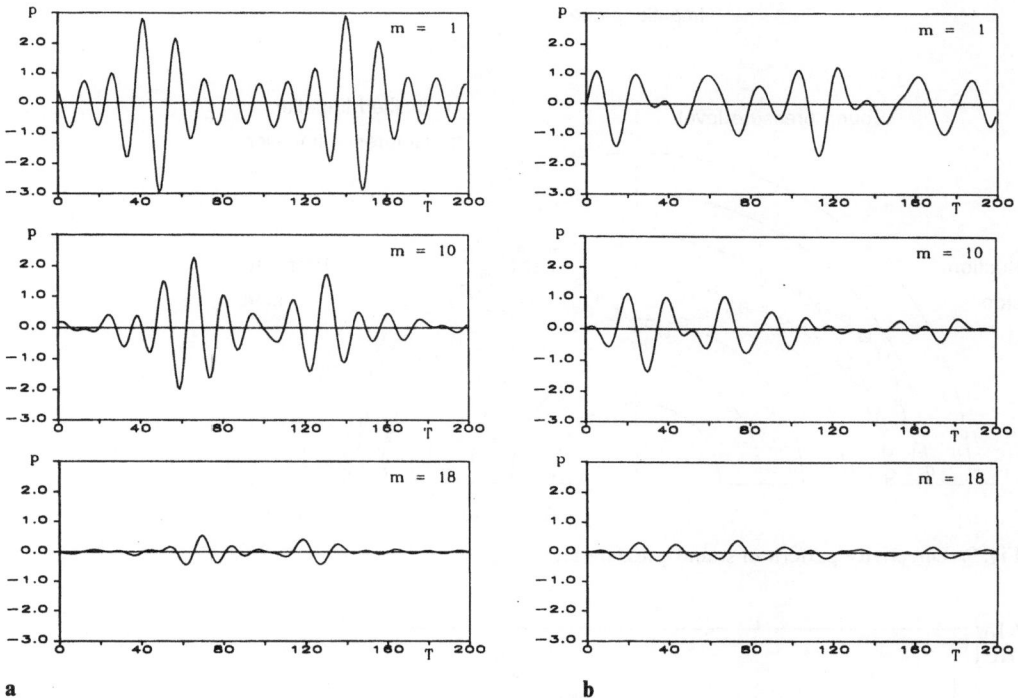

Fig. 8. Decomposed sound pressure field (suction side): **a** referenz impeller, **b** sickled impeller

Between two rib elements the angle range changes in a different way and for $\varphi = 0°$ to $10°$ the cosine similarity is not given. The description of the force function which is different for each radius considered is to be done with 50 Fourier-coefficients. With these coefficients in the harmonic equation the force function can be approximated very well.

In Fig. 7 the sound pressure fields on the suction side of the two fans calculated on a constant radius of 2 metres are compared. m means the position of the observer. $m = 18$ is in the impeller plane and $m = 1$ is near to the axis. The sickled fan shows smaller sound pressure amplitudes for one period (200 time steps) and a relatively smooth pressure field whereas the reference fan evidently has more strong pressure amplitudes. The decomposed sound pressure fields are shown by extract in Fig. 8 for the suction side. This comparison demonstrates the improvements in reducing and smoothing the sound pressure. Looking at the directivity pattern (Fig. 9) improvements of 3 to 5 dB can be noticed for all angles. The integration over the pressure side of the fan yields a difference of 3 dB in sound power. The main regions of radiation at the suction and the pressure side are close to the fan axis. The frequency spectrum of the sound power ΔL_W demonstrates that the tendency of the theory agrees well with the experimental results. Figure 10 shows the difference of the sound power between the reference and sickled fan for the various Strouhal numbers. For smaller Strouhal numbers and for numbers greater than 11 the theoretical calculations and the experiments show the same advantages for the sickle bladed fan. For the medium frequency range the calculated values are higher in the case of the sickled fan. Considering the experimental result, an increase of the sound power level can be stated for the Strouhal number 7 (see Fig. 5). A direct comparison of the experiment and the calculation confirms all tendencies. This is true for the integral field characteristics (sound power level) as well as for

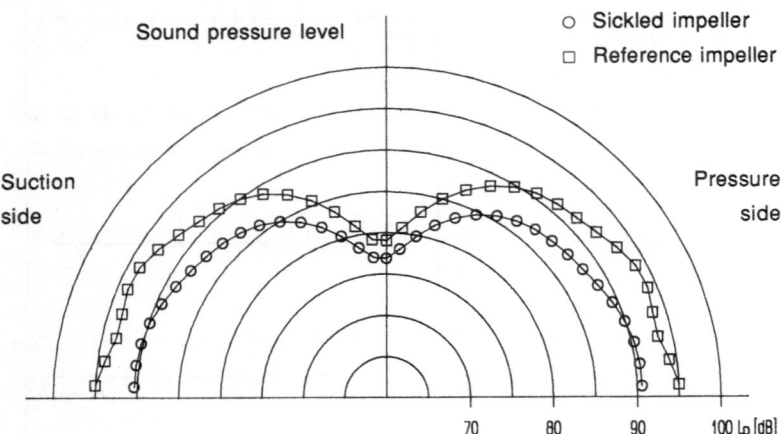

Fig. 9. Directivity pattern of sound pressure level

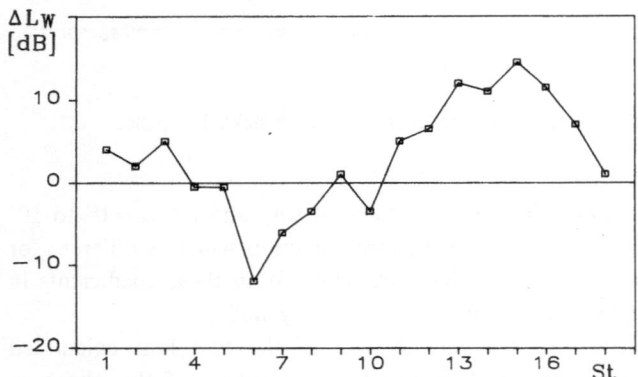

Fig. 10. Difference of sound power levels between sickled and reference impeller ΔL_W at various evaluated Strouhal numbers ($\square$)

the detailled ones (frequency spectrum). In view of the fact that the problem is very complex the obtained result is surprising. If one succeeds in better describing the aerodynamics which are the basis of the whole model better results are to be expected.

4 Nomenclature

a	m/s	speed of sound	t	s	time
D	m	diameter	T	s	$T = 60/n \cdot Z$ period of
f	N/m³; Hz	force per unit volume; frequency			oscillation (blade-passing-frequency tone)
F	N	force	u	m/s	peripheral speed
L_p	dB	sound pressure level	w	m/s	velocity
L_W	dB	sound power level	x, y, z	—	coordinates
m	—	observer position	Z	—	blade number
M	—	Mach number	λ	°	local angle

n	1/min	rotational speed	ϱ	kg/m^3	density
p	Pa	pressure	φ	–	polar coordinate
r	m	radial coordinate	φ_r	–	flow coefficient of fan
St	–	Strouhal number	Ψ	–	pressure coefficient of fan
s	m	clearance	η	–	efficiency of fan

Subscripts

a	blade tip	x	axial component
b	observer	q	source
i, j	indices of summation convention		

References

[1] Stütz, W.: Einfluß der Sichelung auf das aerodynamische und akustische Verhalten von Axialventilatoren. Karlsruhe, Univ. (TH), Dr.-Ing.-Diss. 1991. Also in: Stroemungsmech. Stroemungsmasch. **44**, 1 – 59 (1992).

[2] Wright, T., Simmons, W. E.: Blade sweep for low-speed axial fans. J. Turbomach. **112**, 151 – 158 (1990).

[3] Lighthill, M. J.: On sound generated aerodynamically. I. General theory. Proc. R. Soc. London Ser. **A 211**, 1107, 564 – 587 (1952). II. Turbulence as a source of sound; Proc. R. Soc. London Ser. **A 222**, 1148, 1 – 32 (1954).

[4] Heckl, M.: Strömungsgeräusche. VDI-Z. Fortschr. Ber. **7**, 1 – 125 (1969).

[5] Goldstein, M. E.: Aeroacoustics. New York St Louis San Francisco Auckland: McGraw Hill 1976.

[6] Bronstein, J. N., /Semendjajew, K. A.: Taschenbuch der Mathematik, 17th edn. Zürich Frankfurt a. M. Thun: Harri Deutsch 1977.

[7] Lowson, M. V.: The sound field of singularities in motion. Proc. R. Soc. London Ser. **A 286**, 559 – 572 (1965).

[8] Stütz, W.: Aero-Akustik-Prüfstand – Fortschrittliche Bestimmung der Ventilatorleistungs- und Ventilatorgeräuschdaten. Stroemungsmech. Stroemungsmasch. **42**, 27 – 50 (1990).

Authors' addresses: Professor Dr.-Ing. Dr. h. c. K.-O. Felsch, Institut für Strömungslehre und Strömungsmaschinen Abteilung Strömungsmaschinen, Universität Karlsruhe (TH), D-76128 Karlsruhe, and Dr.-Ing. W. Stütz, Ziehl Abegg GmbH & Co. KG, D-74653 Künzelsau, Federal Republic of Germany

Acta Mechanica (1994) [Suppl] 4: 199–205

Numerical simulation of 3D periodic flow in fluid couplings

A. Kost, N.-K. Mitra, and M. Fiebig, Bochum, Federal Republic of Germany

Summary. Adequate understanding of the flow field in fluid couplings is necessary for the optimized design of such devices. In a fluid coupling torque is transmitted by fluid circulation due to a speed differential between the rotating pump impeller and a matching turbine runner. The structure of the flow field is very complex and detailed studies of the unsteady 3D flow have never been reported. A finite-volume method with non-staggered variable arrangement has been used to solve the unsteady Navier-Stokes equations on boundary-fitted grids and for a rotating frame of reference. The obtained results give insights into the physical process of torque transmission.

1 Introduction

Fluid couplings consists of a pump impeller with radial blades and a similar matching turbine runner, both within the same casing but generally with small differences in blade number (cf. Fig. 1). The casing is partially filled with an oil of low viscosity. The pump impeller is driven by a prime mover, such as an electric motor or an internal combustion engine. Torque is transmitted from the impeller to the turbine by fluid circulation due to pressure differences which are caused by different angular velocities (slip) of the coupling halves. If there is no slip, there is no flow and consequently there will be no torque transmission [1].

The circulating flow in reality is two-phase (oil/air), three-dimensional, unsteady (possibly periodic) and contains separated zones. The complexity of the flow field depends on the design of the working circuit (e.g. effective diameter of the coupling, number and shape of the blades, axial gap between the impeller and the turbine) and dynamical parameters (oil filling, angular velocities and speed differential). Detailed studies of the flow field in fluid couplings have never been reported. Experimental data is extremely expensive to obtain due to the difficulties of measuring flow quantities in rotating narrow passages. A computational analysis is a possible tool to analyze these flows and to optimize the internal design of the working circuit.

In the present study, the simulation of 3D periodic flow in completely filled fluid couplings with rectangular cross-section and equally pitched pump impeller and turbine runner is reported. Laminar flow is assumed for small values of slip between the coupling halves.

2 Basic equations

The equations governing the unsteady 3D single-phase incompressible viscous flow in a fluid coupling are written in non-dimensional form and for a co-ordinate system rotating with

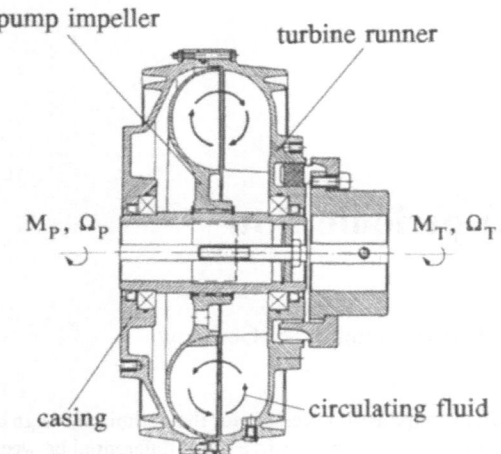

Fig. 1. Cross-sectional view of a fluid coupling

constant angular velocity $\underline{\Omega} = \Omega\underline{k}$ [2]:

$$\nabla \cdot \underline{w} = 0 \tag{1}$$

$$\frac{\partial \underline{w}}{\partial t} + \text{Ro}\, \underline{w} \cdot \nabla\underline{w} + 2\underline{k} \times \underline{w} = -\nabla P - \text{Ek}\, \nabla \times (\nabla \times \underline{w}) \tag{2}$$

Here, $\underline{w}$ is the velocity vector related to the rotating frame of reference and non-dimensionalized by an appropriate velocity scale U. The centrifugal acceleration term is combined with the static pressure in the reduced pressure P. The dimensionless parameters appearing in Eq. (2) are the Ekman number Ek and Rossby number Ro providing ratios of viscous force to Coriolis force and convective inertial force to Coriolis force, respectively:

$$\text{Ek} = \frac{v}{\Omega L^2}; \quad \text{Ro} = \frac{U}{\Omega L} \tag{3}$$

For the flow in fluid couplings it is convenient to use the angular velocity of the pump impeller $\underline{\Omega}_P$ as the rotational speed of the co-ordinate system, the effective diameter D of the coupling as the length scale $L = D$ and the difference of angular velocities between the impeller and the runner as the velocity scale $U = (\Omega_P - \Omega_T) \cdot D$ [3].

With respect to the co-ordinate system rotating with angular velocity $\underline{\Omega}_P$, the boundary conditions at the solid surfaces of the impeller and the turbine are (cf. Fig. 2):

$$\underline{w} = \begin{cases} \underline{0} & \text{, pump impeller} \\ -(\underline{k} \times \underline{r}) & \text{, turbine runner} \end{cases} \tag{4}$$

Here, s is the slip between the impeller and the turbine defined as:

$$s = 1 - \frac{\Omega_T}{\Omega_P}. \tag{5}$$

The analysis of the flow field is limited to one pitch by assuming equal number of blades for the impeller and the turbine and prescribing periodic conditions at the circumferential boundaries of the computational domain (cf. Fig. 2). The blades of the turbine runner are treated

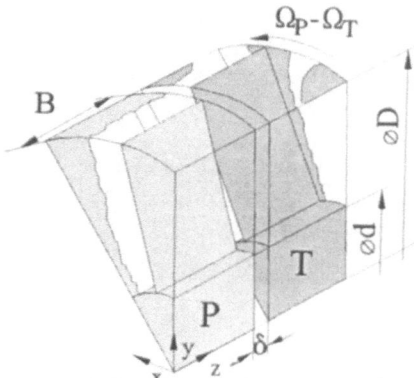

Fig. 2. Sketch of the computational domain; $d/D = 0.326$, $B/D = 0.281$, $\delta/D = 0.006$

as internal obstacles moving relative to the computational domain. This procedure is suitable to avoid problems arising from matching the flow quantities using patched or overlaid grids that move relative to each other [4].

3 Method of solution

Using Cartesian velocity components the basic equations are discretized by employing a finite-volume scheme [5]. The flow domain is subdivided into a finite number of control volumes (CV). All dependent variables are defined in the centrepoint P of the CV (cf. Fig. 3 for grid arrangement and nomenclature). Integration of Eq. (2) for each CV leads to a balance equation of momentum fluxes through the CV faces and volumetric sources. The diffuse part of the momentum fluxes can be obtained by assuming linear variation of the variables between adjacent grid points. Evaluation of the convection fluxes requires discretization schemes for interpolating the variable values at the CV faces from their nodal values. In the present code, the convective fluxes are split into an implicit part which is obtained by first-order upwind differencing, and an explicit part containing the difference between the second-order accurate central differencing scheme and the upwind approximation. This technique originally suggested by Khosla and Rubin [6] is known to enhance the stability of the iterative solution algorithm. A second-order accurate discretization in time has been chosen to obtain time-accurate solutions [5]:

$$\left[\frac{\partial \underline{w}}{\partial t}\right]^{n+1} = \frac{1}{2\Delta t}\left(3\underline{w}^{n+1} - 4\underline{w}^{n} + \underline{w}^{n-1}\right) \tag{6}$$

The volumetric source terms of Eq. (2) are integrated by simply multiplying the specific source at the CV center (which is assumed to be representative for the whole volume) with the cell volume. The resulting finite volume equation for variable ϕ can be written in general form:

$$\frac{a_P\phi_P}{\alpha_\phi} = \Sigma a_{nb}\Phi_{nb} + b_\phi^* + \frac{1-\alpha_\phi}{\alpha_\phi} a_P\phi_P^*; \quad nb = E, W, N, S, T, B \tag{7}$$

where the coefficients a_{nb} represent the combined convection and diffusion effects and b_ϕ^* contains the discretized source terms (i.e. transient term, pressure gradient and Coriolis term)

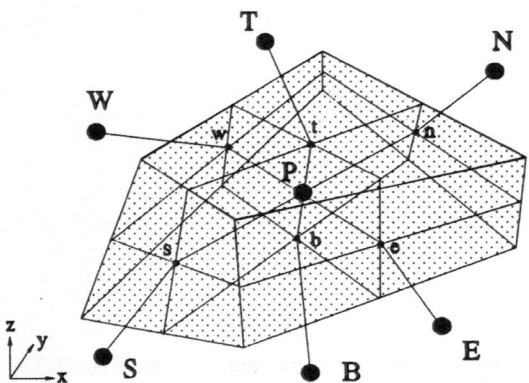

Fig. 3. Grid arrangement and nomenclature

and explicitly treated parts of the convection and diffusion fluxes [5]. The detailed expressions for the coefficients and the source terms are beyond the scope of the present paper and can be found in other publications, e.g. [7]. The equations are non-linear and strongly coupled by the convective and Coriolis term. The convergence of the iterative solution procedure requires under-relaxation of variable changes $(\phi_P - \phi_p{}^*)$ with a factor $0 < \alpha_\phi \leqq 1$. This enhances the diagonal dominance of the coefficient matrix.

For incompressible flows, the convergence of the numerical method for the solution of the momentum and continuity equation depends strongly on an adequate handling of the pressure-velocity coupling [8]. The velocity field obtained by solving the momentum equations using a guessed pressure field in general does not satisfy the continuity equation. The continuity equation serves as an additional constraint for the velocity field to adjust the pressure gradient in the momentum equations. In the present method, the coupling between pressure and velocities is achieved by the well-known SIMPLEC algorithm [9]. In order to avoid an oscillatory pressure field due to the non-staggered variable arrangement a special interpolation has been used to determine the mass fluxes through CV faces from the adjacent CV-centered quantities [10]. Substitution of the mass fluxes obtained from the calculated CV face velocities in the discretized form of the continuity equation will result in a mass imbalance S_m. Flux corrections are needed to annihilate this imbalance. These corrections are based on corresponding velocity corrections at the CV faces. According to the SIMPLEC algorithm [9] these velocity corrections are further related to pressure corrections at the nodal points. Thus, substitution of the mass flux corrections leads to a pressure-correction equation of the final form:

$$a_P p_P{}' = \Sigma a_{nb} p_{nb}' - S_m \tag{8}$$

In the present method, the system of Eqs. (7) and (8) are solved by an incomplete LU decomposition based on the SIP algorithm [11].

4 Results and discussion

The calculations have been performed for $Z = 24$ and $Z = 48$ blades of the impeller and the turbine. The other geometrical quantities are fixed throughout this study and defined in the caption of Fig. 2. The Ekman number is $Ek = 1.88 \cdot 10^{-5}$ and the Rossy number is $Ro = 0.05$. The numerical grid consists of either $36 \times 38 \times 78$ CV for $Z = 24$ or $19 \times 38 \times 78$ CV for $Z = 48$. The convergence of the iterative solution algorithm within each time step has been assumed when the sum of the normalized absolute residues for each dependent variable has fallen

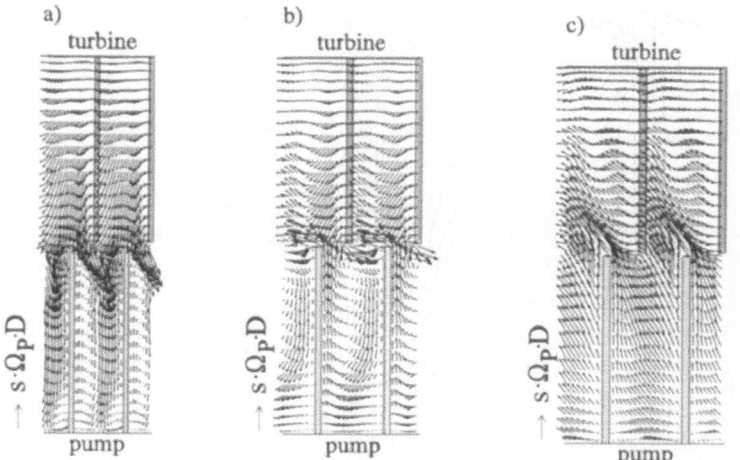

Fig. 4. Instantaneous velocity vectors relative to pump impeller in circumferential ("passage") cross-sections; 24 blades of P and T, **a** $r/D = 0.287$, **b** $r/D = 0.372$, **c** $r/D = 0.441$

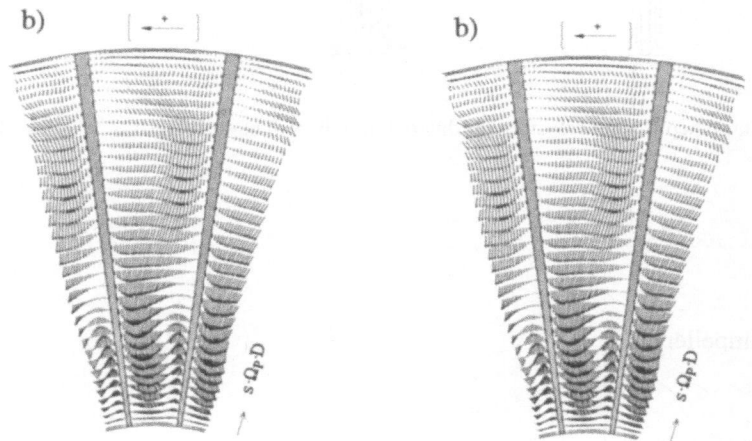

Fig. 5. Instantaneous velocity vectors in axial ("blade-to-blade") cross-sections relative to **a** pump impeller ($z/B = 0.5$) and **b** turbine runner ($z/B = 1.5$), 24 blades of P and T, ([+]: direction of absolute rotation)

below 10^{-3}. In the case of $Z = 24$ the cpu-time required for the passage of 10 blades of the turbine runner through the computational domain was about 78 h on a IBM RS/6000 Workstation.

Figure 4 displays instantaneous velocity vectors relative to the rotating co-ordinate system in different circumferential ("passage") cross-sections for $Z = 24$. Near the hub (Fig. 4 a) the flow is mainly directed from the turbine to the impeller and near the shroud (Fig. 4 c) vice versa, thus indicating the torque transmitting circulating flow between the coupling halves. A neutral zone exists between the pump impeller and the turbine runner (Fig. 4 b). In this region the flow is dominated by leading-edge flows from pressure-side to suction-side of the blades.

Figure 5 shows the corresponding flow field in axial ("blade-to-blade") cross-sections of the coupling. The velocity vectors are displayed relative to the motion of the coupling half they belong to, i.e. for the pump relative to angular velocity Ω_P and for the turbine relative to angular velocity Ω_T. In the pump impeller (Fig. 5 a), the main flow is directed radially outward and a secondary eddy has formed on the suction-sides of the blades near the shroud. The velocity

line	value
1	-1.5
2	-1.25
3	-1.0
4	-0.75
5	-0.5
6	-0.25
7	0
8	0.25
9	0.5
10	0.75
11	1.0
12	1.25
13	1.5

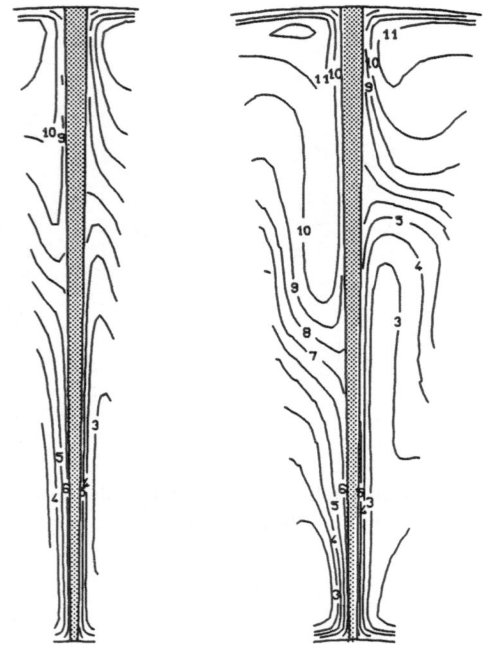

a   b

Fig. 6. Contour-lines of instantaneous axial velocities for different numbers of blades Z ($z/B = 1$); **a** $Z = 48$ and **b** $Z = 24$

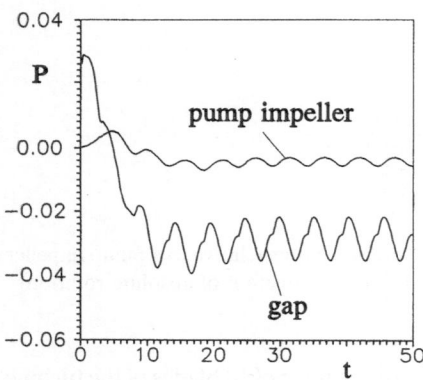

Fig. 7. Time history of reduced pressure at two specific points ($r/D = 0.247$) in the pump impeller ($z/B = 0.321$) and in the gap between the coupling halves ($z/B = 1.015$)

vectors in the turbine runner (Fig. 5 b) show a main flow directed radially inward and a passage vortex near the hub.

The flow structure is strongly influenced by the pitch $2\pi/Z$ of the fluid coupling. Figure 6 shows instantaneous contours of the axial velocity in the pump impeller for $Z = 48$ (Fig. 6 a) and $Z = 24$ (Fig. 6 b). The effects of secondary flow are generally much smaller for higher number of blades. The distribution of axial velocity is almost symmetrical for $Z = 48$ (Fig. 6 a). The contour-line of vanishing axial velocity component indicating the change of axial flow direction shows only small variations in circumferential direction. For $Z = 24$ the axial velocity between the pressure-side and suction-side of the blades exhibits a more unsymmetrical structure. There are regions of positive axial velocities not far from the hub on the pressure-side and negative axial velocities on the suction-side near the shroud, thus indicating the influence of secondary flow between the coupling halves.

Due to the cyclic passage of the turbine blades through the computational domain the flow field in fluid couplings is periodic in time. Figure 7 shows the time history of the reduced pressure at two specific points in the pump impeller and in the gap between the coupling halves. The wake of the turbine blade has a distinct influence on the pressure variation in the gap. With increasing distance from the gap the wake is compensated resulting in much smaller pressure fluctuations in the pump impeller.

Conclusions

A finite-volume method with non-staggered variable arrangement has been developed and used to solve the Navier-Stokes equations for a rotating frame of reference on boundary fitted grids. Results for the 3D periodic flow in fluid couplings show that the flow structure is strongly influenced by the number of blades of the coupling. Secondary flow effects reducing the torque transmission capacity of a fluid coupling are much smaller for higher number of blades. The unsteady effects due to the relative motion of pump and turbine blades decrease with increasing distance from the gap between the coupling halves.

Acknowledgements

This work was sponsored by the Deutsche Forschungsgemeinschaft (DFG) through special research program SFB 278 "Hydrodynamische Leistungsübertragung". The authors gratefully acknowledge this support.

References

[1] Langlois, H. J.: Hydrodynamic adjustable-speed drives. Power Transm. Des. **21** 60−62 (1979).

[2] Greenspan, H. P.: The theory of rotating fluids, p. 6. Cambridge: Cambridge University Press 1968.

[3] Dijkstra, D., van Heijst, G. J. F.: The flow between two finite rotating disks enclosed by a cylinder. J. Fluid Mech. **128**, 123−154 (1983).

[4] Rai, M. M.: Three-dimensional Navier-Stokes simulations of turbine rotor-stator interaction. J. Prop. Power **5**, 305−319 (1989).

[5] Kost, A., Bai, L., Mitra, N. K., Fiebig, M.: Calculation procedure for unsteady incompressible 3D flows in arbitrarily shaped domains. In: Notes on numerical fluid mechanics, vol. 35 (Vos, J. B., Rizzi, A., Rhyming, I. L., eds.), pp. 269−278. 9th GAMM-Conference on Numerical Methods in Fluid Mechanics, Lausanne 1991. Braunschweig: Vieweg 1992.

[6] Khosla, P. K., Rubin, S. G.: A diagonally dominant second-order accurate implicit scheme. Comp. Fluids **2**, 207−209 (1979).

[7] Perić, M.: A finite volume method for the prediction of three-dimensional fluid flow in complex ducts, pp. 58−81. Ph. D. Thesis, University of London 1985.

[8] Patankar, S. V.: Numerical heat transfer and fluid flow, pp. 113−138. New York: McGraw-Hill 1980.

[9] Van Doormaal, J. P., Raithby, G. D.: Enhancement of the SIMPLE method for predicting incompressible fluid flows. Num. Heat Trans. **7**, 147−163 (1984).

[10] Rhie, C. M., Chow, W. L.: A numerical study of the turbulent flow past an isolated airfoil with trailing edge separation. AIAA J. **21**, 1525−1532 (1983).

[11] Stone, H. L.: Iterative solution of implicit approximations of multidimensional partial differential equations. SIAM J. Num. Anal. **5**, 530−558 (1968).

Authors' address: Dr.-Ing. Andreas Kost, Professor Dr. N.-K. Mitra, and Professor Dr.-Ing. M. Fiebig, Institut für Thermo- und Fluiddynamik, Ruhr-Universität Bochum, Universitätsstrasse 150, D-44801 Bochum, Federal Republic of Germany

Acta Mechanica (1994) [Suppl] 4: 207–217
© Springer-Verlag 1994

Passive control of shock-boundary layer interaction in transonic axial compressor cascade flow

S. Yu, Beijing, People's Republic of China, G. H. Schnerr, U. Dohrmann, and O. Sadi, Karlsruhe, Federal Republic of Germany

Summary. The object of this investigation is the ventilated flow through an axial transonic/supersonic compressor cascade. It is a joint research project between the Chinese Academy of Sciences, Institute of Engineering Thermophysics and the University of Karlsruhe (TH), Institute for Fluid Mechanics and Fluid Machinery. Losses in the cascade flow can be reduced by passive control of the shock-boundary layer interaction. The passive control is realized by a cavity covered with a porous plate which is located at the interference region of the shock at the suction side of the blade. The static pressure gradient along the porous surface initiates a flow through the cavity which smoothes the pressure jump across the shock automatically. Experiments are performed in a cascade wind tunnel of the Chinese Academy of Sciences and in a cascade element being installed in the supersonic wind tunnel of the Institute for Fluid Mechanics and Fluid Machinery. Numerical results are obtained at the University of Karlsruhe by solving the 2-D Reynolds averaged Navier-Stokes equation with an explicit, time-dependent finite volume method. Experimental results in the cascade and in the cascade element yield a relative reduction of the total loss of 14–15% and an improvement of the isentropic efficiency of about 2%. In the numerical simulation of stationary viscous flow the total loss reduces by more than 5% and the isentropic efficiency increases by nearly 1%.

1 Introduction

Recent developments of axial compressor cascades lead to an increase of the mass flux and a reduction of the number of stages, even for constant pressure ratios (Bölcs and Suter [1]). The flow velocity increases, the relative inlet Mach number changes from sub- to supersonic flow and leads to complicated shock systems. As a consequence, shock-boundary layer interaction and separation phenomena become more important (Schreiber and Yu [2]). Figure 1 shows a typical schlieren picture of the flow in a cascade with a supersonic inlet Mach number. A bow shock develops in front of the blade which is reflected at the suction side of the neighbouring profile. A nearly normal shock in the passage leads to subsonic outflow. In the rear part of the suction side the boundary layer separates. Both shocks and boundary-layer separation cause additive losses. In this joint research project the method of passive control is applied to transonic internal flow to reduce the losses and separation phenomena. The main idea of this principle is to utilize the pressure rise across the shock to bypass a small secondary flow through a cavity covered by a perforated surface (typical porosity is 8%). The cavity is located at the suction side of the blade at the shock foot point (Fig. 2). Due to the blowing out upstream and the sucking in downstream of the shock, ventilation smoothes the pressure gradient, moves the shock upstream and causes a λ-shock in the interference region. The boundary layer is also influenced by the geometry and location of the cavity. The effect of ventilation on transonic flows in a cascade with given static pressure ratio and mass flux as additional parameters of the internal flow is significantly different

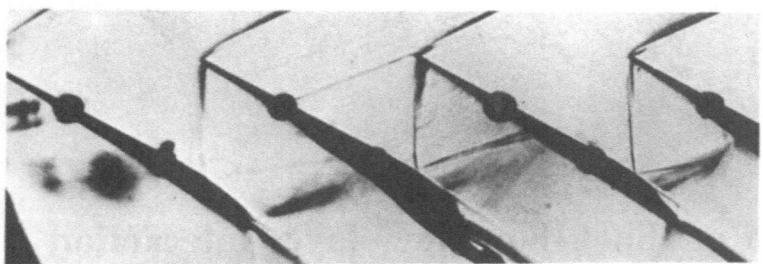

Fig. 1. Schlieren picture of stationary flow in the cascade without passive control, flow from the top; $M_1 = 1.291$, $\pi = 1.56$, chord length $c = 129.7$ mm

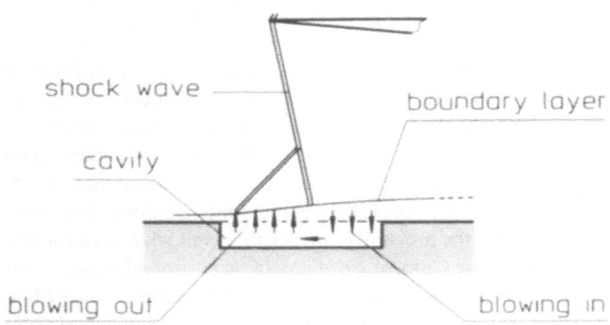

Fig. 2. Bow shock with passive control and normal blowing out; flow from the left

from that around airfoils. Thus the interaction between the ventilation and the inviscid flow is more complicated and the maximum decrease of the losses in the cascade is expected to be less than that in transonic flow around an airfoil.

An axial compressor cascade with wedge-circular arc blades is investigated. The oncoming flow is supersonic ($M_1 = 1.3$), whereas the axial Mach number remains less than unity. The pressure ratio $\pi = p_2/p_1$ is chosen to ensure subsonic flow downstream of the cascade. The subscripts '1' and '2' denote quantities upstream and downstream of the cascade, respectively. The inlet flow angle e β_1 is constant and follows from the 'unique incidence' condition (Bölcs and Suter [1]). Experimental investigations in a cascade wind tunnel with 6 blades are performed by the Chinese group. Further experiments are carried out by the German group in a cascade element consisting of one suction and one pressure side. This has been done to get more detailed information about the mechanism of passive control. For the flow regime without ventilation as shown in Fig. 1, the best improvement of the flow performance with passive control is obtained. Thus the results are discussed for this flow configuration in the subsequence. Experimental results in both wind tunnels lead to a maximum relative reduction of the loss coefficient ($L = (p_{t1} - p_{t2})/(p_{t1} - p_1)$) of $14-15\%$. The isentropic efficiency ($\eta_{is} = (h_{2is} - h_1)/(h_2 - h_1)$) is improved by about $1.5-2\%$. Here h is the enthalpy, the subscripts 't' and 'is' denote total values and the assumption of constant entropy. The zonal method (Breitling [3]) based on the analytical shock-boundary layer interaction model of Bohning and Zierep [4] for transonic flows across airfoils cannot be applied in the numerical simulations because of the boundary layer separation downstream of the shock without reattachment. Therefore the Reynolds-averaged Navier-Stokes equation is used to calculate the viscous stationary cascade flow. Numerical results yield a reduction in the loss coefficient by more than 5% and an improvement of the isentropic efficiency by about 1% which confirms the experimental results in tendency and demonstrate the effectiveness of passive control.

2 Test facility

2.1 Cascade wind tunnel

Experiments of the Chinese cooperation partner are performed in a supersonic intermittent blow down cascade wind tunnel. The cross sectional area of the test section is 220 mm × 290 mm. The axial cascade is simulated by 6 wedge-circular arc blades ($c = 129.7$ mm). A cavity covered by a perforated sheet (porosity 8.2%) is inserted on the suction side (Fig. 3) of the third blade from the bottom. The cavity length is varied by sealing the not required part of the porous area. Two boundary layer scoops eliminate the boundary layers at the upper and the lower wall in the oncoming flow. The back pressure is regulated with valves downstream of the test section.

Traverse measurements of static and stagnation pressure are made downstream of the blades (Fig. 3) to determine the isentropic efficiency and the loss coefficient. Furthermore, the static pressure is measured at the sidewalls upstream of the cascades. The flow can be visualized by a schlieren optical system and a holographic interferometer, pictures are taken by camera or using a video recording system. The measuring time is of the order of one minute.

2.2 Cascade element wind tunnel

The supersonic blow down wind tunnel of the Institute of Fluid Mechanics and Fluid Machinery with a cross sectional area of 50 mm × 230 mm is employed for the experiments of the German cooperation partner. A cascade element is build up by two blades of the wedge-circular arc profile ($c = 187.4$ mm) in order to obtain a great chord length for a high resolution of all details of the

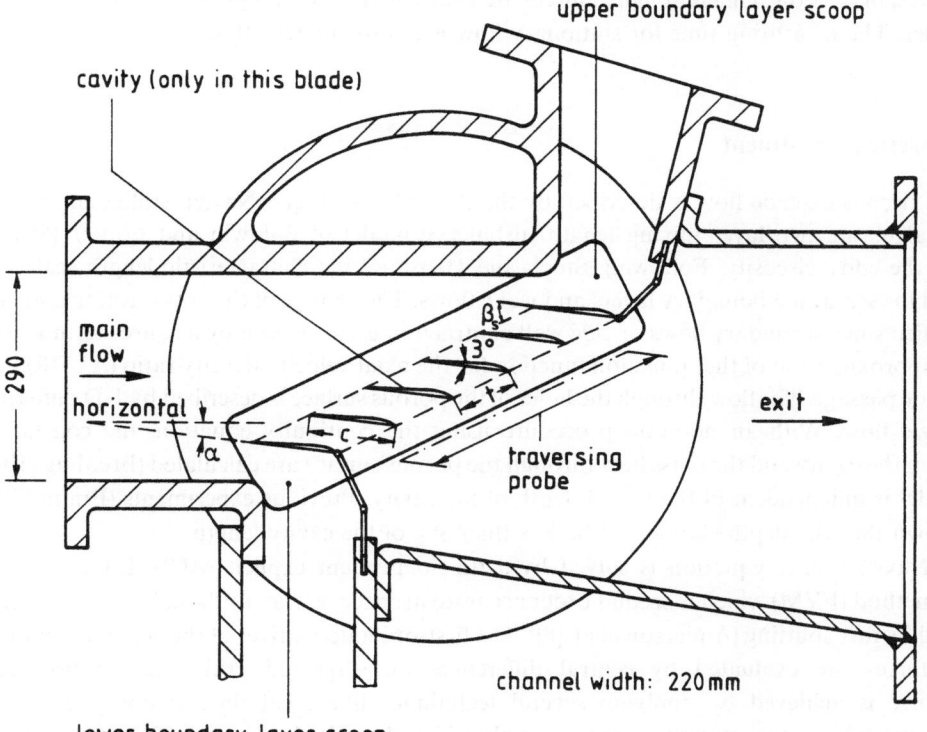

Fig. 3. Cascade wind tunnel, experimental device, length scale in mm; chord length $c = 129.7$ mm, blade spacing $t = 90$ mm, stagger angle $\beta_s = 32.25°$, angle of attack $\alpha = 5.75°$

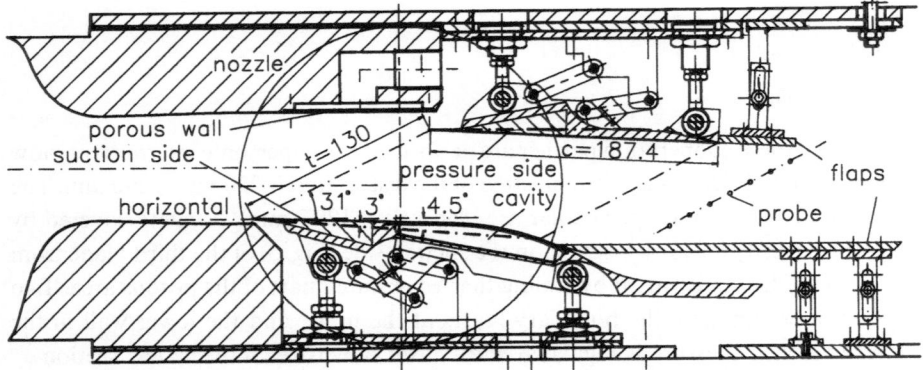

Fig. 4. Cascade element, experimental device, length scale in mm

ventilated shock-boundary interaction (Fig. 4). A plenum chamber for the passive control is inserted in the lower blade element, covered by a perforated sheet (porosity 8.2%). The cavity length is varied by sealing the not required part of the porous area. At the upper nozzle wall the reflection of the bow shock is eliminated by sucking action through a porous wall. The back pressure can be regulated with flaps.

The isentropic efficiency and the loss coefficient are determined by static and total pressure measurements downstream of the cascade passage as indicated in Fig. 4. Moreover, the static pressure is measured at the suction side of the lower blade element. The flow is visualized by a schlieren optical system. A CCD camera and digital image processing are used for registration and analysis of the flow. Different light sources are available, e.g. spark light source (10^{-6} s) and diode laser. The measuring time for stationary flow is approximately 10 s.

3   Theoretical treatment

The 2-D viscous cascade flow is described by the Reynolds-averaged Navier-Stokes equation and the algebraic two layer mixing length turbulence model of Baldwin and Lomax [5] to estimate the eddy viscosity. Following Stock and Haase [6] the characteristic length scale is corrected for separated boundary layers and wake flows. The change of the cross sectional area by 3-D effects like secondary flows or side wall contraction is considered by a source term with a linear approximation of the transition function for the axial velocity density ratio (AVDR) in the cascade passage. The flow through the holes of the porous surface is described by 1-D diffuser and nozzle flow. With an iteration procedure using the continuity equation, the constant pressure in the cavity and the mass flow through the porous surface are calculated (Breitling [3]). This model is independent of the actual depth of the cavity. Previous experiments (Braun [7]) have shown that the depth should not be less than 3% of the cavity length.

The Navier-Stokes equation is solved by a time-dependent explicit MUSCL-type finite volume method (FVM) which is second order accurate in space. Fluxes at the cell interfaces are evaluated by flux splitting (Anderson et al. [8]). The first-order derivatives of the velocities of the viscous fluxes are evaluated by central differences on staggered grids. Acceleration of convergence is achieved by applying several techniques like local time stepping, 3-stage Runge-Kutta scheme for integration in time, multigrid and residual averaging (Baker et al. [9]). The Navier-Stokes equation is solved on a H-grid with a total number of 256×32 cells. Iteration stops when the RMS value of the maximum residual of the conservative variables is less than

5×10^{-7}. A typical CPU-time for inviscid calculations is about 15 minutes. Stationary solutions of the Navier-Stokes equation require a CPU-time of 45 minutes without and about 120 minutes with passive control on a Siemens vector computer VP 600.

4   Results

4.1   Experiments in cascade wind tunnel

Figure 1 shows a schlieren picture of the flow through the compressor cascade without passive control ($M_1 = 1.291$, $\beta_1 = 26.9°$, chord length $c = 129.7$ mm). On the suction side of the blade the shock reflection in the cascade passage is located at $x/c = 0.806$ and the rear shock at $x/c = 0.864$. The loss coefficient is $L = 9.79\%$ and the isentropic efficiency comes to $\eta_{is} = 0.866$.

This performance was investigated with passive control for different geometries and locations of the cavity by nearly constant inlet Mach number and incidence angle. The largest increase of the isentropic efficiency in the ventilated flow is obtained for the cavity located at both the shock reflection and the foot point of the rear shock (cavity length $l_c = 28$ mm, begin of the cavity $x/c = 0.632$, end of the cavity $x/c = 0.848$). The position of the shock reflection relative to the cavity length is $\lambda_1 = 80\%$. Figure 5 shows a schlieren picture of this flow with passive control ($M_1 = 1.294$, $\beta_1 = 26.8°$). The cavity is inserted in the third blade from the left. The pressure ratio is slightly increased in comparison with Fig. 1. At the beginning of the cavity the outflow causes a weak oblique shock [10] which decreases the strength of the impinging bow shock near the surface of the suction side, the rear shock moves upstream. Although separation still exists, the overall performance was improved remarkably. The loss coefficient is reduced to $L = 8.31\%$ which complies with a relative decrease of 15.1% of the original loss coefficient. The isentropic efficiency increases to $\eta_{is} = 0.886$.

4.2   Experiments in the cascade element

To verify the results of the cascade wind tunnel the geometry of the blade passage of Figs. 1 and 5 was enlarged in similarity by a factor of 1.44 (chord length $c = 187.4$ mm). The inlet Mach number M_1 and the inlet flow angle β_1 are almost the same. Figure 6 shows schlieren pictures of the flow with and without passive control ($M_1 = 1.3$, Reynolds number Re $= 2.9 \times 10^6$). The angle 4.5° (Fig. 4) corresponds to $\beta_1 = 26.5°$ and is corrected to compensate boundary layer displacement effects. The pressure ratio $\pi = 1.63$ is chosen to get the same flow regime in the passage without passive control as in the corresponding experiment of the Chinese partner. Different flow exit conditions behind the cascade lead to a higher pressure ratio. Small differences

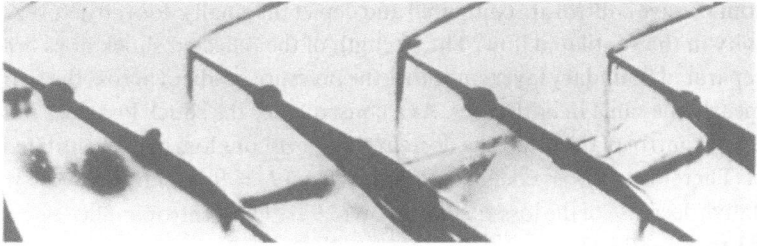

Fig. 5. Schlieren picture of stationary flow in the cascade with passive control, flow from the top; cavity inserted in the 3rd blade from bottom; $M_1 = 1.294$, $\pi = 1.57$, chord length $c = 129.7$ mm, cavity length $l_c = 28$ mm

a

b

Fig. 6. Schlieren pictures of stationary flow in the cascade element with **(b)** and without **(a)** passive control, flow from the left; $M_1 = 1.3$, Re $= 2.9 \times 10^6$, $\pi = 1.63$, chord length $c = 187.4$ mm, cavity length $l_c = 40$ mm

in the flow parameters M_1, β_1 and π compared with the experiment in the cascade wind tunnel cause small differences in the location of the shocks. On the suction side of the blade the shock reflection in the cascade passage is located at $x/c = 0.758$ and the rear shock at $x/c = 0.880$.

Experiments with different geometries and locations of the cavity reveal the best improvement of the overall performance for a cavity located under both the shock reflection and the shock foot point of the rear shock (cavity length $l_c = 40$ mm, begin of the cavity $x/c = 0.715$, end of the cavity $x/c = 0.928$). This confirms the results in the cascade wind tunnel. For the cascade element the position of the shock reflection relative to the cavity length is $\lambda_1 = 20\%$. At the leading edge of the cavity the outflow causes a weak oblique shock which interacts with the impinging bow shock near the surface of the suction side (Fig. 6). The boundary layer separates behind the shock reflection as in the case without passive control. In the rear part of the cavity the sucking in leads to a smaller extension of the separated boundary layer along the porous surface in the ventilated case. Therefore the viscous losses are reduced. The rear shock becomes curved and is reflected several times until subsonic flow is reached downstream of the passage. In Fig. 7 the static pressure distributions at the suction side with and without passive control are compared and depict the small λ-foot structure at the leading edge of the cavity in the ventilated flow. The strength of the reflected shock does not change remarkably. The separated boundary layer smoothes the pressure gradient across the rear shock, the distribution is nearly the same in both cases. As a consequence the shock loss does not change decisively with passive control of the flow. The decrease of the viscous loss in the ventilated flow is the dominant effect. Therefore the loss coefficient reduces from $L = 9.12\%$ to $L = 7.85\%$ which corresponds to a relative decrease of the loss coefficient by 13.9%. The isentropic efficiency is improved from $\eta_{is} = 0.883$ to $\eta_{is} = 0.897$.

The improvement of the overall performance of the flow with passive control and the adequate location of the cavity in the one element cascade passage agrees very well with the experiment in the cascade wind tunnel.

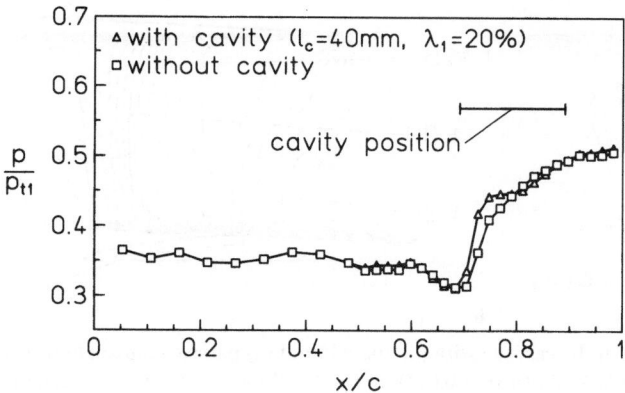

Fig. 7. Normalized static pressure distribution at the suction side of the blade element-experiment; $M_1 = 1.3$, $\text{Re} = 2.9 \times 10^6$, $\pi = 1.63$, chord length $c = 187.4$ mm, cavity length $l_c = 40$ mm

4.3 Numerical investigation

Numerical results are calculated with $M_1 = 1.3$, $\text{Re} = 3 \times 10^6$ and the same chord length $c = 187.4$ mm as in the corresponding experiments in the cascade element. Figure 8 shows constant Mach number contours in the supersonic region of the cascade for viscous flows with and without passive control ($\beta_1 = 25.1°$, $\pi = 1.721$). The pressure ratio is chosen to get nearly the same flow regime in the non-ventilated flow as in the experiments. The smaller inlet flow angle β_1 follows from the unique incidence condition and the chosen transition function for the AVDR. Thus the reflection of the bow shock on the suction side of the blade is located more upstream at $x/c = 0.731$. Because of this and as a consequence of the weakening of the pressure jump across the reflected shock near the surface due to the numerical method, separation does not start at the

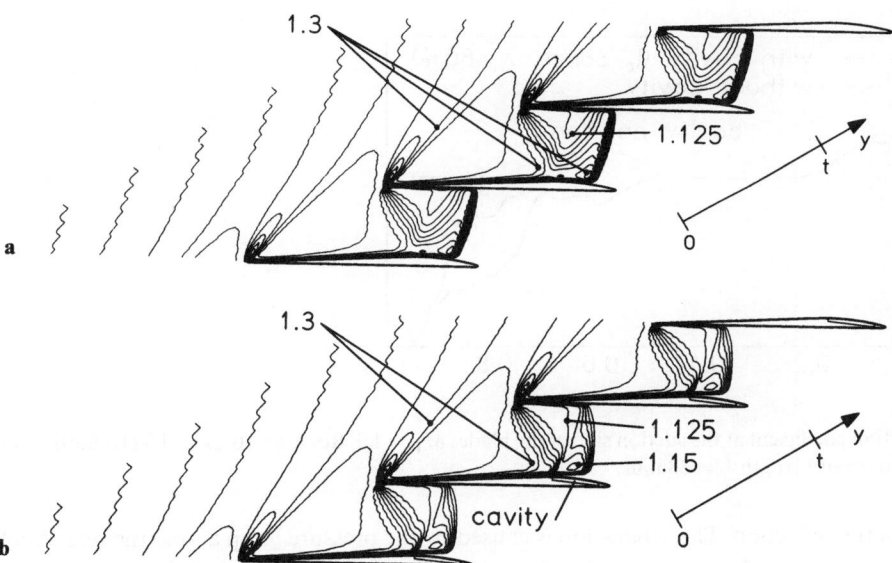

Fig. 8. Viscous flow through a compressor cascade with (**b**) and without (**a**) passive control, flow from the left; supersonic Mach number contours $M \geqq 1$ ($\Delta M = 0.025$); $M_1 = 1.3$, $\text{Re} = 3 \times 10^6$, $\pi = 1.721$, chord length $c = 187.4$ mm, cavity length $l_c = 38$ mm

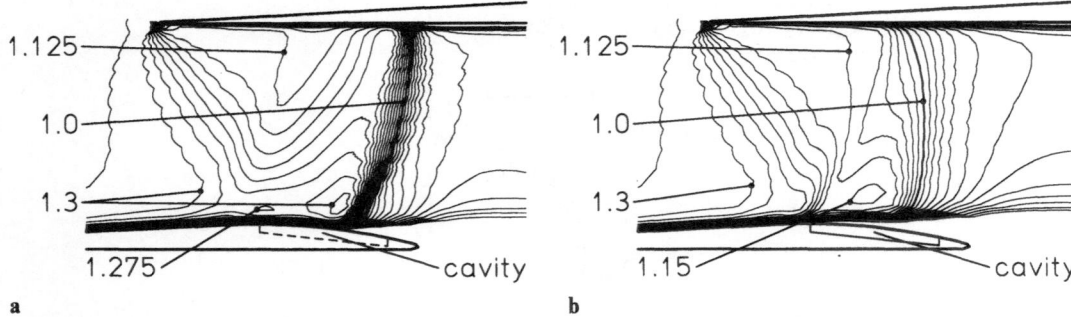

a   **b**

Fig. 9. Enlargements of the shock region in the cascade with **(b)** and without **(a)** passive control, flow from the left; constant Mach number contours $M \geqq 0.7$ ($\Delta M = 0.025$); $M_1 = 1.3$, Re $= 3 \times 10^6$, $\pi = 1.721$, chord length $c = 187.4$ mm, cavity length $l_c = 38$ mm

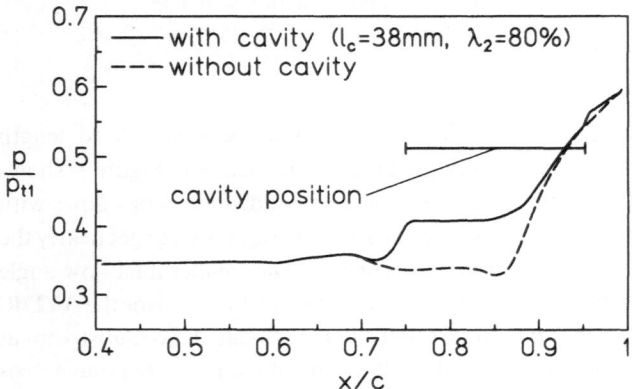

Fig. 10. Normalized static pressure distribution at the suction side of the blade; $M_1 = 1.3$, Re $= 3 \times 10^6$, $\pi = 1.721$, chord length $c = 187.4$ mm, cavity length $l_c = 38$ mm

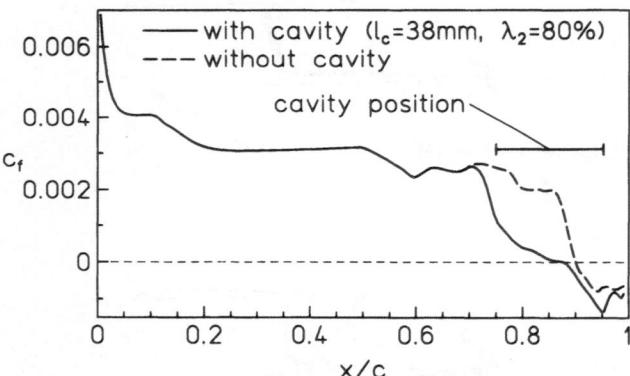

Fig. 11. Friction coefficient at the suction side of the blade; $M_1 = 1.3$, Re $= 3 \times 10^6$, $\pi = 1.721$, chord length $c = 187.4$ mm, cavity length $l_c = 38$ mm

foot point of the reflection. The separation is caused by the pressure jump across the rear shock. Thus the pressure ratio for a comparable flow regime is higher in the numerical investigation than in the experiments.

Several calculations with passive control are performed by varying geometry and location of the cavity. The ventilated flow in Fig. 8 (cavity length $l_c = 38$ mm, begin of the cavity $x/c = 0.750$,

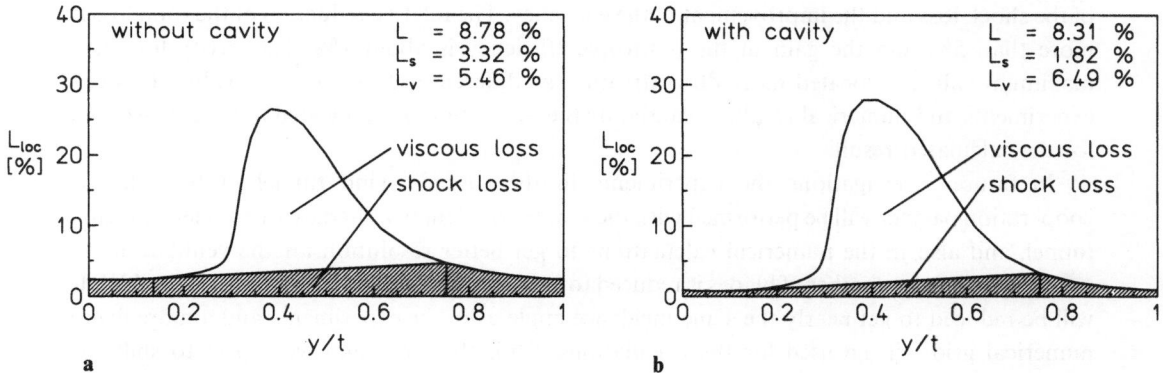

Fig. 12. Local loss coefficient calculated downstream of the cascade with **(b)** and without **(a)** passive control; $M_1 = 1.3$, Re $= 3 \times 10^6$, $\pi = 1.721$, chord length $c = 187.4$ mm, cavity length $l_c = 38$ mm

end of the cavity $x/c = 0.952$) shows the Mach number contours for the largest reduction of the loss coefficient. The cavity is located below the foot point of the rear shock ($x/c = 0.911$); the position relative to the cavity length is $\lambda_2 = 80\%$. Figure 9 shows enlargements of the passage flow with and without ventilation. Constant Mach number contours $M \geq 0.7$ are presented, the sonic line is plotted thicker. The ventilation causes an oblique shock at the leading edge of the cavity, the rear shock is shifted upstream. The Mach number in front of the rear shock decreases and as a consequence the shock strength is weakend. The boundary layer is thickened along the porous surface and separation is started more upstream. Figure 10 depicts the normalized static pressure distribution at the suction side of the blade. The pressure distribution of the flow with passive control shows a slight pressure increase across the oblique shock at the beginning of the cavity and a weaker pressure increase across the rear shock. The friction coefficient c_f decreases along the cavity, and negative values indicate an upstream shift of the boundary layer separation for the ventilated flow (Fig. 11). The loss coefficient is determined in the wake of one cascade element in a plane downstream of the cascade marked by the y-axis in Fig. 8. Figure 12 depicts the variation of the local loss coefficient L_{loc} for the flow with and without passive control. To separate shock loss and viscous loss we use the empirical model of Schreiber [11]. The shaded area denotes the shock loss, while the unshaded area shows the viscous loss. Ventilation reduces the wave drag and increases the viscous loss. As net effect the loss coefficient L decreases about 5.4%. The isentropic efficiency increases from $\eta_{is} = 0.901$ to $\eta_{is} = 0.908$.

The numerical results for the flow with passive control confirm the decrease of the loss coefficient and the increase of the isentropic efficiency being observed in the experiments (Schnerr et al. [12]). The difference in the absolute values and in the location of the cavity for the optimum improvement results from the different separation behavior.

5 Conclusions

Experiments and numerical calculation show the effectiveness of the passive control of the shock-boundary layer interaction in internal flows. The experimental results in the cascade element and in the cascade wind tunnel yield a relative reduction of the total loss of $14 - 15\%$ and an improvement of the isentropic efficiency by about 2%. For nearly the same relative cavity length, the agreement of these experiments is very good. Numerical results confirm the reduction of the shock loss and the improvement of the efficiency. The relative reduction of the total loss is more than 5% and the gain in the isentropic efficiency is about 1%. The cavity for these maximum values is located more downstream. The difference of the absolute values between experiments and numerical results is caused by the separation being located more downstream for the calculated results.

In future investigations the experiments in the cascade wind tunnel of the Chinese cooperation partner will be performed with the same chord length as in the cascade element wind tunnel, and also in the numerical calculations to get better resolution for the ventilation. As a consequence, the number of blades is reduced to four. In the numerical investigation the AVDR will be reduced to get nearly the same incidence angle as in the experiments and a more dense numerical grid will be used for the calculations. With these changes we expect to shift the separation upstream to the location of the shock reflection in the non-ventilated case and to get an improvement of the characteristic values (loss coefficient, isentropic efficiency) in the order of the experiments. Further improvement may be expected from a variation of the cavity in- and outflow angle.

Experiments will also be made for unsteady cascade flows to investigate the influence of passive control on oscillating flows. First results show a significant stabilizing of shock oscillations for frequencies up to 100 Hz, the amplitudes are reduced up to 40% (Schnerr et al. [13]) and more (Yu [14]).

Acknowledgements

This joint research project is financially supported by DFG (Deutsche Forschungsgemeinschaft) and NSFC (National Natural Science Foundation of China). The authors dedicate this paper to Prof. Dr.-Ing. Dr. techn. E. h. J. Zierep on the occasion of his 65th birthday; his guidance and proposal has been decisive for the success of this investigation. The authors acknowledge the experimental work of the whole cascade tunnel group in China.

References

[1] Bölcs, A., Suter, P.: Transsonische Turbomaschinen. Karlsruhe: G. Braun 1986.
[2] Schreiber, H. A., Yu, S.: Experiments on strong shock wave turbulent boundary layer interaction in a supersonic compressor cascade. DLR-IB-325-02/89 (1989).
[3] Breitling, T.: Berechnung transsonischer reibungsbehafteter Kanal- und Profilströmungen mit passiver Beeinflussung. Dissertation Universität Karlsruhe (TH) 1989.
[4] Bohning, R., Zierep, J.: Der senkrechte Verdichtungsstoß an einer gekrümmten Wand unter Einfluß der Reibung. ZAMP 27, 225−240 (1976).
[5] Baldwin, B. S., Lomax, H.: Thin layer approximation and algebraic model for separated turbulent flow. AIAA 78−257, AIAA 16th Aerospace Sciences Meeting, Huntsville, Alabama 1978.
[6] Stock, H. W., Haase, W.: The determination of turbulent length scales in algebraic turbulence models for attached and slightly separated flows using Navier-Stokes methods. AIAA 87−1302 (1987).
[7] Braun, W.: Experimentelle Untersuchungen der turbulenten Stoß-Grenzschicht-Wechselwirkung mit passiver Beeinflussung. Dissertation Universität Karlsruhe (TH) 1990.
[8] Anderson, W. K., Thomas, J. L., van Leer, B.: Comparison of finite volume flux vector splittings for the Euler equations. AIAA J. 24, 1453−1460 (1986).
[9] Baker, T. J., Jameson, A., Schmidt, W.: A family of fast and robust Euler codes. MAE Report 1652 (1984).
[10] Yu, S., Li, J. Y., Yu, B., Fan, C. F.: Some preliminary experimental results on passive control of shock wave boundary layer interaction in compressor cascades. In: Proceedings of the 1st International

Symposium on Experimental and Computational Aerothermodynamics of Internal Flows (Chen, N. X., Jiang, H. D., eds.) pp. 217–219. Beijing: World Publishing Corporation 1990.

[11] Schreiber, H. A.: Shock losses in transonic and supersonic compressor cascades. DLR-IB-325-03/89 (1989).

[12] Schnerr, G. H., Dohrmann, U., Zierep, J.: Passive Control of the Shock-Boundary Layer Interaction in a Transonic Compressor Cascade. Proceedings of the 2 ISAIF, Prague, July 12–15 (eds. R. Dvořák, J. Kvapilová), Society of Czech Mathematicians and Physicists, Prague, pp. 491–498 (1993)

[13] Schnerr, G. H., Dohrmann, U., Sadi, O., Zierep, J.: Numerical and experimental investigation of passive control of the shock-boundary layer interaction in a transonic compressor cascade. Proceedings ICFM-II, Beijing, China, July 7–10 (eds. Zhao Dagang, Zhang Zhixin), Peking University Press, Beijing, pp. 504–510 (1993).

[14] Yu. S.: Suppression of shock oscillations in a compressor cascade by using passive control. Proceedings of the 2 ISAIF, Prague, July 12–15 (eds. R. Dvořák, J. Kvapilová), Society of Czech Mathematicians and Physicists, Prague, pp. 609–611 (1993).

Authors' addresses: Professor S. Yu, Institute of Engineering Thermophysics, Chinese Academy of Sciences, P.O. Box 2706, Beijing, 100 080, People's Republic of China, Professor Dr.-Ing. habil. G. H. Schnerr, Dr.-Ing. U. Dohrmann, Dipl.-Ing. O. Sadi, Institut für Strömungslehre und Strömungsmaschinen, Universität Karlsruhe (TH), D-76128 Karlsruhe, Federal Republic of Germany

Acta Mechanica (1994) [Suppl] 4: 219−231

Part 6: Computational Fluid Dynamics

Numerical simulation of viscous transonic airfoil flows with passive shock control

R. Bohning, Karlsruhe, **P. Thiede** and **G. Dargel**, Bremen, Federal Republic of Germany

Summary. An interactive zonal method for the numerical simulation of viscous transonic airfoil flows with passive shock control is presented. The approach is based on a global viscous-inviscid interaction procedure including a local shock boundary layer interaction solution. The predicted shock control effects on the airfoil characteristics result not only in wave and viscous drag reductions but also in a lift increase. The results show that the transonic performance of a typical supercritical airfoil can be substantially improved by a passive shock control device consisting of a perforated surface with a cavity underneath. Furthermore, it is demonstrated that a perforation with inclined holes is more effective than that with normal ones.

List of symbols

c	chord length	u, U, v, V	mean flow velocity components
c_D	drag coefficient	x, y	cartesian coordinates
c_{Dv}	viscous drag coefficient		
c_{Dw}	wave drag coefficient	α	angle of attack
c_f	friction coefficient	β	wake angle
c_l	lift coefficient	δ	boundary layer thickness
c_p	pressure coefficient	δ_1	displacement thickness
M	Mach number	δ_2	momentum thickness
p	static pressure	$\varkappa^*$	surface curvature
q	total velocity, $q^2 = u^2 + v^2$	ν	viscous-inviscid iteration number
Re	Reynolds number	ϱ	density
s	coordinate along the surface	ω	relaxation parameter

Subscripts and superscripts

e	boundary layer edge	U, L	upper, lower surface
i	equivalent inviscid flow	w	wall
n, t	normal, tangential	∞	free stream conditions

1 Introduction

The aerodynamic performance of transonic wings can be substantially improved by passive control of shock boundary layer interaction (SBLIC). Passive shock control (SC) can be accomplished by boundary layer ventilation in the shock region through a porous surface with a cavity located underneath, Fig. 1. Passive control has the potential to weaken the shock wave, to improve the flow recovery behind the shock and to reduce the shock-induced separation

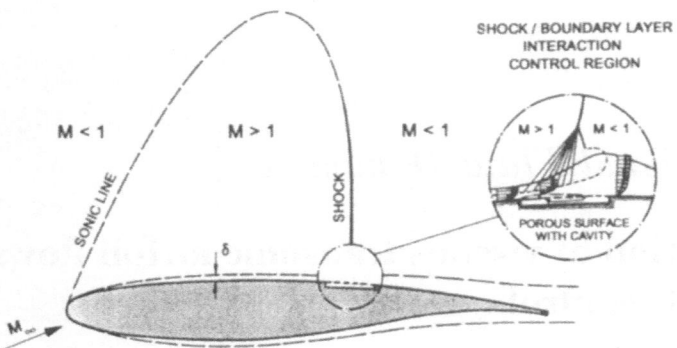

Fig. 1. Transonic airfoil flow with passive shock control

tendency, resulting in wave and viscous drag reductions. In addition, passive shock control offers a means to postpone the transonic drag rise and buffet boundaries as demonstrated in several 2D experiments, [1 – 3].

This paper aims at the numerical simulation of transonic airfoil flows with passive shock control. A first approach to compute viscous transonic flows over porous airfoils has been made by Chen et al. [4] on the basis of the Navier-Stokes equations. The present prediction method is based on the viscous-inviscid interaction (VII) code of Deutsche Aerospace Airbus (DA) [5] allowing an efficient simulation of viscous transonic airfoil flows. For the simulation of shock control, the local triple-deck solution described in [6, 7] extended to shock control conditions has been implemented. The global VII method based on the defect formulation concept of Le Balleur [8] including the local SBLIC solution will be described in the following. Finally, some effects of shock control on transonic airfoil flow characteristics are computed using the present method and will be presented.

2 Global viscous inviscid interaction (VII) method

The global interaction method described in [5] is widely used by DA for the prediction of airfoil flows without SC. Based on the defect formulation concept of Le Balleur [8], the real viscous flow is split into a viscous defect flow and an equivalent inviscid flow which is extended to the airfoil surface and to the wake streamline as the location of the prescibed viscous boundary conditions.

2.1 Inviscid transonic full potential (TFP) method including Prandtl shock operator

The equivalent inviscid flow with prescribed boundary conditions is computed using the TFP method of Jakob [9] with an entropy correction in the shock region from Mertens et al. [10]. The full potential equation for transonic flows is transformed from the physical domain into the incompressible potential-line and streamline plane. Using Murman's finite difference scheme the finite difference equation is solved by the SLOR-method. Because the potential flow solution is only valid for isentropic shock conditions, a shock operator is introduced satisfying the Rankine-Hugoniot relation. Hereby, a controlled mass increase simulates the entropy rise in the potential equation. The interaction of the inviscid flow with

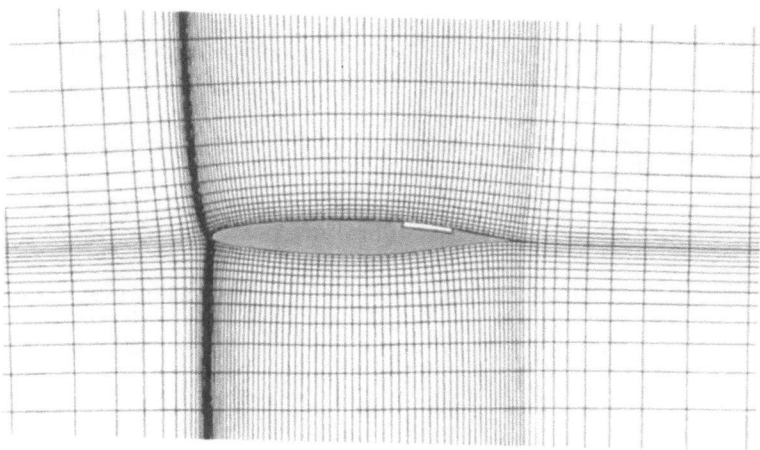

Fig. 2. Grid for SC airfoil flow computation, DA airfoil LVA-1A-SC, $\alpha = 1°$

the boundary layer, resulting in a weakening of the shock, can be reasonably approximated by prescribing the oblique shock relation with maximum stream deflection in the Prandtl shock operator. The entropy jump calculation across the shock leads to a precise wave drag prediction.

The H-type grid, Fig. 2, is generated from the incompressible potential-lines and streamlines of the inviscid flow past the airfoil at given incidence. The standart grid consists of 101 points in streamwise direction (60 points along the airfoil surface) with a refinement in the ventilation region.

2.2 Viscous boundary conditions for the inviscid flow

Normal velocities v_{iw} on the surface and normal velocity jumps $\langle v_{iw} \rangle$ across the wake simulate the displacement effect of the viscous flow, Fig. 3. In the shock control region the ventilation velocities v_w are superimposed. The viscous curvature effects result in a correction between the inviscid and viscous wall pressure $(p_w - p_{iw})$ which can be related to tangential velocity jumps $\langle u_{iw} \rangle$ across the wake effecting the Kutta-condition. Due to the viscous-inviscid interaction the position of the wake streamline and its curvature has to be iteratively determined in the equivalent inviscid flow by the condition of equal normal velocities of the inviscid (I) and viscous (V) solution giving a correction for the wake angle.

2.3 Interactive shear layer integral method

The viscous defect flow is efficiently computed by solving the streamwise momentum and the mean flow kinetic energy integral equations, written both in direct and inverse form. The inverse mode overcomes the breakdown of the direct mode at separation and predicts the inviscid Mach number by a given displacement thickness. The compressibility transformations and the laminar and turbulent closure relations are deduced from the corresponding laminar and turbulent boundary layer relations described in [11]. An empirical dissipation correlation for non-equilibrium effects is taken into account. The attached laminar and turbulent boundary layer

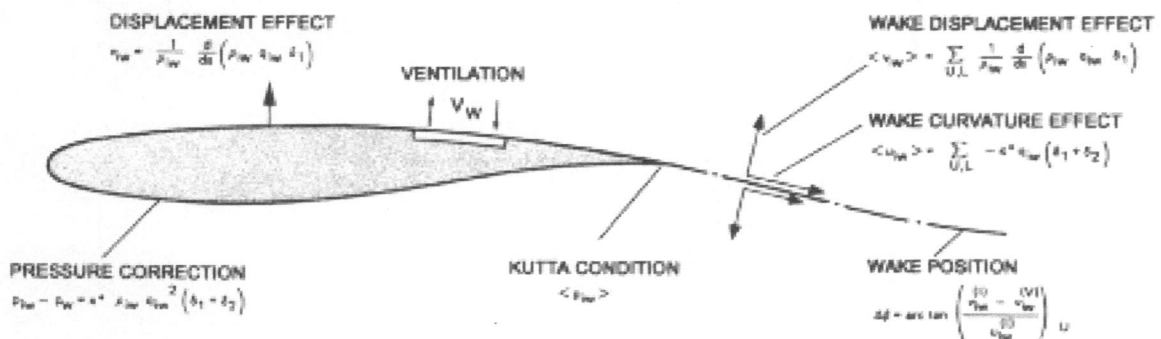

Fig. 3. Viscous boundary conditions for equivalent inviscid flow

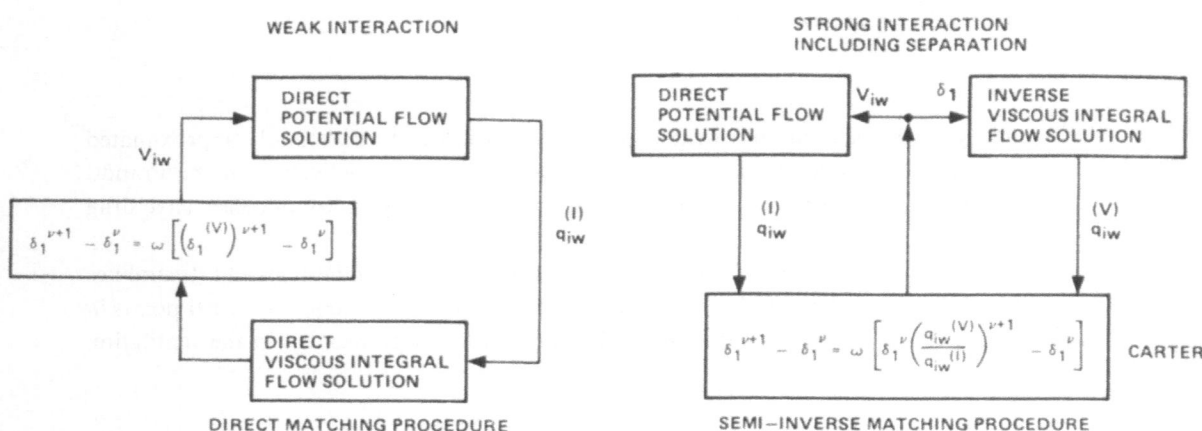

Fig. 4. Viscous-inviscid matching procedures

equations are solved by a modified Walz-method [12]. The transition point location can be prescribed or calculated from a N-factor law.

For the lower and upper half wakes the integral equations are solved setting the skin friction coefficient to zero. Concerning the turbulent closure relations the dissipation law is only changed due to the wake turbulence structure. The initial values of the wake calculation can be modified to consider a thick trailing edge. The viscous drag is calculated from the far wake momentum thicknesses.

2.4 Viscous-inviscid matching procedures

In case of attached flow, that means for weak interaction, the classical direct matching procedure is applied for coupling the inviscid and viscous flow solution, Fig. 4. Within strong interaction areas as the shock and trailing edge regions with incipient flow separation, the semi-inverse matching scheme, proposed by Carter [13] is applied coupling the direct inviscid solution and the inverse viscous solution iteratively, Fig. 4. Both solutions are separately computed with a given displacement thickness, and the resulting velocities are used to update the displacement thickness distribution. Underrelaxation is needed to assure convergence of the viscous-inviscid interaction procedure. The switch between both matching procedures is controlled by a shape parameter criterion.

The VII method described has been validated by numerous test cases as shown in [14]. Reliable results are obtained for cruise conditions with attached flow and moderate off-design conditions with incipient flow separation.

3 Local shock boundary layer interaction solution with passive shock control (SBLIC)

For the computation of the ventilated shock region, a local method has been developed based on the SBLI model described in [6, 15]. This triple-deck model, Fig. 5, consisting of

— an outer layer: inviscid flow outside of the boundary layer
— a middle layer: outer part of the boundary layer, in which the viscous terms are neglegted, but the shear flow vorticity terms are important
— and a wall layer: thin sublayer of the boundary layer, in which the viscous terms are dominant

has been extended in [7] and [16] to simulate mass flux through the porous wall due to the passive ventilation effect. The computation of the inviscid outer transonic flow has been derived from the numerical solution of the global VII method. The resulting pressure distribution and streamwise velocity distribution at the boundary layer edge serve as boundary conditions for the analytical perturbation solution in the middle deck respectively for the Navier-Stokes solution in the interference region.

3.1 Analytical solution

The outer part of the boundary layer, in which according to Lighthill the viscous terms can be neglected but the shear flow vorticity terms have to be considered, is calculated using the analytical perturbation method described in [7, 15, 16]. The analytical solution depends on the thickness of the viscous sublayer which is determined requiring a minimum shock-induced wall shear stress increase. The thickness parameter is a function of the local Reynolds number and the shape parameter of the incoming undisturbed velocity profile.

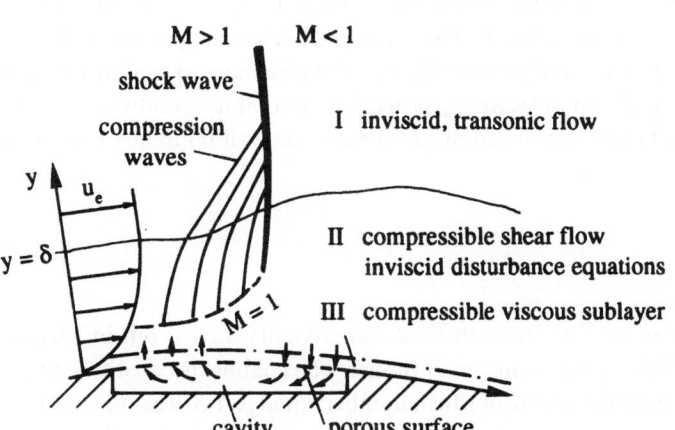

Fig. 5. SBLIC triple-deck model with passive shock control

The boundary conditions of the analytical calculation are:

— the velocity at the beginning of the interference area upstream of the the shock and cavity position approximated through a power law function in the outer part of the boundary layer,
— a constant pressure distribution normal to the main flow direction at the beginning and the end of the interference area,
— the pressure distribution at the boundary layer edge given from the global VII method
— and the normal velocity distribution at the limit between shear layer and viscous sublayer obtained from the mass flux calculation through the porous surface.

As a result of the perturbation solution the insensitive quantities as the pressure, the density and the normal velocity are obtained with sufficient accuracy. The tangential velocity is more sensitive even near the onset of separation. In this case the disturbance model is no longer valid.

Therefore, the tangential velocity is calculated in a subsequent step using the full Navier-Stokes equation in streamwise direction.

3.2 Navier-Stokes solution

The Navier-Stokes equation in streamwise direction is solved using a time stepping algorithm based on the "Implicit Factored Scheme" developed by Beam and Warming [17] and introducing the turbulence model of Chen [18]. Density, static pressure and normal velocity are given quantities known from the analytical perturbation solution of the boundary layer. The required boundary conditions are:

— the tangential velocity profile at the inflow area,
— the tangential velocity distribution at the wall, different from zero in the case of ventilation through inclined holes,
— the tangential velocity distribution at the upper edge of the interference area
— and vanishing second derivative of the tangential velocity in streamwise direction at the downstream end of the interference region.

3.3 Shock control boundary conditions

The ventilation velocities are dependent on the pressure difference and losses across the porous wall. The approximately constant pressure inside of the cavity is iteratively determined fulfilling the mass conservation law. For the calculation of the mass flux across the porous wall, loss coefficients determined from experiments and predicted from a data base are taken into account. The resulting normal component of the velocity v_w is used as boundary condition for the analytical perturbation solution. The tangential component serves as wall boundary condition for the Navier-Stokes solution.

3.4 Experimental verification

The local SBLIC solution has been verified through LDA-experiments carried out by Délery, Bur and Pot [19, 20] in the ONERA-S8 transonic wind tunnel. The channel is constituted by a half contoured nozzle producing a uniform flow of Mach number equal to 1.3. The nozzle block is installed on the upper wall, the cavity being placed on the lower flat wall at the origin of the

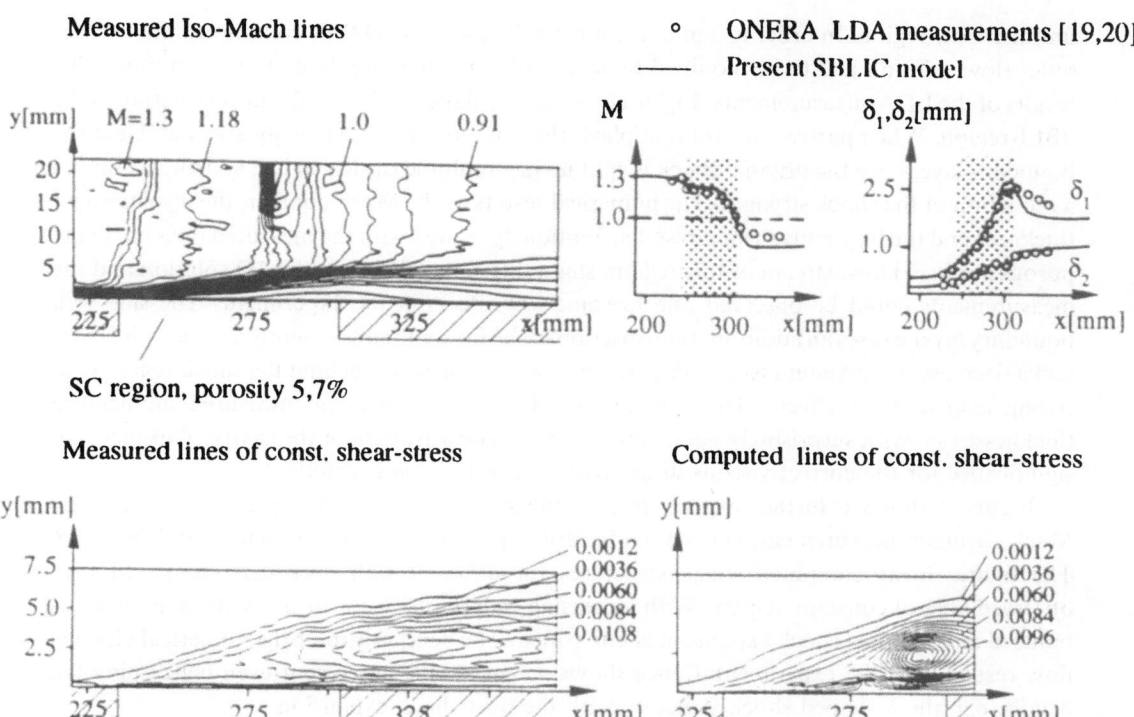

Fig. 6. Comparison of the SBLIC triple-deck solution with LDA measurements

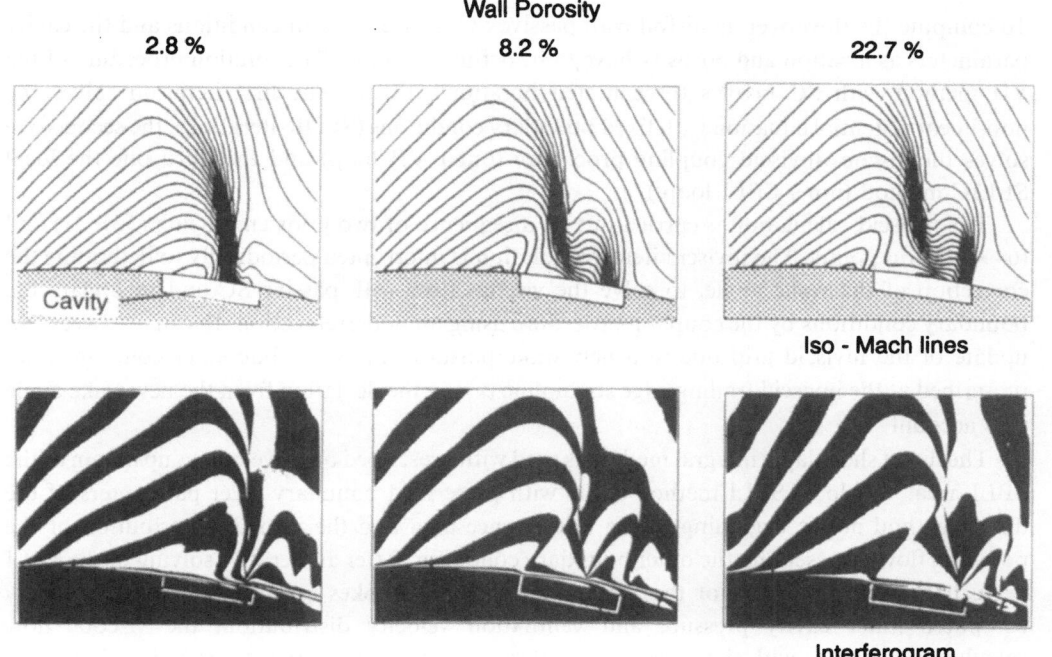

Fig. 7. Calculated/measured shock structure with shock control. Pre-shock Mach number of the solid-wall case $M = 1.34$

uniform flow region. In order to simulate numerically the channel flow conditions, the inviscid outer flow solution had to be modified by means of corresponding boundary conditions. The results of the LDA-measurements, Fig. 6, show the ventilation effect on the flow structure in the SBLI-region. When passive control is applied, the flow structure is strongly affected. Near the boundary layer edge the original shock is split up in a multiple iso-Mach line system, leading to a reduction of the shock strength. The numerical results as the Mach number, the displacement thickness and the Reynolds shear stress distribution agree well with the measured ones along the porous surface. Downstream of this region, slight differences between the 2D solution and the measurements could be observed due to some 3D effects in the experiment: The side wall boundary layer causes an additional constriction of the flow and subsequently a velocity increase and a decrease of maximum Reynolds shear stresses, in particular behind the shock region with strong displacement effects. The comparison of the calculated and measured momentum thicknesses shows a surprisingly good agreement even downstream of the cavity. This fact is of significance for the correct viscous drag prediction of transonic airfoils.

Figure 7 shows a further experimental verification of the SBLI triple-deck solution by Mach-Zehnder measurements carried out by Braun [21] in the transonic channel in Karlsruhe. The figure shows computed shock structures for different wall porosities compared with observed lines of constant density. With increasing wall porosity, the primary shock intensity is reduced and the post shock expansion is intensified. The comparison of the numerical channel flow results with the experimental ones shows a remarkable agreement even concerning the details, e.g. the λ-shaped shock structure with the post shock expansion.

4 Numerical solution procedure

To compute the flow over an airfoil with passive SC the free-stream conditions and the cavity parameters as location and porosity have to be defined as input. The solution procedure of the VII method with SC involes three nested iterations, Fig. 8. The first iteration solves the non-linear potential equation with fixed boundary conditions (SLOR-iteration), the second one solves the viscous-inviscid coupling procedure (outer VII loop) and the third one the local SBLIC-method (inner SBLI loop.)

The inviscid calculation is carried out on a sequence of two grids employing (32×21) and (64×41) grid points. The inviscid iterative solution is interrupted periodically to determine the correction of the wake angle, to solve the viscous flow with passive SC and to update the boundary conditions by the coupling procedure using an underrelaxation. In order to avoid an update of the inviscid grid due to a new wake position, the wake boundary conditions are prescribed at the inviscid trailing edge streamline taking the deviation from the new wake angle into account.

The direct shear layer integral method is used with prescribed outer velocities upstream of the SBLI area. The local SBLI method starts with prescribed boundary layer parameters of the direct method at the beginning of the interference area and the velocity distribution of the potential flow representing the outer boundary condition. After an iterative solving of the SBLI perturbation equation and of the u-component Navier-Stokes equation with an iteratively computed inner cavity pressure and ventilation velocity distribution, the viscous flow calculation proceeds with the inverse shear layer integral method for the boundary layer and wake flow downstream of the interference area. Depending on the viscous-inviscid interaction the direct coupling procedure is applied to the direct shear layer and the local SBLI calculation

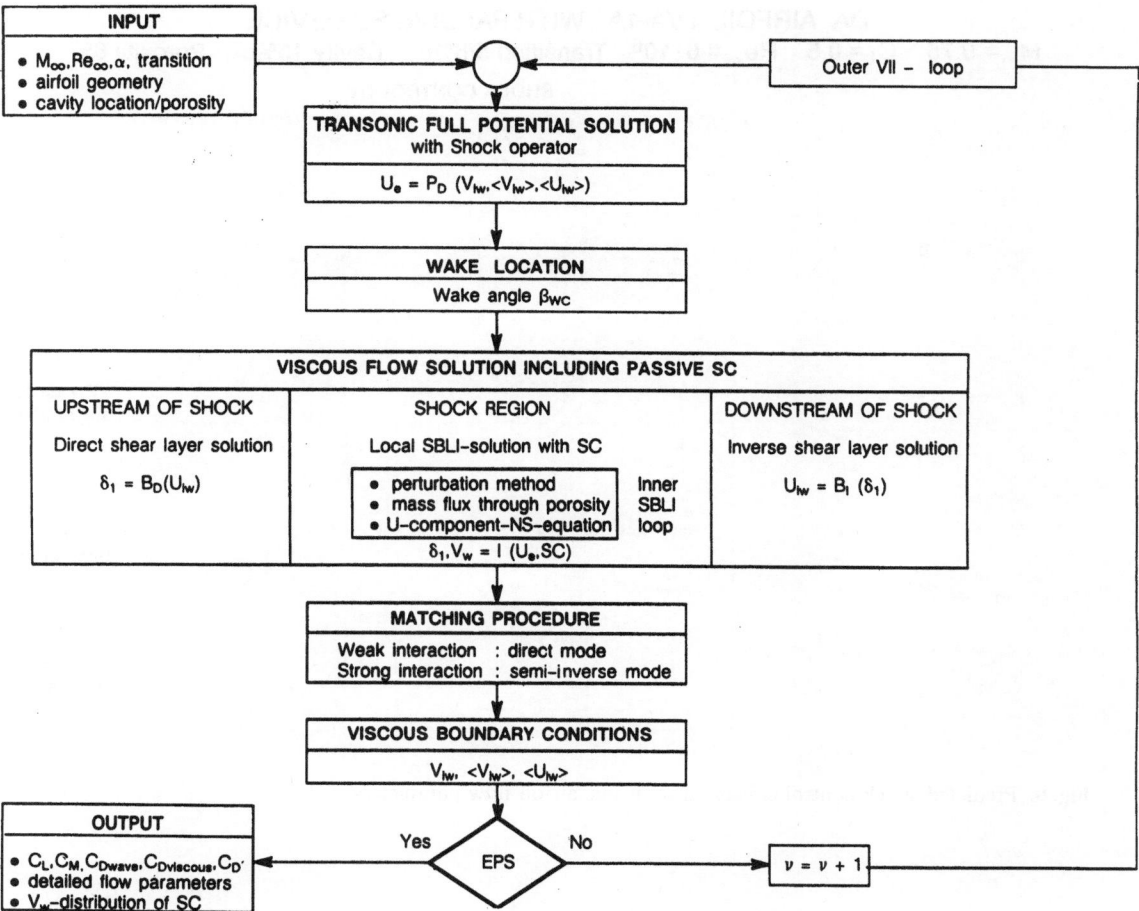

Fig. 8. Structure of computational method

and the semi-inverse one to the inverse shear layer and wake calculation. The viscous parameters including the ventilation velocity are used to predict the new viscous boundary conditions for the next VII iteration loop.

The viscous-inviscid iteration is controlled by an average potential correction in the SLOR-method and a relative displacement correction in the matching procedure. Numerical investigations have shown that it is reasonable to start the passive SC calculation after obtaining a converged solution of the VII-method without SC in the finest grid. Lift and moment characteristics, integrated from the surface pressure and from the skin friction distributions and the total drag predicted from the moment and wave drag, are given in the output.

5 Results

The method described above has been used to compute shock control effects on transonic airfoil flow characteristics. In this scope, the laminar-type transonic airfoil LVA-1A equipped with two different passive SC devices with the same cavity and porosity but different hole inclination (0° and 45°) has been investigated.

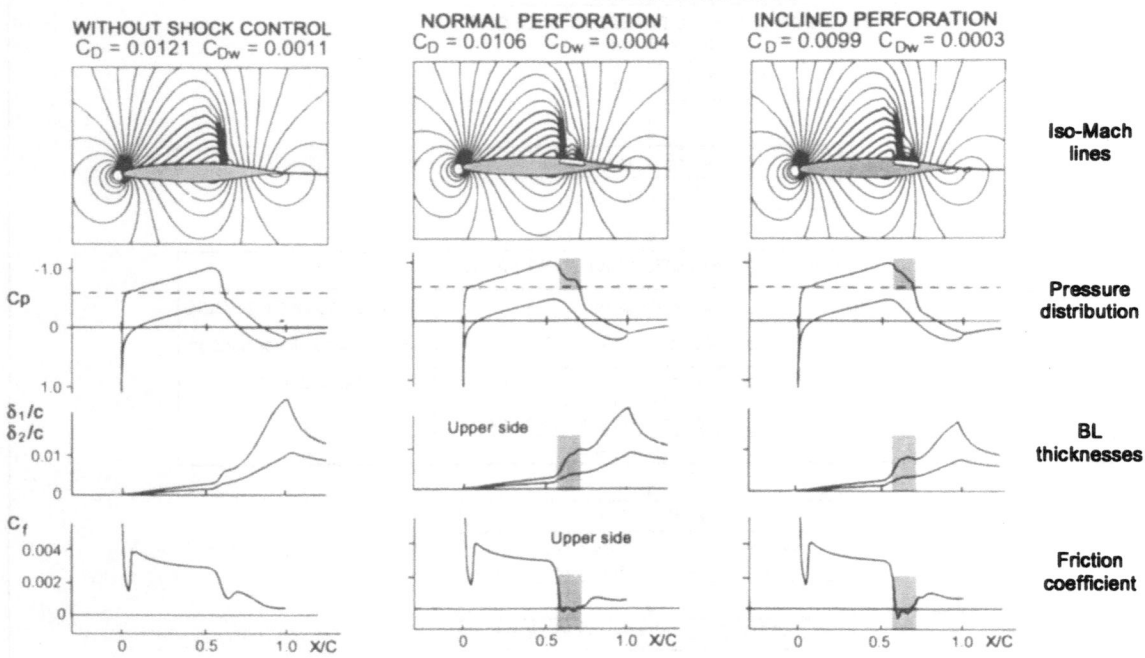

Fig. 9. Predicted shock control effects on transonic airfoil flow parameters

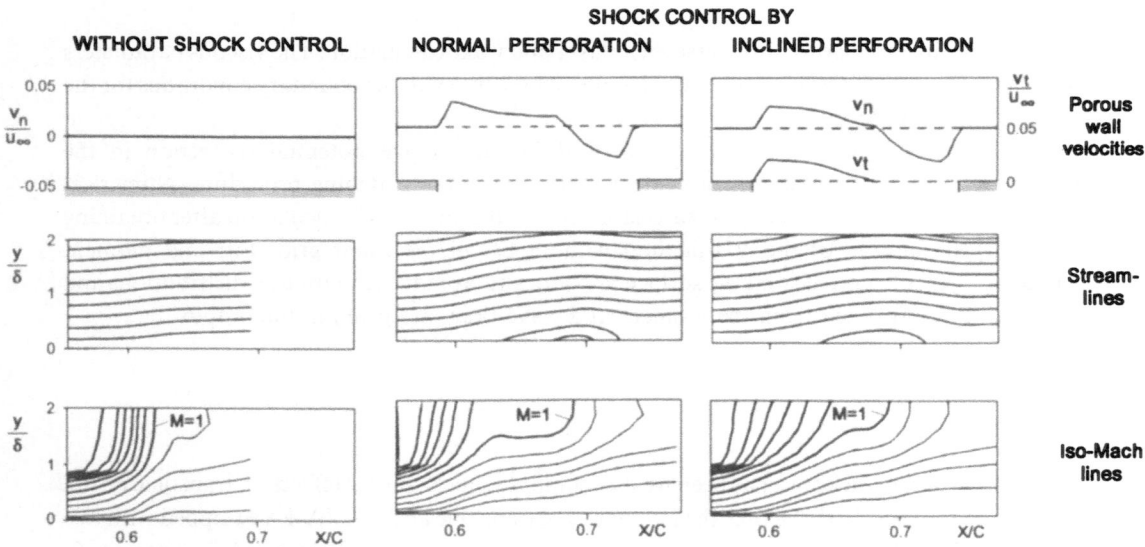

Fig. 10. Predicted shock control effects on SBLI flow field details

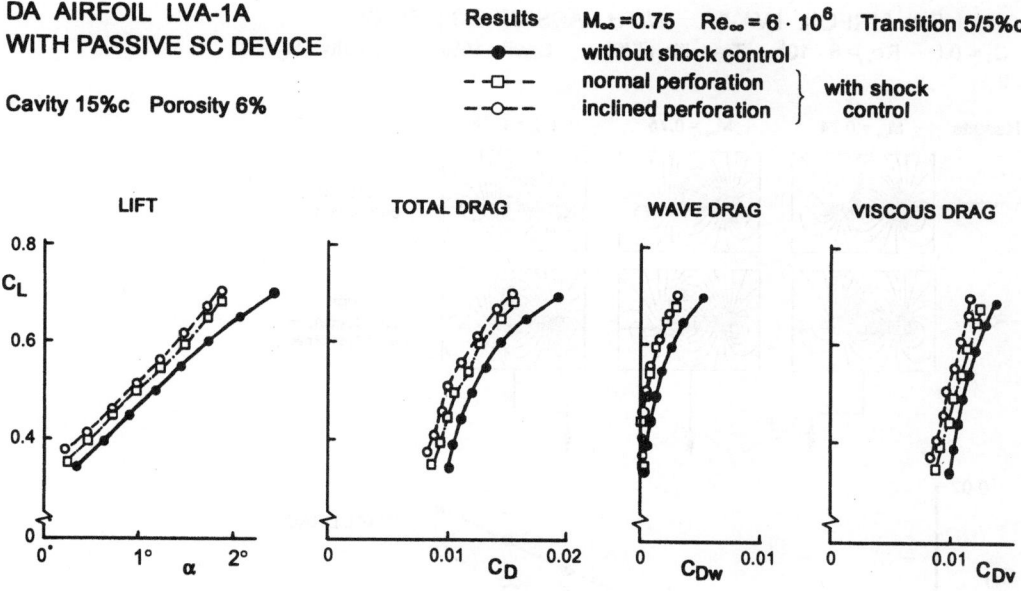

Fig. 11. Predicted shock control effects on transonic airfoil characteristics

Figure 9 shows the computed effects of both control devices on the Iso-Mach lines, the pressure distribution and the boundary layer parameters at the same lift coefficient. From the Iso-Mach plots the influence of passive shock control on the shock structure can be seen. By ventilation the primary shock is reduced and the flow expansion behind the shock is increased, leading to a double shock system with a reduced wave drag. The typical pressure plateau caused by normal perforation in the shock region disappears in the inclined perforation case. The flow recovery behind the shock is more improved by inclined perforation than by normal perforation, resulting in a viscous drag reduction. Computed SBLI flow field details such as porous wall velocities, streamlines near the wall, as well as Iso-Mach lines of the corresponding cases are plotted in Fig. 10. The stronger streamline displacement and the larger extension of the SBLI region in the ventilation case responsible for the shock-weakening can be seen from the plots.

Figure 11 shows predicted shock control effects on the transonic characteristics of the LVA-1A airfoil. There is not only a positive ventilation effect on the wave and viscous drag but also on the lift. The results show that the transonic performance of the LVA-1A airfoil can be substantially improved by a passive SC device. Furthermore, they show that the inclined perforation is more effective than the normal one.

Moreover, the effect of passive shock control on the transonic drag rise has been predicted, Fig. 12. The computations indicate that the drag rise of the LVA-1A airfoil can be considerably reduced by a passive SC device with inclined perforation.

6 Conclusions

A method for the numerical simulation of viscous transonic airfoil flows with passive shock control based on a global VII approach and including a local SBLIC solution has been described. For the simulation of shock control, the local triple-deck solution has been verified by LDA-measurements of Délery et al. and by channel flow experiments of Braun. Both test cases

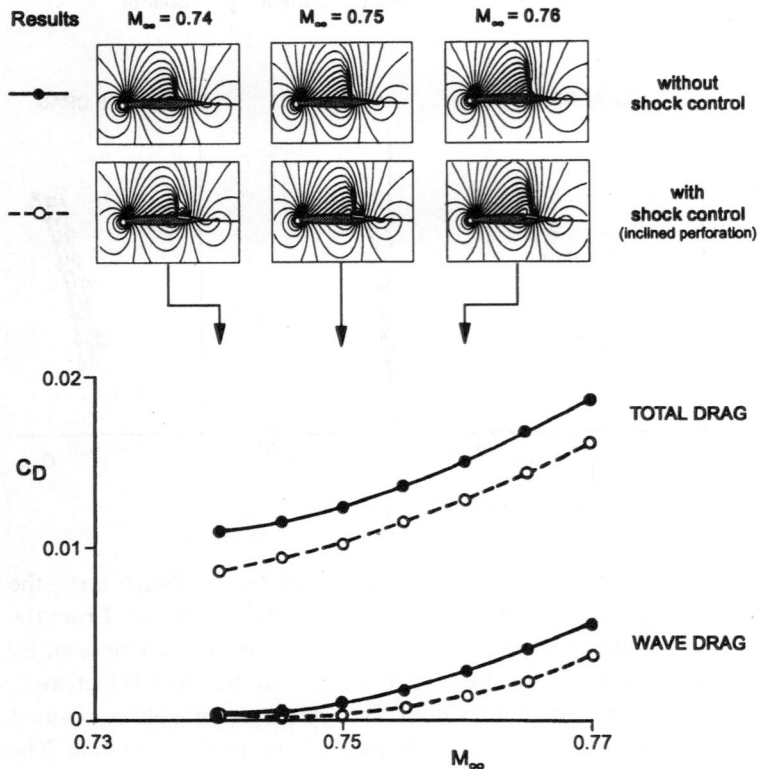

Fig. 12. Predicted shock control effects on transonic airfoil drag rise

demonstrate that the local triple-deck solution is able to simulate not only the main shock control effects but also the details of the flow. The shock control boundary conditions determined from a data base need more experimental verification.

For the validation of the present transonic airfoil flow method with shock control, no reliable test cases are up to now available. However, suitable transonic airfoil experiments with passive shock control are under preparation now.

Finally, promising shock control effects on transonic airfoil characteristics have been predicted using the present method.

References

[1] Bahi, L., Ross, J. M., Nagamatsu, H. T.: Passive shock wave/boundary layer control for transonic airfoil drag reduction. AIAA-Paper 83-0137 (1983).
[2] Thiede, P., Krogmann, P., Stanewsky, E.: Active and passive shock/boundary layer interaction control on supercritical airfoils. AGARD-CP-365, Brussels 1984.
[3] Krogmann, P., Stanewsky, E., Thiede, P.: Transonic shock/boundary layer interaction control. ICAS-Paper 84-2.3.3, Toulouse 1984.
[4] Chen, C. L., Chow, C. Y., van Dalsem, W. R., Holst, T. L.: Computation of viscous transonic flow over porous airfoils. AIAA Paper 87.0359 (1987).

[5] Dargel, G., Thiede, P.: Viscous transonic airfoil flow simulation by an efficient viscous-inviscid interaction method. AIAA Paper 87-0412 (1987).

[6] Bohning, R., Zierep, J.: Normal shock-turbulent boundary-layer interaction at a curved wall. AGARD CP No. 291, Computation of Viscous-Inviscid Interactions, 17-1 — 17-8 (1980).

[7] Bohning, R., Zierep, J.: Calculation of 2D turbulent shock/boundary-layer interaction at curved surfaces with suction and blowing. Turbulent Shear Layer/Shock Wave Interactions IUTAM Symposium Palaiseau 1985 (Délery, J., ed.) pp. 105—112. Berlin Heidelberg New York Tokyo: Springer 1985.

[8] Le Balleur, J. C-.: Strong matching method for computing for transonic viscous flows including wakes and separations. Lifting airfoils. La Recherche Aerospatiale 1981—3, pp. 21—45 (1981).

[9] Jakob, H.: Ein Verfahren zur Berechnung der ebenen transsonischen Strömung in Stromlinienkoordinaten. MBB LFK 81117 (IFAS 11), 1984 (not published).

[10] Mertens, J., Klevenhusen, K. D., Jakob, H.: Accurate transonic wave drag prediction using simple physical models. AIAA Paper 86-0512 (1986).

[11] Thiede, P.: Ein inverses Integralverfahren zur Berechnung abgelöster turbulenter Grenzschichten. DLR-FB 77-16 (1977).

[12] Otte, F., Thiede, P.: Berechnung ebener und rotationssymmetrischer kompressibler Grenzschichten auf der Basis von Integralbedingungen. Fortschr.-Ber. VDI-Z, Reihe 7, 33 (1973).

[13] Carter, J. E.: A new boundary layer inviscid iteration technique for separated flow. AIAA Paper 79-1450 (1979).

[14] Holst, T. L.: Viscous transonic airfoil workshop compendium of results. AIAA Paper 87-1460 (1987).

[15] Bohning, R.: Die Wechselwirkung eines senkrechten Verdichtungsstoßes mit einer turbulenten Grenzschicht an einer gekrümmten Wand. Habilitationsschrift, Universität Karlsruhe 1982.

[16] Breitling, Th.: Berechnung transsonischer, reibungsbehafteter Kanal- und Profilströmungen mit passiver Beeinflussung. Dissertation, Universität Karlsruhe (TH) 1989.

[17] Beam, R. M., Warming, R. F.: An implicit factored scheme for the compressible Navier-Stokes equations. AIAA J. 16, 393—402 (1978).

[18] Chen, C. L.: Computation of transonic flow over porous airfoils. Ph. D. Thesis, University of Colorado 1986.

[19] Délery, J., Bur, R., Pot, T.: Basic experiments on passive control of shock-wave/boundary-layer interaction in transonic flow. 4th STAB-Workshop, DLR Göttingen, 1989.

[20] Bur, R.: Etude fundamentale sur le contrôle passif de l'interaction onde de choc-couche limite turbulente en écoulement transsonique. Ph. D. Dissertation, Université Pierre et Marie Curie, Paris 1991.

[21] Braun, W.: Experimentelle Untersuchung der turbulenten Stoß-Grenzschicht-Wechselwirkung mit passiver Beeinflussung. Dissertation, Universität Karlsruhe (TH) 1990.

Authors' addresses: Professor Dr.-Ing. habil. R. Bohning, Institut für Strömungslehre und Strömungsmaschinen, Universität Karlsruhe (TH), D-76128 Karlsruhe; Professor Dr. P. Thiede and Dipl.-Ing. G. Dargel, Deutsche Aerospace Airbus GmbH, Abt. EF 11, D-28078 Bremen, Federal Republic of Germany

Acta Mechanica (1994) [Suppl] 4: 233–240
© Springer-Verlag 1994

Zonal computational method for turbulent plane cascade flow

G. Fleberger, W. Schneider, Vienna, Austria, and **H. Keck,** Zürich, Switzerland

Summary. A zonal method is presented for computing the incompressible flow through plane cascades with blunt trailing edges. The flow field is divided into three distinct regions, i.e. the potential flow region, the turbulent boundary layers, and the region of the turbulent flow near the trailing edge and in the near wake. The potential flow is computed by a panel method that accounts for the periodicity of the cascade flow in an exact manner. A finite-difference scheme is applied to solve the boundary-layer equations together with a classical mixing-length formula. In the trailing edge and near wake region, a local solution of the time-averaged Navier-Stokes equations is obtained by a finite-element method, which incorporates a k-ε model for the Reynolds stresses. To match the potential-flow solution to the Navier-Stokes solution, the circulation is determined iteratively such that the pressure difference across the wake at the outflow boundary becomes zero. A relaxation procedure is applied to stabilize the iteration process. Numerical results for the flow through a plate cascade, as a test case, show fair agreement with experimental data for the static pressure difference. The computed outflow angle, however, is smaller than the value given in the literature. The discrepancy seems to be due to the fact that in recirculation zones the k-ε model predicts values of the turbulent normal stresses that are too large.

1 Introduction

In the present work the incompressible, turbulent flow through plane cascades is computed. The aim of this study is to determine outflow angle and pressure loss as a basis for predicting the lift and drag forces acting on the profiles. As shown in Fig. 1, the cascade with blade spacing t and stagger angle β_s consists of profiles of length l. Note that the profiles have blunt trailing edges of thickness d (for reasons of manufacturing, mechanical stresses and fatigue). To avoid Kármán excitation the blunt trailing edge may have special contours, e.g. asymmetric fairing, which is neglected here. However, the assumption of a stationary, non-oscillating wake flow behind the blunt trailing edge conforms to the real flow pattern of practical interest.

Since the Kutta condition cannot be applied to profiles with blunt trailing edges, the circulation is determined, in this investigation, in conjunction with a local solution of the time-averaged Navier-Stokes equations. Thus, various effects having a strong influence on the outflow angle, such as the strong interaction between the viscous flow in the vicinity of the trailing edge and the inviscid flow, can be taken into account. Following the approach outlined by Schmatz[11], who investigated the turbulent flow past a single airfoil, the flow field is divided into three regions, cf. Fig. 1. However, whereas Schmatz solved Euler equations to determine the inviscid flow, in the present paper the inviscid flow in region I is assumed to be a potential flow. The flow upstream of the profiles is assumed to be uniform with inflow velocity w_1, incidence angle β_1 and pressure p_1. The turbulent boundary layers are supposed to have no separation points in region II. In region III the boundary layers are separating at the trailing edge forming

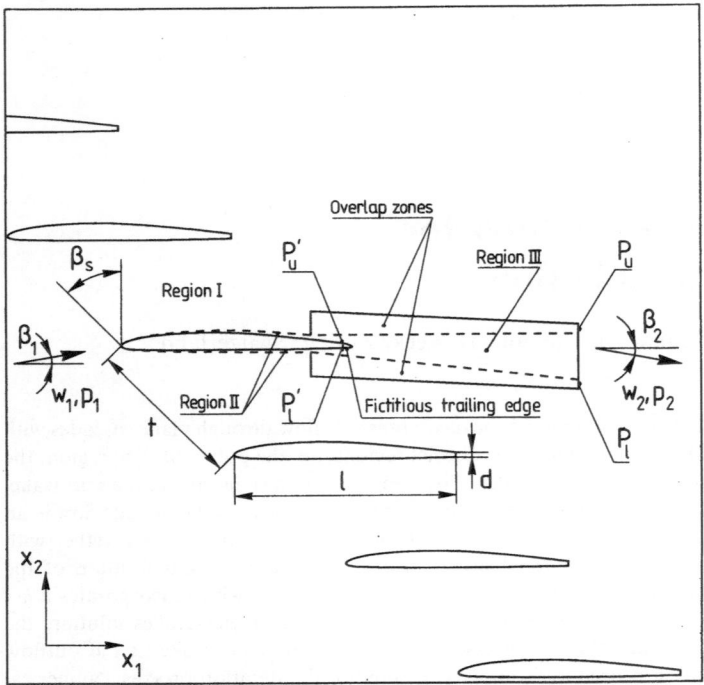

Fig. 1. Flow regions in the plane cascade flow: region I, potential flow; region II, turbulent boundary layers; region III, trailing edge and near wake region (dashed line: source distribution)

asymmetric, turbulent, free shear layers. Downstream of a small recirculation zone the shear layers merge, and a near wake flow develops. The flow field in the trailing edge and near wake region is represented by a local solution of the Navier-Stokes equations. The flow far downstream of the trailing edge, where the wakes of adjacent profiles merge, is not considered in this paper.

The solutions for the three regions are coupled iteratively by imposing appropriate boundary and coupling conditions. The uniform flow properties far downstream of the trailing edge, e.g. outflow velocity w_2, outflow angle β_2 and pressure p_2, are determined by means of mass and momentum balances in the near wake region. In addition, loss coefficients are determined to estimate the energy losses in the viscous flow.

The method can also be of interest for profiles with sharp trailing edges if rather thick, and markedly different, boundary layers merge at the trailing edge. With the displacement thickness to be taken into account, the Kutta condition becomes meaningless, since displacement effects at the trailing edge render the profile "blunt-edged" from the point of view of the outer potential flow.

2 Governing equations

The incompressible, inviscid, irrotational flow in region I is described by the Laplace equation. A panel method incorporating an exact treatment of the periodicity condition is used to compute the potential flow.

To compute the turbulent flow in the boundary layers at the profile surface, the Reynolds shear stress is modelled with an eddy viscosity μ_t, which is determined from Prandtl's mixing-length formula. The eddy viscosity is combined with the constant laminar viscosity, μ_l, to form the effective viscosity $\mu_e = \mu_l + \mu_t$. Using this effective viscosity the boundary-layer equation together with the continuity equation are solved numerically with the finite-difference code STAN 5 [2].

The turbulent flow in the trailing edge and near wake region is described by a local solution of the time-averaged Navier-Stokes equations. The Reynolds stresses are determined with a k-ε turbulence model [6], in which the eddy viscosity μ_t is related to the turbulent kinetic energy, k, and the turbulent dissipation rate, ε, according to the relationship $\mu_t = \varrho c_\mu k^2/\varepsilon$, where ρ is the constant density of the fluid and $c_\mu = 0.09$ is an empirical constant. Hence, the stress tensor σ_{ij} ($i,j = 1, 2$), comprising static pressure p as well as viscous and turbulent stresses, is given by the relationship

$$\sigma_{ij} = -\left(p + \frac{2}{3}\varrho k\right)\delta_{ij} + \mu_e\left(\frac{\partial u_i}{\partial x_j} + \frac{\partial u_j}{\partial x_i}\right). \tag{1}$$

The tensor notation is used, with u_1, u_2 denoting the velocity components in the direction of the Cartesian coordinates x_1, x_2, respectively, and δ_{ij} the Kronecker symbol. To determine k and ε, well-known transport equations [6] are used. In regions close to solid walls, where the k-ε model ceases to be valid, the wall-function method [7] is applied in the version described in [3]. The nonlinear system of governing equations subject to boundary and coupling conditions, which are discussed in the next Section, is solved numerically with the finite-element program FIDAP [3].

3 Boundary conditions and coupling procedure

The solutions for the three regions are coupled according to the displacement effect of the boundary layers, the interaction between the separated shear layers and the inviscid flow as well as the deflection of the wake downstream of the trailing edge.

To compute the potential flow, the source distribution is placed at the edges of the boundary layers outside the trailing edge and near wake region (Navier-Stokes region) (Fig. 1). In the latter region the source distribution is placed along two straight line segments, located in the vicinity of the separated shear layers and the wake and denoted as SD in Fig. 2. Between

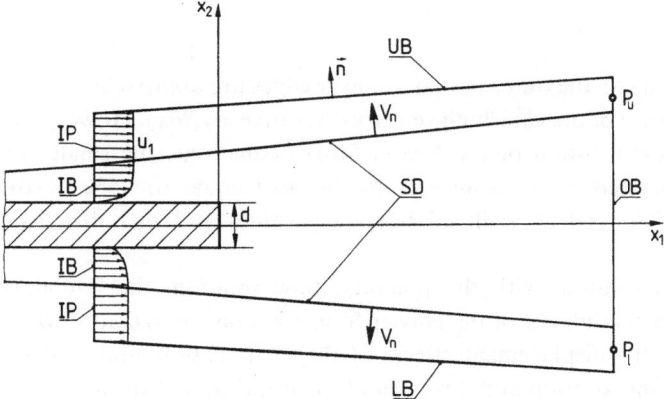

Fig. 2. Zonal boundaries of Navier-Stokes solution: inflow boundaries IB and IP, upper and lower boundaries UB and LB, outflow boundary OB, source distribution SD in the trailing edge and near wake region, points P_u and P_l to determine the circulation in the potential flow

the latter source distribution and the upper and lower boundaries, *UB* and *LB*, respectively, of the Navier-Stokes region, there are two overlap regions, in which both the potential flow and the Navier-Stokes solutions must be valid (Fig. 2). On the source-distribution line the velocity component V_n directed normal to the panels is prescribed as a boundary condition for the potential flow. In the boundary-layer region, on the other hand, V_n is determined from the velocity at the boundary-layer edge as obtained from the boundary-layer solution. In the trailing edge and near wake region, V_n can be taken directly from the Navier-Stokes solution.

To obtain the boundary-layer solution, the no-slip condition is prescribed at the wall. Furthermore, the tangential velocity component obtained from the potential flow solution at the wall is prescribed at the outer boundary-layer edge.

At the inflow boundaries of the Navier-Stokes region, which are located a few trailing edge thicknesses upstream of the trailing edge (Fig. 2), the velocity component u_1, as obtained from the boundary-layer solution at *IB* and from the potential solution at *IP*, is prescribed as a boundary condition for the Navier-Stokes equations. In accordance with the boundary-layer concept, a further boundary condition for the Navier-Stokes equations is to be specified such that the velocity component u_2 is not prescribed a priori. If the finite-element code FIDAP [3] is used to solve the Navier-Stokes equations, an appropriate boundary condition can be obtained by prescribing the stress σ_{n2}, which is the x_2 component of the stress vector $\sigma_{ni} = \sigma_{ij} n_j$ acting on a boundary with outward directed unit normal vector n_j (Fig. 2). The stress tensor σ_{ij} is obtained from Eq. (1) with the velocity derivatives taken from the boundary-layer solution on *IB* and from the potential-flow solution on *IP*, respectively. Since $\partial u_2/\partial x_1 \ll \partial u_1/\partial x_2$ in the boundary layer, the term $\partial u_2/\partial x_1$ is neglected at *IB*. The boundary conditions for the turbulent kinetic energy k and the turbulent dissipation ε at *IB* have to account for the fact that in the boundary-layer solution the eddy viscosity is determined with the mixing-length model, whereas in the Navier-Stokes solution the k-ε model is applied. Combining the eddy-viscosity relations for these models and making use of the eddy-viscosity formula of Prandtl's one-equation turbulence model [9], the boundary conditions for k and ε at *IB* can be written as follows.

$$k = c_\mu^{-1/2} \left(l_m \frac{\partial u_1}{\partial x_2} \right)^2, \qquad \varepsilon = c_\mu k^2 \left(l_m^2 \frac{\partial u_1}{\partial x_2} \right)^{-1}, \tag{2}$$

where l_m denotes the mixing-length. As the outer boundary-layer edges are approached, k and ε would tend to zero as a consequence of the vanishing velocity derivative, $\partial u_1/\partial x_2$. However, to avoid instabilities in the numerical solution of the Navier-Stokes equations, the small, but non-zero values of k and ε obtained at a distance from the wall equal to 95% of the boundary-layer thickness are prescribed as artificial boundary values at larger distances, including the boundary *IP*.

To couple the Navier-Stokes solution with the potential flow solution, the boundary conditions on the upper and lower boundaries of the Navier-Stokes region, i.e. *UB* and *LB*, cf. Fig. 2, are formulated such that the displacement effects of the recirculation zone and the separated shear layers, as well as the location and direction of the wake are not fixed a priori. Coupling is achieved by prescribing the stress components σ_{n1} and σ_{n2} at *UB* and *LB*. To determine the stress tensor at *UB* and *LB*, the pressure and the velocity derivatives as obtained from the potential-flow solution are introduced into Eq. (1). Furthermore, the normal derivatives of both k and ε are assumed to vanish at UB and LB, i.e.

$$\frac{\partial k}{\partial n_j} = \frac{\partial \varepsilon}{\partial n_j} = 0. \tag{3}$$

Finally, outflow boundary conditions have to be prescribed to solve the elliptic system of the Navier-Stokes equations and the transport equations for k and ε. The outflow boundary OB (cf. Figs. 1 and 2) is placed at a distance of about one and a half blade spacings downstream of the trailing edge in a region where the pressure is almost uniform over the cross-section. Following common practice with finite-element methods, outflow boundary conditions are obtained by requiring the stress components σ_{n1} and σ_{n2} to vanish at the outflow boundary ("traction-free conditions") [8]. Furthermore, Eq. (3) provides outflow boundary conditions for k and ε that are consistent with the traction-free conditions.

To obtain a unique solution for the potential flow, the circulation has to be fixed by means of an appropriate coupling condition. In this work, the circulation is determined such that in the vicinity of the outflow boundary the potential flow matches the wake with respect to both location and direction. This condition can be formulated in a simple and convenient way if it is assumed that the flow angle is uniform in the potential flow near the outflow boundary. This assumption can be supported by the argument that the displacement thickness in the far wake is known to be approximately constant. Hence, the potential flow between the wakes of adjacent profiles can be regarded as a one-dimensional channel flow with constant cross-section, where the flow angle as well as the pressure are uniform. The circulation, which is induced by a vortex panel located near the leading edge, is therefore determined such that the pressure difference across the wake vanishes, i.e.

$$p_u - p_l = 0, \tag{4}$$

where p_u and p_l, respectively, are the pressures at properly chosen points P_u and P_l, cf. Fig. 2. Regarding the location of the points P_u and P_l, a series of test computations was performed. As a result of this numerical study, the points P_u and P_l are recommended to be placed, as shown in Fig. 2, downstream of the trailing edge at the outflow boundary OB, in the middle of the upper and lower overlap zones, respectively.

Coupling of the solutions in the three flow regions is performed with the following iteration procedure. First, a potential-flow solution is obtained with the source distribution located at the profile surface including a fictitious trailing edge of semi-elliptical shape, which is an approximate substitute for the displacement effect of the recirculation zone (Fig. 1). The normal velocity components V_n are set equal to zero in this first step of the iteration. To determine the circulation for the start solution, the outflow angle has to be estimated or other ad-hoc assumptions have to be made, cf. also the example considered in Section 4. Starting from this first potential-flow solution, the boundary-layer and Navier-Stokes solutions, subject to the boundary conditions given above, are computed. For all subsequent iterations the source distribution is located on the boundary-layer edges and in the vicinity of the separated shear layers and the wake, while the velocity components V_n are determined from the boundary-layer and Navier-Stokes solutions. The potential-flow solutions are obtained with the updated velocity components V_n, together with Eq. (4) to fix the circulation. Beginning with the second iteration, a relaxation procedure is employed for the velocity components V_n at SD in order to avoid physically unrealistic, large deflections of the wake in the next iteration for the Navier-Stokes solution. For the relaxation factor, values between 0.7 and 0.85 were found satisfactory.

4 Numerical results and discussion

As a test case, the flow through a cascade of plane plates with finite thickness was computed. For this case, the flow turning and the static pressure difference across the cascade are solely due to viscous effects, which take place in the boundary-layers, in the separated shear layers and in the wake. Furthermore, the numerically obtained results for the flow-turning angle and the static pressure difference can be compared with analytical and experimental data due to Gersten [5]. The cascade investigated in [5] consisted of plane plates having a length $l = 250$ mm and a thickness $d = 8$ mm. The leading edge was of semi-circular cross section, the trailing edge was square-cut. The blade spacing was $t = 125$ mm and the stagger angle was $\beta_s = 45°$. The experiments were conducted with air at a temperature of $20°C$. The inflow velocity was $w_1 = 30$ m/s at an incidence angle $\beta_1 = 0$. The Reynolds number based on inflow velocity and plate length was $Re_l = 5 \times 10^5$.

The iteration procedure of our numerical solution was started with a potential flow solution, for which the circulation was determined by assuming a vanishing pressure difference between the two points P_u' and P_l' near the trailing edge, cf. Fig. 1. Ten iterations were found to be sufficient for matching the solutions of the three regions to each other.

The outflow velocity w_2, the outflow angle β_2 and the pressure p_2 far downstream of the trailing edge were determined by means of mass and momentum balances in the wake [12]. As shown in Table 1, the present result for the static pressure difference compares quite well with the wake measurements given in [5]. The corresponding value of the boundary-layer measurement obtained in [5] is slightly higher. On the other hand, the present method predicts too small a flow-turning angle. This discrepancy is most likely due to the well-known fact that in recirculation zones the k-ε model predicts values of the turbulent normal stresses that are too large. As a consequence, recirculation zones predicted on the basis of k-ε models are consistently shorter than those measured. In the present case, a recirculation-zone length of about 2.0 plate thicknesses is predicted, cf. Fig. 3, for a Reynolds number $Re_d = 1.6 \times 10^4$ based on the plate

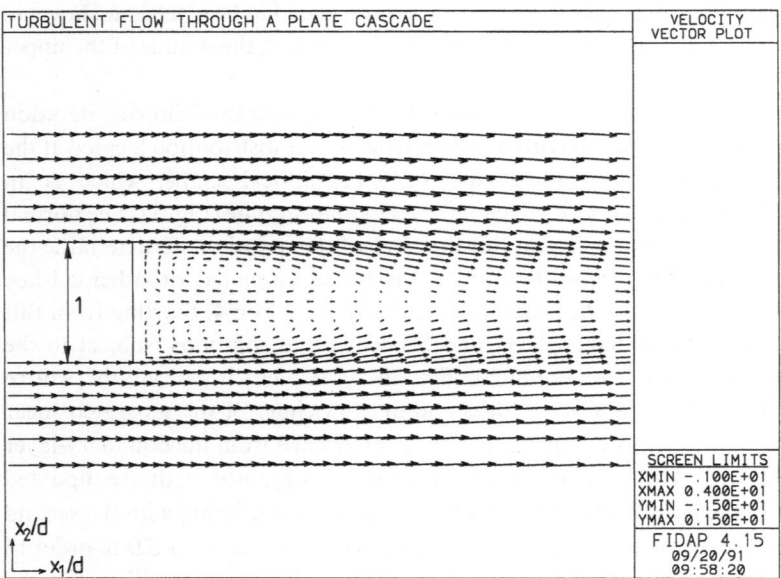

Fig. 3. Flow through a plate cascade: velocity vectors of the Navier-Stokes solution in the vicinity of the trailing edge and in the recirculation zone (plate thickness d)

Table 1. Comparison of results for the flow through a plate cascade ($l/t = 2$, $d/l = 0.032$): dimensionless pressure difference $2(p_1 - p_2)/\rho w_1^2$; flow-turning angle $\beta_1 - \beta_2$

	$2(p_1 - p_2)/\rho w_1^2$	$\beta_1 - \beta_2$
Boundary-layer measurements [5]	0.15	2.7°
Wake measurements [5]	0.12	2.5°
Approximate analysis [5]	0.15	2.9°
Present computation	0.121	2.0°

thickness. On the other hand, in Bearman's experiments [1] with a single plate at a similar Reynolds number, $\mathrm{Re}_d = 2.4 \times 10^4$, a recirculation-zone length of about 2.9 plate thicknesses was observed. Analogous discrepancies were reported by Simpson [13], Speziale [14] and Taulbee [15] for turbulent flows over a backward-facing step or through a channel with a sudden expansion. Possible improvements of the turbulence models for predicting recirculation zones were discussed by Rodi [10].

The pressure-loss coefficient ζ_P, the drag coefficient ζ_D and the drag-to-lift ratio λ were also determined from the computed flow field by means of mass and momentum balances. Based on the usual definitions [12], the results for the particular plate cascade are $\zeta_P = 0.167$, $\zeta_D = 0.029$, and $\lambda = 0.521$. For these coefficients no data are given in [5].

To show the relevance of the method to industrial applications, the flow through a profile cascade was also computed with the present method. The numerical results and further details of the method may be found in [4].

5 Conclusions and outlook

The zonal method presented in this paper is well suited to compute the turbulent flow through plane cascades. Viscous effects, including apparent turbulent viscosities, are taken into account with a computational effort that is much less than for a non-zonal solution of the Navier-Stokes equations in the whole flow field. The present method accounts for the interaction of the potential flow, the boundary layers, and the wake behind the (possibly blunt) trailing edge. Pressure losses determined with the present method are found to be in reasonable agreement with experimental data. Discrepancies in the prediction of the outlet angle are attributed to well-known drawbacks of the k-ε model, which was incorporated in the available finite-element programm. In future work, more advanced turbulence models should be tested with respect to their performance in the framework of the present method.

Possible applications of the method include guide vane configurations of hydraulic turbines, and cascades of spiral casings of Kaplan, Francis or pump turbines. The method would also allow studies for optimizing the shape of profile trailing edges. Furthermore, it seems possible to extend the current method to the computation of the three-dimensional flow through complete stator or rotor configurations.

Acknowledgements

This research was initiated and financially supported by Sulzer-Escher-Wyss AG, Zürich, Switzerland. Additional funding was provided by the Fonds zur Förderung der wissenschaftlichen Forschung in Österreich (FWF), Project No. P5716. The authors are grateful to Dr. P. E. Roach and two anonymous referees for valuable comments on this work.

References

[1] Bearman, P. W.: Investigation of the flow behind a two-dimensional model with a blunt trailing edge and fitted with splitter plates. J. Fluid Mech. **21**, 241−255 (1965).

[2] Crawford, M. E., Kays, W. M.: STAN5 — a program for numerical computation of two-dimensional internal and external boundary-layer flows. NASA CR 2742 (1976).

[3] Engelman, M. S.: FIDAP — Theoretical manual. Rev. 4.15, Fluid Dynamics International Inc., Evanston 1987.

[4] Fleberger, G.: Berechnung der turbulenten Gitterströmung mit Hilfe eines iterativen Kopplungsverfahrens. Diss. T. U. Wien 1992.

[5] Gersten, K.: Experimenteller Beitrag zum Reibungseinfluß auf die Strömung durch ebene Schaufelgitter. Abhandl. Braunschweig. Wissenschaftl. Ges. **7**, 93−99 (1955).

[6] Launder, B. E., Spalding, D. B.: The numerical computation of turbulent flows. Comput. Meth. Appl. Mech. Eng. **3**, 269−289 (1974).

[7] Patel, V. C., Rodi, W., Scheurer, G.: Turbulence models for near-wall and low Reynolds-number flows: a review. AIAA J. **23**, 1308−1319 (1985).

[8] Pelletier, D., Schetz, J. A.: A Navier-Stokes calculation of 3-D turbulent flow near a propeller in a shear flow. AIAA Paper 85-0365 (1985).

[9] Prandtl, L.: Über ein neues Formelsystem für die ausgebildete Turbulenz. Nachr. Akad. Wiss. Göttingen, Math.-Phys. Klasse, 6−19 (1945).

[10] Rodi, W.: Recent developments in turbulence modelling. Proc. 3rd Int. Symp. on Refined Flow Modelling and Turbulence Measurements, Tokyo 1988.

[11] Schmatz, M. A.: Simulation of viscous flows by zonal solutions of Euler, boundary-layer and Navier-Stokes-equations. Z. Flugwiss. Weltraumforsch. **11**, 281−290 (1987).

[12] Scholz, N.: Aerodynamik der Schaufelgitter. Bd. I. Karlsruhe: G. Braun 1965.

[13] Simpson, R. L.: A review of some phenomena in turbulent flow separation. ASME J. Fluids Eng. **103**, 520−533 (1981).

[14] Speziale, C. G.: Analytical methods for the development of Reynolds-stress closures in turbulence. Annu. Rev. Fluid Mech. **23**, 107−157 (1991).

[15] Taulbee, D. B.: Physical and computational issues for turbulent flow modeling. Proc. 3rd Int. Congress Fluid Mech., Cairo 1990 (A. Nayfeh, et al., eds.), pp. 1565−1574.

Authors' addresses: Dipl.-Ing. G. Fleberger and o. Professor Dipl.-Ing. Dr. techn. W. Schneider, Institut für Strömungslehre und Wärmeübertragung, Technische Universität Wien, Wiedner Hauptstrasse 7, A-1040 Wien, Austria, and Dipl.-Ing. Dr. techn. H. Keck, Sulzer-Escher-Wyss AG, CH-8023 Zürich, Switzerland

Acta Mechanica (1994) [Suppl] 4: 241–249

Simulations of compressible inviscid flows over stationary and rotating cylinders

M. Hafez and **A. C. B. Dimanlig,** Davis, California

Summary. Solutions of potential and Euler equations for compressible flows over a cylinder are obtained based on simple artificial viscosity forms using standard numerical techniques. The equations, written in cylindrical coordinates are discretized on an orthogonal grid via central differences with special treatment of boundary conditions. The resulting nonlinear algebraic equations are solved via Newton's method. A direct solver based on an efficient Gaussian elimination procedure for banded matrices, is employed at each iteration. Preliminary results demonstrate multiple solutions of both systems for the transonic cases of non-zero circulation in the far field. Finally, some remarks about the non-uniqueness problem are discussed.

1 Introduction

The analytical solution of an incompressible potential flow over a cylinder is studied in almost every textbook on elementary fluid mechanics. The solution is obtained via separation of variables and the superposition principle. For compressible flows, the potential equation is nonlinear and no analytical solution is available. Numerical results can be easily obtained for subsonic flows, since the equation is elliptic. For transonic flows, the equation is of mixed type and numerical solutions can be obtained if artificial viscosity is introduced (either explicitly or implicitly) in the supersonic region.

The Euler equations are, in general, more complicated since flows with vorticity are admissible solutions. Even for incompressible flows, Euler equations admit, beside the potential solution, solutions with free streamlines and recirculating zones. Irrotational solutions of the Euler equations are not easily calculated due to the presence of numerical vorticity generated by most of the existing schemes. For transonic flows with shocks, vorticity is generated from curved shocks, and bubbles with closed streamlines appear in the solution. Again, special care is required in such numerical simulation since the real vorticity is of the same order of the numerical vorticity in most of the existing codes.

One motivation to study flows over cylinders is to explain the generation of lift. If the cylinder is rotating, and assuming the flow around the cylinder will have the same circulation (we exclude the trivial case of a rotating cylinder in a stand-still air), the speed, and hence the pressure, on the top and bottom will be different thereby producing a lifting force. To obtain an analytical solution, a cut is introduced to reduce the doubly connected domain around the cylinder to a simply connected one. Applying Stokes' theorem, it is readily seen that the circulation around any curve enclosing the cylinder is the same.

For compressible flows, one must resort to numerical techniques. In the present paper, numerical solutions of both potential and Euler equations, with circulation imposed in the far field, will be calculated. It should be mentioned that the present study is relevant to flow analysis over airfoils. It is well known [1]–[3] that airfoils can be mapped onto circles, and flows over lifting airfoils correspond to flows over rotating cylinders. The mapping is a function of the geometry. To study the anomaly of the formulations, it is instructive to consider the cylinder case and to exclude the geometrical effects.

In the following, the details of the numerical calculations are presented and some concluding remarks of these preliminary results are discussed.

2 Governing equations

Consider a steady two-dimensional inviscid flow over a cylinder. Using standard notations, conservation of mass, momentum, and energy can be written (for a perfect gas) in the form:

$$\nabla \cdot \varrho \boldsymbol{q} = 0, \tag{1}$$

$$\nabla \varrho \boldsymbol{q} \boldsymbol{q} = -\nabla p, \tag{2}$$

$$\nabla \cdot \varrho \boldsymbol{q} H = 0, \tag{3}$$

where:

$$H = \gamma/(\gamma - 1)\,(p/\varrho) + (1/2)\,q^2.$$

From Eq. (3), it is clear that the total enthalpy, H, remains constant along a streamline. Moreover, if the incoming flow has uniform H, then H is constant everywhere. Taking the curl of Eq. (2), one can obtain the vorticity transport equation and again it can be shown that if the incoming flow has zero vorticity, the vorticity vanishes everywhere. In this case, a potential function can be introduced such that

$$\boldsymbol{q} = \nabla \phi. \tag{4}$$

Using the vector identity

$$(\boldsymbol{q} \cdot \nabla)\,\boldsymbol{q} = (1/2)\,\nabla q^2 + \omega \times \boldsymbol{q} \tag{5}$$

Eq. (2) can be rewritten (for smooth flows) as

$$\omega \times \boldsymbol{q} = -\nabla p/\varrho - (1/2)\,\nabla q^2 = -\nabla H + T\nabla S. \tag{6}$$

Hence, a steady irrotational ($\omega = 0$) and isoenergetic ($H = $ constant) flow is also isentropic ($S = $ constant).

Normalizing the velocity, $\boldsymbol{q}$, and the density, ϱ, by the speed and the density of the incoming flow, q_∞ and ϱ_∞, respectively, the isentropic relation for a perfect gas is given by

$$\bar{p} = \bar{\varrho}^\gamma/(\gamma M_\infty^2), \tag{7}$$

where p is normalized by $\varrho_\infty q_\infty^2$.

The classical potential flow formulation is then

$$\nabla \cdot \bar{\varrho} \nabla \bar{\phi} = 0, \tag{8}$$

where:

$$\bar{\varrho}^{\gamma-1} = 1 - (1/2)\,(\gamma - 1)\,M_\infty^2(|\nabla \bar{\phi}|^2 - 1). \tag{9}$$

Equation (9) is obtained by substituting Eq. (7) in the definition of H. Equations (8) and (9) can be combined in a single, nonlinear partial differential equation for $\bar{\phi}$. For smooth flows, the potential solution is also a solution of the Euler equations. This is not true, however, once a shock appears in the flow.

3 Boundary conditions

The flow must be tangent to a solid surface. Hence, the no penetration condition implies

$$q \cdot n = 0, \tag{10}$$

where n is the normal vector to the surface. In the far field, q approaches q_∞ (except in the wake of a shock wave). The ambient pressure is always recovered downstream of any obstacle.

For potential flows, Eq. (10) is equivalent to

$$\partial\phi/\partial n = 0. \tag{10.1}$$

The density is calculated from Bernoulli's law. Notice, $\partial\varrho/\partial n$ does not vanish at the wall since $\partial\varrho/\partial n = -\varrho M^2/R$. An asymptotic solution in the far field includes the contribution of a uniform flow, a doublet, and an irrotational vortex. The potential function is discontinuous across a line in the doubly connected domain around the body, and the jump in potential across such a line is equal to the circulation

$$\Gamma = \oint q \cdot dt.$$

The circulation is the same for any closed curve around the body.

4 Numerical methods

In general, to obtain a numerical solution of the Euler or potential equations, artificial viscosity terms are added to guarantee numerical stability and/or capture discontinuities (shocks and wakes). The following forms are adopted

$$\nabla \cdot \varrho q = \varepsilon_c(\nabla^2 p + g), \tag{11}$$

$$\nabla\varrho qq = -\nabla p + \varepsilon_m \nabla^2 q, \tag{12}$$

$$H = H_\infty, \tag{13}$$

where ε_c and ε_m are small parameters and g in Eq. (11) is given by

$$g = \nabla \cdot \big(\varrho(\nabla \cdot q)\, q\big).$$

On the differential level, the right-hand side of Eq. (11) vanishes for smooth flows since it is the divergence of the momentum equations. The right-hand side of Eq. (12) vanishes for irrotational incompressible flows since

$$\nabla^2 q = \nabla(\nabla \cdot q) - \nabla \times \omega. \tag{14}$$

In general, an estimate of ω is needed to construct a higher order viscosity based on Eq. (14).

For potential flows, the following augmented continuity equation is used

$$\nabla \cdot \bar{\varrho}\nabla\bar{\phi} = \varepsilon\nabla^2\bar{\varrho}, \tag{15}$$

while Eq. (9) (Bernoulli's law) is unchanged.

Other forms of artificial viscosity ([4]) have been tested and will be discussed elsewhere. In passing, the Laplacian of the velocity vector is not simply the Laplacian of its components, in a general coordinate system.

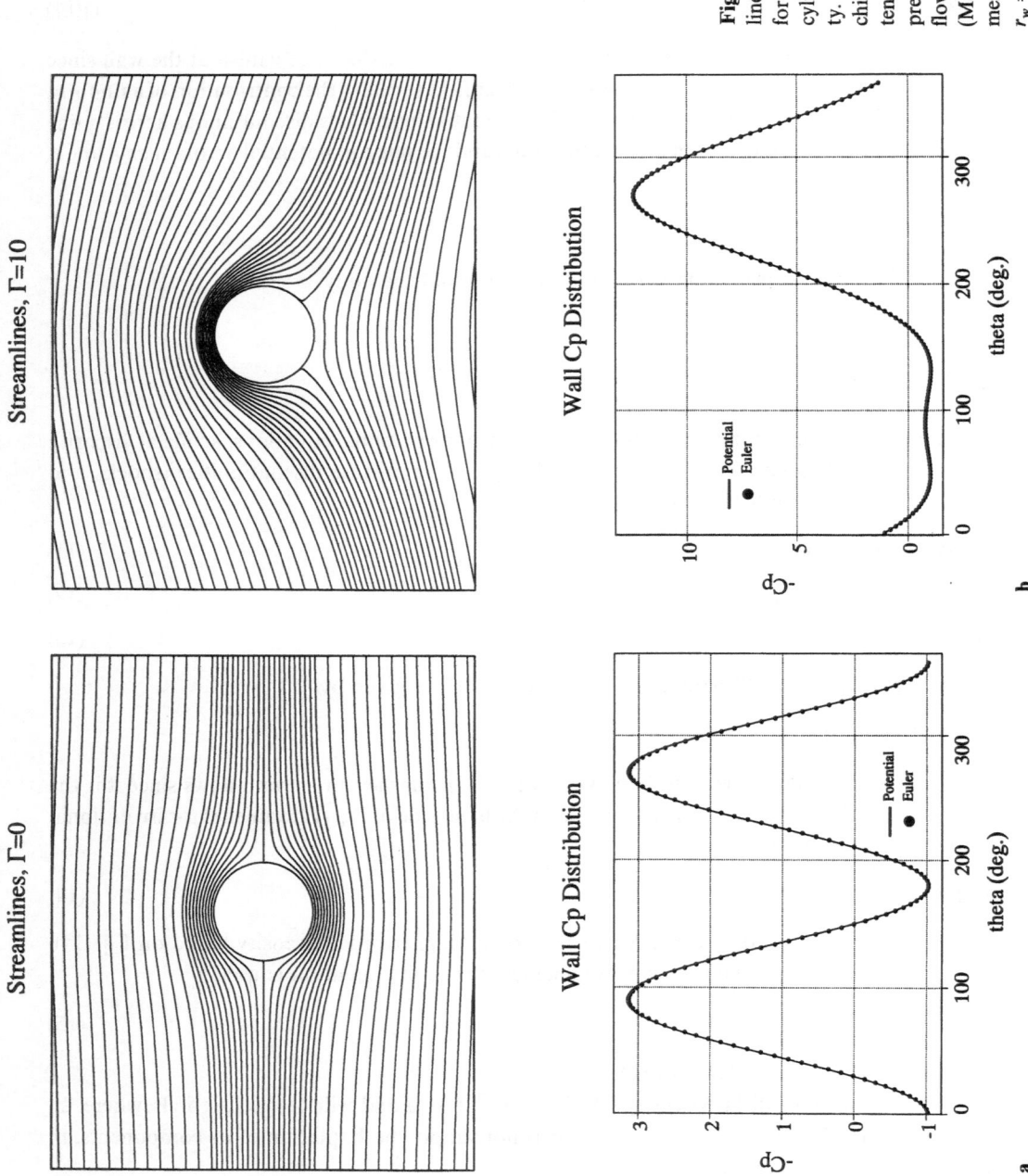

Fig. 1. a Potential and Euler streamlines and wall pressure distributions for subsonic flow around a stationary cylinder ($Ma = 0.2$, $\Gamma = 0$, no viscosity. Fine mesh: 136×34; 15% stretching from $r_w = 1$ to $r_0 = 12$). **b** Potential and Euler streamlines and wall pressure distributions for subsonic flow around a rotating cylinder ($Ma = 0.2$, $\Gamma = 10$, no viscosity. Fine mesh: 136×34; 15% stretching from $r_w = 1$ to $r_0 = 12$)

A careful treatment of numerical boundary conditions is necessary to obtain acceptable results. The normal velocity component at the nodes on the surface is set to zero, while the pressure and the tangential velocity component are obtained by enforcing the normal and tangential momentum equations, respectively, using a Crank-Nicholson scheme. The scheme includes both the rings on and next to the body (no artificial viscosity is used for the normal momentum equation at the surface, while the tangential momentum equation includes artificial viscosity terms only in the tangential direction). For subsonic flows, results can be obtained without any use of artificial viscosity terms ([5]). In this case, the tangential velocity component is calculated from the zero vorticity condition. The density is always calculated from the constant total enthalpy condition. In the far field, an analytical asymptotic solution is used including the contribution of the circulation.

The above equations, in cylindrical coordinates, are discretized on an orthogonal grid, via central differences. The resulting nonlinear algebraic equations are solved by Newton's method. At each iteration, a direct solver (LAPACK) based on an efficient Gaussian elimination procedure of banded matrices is employed to calculate the corrections. The solution is updated and the process is repeated until convergence to machine accuracy.

5 Numerical results

The results of numerical simulations of a compressible flow over a cylinder based on potential and Euler equations are compared. Streamlines and surface pressure distributions are plotted in Fig. 1 for a subsonic flow, not aligned with the grid. Unlike the work of Pulliam [6] and others [7] – [9], the potential and Euler solutions are indistinguishable for subsonic flows on a fine grid. In Fig. 2, the circulation (Γ) is plotted versus the stagnation point location. Again, the potential and Euler calculations (with and without artificial viscosity) are in good agreement. Notice, the present work is different from that of [10] where special schemes on staggered grids were used to guarantee the existence of a discrete potential.

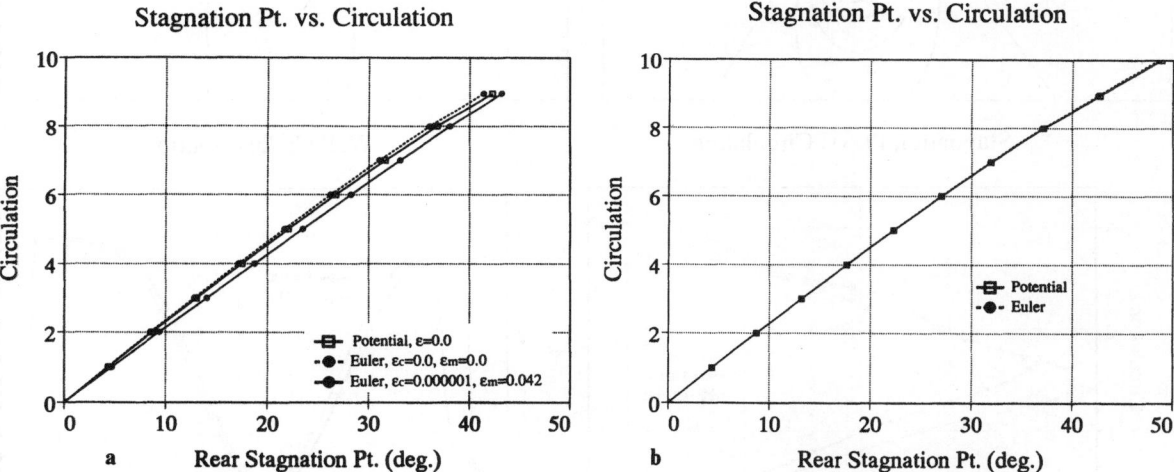

Fig. 2. a Variation of rear stagnation point with circulation for potential and Euler flow (Ma = 0.2. Course mesh: 96 × 22; 25% stretching from $r_w = 1$ to $r_0 = 12$). **b** Variation of rear stagnation point with circulation for potential and Euler flow (Ma = 0.2, no viscosity. Fine mesh: 136 × 34; 15% stretching from $r_w = 1$ to $r_0 = 12$)

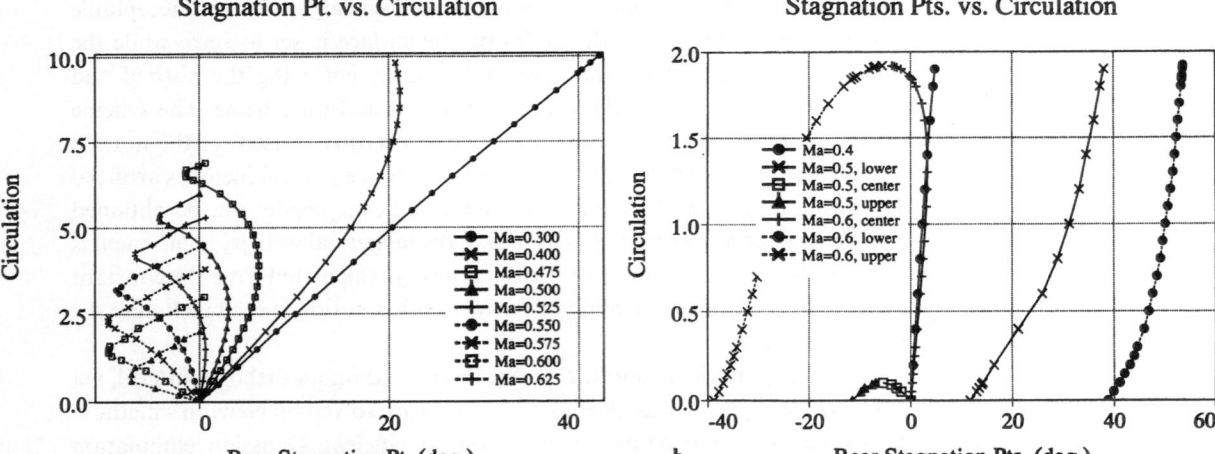

Fig. 3. a Variation of rear stagnation point with circulation for potential flows ($\varepsilon = 0.11$. Fine mesh: 136×34; 15% stretching from $r_w = 1$ to $r_0 = 12$). **b** Variation of stagnation points with circulation for Euler flows ($\varepsilon_c = 0.00001$, $\varepsilon_m = 0.019$. Unif. mesh: 136×34; from $r_w = 1$ to $r_0 = 6.25$)

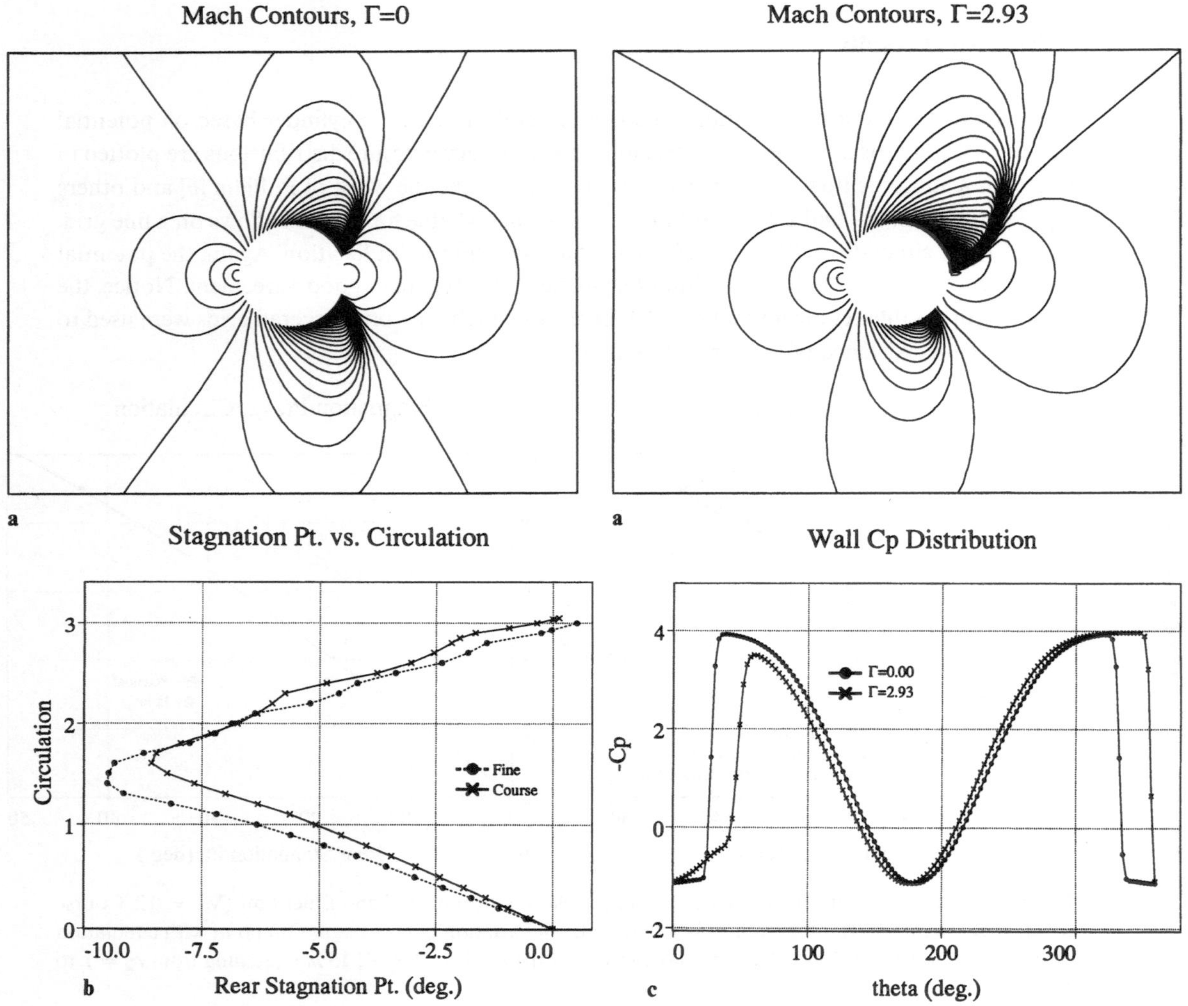

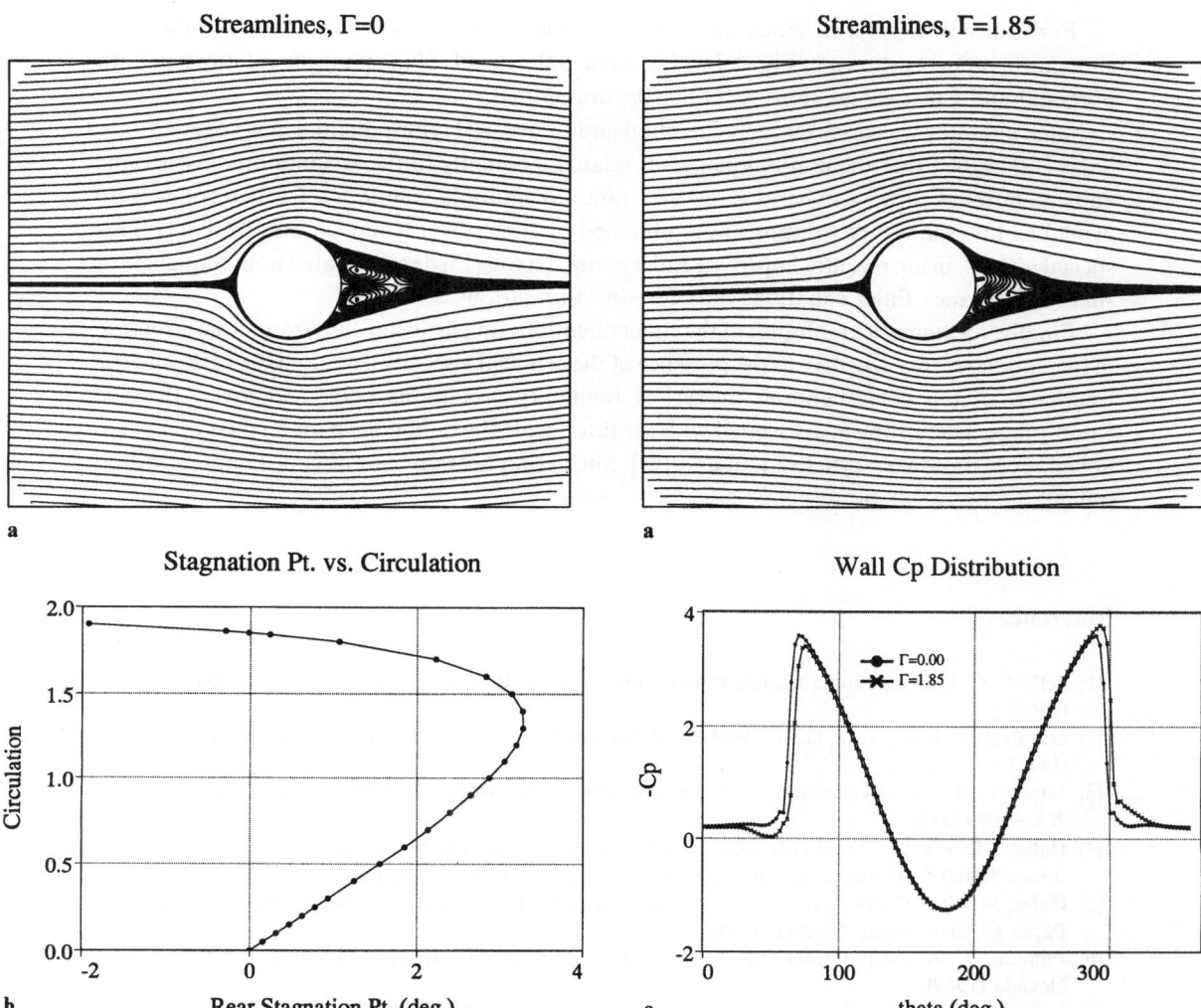

Fig. 5. a Streamlines of multiple solutions of Euler calculations (Ma $= 0.6$, $\varepsilon_c = 0.00001$, $\varepsilon_m = 0.019$. Unif. mesh: 136×34; from $r_w = 1$ to $r_0 = 6.25$). **b** Variation of rear (center) stagnation point with circulation of Euler calculations (Ma $= 0.6$, $\varepsilon_c = 0.00001$, $\varepsilon_m = 0.019$. Unif. mesh: 136×34; from $r_w = 1$ to $r_0 = 6.25$). **c** Multiple solutions (wall pressure distributions) of Euler calculations (Ma $= 0.6$, $\varepsilon_c = 0.00001$, $\varepsilon_m = 0.019$. Unif. mesh: 136×34; from $r_w = 1$ to $r_0 = 6.25$)

Fig. 4. a Mach contours of multiple solutions of potential calculations (Ma $= 0.6$, $\varepsilon = 0.11$. Fine mesh: 136×34; 15% stretching from $r_w = 1$ to $r_0 = 12$). **b** Variation of stagnation points with circulation for potential flows on a course and fine mesh (Ma $= 0.6$, $\varepsilon = 0.11$. Fine mesh: 136×34; 15% stretching from $r_w = 1$ to $r_0 = 12$. Course mesh: 96×22; 25% stretching from $r_w = 1$ to $r_0 = 12$). **c** Multiple solutions (wall pressure distributions) of potential calculations (Ma $= 0.6$, $\varepsilon = 0.11$. Fine mesh: 136×34; 15% stretching from $r_w = 1$ to $r_0 = 12$)

For transonic flows with shock waves, it is possible to find multiple circulations for the same stagnation point, as shown in Fig. 3. In Figs. 4 and 5 the details of the multiple solutions of both potential and Euler calculations at $M_\infty = 0.6$ are plotted.

Such anomaly was discovered by Steinhoff and Jameson [11] for potential flows over airfoils. Salas [12] argued that the non-uniqueness is related inherently to the assumption of isentropic and irrotational flow, just because at that time no multiple solutions of Euler codes were available. Later, multiple solutions were obtained by Jameson [13] and by Garabedian [14] for special airfoils, using recently improved Euler codes. Nixon [15] demonstrated nonuniqueness of small disturbance Euler equations only for very thin airfoils.

Finally, the numerical solution of the Euler equations (in particular the size and the structure of the bubble) is very sensitive to the presence of the artificial viscosity (see also [16] – [18] and the treatment of the corresponding numerical boundary conditions [19]). Moreover, there is a non-uniqueness problem associated with the flow inside the bubble even for an incompressible flow over a stationary cylinder [20] and [21]. Such complications are obviously absent in the potential flow calculations.

References

[1] Sells, C. C. L.: Plane subcritical flow past a lifting airfoil. Proc. R. Soc. London Ser. A **308**, 377 – 401 (1968).

[2] Garabedian, P. R., Korn, D. G.: Analysis of transonic airfoils. Comm. Pure Appl. Math. **24**, 841 – 851 (1971).

[3] Jameson, A.: Iterative solution of transonic flows over airfoils and wings. Comm. Pure Appl. **27**, 283 – 309 (1974).

[4] Hafez, M., Kinney, D.: Finite element simulation of transonic flows with shock waves. In: 13th International Conference on Numerical Methods in Fluid Dynamics, Italy 1992.

[5] Hafez, M., Yam, C., Tang, K., Dwyer, H.: Calculations of rotational flows using stream function. AIAA Paper 89-0474, Reno, Nevada (1989).

[6] Pulliam, T.: A computational challenge: Euler solutions for ellipses. AIAA Paper 89-0464, Reno, Nevada (1989).

[7] Lottati, I., Eidelman, S., Drobot, A.: A fast unstructured grid second order Godunov solver. AIAA Paper 90-0699, Reno, Nevada (1990).

[8] Ramakrishnan, S. V., Ota, D. K., Chakravarthy, S. R.: Numerical simulation of inviscid flow past an ellipse at incidence. In: Fourth International Symposium on CFD, University of California, Davis 1991.

[9] Baruzzi, G., Habashi, W., Hafez, M.: Finite element solutions of the Euler equations for transonic external flows. AIAA Paper 90-0405, Reno, Nevada (1990).

[10] Hafez, M., Brucker, D.: Effects of artificial vorticity on the discrete solution of Euler equations. AIAA Paper 91-1553, AIAA 10th CFD Conference, Honolulu, Hawaii (1991).

[11] Steinhoff, J., Jameson, A.: Multiple solutions of the transonic potential flow equation. AIAA Paper 81-1019, AIAA CFD Conference, Palo Alto, California (1981).

[12] Salas, M. D., Jameson, A., Melnik, R. E.: A comparative study of the nonuniqueness of the potential equation. AIAA Paper 83-1888, AIAA CFD Conference, Danvers, Massachusetts (1983).

[13] Jameson, A.: Nonunique solutions to the Euler equations. AIAA Paper 91-1625, Honolulu, Hawaii (1991).

[14] Garabedian, P.: Comparison of numerical methods in transonic aerodynamics. In: Fourth International Symposium on CFD, University of California, Davis 1991.

[15] Nixon, D.: The occurrence of multiple solutions for the transonic small disturbance Euler equation. Acta Mech. **80**, 191 – 199 (1989).

[16] Salas, M. D.: Recent development in transonic Euler flow over a circular cylinder. Math. Comput. Simul. **25**, 232 – 236 1983.

[17] Buning, P. G., Steger, J. L.: Solution of the two-dimensional Euler equations with generalized coordinate transformation using flux vector-splitting. AIAA Paper 82-0971 (1982).

[18] Pandolfi, M., Larocca, F.: Transonic flow about a circular cylinder. Comput. Fluids 17, 205 – 220 (1989).

[19] Dadone, A., Grossman, B.: Surface boundary conditions for the numerical solution of the Euler equations. ICAM Report 92-10-04, Virginia Polytechnic Institute and State University, Blacksburg, VA (1992).

[20] Bruneau, C. H., Chattot, J. J., Laminie, J., Temam, R.: Numerical solutions of the Euler equations with separation by a finite element method. In: Ninth International Conference on Numerical Methods in Fluid Dynamics, France 1984.

[21] Garabedian, P.: Nonparametric solution of the Euler equations for steady flow. Comm. Pure Appl. Math. 36, 529 – 536 (1983).

Authors' address: Professor M. Hafez Ph. D. and Mr. A. C. B. Dimanlig, Department of Mechanical, Aeronautical and Materials Engineering, University of California, Davis, California 95616, U.S.A.

Acta Mechanica (1994) [Suppl] 4: 251–257
© Springer-Verlag 1994

Computation of fictitious gas flow with Euler equations

P. Li, Karlsruhe, and **H. Sobieczky**, Göttingen, Federal Republic of Germany

Summary. The Fictitious Gas Concept supports some computational design methods to construct shock-free transonic flows. It was originally developed for potential flows, here it is introduced to the Euler equations for more general applications. A new equation of state needs to be defined in order to simulate results of the simpler potential approach. An operational numerical Euler code was chosen for the modifications and tested on a basic boundary value problem: The inviscid flow past a circular cylinder with a local supersonic flow region. The numerical computation is based on a finite volume method to solve the time dependent Euler equations in integral form. Conclusions are drawn for a physical explanation of the hitherto abstract "fictitious" gas: an internal momentum and energy supply/removal is modelled and the results for locally non-isentropic flow may be interpreted as an internal cooling/heating process controlled by the flow velocity.

1 Introduction

The Fictitious Gas Concept (FGC) and its applications to transonic design of shock-free aerodynamic configurations were proposed by Sobieczky and extensively investigated by Sobieczky and Seebass [1], [2]. The main purpose of the FGC is to design airfoil and wing shapes as well as turbomachinery components which exhibit shock-free supercritical flows at operation conditions. In a first step of the numerical solution the subsonic part of the flow is solved with an altered perfect gasdynamic law in the supersonic domain, which results in a preliminary elliptic replacement within this domain – a "fictitious" part of the flow. Because of its elliptic type, this flow is shock-free and carries conditions to support also shock-free "real" transonic flow, which in a second step and compatibly with the sonic surface found in the first step still has to be computed. The correct mixed type structure of the transonic field is recovered in this second step: real supersonic flow calculation uses the sonic surface as initial data and yields the boundary shape wetted by supersonic flow compatible with the combined/supersonic flow domains.

During the 1980's many operational computer flow analysis programs based on potential theory have been extended to be design tools by the use of the FGC. With new inverse design tools available today for airfoils and simple wings as reviewed in [3], the value of the concept may be reduced to applications in complex configuration optimization, where use of the FG in combination with flexible direct geometry generators allows for control of complex local supersonic flow domains. It has been shown by Zhu and Sobieczky [4] that for engineering applications the above-mentioned second step of computing the real supersonic domain can be replaced by a simple surface alteration within the sonic bubble, with shape parameters defined by the extent and size of this bubble resulting from a FG calculation. Nearly shock-free flow results from this approach, as an easily obtained precondition for an accelerated aerodynamic optimization strategy.

With faster computers and accelerated CFD software available, the use of Euler solvers for practical aerodynamic design becomes feasible and all techniques successfully developed with potential theory should be available now also for the use of these improved CFD methods in applied aerodynamics. In this situation we may ask the question:

How can the FGC be implemented to the Euler equations and is there an explanation from gasdynamics or flow physics supporting the concept? In this article we duplicate FG models developed for potential theory with an Euler code. Illustrations are shown for the simple but computationally non-trivial example of transonic flow past the circular cylinder. The computer code of Kroll and Rossow [5], [6], [7] is applied to solve the two-dimensional Euler equations using a finite volume spatial discretization and Runge-Kutta time stepping schemes as developed by Jameson, Schmidt and Turkel [8].

2 Euler equations and fictitious gasdynamic relations

The two-dimensional Euler equations describing conservation of mass, momentum and energy for unsteady inviscid flows are written in the following conservative form

$$\frac{\partial U}{\partial t} + \frac{\partial F}{\partial x} + \frac{\partial G}{\partial y} = RHS, \tag{1}$$

where

$$U = \begin{bmatrix} \varrho \\ \varrho u \\ \varrho v \\ \varrho e \end{bmatrix}, \quad F = \begin{bmatrix} \varrho u \\ \varrho u^2 + p \\ \varrho uv \\ (\varrho e + p)\, u \end{bmatrix}, \quad G = \begin{bmatrix} \varrho v \\ \varrho uv \\ \varrho v^2 + p \\ (\varrho e + p)\, v \end{bmatrix}, \quad RHS = 0.$$

For the perfect gas the equation of state can be described as

$$p = (\gamma - 1)\, \varrho \left(e - \frac{u^2 + v^2}{2} \right), \tag{2}$$

with symbols u, v, p, ϱ, and e for flow velocity components, pressure, density and energy, respectively; and γ for the ratio of specific heats.

For the FG gasdynamic relations are asked for to provide a change from a locally hyperbolic to an elliptic or parabolic behavior. To acquire this an analytical fictitious relation for the speed of sound

$$A_f(Q) \geqq Q \tag{3}$$

needs to be prescribed, which gives with a relation for the compatible equation of state

$$A_f{}^2 = \frac{1}{\gamma} \cdot \frac{dP_f}{dD_f}. \tag{4}$$

Here the non-dimensional variables are defined by sonic (critical) conditions of perfect gas flow:

$$A = \frac{a}{a^*}, \quad Q = \frac{q}{a^*}, \quad D = \frac{\varrho}{\varrho^*}, \quad P = \frac{p}{p^*}.$$

Model functions for fictitious speed of sound and fictitious density implemented respectively into various non-conservative and conservative potential codes are

$$A_f{}^2 = Q \left[1 + \frac{1}{\lambda} (Q - 1) \right], \tag{5}$$

$$D_f = \left[1 + \frac{1}{\lambda} (Q - 1) \right]^{-\lambda}. \tag{6}$$

With Eqs. (4)–(6) the relation for fictitious pressure, to be implemented into the Euler equations, can be obtained

$$P_f = 1 + \frac{\lambda \gamma}{2 - \lambda} \left[2 - \lambda D^{\left(\frac{\lambda - 2}{\lambda} \right)} - (2 - \lambda) D^{\left(\frac{\lambda - 1}{\lambda} \right)} \right], \tag{7}$$

or

$$P_f = 1 + \frac{2 \lambda \gamma}{2 - \lambda} - \frac{\lambda \gamma}{2 - \lambda} (Q + 1) \left[1 + \frac{1}{\lambda} (Q - 1) \right]^{1 - \lambda}. \tag{8}$$

These relations constitute a one-parametric $(0 < \lambda \leqq 1)$ family of fictitious gases which includes parabolic sonic flow $(\lambda = 1)$ and approaches an incompressible limit for the fictitious gas flow $(\lambda \to 0)$, viz. $\varrho = \varrho^*$.

In the present computation the fictitious pressure-density relation (7) is used as an equation of state within isolated supercritical domains where $Q \geqq 1$. At the sonic line $(Q = 1)$ the real subsonic flow and the FG flow have a smooth continuation. Even though the total energy (e) is independent of the fictitious pressure, the total enthalpy $(h = e + p/\varrho)$ as well as the entropy $(s = p/\varrho^\gamma)$ in fictitious gas flow deviate from that in real flow. By contrast to the real subsonic isentropic flow the FG flow is a non-isentropic one. The fictitious density-velocity relation (6) will be proven as a numerical result of the manipulated equations.

3 Discussion of fictitious gas models

The Euler equations for the FG flow can be transferred into another form. Assuming that p at the left hand side of Eq. (1) is a perfect gas pressure, the *RHS* for the FG model has then the following form:

$$RHS = \begin{bmatrix} 0 \\ \dfrac{\partial}{\partial x} (\varDelta p) \\ \dfrac{\partial}{\partial y} (\varDelta p) \\ \dfrac{\partial}{\partial x} (\varDelta p \cdot u) + \dfrac{\partial}{\partial y} (\varDelta p \cdot v) \end{bmatrix}, \tag{9}$$

where

$$\varDelta p = p - p_f.$$

p and p_f are respectively determined by Eq. (2) and Eq. (7) ($p_f = P_f \cdot p^*$). In this case the Euler equations for the FG model are equal to those for the real flow with additional source terms in the momentum and energy equations. Therefore the FG flow can be interpreted as a flow with internal momentum and energy supply/removal.

4 Numerical algorithm

In the numerical computation the time dependent Euler equations in integral form are discretized by the finite volume approach based on the method of Jameson et al. [8]. The physical domain is covered with a body-fitted grid using curvilinear coordinates. In the present work an O-grid was generated by conformal mapping. In the finite volume spatial discretization a cell vertex scheme is applied, in which the flow variables are defined at the vertices of the cells and a volume weighted distribution formula of Hall [9] is used for the rate of change of the flow variables at a vertex. According to Kroll and Rossow in the case of the cell vertex formulation, a solid wall boundary condition is implemented by the projection of velocities. The treatment of the far field boundary conditions is based on the characteristic variables for one-dimensional flow normal to the boundary.

In order to damp high frequency oscillations caused by the finite volume discretization with central averaging, artificial dissipative terms are introduced by a blend of first and third order dissipative fluxes. For the integration of the discrete equations to steady state an explicit 5-stage time stepping scheme of Runge-Kutta type is used with two evaluations of the dissipative terms. To accelerate the convergence of the solution of unsteady Euler equations to steady state, the techniques including local time stepping, enthalpy damping and residual averaging are applied in the present code extended to allow calculations with the fictitious gas model. As convergence criterion the relative rate change of density is used and the accuracy is 10^{-5}.

5 Results and discussion

The Euler code has been used for various airfoils as well as for the simple test case of a circular cylinder in transonic flow. The critical Mach number for this boundary value problem without circulation has been determined by Van Dyke and Guttmann [10] as $M^* = 0.3982$. The present investigation is focused on supercritical flow at $M_\infty = 0.45$. The parameter λ in the fictitious gas laws is set to 0.8. The local Mach number contours for perfect gas flow and the O-grid around the circle are plotted in Fig. 1a. In this transonic flow because of a strong shock, the flow field is not symmetrical about the vertical axis. On the contrary the shock-free flow fictitious gas within the sonic bubble is completely symmetrical about the vertical axis (Fig. 1b). In Fig. 2 the fictitious density-velocity relation Eq. (6) is satisfactorily verified by the values for density and velocity on all grid points within the sonic bubble. The corresponding isentropic relation is also depicted.

We ask now for a perfect gas flow which exhibits the real subsonic part as computed with using the fictitious gas and a compatible real supersonic shock-free flow pattern within the sonic bubble. Here we skip the previously mentioned "second step" method of characteristics calculation and use instead the simpler optimization approach as developed by Zhu and Sobieczky [4]. The resulting flow field past the locally flattened circle is found by a final perfect gas flow analysis, it is found to be practically shock-free as shown in Fig. 1c. The pressure distributions for the transonic perfect gas flow past the original circle with strong shock and the

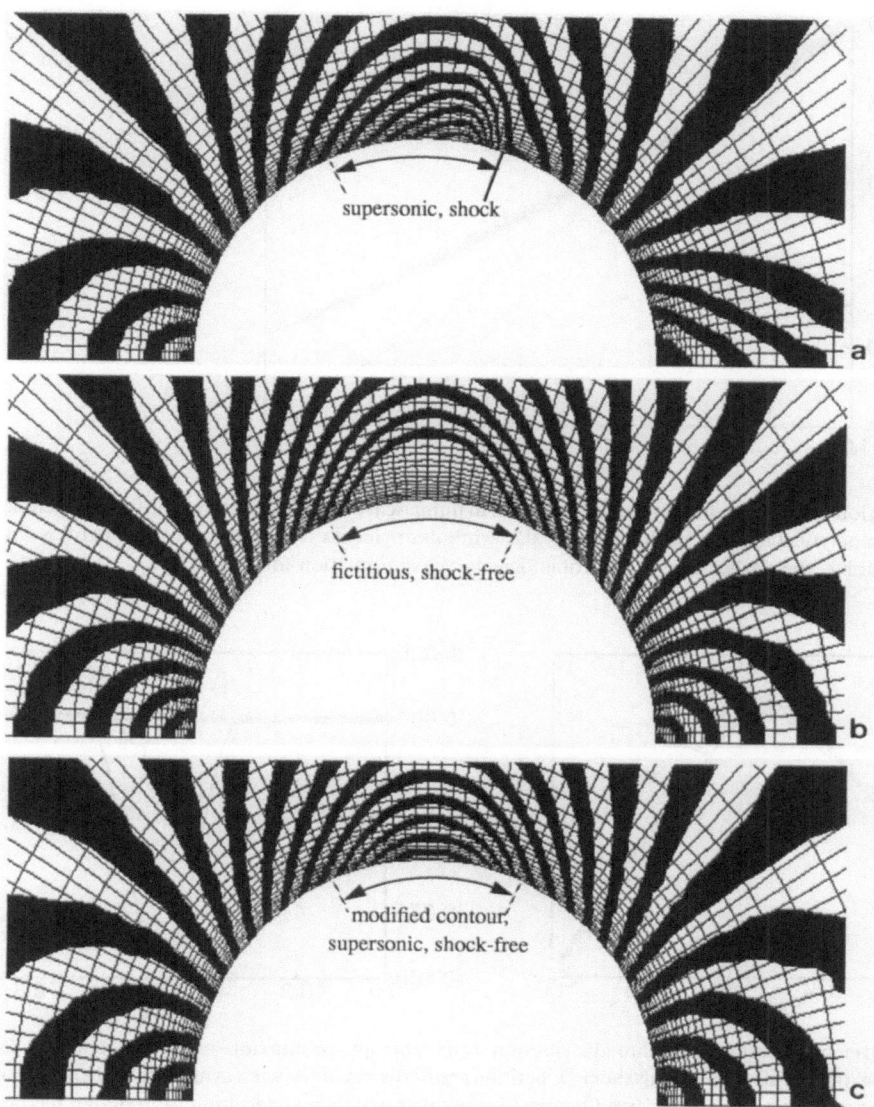

Fig. 1. Local Mach number contours and grid (*O*-grid with 160 × 32 cells) for transonic flow past circular cylinder, $M_\infty = 0.45$ and **a** perfect gas with recompression shock, **b** fictitious gas ($\lambda = 0.8$) within sonic bubble, and **c** perfect gas and contour modified by FG shock-free design technique

fictitious gas flow showing symmetrical distribution as well as the transonic practically shock-free flow past the locally deformed circle are compared in Fig. 3. Also, the entropy productions along the contour of the original circle in transonic perfect gas and fictitious gas flow and of the modified circle in transonic flow are shown. In fictitious gas flow the entropy is first decreased and subsequently raised to its upstream value, which can be interpreted as heat removal and subsequent equivalent heat addition within the sonic bubble. Compared with the increment of entropy caused by the strong shock in transonic flow past the original circle, the flow past the locally modified circle indicates practically no entropy inrease which verifies the shock-free design of the contour, resulting here in a thickness reduction of about half a percent of the circle diameter.

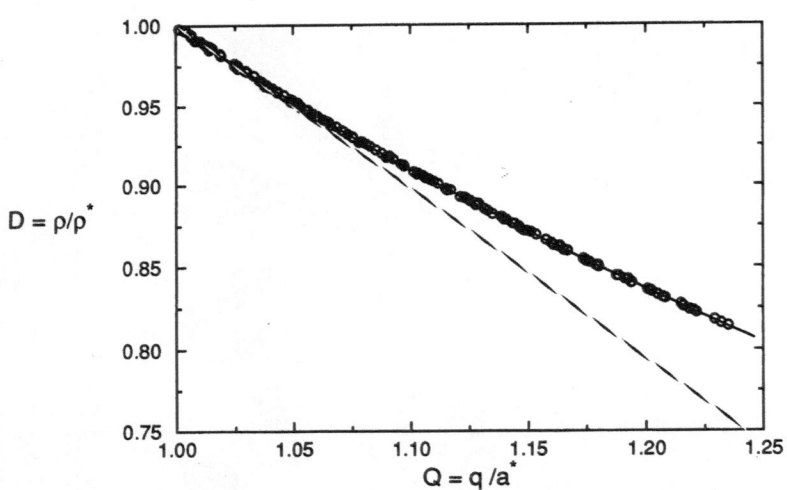

Fig. 2. Verification of fictitious density-velocity relation in numerical results for fictitious gas flow past circle: the solid line is analytical relation Eq. (6) with $\lambda = 0.8$, symbols are local values for density on all grid points within sonic bubble, the dashed line is isentropic density-velocity relation for perfect gas

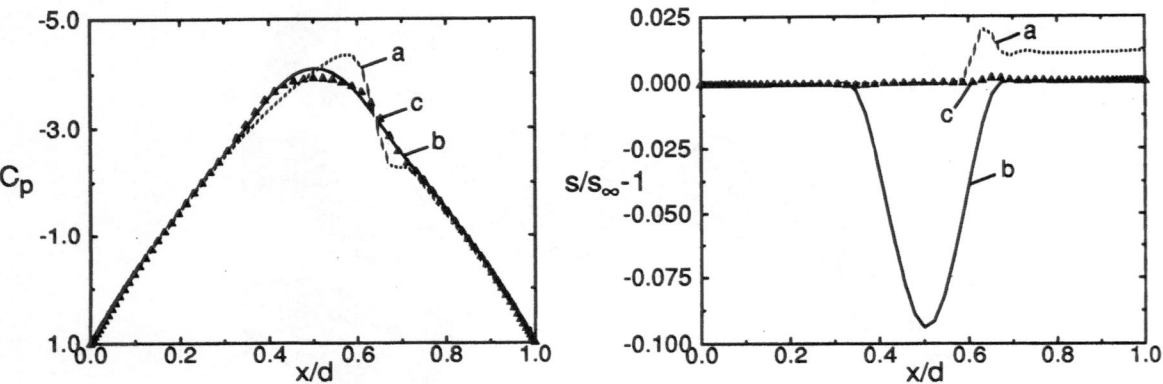

Fig. 3. Comparison of surface pressure distributions and entropy production along the contour for *a* transonic flow past circle with strong shock, *b* fictitious gas flow circle showing symmetrical distribution, and *c* for transonic practically shock-free flow past locally deformed circle, resulting from design method based on fictitious gas concept

6 Conclusions

The two-dimensional Euler equations in association with the fictitious gas laws are numerically solved by a finite volume spatial discretization and Runge-Kutta time stepping scheme. Numerical tests have shown that the fictitious pressure-density relation – a fictitious equation of state – is the appropriate function to be implemented into an Euler code in order to replace previous potential solver results which used the fictitious density-velocity relation. The latter agrees with numerical results of the manipulated equations. A simple surface alteration within the sonic bubble, resulting from an engineering design approach based on the fictitious gas concept, yields practically shock-free aerodynamic configurations. The fictitious gas flow is non-isentropic, which can be physically explained as an internal cooling/heating process controlled by the local flow velocity. The introduction of the fictitious gas model into the Euler equations provides a more general method of designing transonic flow fields with favorable aerodynamic characteristics.

Acknowledgements

This research is funded by Max-Planck-Forschungspreis 1991, awarded to H. Sobieczky and A. R. Seebass. Supply of computational facility by DLR Institute for Theoretical Fluid Mechanics, Göttingen, is gratefully acknowledged.

References

[1] Sobieczky, H.: Verfahren für die Entwurfsaerodynamik moderner Transportflugzeuge. DFVLR-FB **85–43** (1985).

[2] Sobieczky, H., Seebass, A. R.: Supercritical airfoil and wing design. Ann. Rev. Fluid Mech. **16**, 337–363 (1984).

[3] Sobieczky, H.: Progress in inverse design and optimization in aerodynamics. Conference on Computational Methods for Aerodynamic Design (Inverse) and Optimization, AGARD CP **463**, pp. 1.1.–1.10 (1989).

[4] Zhu, Z., Sobieczky, H.: An engineering approach for nearly shock-free wing design. Chin. J. Aeronautics, **2**, 81–86 (1989).

[5] Kroll, N., Jain, R. K.: Solution of two-dimensional Euler equations – experience with a finite volume code. DFVLR-FB **87–41** (1987).

[6] Rossow, C.: Berechnung von Strömungsfeldern durch Lösung der Euler-Gleichungen mit einer erweiterten Finite-Volumen Diskretisierungsmethode. DLR-FB **89–38** (1989).

[7] Kroll, N., Rossow, C.: Foundations of numerical methods for the solution of Euler equations. Carl-Crantz-Gesellschaft Lecture Series **F6. 03**, Braunschweig, Germany 1989.

[8] Jameson, A., Schmidt, W., Turkel, E.: Numerical solutions of the Euler equations by finite volume methods using Runge-Kutta time stepping schemes. AIAA Paper **81–1259** (1981).

[9] Hall, M. G.: Cell-vertex multigrid scheme for solution of the Euler equations. Proceedings of the Conference on Numerical Methods for Fluid Dynamics, Reading, 1985.

[10] Van Dyke, M., Guttmann, A. J.: Subsonic potential flow past a circle and the transonic controversy. J. Aust. Math. Soc. Ser. **B24** (1983).

Authors' address: Dr.-Ing. P. Li, Institut für Strömungslehre und Strömungsmaschinen Universität Karlsruhe (TH), D-76128 Karlsruhe, and Professor Dr. habil. H. Sobieczky, DLR-Institut für Theoretische Strömungsmechanik, Bunsenstrasse 10, D-37073 Göttingen, Federal Republic of Germany

Acta Mechanica (1994) [Suppl] 4: 259–267
© Springer-Verlag 1994

Calculation of three-dimensional flows with complex boundaries using a multigrid method

A. Orth* and **W. Rodi**, Karlsruhe, Federal Republic of Germany

Summary. The paper describes a multigrid method for calculating incompressible laminar and turbulent 3D flows with complex geometries. The flow solver is based on an iterative finite-volume method employing non-staggered, general curvilinear grids. The convergence of the basic solver is accelerated by a Full-Approximation-Scheme/Full Multigrid Method. The iterative SIMPLE scheme for solving the coupled momentum and continuity equations and the strongly implicit procedure (SIP) for solving iteratively the linear algebraic equations are found to have good smoothing properties for use in the multigrid method. Applications of the method to a few laminar and turbulent flows are presented. The performance is compared with that of single-grid calculations and demonstrates that considerable computing time can be saved with the multigrid method also for geometrically complex flows and when turbulence-model equations are solved.

1 Introduction

Numerical calculations of fluid flow processes play an increasingly important role in practical flow analysis and are used more and more for design purposes. Rapid advances in computer technology allow more complex problems to be tackled with increasingly realistic computational models. In general, the flows in engineering practice are three-dimensional, turbulent and have complex boundaries and therefore require a fine numerical resolution. Conventional iterative schemes for solving the strongly coupled non-linear differential equations governing the flow processes have convergence rates that decrease exponentially with increasing number of grid points, the exponent being 2–3. In spite of all the advances in computer speed, this leads quickly to unacceptably large computing times for complex flow problems. Hence, there is a great need for convergence acceleration methods, and the multigrid method is one of the most promising techniques because of its property of theoretical independence of the number of iterations on the grid fineness so that the computing effort increases only linearly with the number of grid points. In the multigrid technique, the solution is obtained by switching between finer and coarser grids during the iteration process. On the coarser grids, low-frequency errors are eliminated faster, while the final solution has the accuracy obtained on the finest grid.

Originally, the multigrid method was developed for linear systems of elliptic equations [1, 2], but recently the technique was extended for solving the Euler and Navier-Stokes equations (see e.g. [3–5]). However, this work was restricted so far mainly to simple geometries and laminar flow, where impressive speed-up factors over a solution with a single grid could be obtained. The present paper reports on the incorporation of a multigrid method into a 3D finite-volume method for solving geometrically complex laminar and turbulent flow problems on curvilinear boundary-fitted grids and presents some calculation examples.

* Present address: Lurgi AG, Hauptabteilung Prozeßentwicklung, D-60295 Frankfurt, Federal Republic of Germany.

2 Multigrid finite-volume method

The aim of the method is to solve efficiently and accurately the three-dimensional momentum and continuity equations as well as turbulence model equations written in arbitrary, non-orthogonal curvilinear coordinates. To this end, the flow field is discretized with a boundary-fitted grid, i.e. it is subdivided into small control volumes. Formal integration of the differential equations over the control volume yields finite-difference equations which express the balance of convective and diffusive fluxes through the control volumes faces and of source terms and pressure gradients. All variables are stored at the cell centres (non-staggered arrangement) and Cartesian velocity components are used. The fluxes need to be determined at the control volume faces from variables stored at the centres. A special momentum interpolation procedure is used for the mass fluxes, hybrid central/upwind discretization is used for the convection fluxes, central differencing for the normal diffusion fluxes, and the cross-derivative diffusion is treated explicitly. Pressure velocity coupling is handled via the iterative SIMPLE guess-and-correct procedure. The resulting system of linear equations is solved with an extended version [6] of Stones' Strongly Implicit Procedure (SIP). This basic finite-volume method is described in detail in [7]. When turbulent flows are calculated, the effects of turbulence are simulated with the k-ε model using wall functions for bridging the viscosity-affected near-wall layers [8].

The basic finite-volume method outlined above has the undesirable feature that the number of iterations necessary to reach convergence increases with the number of grid points. Incorporation of a multigrid procedure has the potential of removing this feature to a large extent. A necessary requirement for the success of the multigrid technique are good smoothing properties of the iterative algorithms used, where smoothing means the fast elimination of high frequency errors on a given mesh. The two iterative algorithms involved in the basic finite-volume method have been found to have good smoothing properties and could therefore be retained: They are the strongly implicit procedure (SIP) for solving iteratively the sets of linear algebraic equations and the pressure-correction algorithm SIMPLE for solving the coupled momentum and continuity equations. The good smoothing properties of the SIMPLE algorithm have been confirmed theoretically as well as in test calculations [9].

The features of the multigrid method incorporated have been described in detail in [10] and are now summarised briefly. The flow domain is discretized by several levels of grids. The coarsest grid is subdivided such that one coarse-grid control volume yields eight fine-grid control volumes and these are then subdivided again and so on. Up to five grid levels are used. The solution basically proceeds in V-cycles as illustrated in Fig. 1. The first step is relaxation (smoothing) on the finest grid, i.e. mainly the high-frequency errors are smoothed out by the iterative algorithms. The number of iterations for the SIMPLE algorithm is prescribed $(1-3)$.

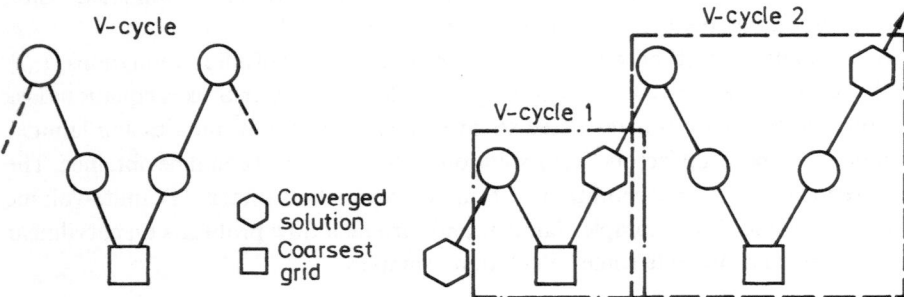

Fig. 1. V-cycle and typical FMG cycle

The next step is a so-called restriction, which involes the transfer of information from the fine to the next coarser grid, namely the transfer of residua (remaining solution errors), mass fluxes through the control-volume faces and solution variables. With these as starting values, equations on the coarse grid are solved for the sum of coarse-grid variables and solution errors, thereby accounting for non-linear source terms in the differential equations. This is called Full Approximation Scheme (FAS); in contrast the so-called Correction Scheme solves only equations for the error and cannot account for nonlinear source terms. Smoothing is then carried out again by the iterative algorithms. Continuity in the coarser grids is satisfied by solving the pressure-correction equation for the errors only. In the same way as just described the procedure then moves on to the next coarser grid and so on until the coarsest grid is reached (bottom of the V-cycle). The next step is called prolongation: The corrections, i.e. the errors calculated on the coarse grid, are transferred back to the next fine grid. First the corrections have to be calculated from the coarse-grid solution for the sum of errors and variables by subtracting the restricted fine-grid variable values. Then the corrections are interpolated linearly to yield the fine-grid corrections with which the original fine-grid values (before proceeding to the next coarser grid) are improved. A prescribed number of relaxation steps $(1-2)$ follows as so-called "post-smoothing". This procedure is repeated until the finest grid in the V-cycle is reached.

The method incorporated does not use straightforward V-cycles, but V-cycles with increasing number of grid levels, as sketched in Fig. 2 b. In the first cycles, the fine grid starting values are obtained by interpolation of a converged solution on the finest mesh of the previous V-cycle. The whole procedure starts with a converged solution on the coarsest grid. This is the so-called Full Multigrid (FMG) procedure which leads to faster convergence. Finally, it should be mentioned here that the complete computer programme has been vectorised to a large extent (98%), including the recursive SIP algorithm [10].

3 Application examples

Applications of the multigrid method described above to various 3D laminar and turbulent flows will now be presented. All calculations were carried out on the Siemens VP 400-EX vector computer. The calculation results were declared converged when the maximum normalised residue of the variables was less than a prescribed value ε. First, the multigrid method was

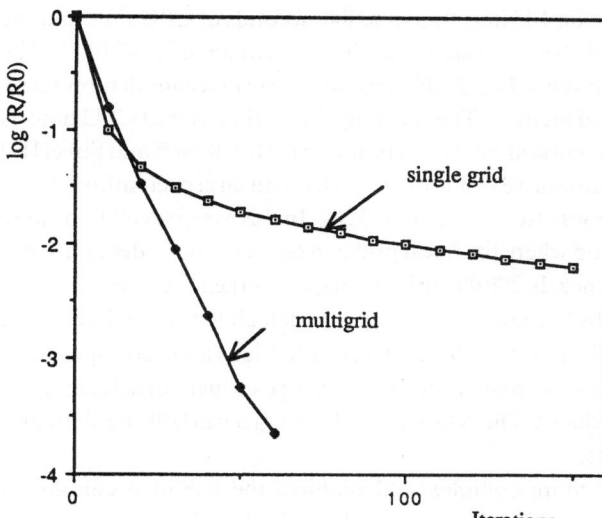

Fig. 2. Convergence history for cubical cavity calculations on $66 \times 66 \times 66$ grid

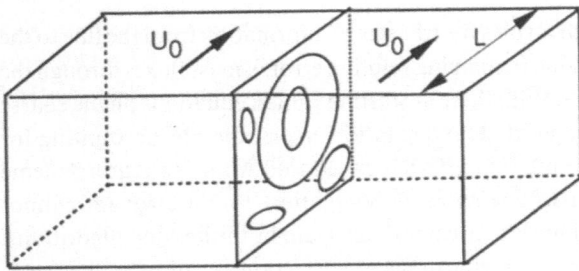

Fig. 3. Flow configuration of cavity with aspect ratio 3:1

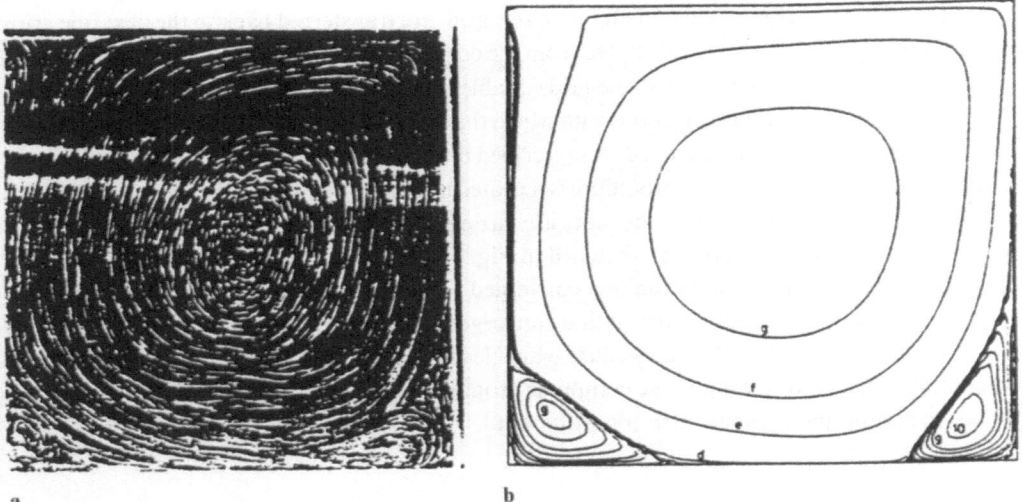

a   b

Fig. 4. Flow behaviour in symmetry plane of cavity with aspect ratio 3:1. **a** Flow visualisation picture [11]; **b** calculated streamlines

tested for 3D laminar cavity flow driven by a moving lid. Flow in a cubical cavity at Re = 1 000 (based on cavity depth and lid velocity) was calculated with a single-grid and the multigrid method on a $66 \times 66 \times 66$ grid; in the multigrid case this was the finest grid and three levels of coarser grids were used. The single-grid calculations took 105 iterations and 840 seconds of CPU time and the multigrid calculations 29 iterations and 216 seconds to reach an accuracy of $\varepsilon = 10^{-3}$. The convergence history of the calculations is given in Fig. 2 which shows clearly that considerably faster convergence is achieved with the multigrid method. The speed-up factor (in terms of CPU time) is 3.89. The results of the calculations for the cubical cavity can be found in [10]. Koseff and Street [11] investigated experimentally the laminar flow in a 3D lid-driven cavity with an aspect ratio of 3:1 as sketched in Fig. 3. The Reynolds number was Re = U_0L/v = 3200. In this case, calculations were carried out only with the multigrid method where the finest grid had $66 \times 66 \times 66$ nodes and 4 grid levels were used. In this calculation with nearly 290 000 grid points, convergence to $\varepsilon = 10^{-4}$ was reached after 13 fine-grid iterations in 118.6 seconds CPU time. The calculated streamlines in the mid-symmetry plane are compared with the flow visualisation picture in Fig. 4 and good agreement can be seen. In particular, the evolution of one main central vortex, two corner vorticies and one vortex along the left side wall is well reproduced. The velocity profiles are given in [10] and show also good agreement with the measurements.

   The next example is geometrically more complex and requires the use of a curvilinear boundary-fitted grid. It is the 3D laminar flow past a circular cylinder placed in a duct at

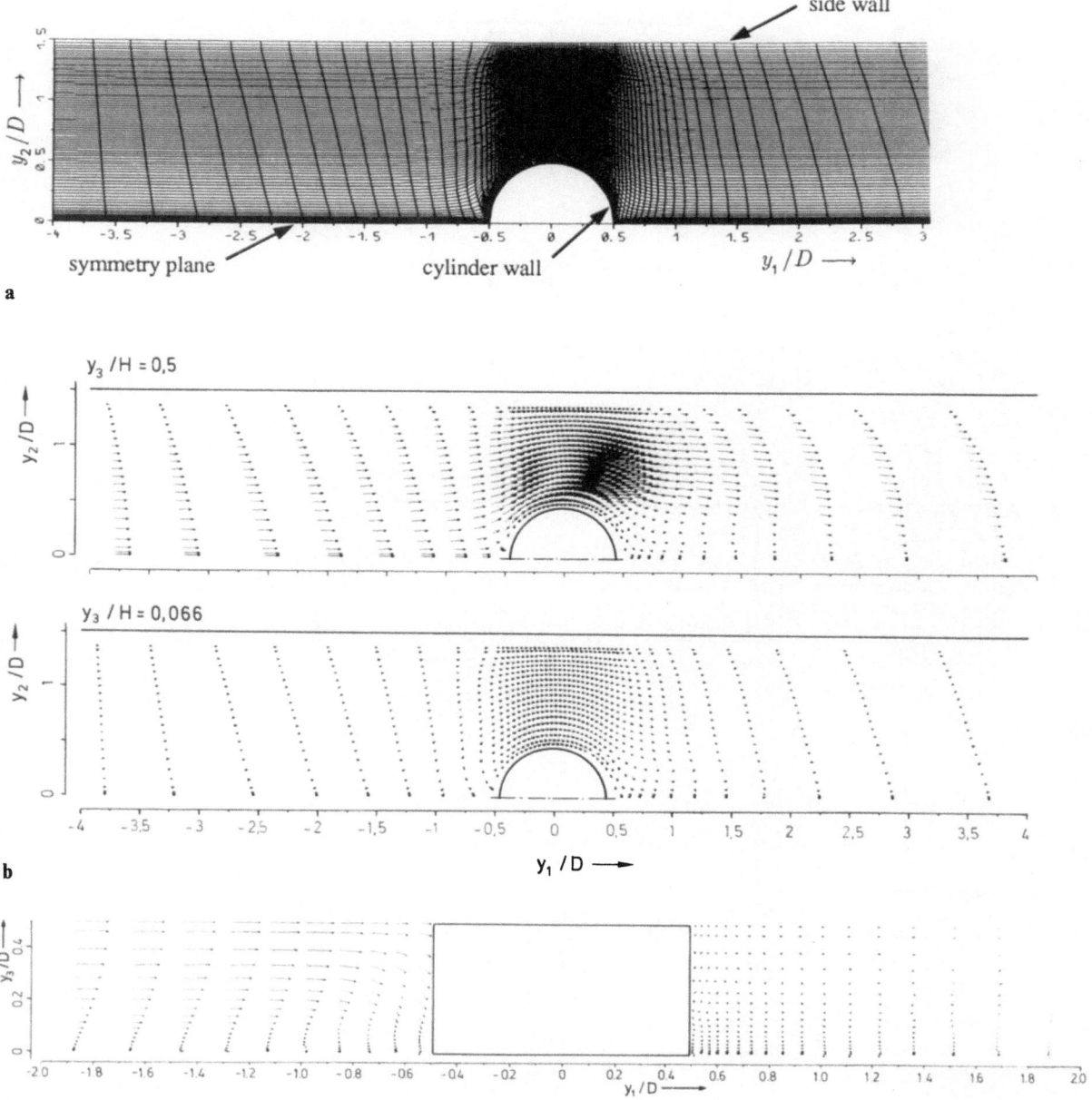

Fig. 5. Flow around cylinder placed in duct. **a** Geometry in horizontal plane and grid; **b** velocity vectors in horizontal $y_1 - y_2$-planes; **c** velocity vectors in vertical $y_1 - y_3$ symmetry plane

Re = 67 (based on cylinder diameter D and approach velocity). In a numerical study covering a wider parameter range Kiehm et al. [12] have shown that at this Reynolds number periodic vortex shedding, which would be expected for the case of flow past a long cylinder without the influence of channel walls, does not occur. Figure 5a shows the geometry in the horizontal plane and the 98 × 82 grid used. The channel depth H in the vertical direction was 1.08 D and 18 grid

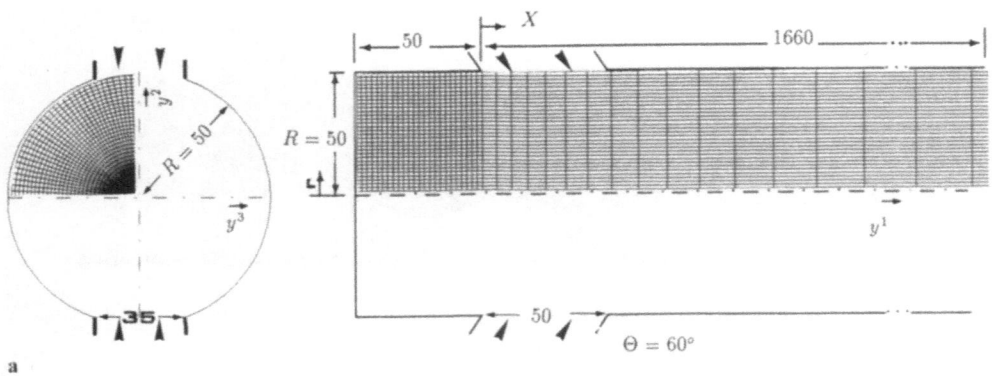

a

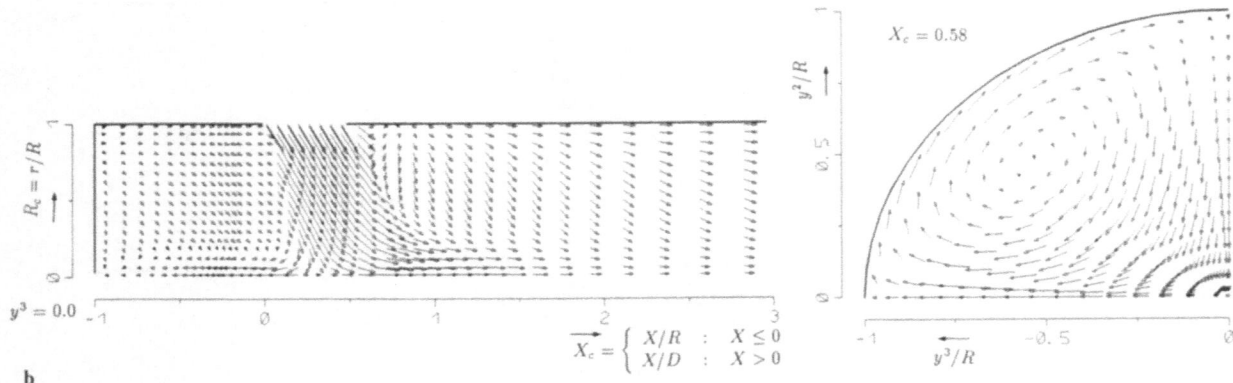

b

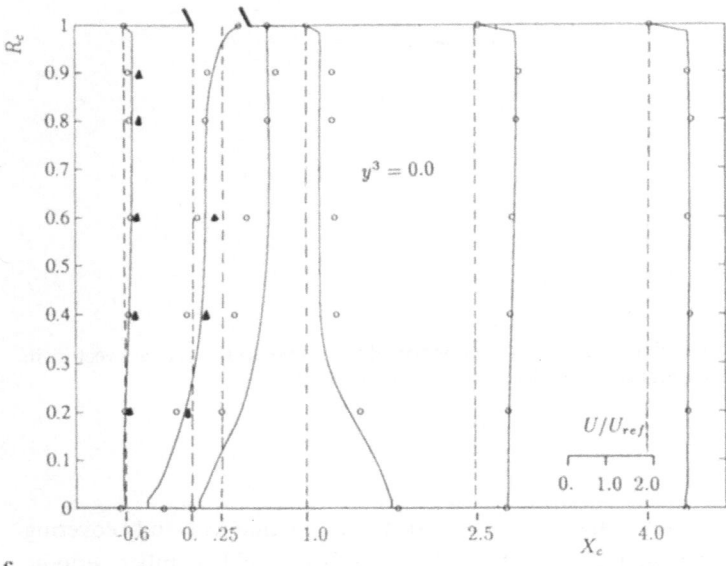

c

Fig. 6. Flow in a one-side closed cylindrical duct with injections. **a** Flow geometry and numerical grid; **b** predicted velocity vectors; in vertical symmetry plane; in $y_2 - y_3$-plane; **c** velocity profiles in mean-flow direction, ○ measurements Liou et al. [13], ● calculations Liou et al. [13], — present calculations

Table 1. Iterations and computing times for calculation of flow in pipe with injections

Gridpoints	$42 \times 18 \times 18$			$82 \times 34 \times 34$		
	Single grid	Multigrid	Ratio	Single grid	Multigrid	Ratio
Iterations	932	49	19.0	> 1 000	45	> 22.0
Comp.time	417 s	48 s	8.75	> 3 000 s	179 s	> 16.7

points were placed over the half-depth. For symmetry reasons, only one fourth of the total flow domain was calculated. The resulting $98 \times 82 \times 18$ grid was considerably finer than the $50 \times 20 \times 12$ grid used by Majumdar et al. [7] in their previous single-grid calculations of the same problem. At the inflow boundary 4D upstream of the cylinder, a fully developed parabolic velocity profile was prescribed and the exit boundary with zero gradient conditions was placed 8D downstream of the cylinder. Convergence to reach $\varepsilon = 5 \times 10^{-3}$ required 295 iterations and 17.5 min. with the single-grid method and 33 iterations and 2.75 min. with the multigrid method using three grid levels. The speed-up factor therefore is 6.36 in this case. Calculated velocity vectors are shown in Fig. 5b for two horizontal planes and in Fig. 5c for the vertical symmetry plane. Close to the bottom wall ($y_3/H = 0.066$), the oncoming boundary layer can be seen to separate in front of the cylinder. This separation, leading to the formation of the well known horseshoe vortex, is also obvious from Fig. 5c. In the lee of the cylinder, the flow separates approximately at 125° from the forward stagnation point and the separation region extends to about 1D downstream of the rearward stagnation point. At the horizontal symmetry plane, the flow does not separate in front of the cylinder but separates at the rear at an angle of 134° and the length of the separation zone is again about 1D. Figure 5c shows clearly the formation of the horseshoe vortex in front of the cylinder and the separated region in the rear. This calculation example demonstrates that the multigrid method allows to accelerate convergence also in the case of strongly non-orthogonal grids.

The final test case concerns turbulent flow in a cyclindrical duct with one side closed and oblique injections from top and bottom. This flow situation can be found often in mixing and combustion chambers as well as in chemical reactors. The oblique jets collide and mix into each other in the middle of the tube, a dead-water zone develops at the closed end and a significant secondary flow is present downstream of the injection where the injected fluid leaves through the open end. A situation was simulated which was investigated experimentally by Liou et al. [13]. The geometry and the numerical grid are shown in Fig. 6a, which also gives the locations of the inlet ports. For symmetry reasons only one fourth of the flow domain was covered in the calculations. The Reynolds number based on the average exit velocity and pipe diameter was 26 000. In this case with turbulent flow, the molecular viscosity in the momentum equations was replaced by the eddy viscosity calculated with the aid of the k-ε-model. The viscous near-wall layers were not resolved but bridged with the aid of wall functions [8]. The inlet conditions are given in [10]. The calculations were carried out both with the single- and the multigrid method on two different grids, and details on the number of grid points, iterations and computing times required to reach an accuracy of $\varepsilon = 5 \times 10^{-3}$ can be found in Table 1. On the coarser grid the speed-up factor was 8.75 while on the finer grid it was more than 16. This shows convincingly that convergence acceleration through the multigrid method improves as the grid gets finer and that the method also works for turbulent flow calculations. In this test case the results obtained with the two different grids were nearly the same. Figure 6b shows calculated velocity vectors in the symmetry plane and in a cross-sectional plane 0.5 pipe diameters downstream of the injection. It should be noted that, for better visualisation, the x-scale in the dead-end region is twice as large as in the downstream region. The predicted velocity vectors clearly show the complex flow field evolving. Figure 6c compares calculated and measured profiles of velocity in the mean-flow direction in the vertical symmetry plane. The agreement is generally good and somewhat better than that between the data and the calculations of Liou et al. [13] also included in the figure. The figure supports quantitatively the picture already given in Fig. 6b that most of the injected fluid is deflected downstream while only a small amount is deflected towards the closed end which is then reentrained into the jet. Also, two diameters downstream of the injection, the velocity in the pipe is already nearly uniform.

4 Concluding remarks

The paper has shown that considerable computing time can be saved by use of the multigrid technique also for solving geometrically complex flow problems on non-orthogonal curvilinear grids and when turbulence models introducing additional differential equations are employed. With conventional single-grid methods, the computing effort increases exponentially with the number of grid points, while with the multigrid method the increase is only linear. Per multigrid cycle, $3-6$ iterations are usually performed on the finest grid. Therefore the multigrid method is effective only when many more iterations would have been necessary with the single-grid method to reach a prescribed accuracy. Hence, the multigrid method may be not be effective when the accuracy demands are low or for solving unsteady flow problems where only few iterations are necessary per time step. For the steady problems considered here, with 30 grid points in each direction and solution to an accuracy of $\varepsilon = 10^{-3}$, typical speed-up factors of 2 were achieved. With finer grids the speed-up factors were much higher so that more than 90% of the computing time could be saved. Hence the use of the multigrid method is recommended mainly when high resolution is necessary. A disadvantage of the method is that it is not easy to implement in an existing code; this is possible only when the code is highly modular. However, when new codes are developed, the method offers great potential in saving computing time, particularly when fine resolution is required for complex flow problems. This opens up the application of CFD methods for solving practical flow problems also on smaller, widespread computers like workstations.

References

[1] Brandt, A.: Multi-level adaptive solutions to boundary-value problems. Math. Comp., 31, 333−390 (1977).
[2] Hackbusch, W., Trottenberg, U. (eds.): Multigrid methods. Lectures Notes in Mathematics, Vol. 960. Berlin Heidelberg New York: Springer 1982.
[3] Vanka, S. P.: A calculation procedure for three-dimensional steady recirculating flows using multigrid methods, Comp. Meth. Appl. Mech. Eng. 55, 321−338 (1986).
[4] Barcus, M., Perić, M., Scheuerer, G.: A control-volume-based full multigrid procedure for the prediction of two-dimensional, laminar, incompressible flows. In: Notes on numerical fluid mechanics, Vol. 20 (Aeville, M., ed.), pp. 9−16. Braunschweig: Vieweg 1988.
[5] Thompson, M. C., Ferziger, H. J: An adaptive multigrid solution technique for the steady-state incompressible Navier-Stokes equations. In: Computational fluid dynamics (de Vahl Davies, G., Fletcher, C., eds.), pp. 715−724. Amsterdam: Elsevier 1988.
[6] Perić, M.: Efficient semi-implicit solving algorithm for nine-diagonal coefficient matrix, Num. Heat Transfer 11, 251−279 (1987).
[7] Majumdar, S., Rodi, W., Zhu, J.: Three-dimensional finite-volume method for incompressible flows with complex boundaries. J. Fluids Eng. 114, 496−503 (1992).

[8] Launder, B. E., Spalding, D. B.: The numerical computation of turbulent flows. Comp. Meth. Appl. Mech. Eng. **3**, 269−289 (1974).

[9] Sivaloganathan, S., Shaw, G. J.: On the smoothing properties of the SIMPLE pressure-correction algorithm, Int. J. Num. Meth. Fluids **8**, 441−461 (1988).

[10] Orth, A.: Mehrgittermethode zur Berechnung inkompressibler, stationärer Strömungen mit krummlinigen Berandungen. Dissertation, Universität Karlsruhe 1991.

[11] Koseff, J. R., Street, R. L.: Visualisation studies of a shear-driven 3D recirculating flow. J. Fluids Eng. **106**, 21−29 (1984).

[12] Kiehm, P., Mitra, N. K., Fiebig, M.: Numerical investigation of two- and three-dimensional confined wakes behind a circular cylinder in a channel. AIAA Paper 86-0035 (1986).

[13] Liou, T.-M., Hwang, Y.-H., Wu, S.-M.: The tree-dimensional jet-jet impingement flow in a closed-end cylindrical duct. J. Fluids Eng. **112**, 171−178 (1990).

Authors' address: Dr.-Ing. A. Orth and Professor Dr. W. Rodi, Institut für Hydromechanik, Universität Karlsruhe (TH), D‑76128 Karlsruhe, Federal Republic of Germany

Acta Mechanica (1994) [Suppl] 4: 269—277

Computational analysis of viscous hypersonic flow over delta wings

A. Rizzi and **P. Eliasson,** Bromma, Sweden

Summary. Numerical simulations for the laminar hypersonic flow past blunt edged delta wings have been performed by solving the Navier-Stokes equations for four different angles of attack ($0°$, $15°$, $25°$, and $30°$) and flow conditions $M_\infty = 7.15$, $\text{Re}_c = 5.85 \times 10^6$, and for three different chord Reynolds numbers (5.625×10^5, 5.625×10^6, and 10.0×10^6), and flow conditions $M_\infty = 8.7$, $\alpha = 30°$. It is known that for blunt edged delta wings at high angles of attack, the leeside hypersonic flow is dominated by a shear layer that separates just past the blunt leading edge and forms a more distributed vortical region over the wing, rather than a concentrated vortex structure as observed at lower speeds. Our calculations here show that the size of the vortical regions increases with increasing angle of attack, and as the Reynolds number increases, the character of the separation on the leeside of the wing changes from a primary separation inboard to one closer to the leading edge but with the addition of a secondary separation near midspan. The shape of the wing apex determines the location where the separation begins. The solutions are analyzed and compared with available experimental data.

1 Introduction

In the last few years there has been a renewed interest in hypersonic flow research. Large scale research programs have been initiated for both reentry vehicles and hypersonic transport aircraft. Current interest in developing advanced space planes like the *NASP, HERMES, SANGER* and the super Concorde has brought forth the problem of understanding the leeside vortical flow over these vehicles when they travel at hypersonic speeds. The vortical structures are characterized by shock waves, separated flow and shear layers and have important consequences on heating and local effects like a shock wave or shear layer impinging on a configuration detail, e.g. a flap or other protuberance. Vortex phenomena for example have been found responsible for intense local heating over the lee meridian of the delta wings. Due to the near vacuum pressures on the leeward side of the wing, the effectiveness of the control surfaces and lift performance are little affected by these vortical structures, but the flow features of this low-pressure region, its structure and interactions are of primary interest. The occurence of vortical regions in hypersonic flow at high angle of attack has been investigated by Rizzi et al. [1]. Srinivasan et al. [2] studied the change brought about in the vortex structure by varying the angle of attack. Preliminary results have suggested the effect of varying the chord Reynolds number on the separation characteristics of the leeside flow [3], and we present here more refined computations together with a discussion of these effects.

1.1 Features of hypersonic wing flow

Current understanding of hypersonic flow over a delta wing at various angles of attack follows from classical analysis based on flow past a flat thin delta wing of zero thickness. Squire [4] and later Miller and Wood [5] among others have studied the various types of structures that may occur. At moderate supersonic speed the bow shock is detached, and the flow attaches on the windward side inboard of the leading edge. It cannot expand smoothly around the sharp leading edge but separates and develops into a vortex. At higher speed the attachment points shifts outboard to the leading edge. Now the flow can deflect smoothly around the leading edge through a Prandtl-Meyer expansion fan. But at some inboard point on the leeward side it must turn in the freestream direction through a cross-flow shock which usually causes the boundary layer to separate. At still higher speeds the bow shock attaches at the leading edge, but the expansion and crossflow shock remain as at lower speed.

There are, however, limits to each of the above processes, as the usual classical analysis of flow normal to the leading edge shows. When expanding around the leading edge, the flow can reach vacuum conditions, if the Mach number is high enough, before it becomes parallel to the upper surface. For a 70° swept delta wing, vacuum is reached in the flow at $M_\infty = 7$ and $\alpha = 40°$. The conclusions from this analysis lose precision when the wing has a round leading edge and thickness. The analysis must then be carried out by numerical solution.

1.2   Outline of this paper

The paper reviews our work on the numerical simulation of a hypersonic vortical flow over the leeside of blunt delta wings. The computations have been performed by solving the full Navier-Stokes equations to evaluate the inviscid and viscous mechanisms of leeside vortex formation. Two different blunt-edged wings are treated, each with 70° sweep and constant leading edge radius, i.e. non-conical. The results for the first wing to be presented are for flow at $M_\infty = 7.15$, $Re_c = 5.85 \times 10^6$ and $\alpha = 0°$, 15°, 25°, and 30° incidence, and $M_\infty = 8.7$ and $\alpha = 30°$ incidence for three different chord Reynolds number, viz. 5.625×10^5, 5.625×10^6, and 10.0×10^6 for the second wing. Some experimental measurements are available for comparison. In the spirit of Prof. Zierep's research over his long and productive career, the paper emphasizes the analysis of the results, through which a sound description and reasonable understanding of the computed flow structures, including flow separation, shear layers, vortices, shock waves, and entropy losses can be obtained. Leeside values of surface pressure and heat transfer completes the description of the leeside flow.

2   Numerical method

Several numerical schemes have been developed over the years to solve the three dimensional Navier-Stokes equations. The results presented here have been obtained using a finite-volume code that originated for solving transonic flow [6], but has been extensively modified and adapted for hypersonic flows, by B. Winzell at SAAB [7]. The finite-volume method uses central discretization in space with an artificial smoothing term added that is a blend of second and fourth differences to stabilize the discretization. The viscous terms are computed using a staggered grid to obtain the gradients in the cell vertices. This gives a compact discretization of the viscous terms. The resulting semidiscrete system is then integrated over time with an explicit Runge-Kutta scheme; [1] describes the finite-volume approach taken to solve the Navier-Stokes equations.

With appropriate grid clustering, a C−O type grid resolves the boundary layer around the delta wing. A grid size of $33 \times 129 \times 97$ in the streamwise, normal, and spanwise directions, respectively has been used for the first wing, and $41 \times 97 \times 97$ for the second. The grids in the streamwise direction extend only up to the trailing edge, and the wake is not simulated in the computations. The C−O grid topology introduces a polar singular line at the apex of the wing that degrades the convergence.

3 Case 1: angle of attack effects

In January 1990 INRIA and GAMNI organized a workshop *Hypersonic Flows for Reentry Problems* to investigate the suitability of computational techniques for hypersonic flow analysis applied to vehicle design [8]. One test case was to calculate the laminar flow over a 70° swept blunt delta wing at $\alpha = 30°$, $M_\infty = 7.15$, $Re_c = 5.85 \times 10^6$, with an isothermal wall temperature $T_w = 288$ K, and freestream temperature $T_\infty = 74$ K. Since then we have carried out further calculations at different angles of attack in order to understand how the flow structures change with α [2]. The windtunnel model of the delta wing is 0.15 m long, 0.015 m thick at the center of the trailing edge, and has a constant leading edge radius of 0.001 95 m. A sting attached to the base has been used to hold the model during windtunnel tests. Linde [9] gives the details of the experiment and the geometry.

3.1 Crossflow plane

Figure 1 a presents isoMach contours of the flowfield at the 50% chord section for the results at 0°, 15°, 25°, and 30° incidence. At nonzero α, the flow negotiates the leading edge and separates just past the maximum expansion forming a shear layer. This shear layer extends inboard to where it meets the crossflow shock over the wing and then turns downward towards the wing

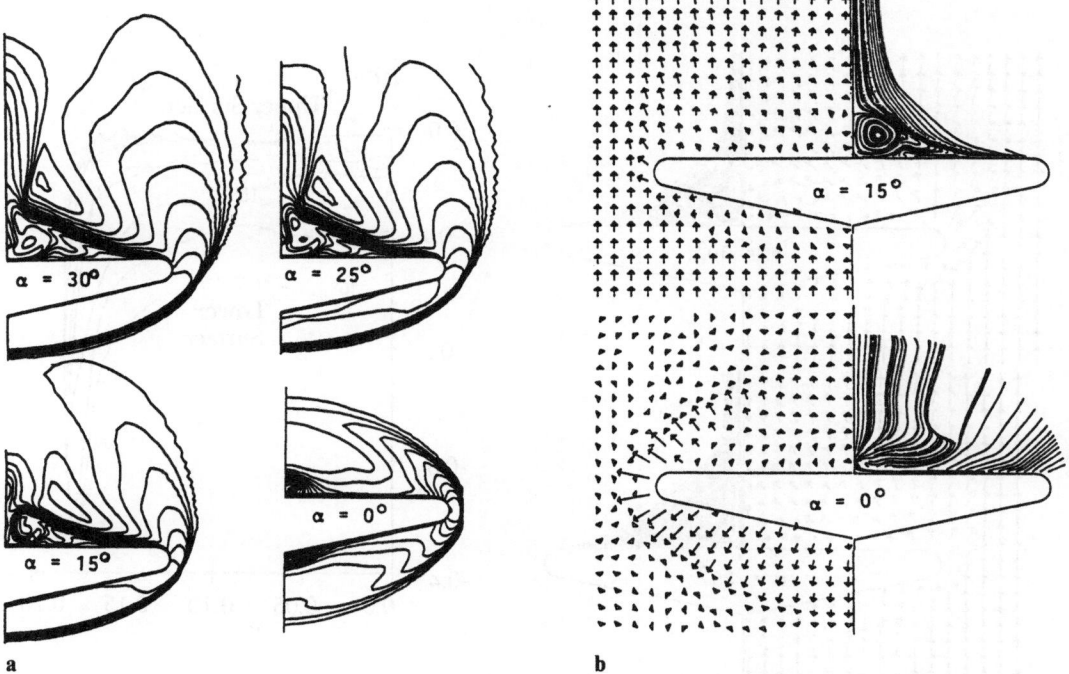

a   **b**

Fig. 1. a IsoMach contours in 50% chord plane; **b** velocity vectors and in-plane streamlines, 50% chord plane

surface where a small embedded shock turns it once again outboard towards the leading edge. It finally rolls up to form a vortical layer. Between the wing and the shear layer a secondary vortex develops from the boundary layer separating from the wing. The direction of the flow motion in this plane is given by velocity vectors and their integrated path lines (i.e. in-plane streamlines) as displayed in Fig. 1b. In each case the vortex is clearly identified. The flow features are more or less similar for the 15°, 25°, and 30° cases, but the vortical features grow larger with increasing α. In the 0° case, there is very little expansion around the leading edge and the flow remains attached on most of the leeward side. However, it separates close to the symmetry plane. As the angle of attack is increased, the flow tends to separate forming a shear layer, closer to the leading edge and the vortex structure close to the symmetry plane is well defined. In other words, the primary separation line moves outboard as the angle of attack increases.

3.2   Upper surface

The Stanton number on the wing at 50% chord position for all the four cases is displayed in Fig. 2. The negative values indicate that on both the upper and lower surface, heat is being transferred from the surroundings to the wing. The peak value for the case of $\alpha = 30°$ is about 3.1 times larger in magnitude than that at the windward symmetry plane. It is noteworthy that this factor agrees well with the value 3.2 given by the infinite yawed-cylinder analysis of Reshotko and Beckwith [10]. As α increases, there is an increase in the heat transfer at the leading edge and the windward side. However, the heat transfer is approximately the same for the three angles of attack on the leeward side. This is to be expected, because the flow that passes through the strongest part of the bow shock undergoes rapid expansion around the leading edge but does not come near to the leeside surface.

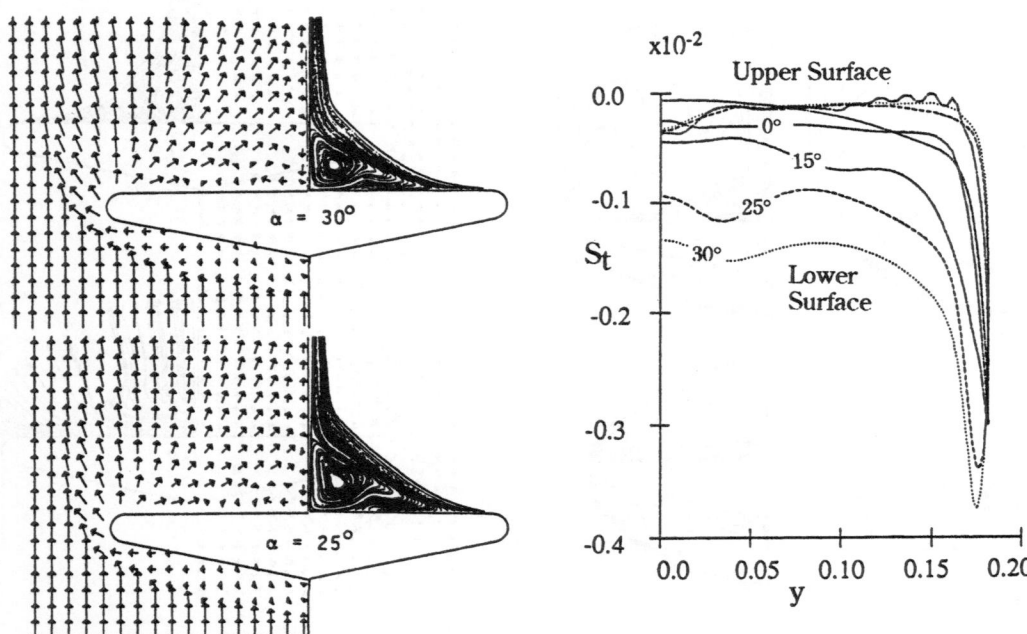

Fig. 2. Stanton number distribution, 50% chord plane

3.3 Comparison with experiments

The oil flow patterns and the corresponding computed skin friction plots displayed in Fig. 3 confirm the flow structure of one primary and one secondary separation on the upper surface of the wing for all the three angles of attack cases. The oil flow patterns also show a very substantial trailing edge effect on the boundary layer that extends almost up to midchord. The reasons for this are still unclear, but it may be due to 1) upstream influence of the trailing edge through the boundary layer, 2) vortex breakdown, or 3) nitrogen liquefaction effects in the tunnel. In any case the abrupt kink in the secondary separation line suggests that laminar to turbulent transition has taken place. However, up to the transition location, the computed skin friction agrees very well with the experimental oil flow pattern. Beyond the transition location, the computed results differ significantly and the computed results do not show any signs of a large

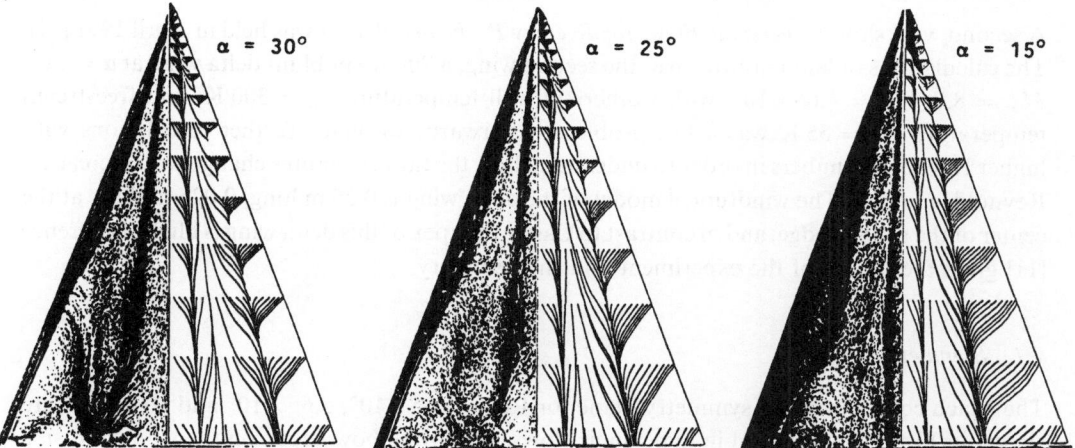

Fig. 3. Comparison of computed skin friction lines with oil flow pattern

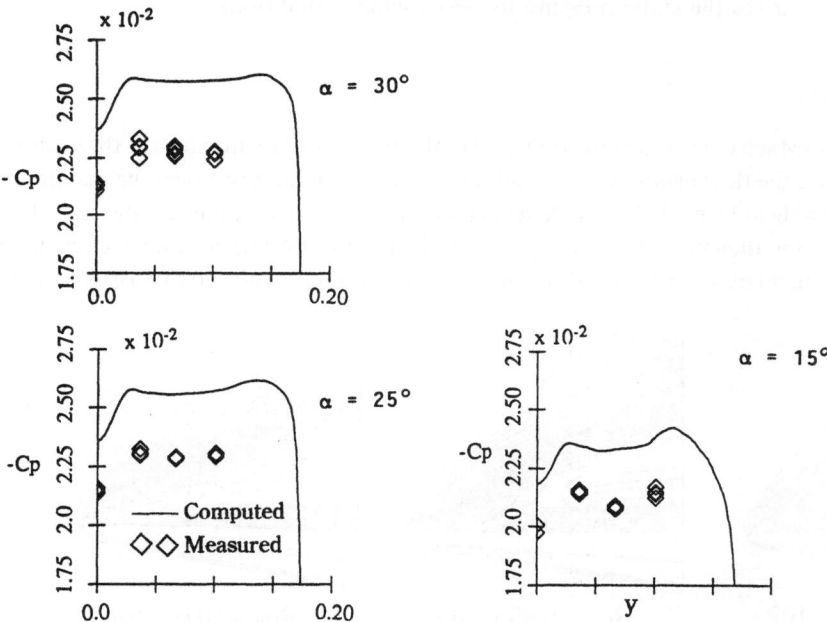

Fig. 4. Comparison between computed and measured C_p on upper surface, 50% chord plane

recirculation region near the trailing edge as seen in the experiments. Since the calculations are purely laminar, some differences are to be expected.

Four pressure taps are located at the mid-chord station, just ahead of the separation bubble. Figure 4 presents the comparison between the computed and experimental surface Cp values for the 15°, 25°, and 30° incidence cases. The computed results underpredict the pressure on the leeward side in all the three cases. Although there is a shift in the levels of the two sets of results, the trend to a higher pressure at the symmetry plane, where the shear layer meets the wing, is similar in each result. The data is rather sparse, but what little there is, seems to confirm our picture of a shear-layer dominated vortical flow structure.

4 Case 2: Reynolds number effects

A second workshop *Hypersonic Flows for Reentry Problems, Part II* was held in April 1991 [11]. The calculations of laminar flow over the second wing, a 70° swept blunt delta wing at $\alpha = 30°$, $M_\infty = 8.7$, $Re_c = 5.65 \times 10^5$, with isothermal wall temperature $T_w = 300$ K, and freestream temperature $T_\infty = 55$ K was a test problem. Afterwards we made further calculations with higher Reynolds numbers in order to understand how the flow structures change with increasing Reynolds number. The windtunnel model of the delta wing is 0.25 m long, 0.025 m thick at the center of the trailing edge, and in contrast to Case 1 the apex of this delta wing is blunt. Reference [11] gives the details of the experiment and the geometry.

4.1 Symmetry plane

The Mach contours in the symmetry plane for $Re_c = 5.65 \times 10^5$, 5.65×10^6 and 10.0×10^6 are shown in Fig. 5. As expected in hypersonic flows, the strong bow shock on the windward side and the flow expansion on the upper surface are clearly visible. The expansion on the leeward side is strong which drives the static pressure to near vacuum conditions.

4.2 Crossflow plane

Figure 6 presents isoMach contours of the flowfield at the 50% chord section for the three cases. In all the three cases, the flow negotiates the leading edge and separates just past the maximum expansion forming a shear layer. This shear layer extends inboard to where it meets the crossflow shock over the wing and then turns downward towards the wing surface where a small embedded shock turns it once again outboard towards the leading edge. It finally rolls up to form a vortical

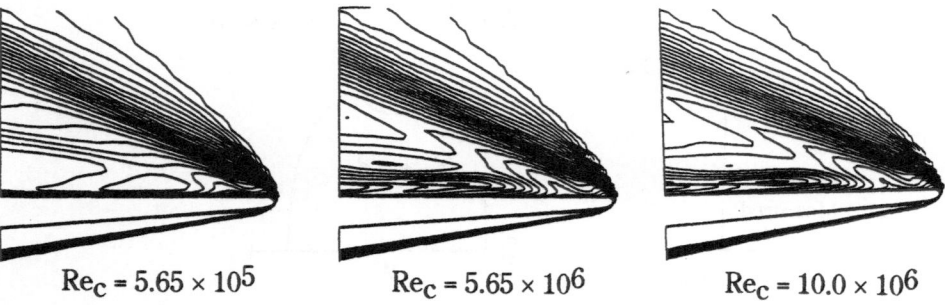

$Re_c = 5.65 \times 10^5$ $Re_c = 5.65 \times 10^6$ $Re_c = 10.0 \times 10^6$

Fig. 5. IsoMach contours in symmetry plane

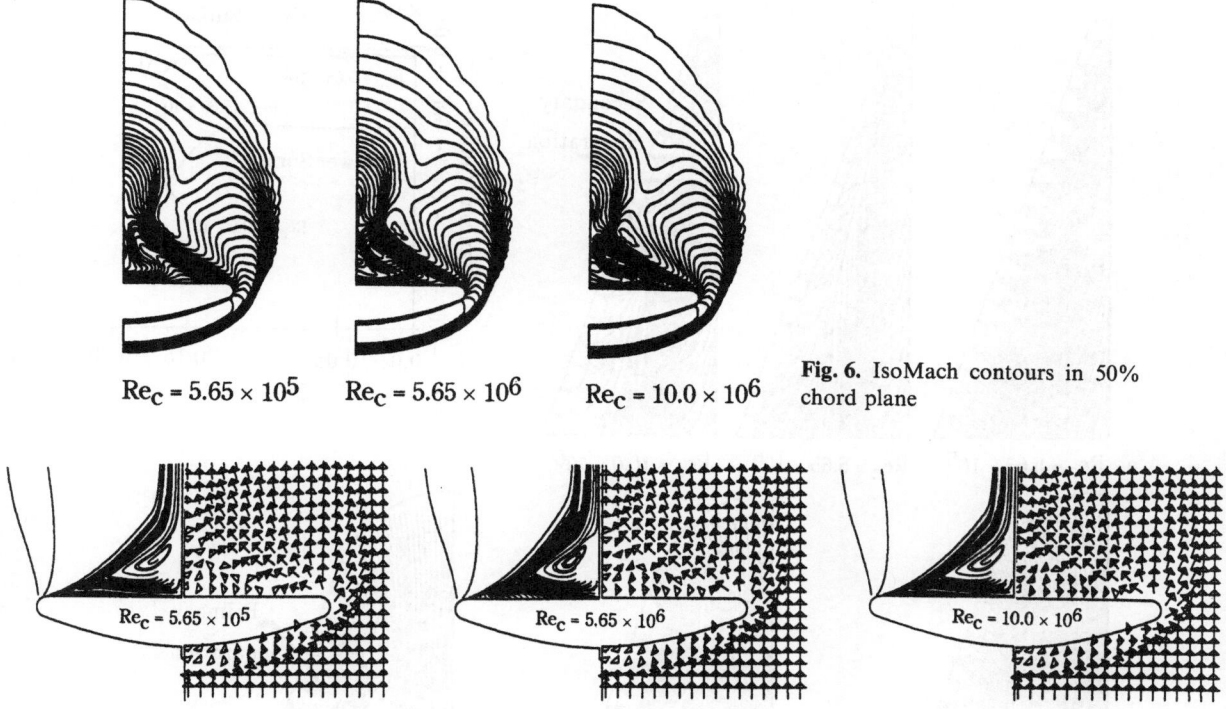

$Re_C = 5.65 \times 10^5$ $Re_C = 5.65 \times 10^6$ $Re_C = 10.0 \times 10^6$

Fig. 6. IsoMach contours in 50% chord plane

Fig. 7. Velocity vectors and inplane streamlines, 50% chord plane

layer. At the two higher Reynolds numbers a secondary vortex develops from the boundary layer separating from the wing, but it does not occur at the lowest Reynolds number. However, for the lowest Reynolds number case, the shear layer separates farther inboard of the leading edge. As the Reynolds number increases, both the crossflow shock and the small embedded shock are more pronounced. The direction of the flow motion in this plane is given by velocity vectors and their integrated path lines (i.e. in-plane streamlines) as displayed in Fig. 7. In each case the vortex is clearly identified. The flow features for the two higher Reynolds number cases are more or less similar. As the Reynolds number increases, the flow tends to separate and to form a shear layer closer to the leading edge.

4.3 Upper surface

The skin friction lines on the upper surface are shown in Fig. 8a for all the three cases. It is clear that for the lowest Reynolds number case there is one separation line, whereas for the two higher Reynolds number cases, the flow separates nearer the leading edge and reattaches inboard. There is a secondary separation close to the midspan of the wing. As the Reynolds number increases, the shear layer tends to separate closer to the leading edge. In the case of the highest Reynolds number, the secondary separation line is curved and moves a little outboard. The zoom view of the skin friction lines near the apex region is displayed in Fig. 8b. The shape of the wing apex has some influence on the streamwise location where the separation begins.

The Stanton number on the wing at 50% chord position for all the three cases is displayed in Fig. 8c. The negative values indicate that on both the upper and lower surface, heat is being transferred from the surroundings to the cold wing. As expected, because of the proximity to the

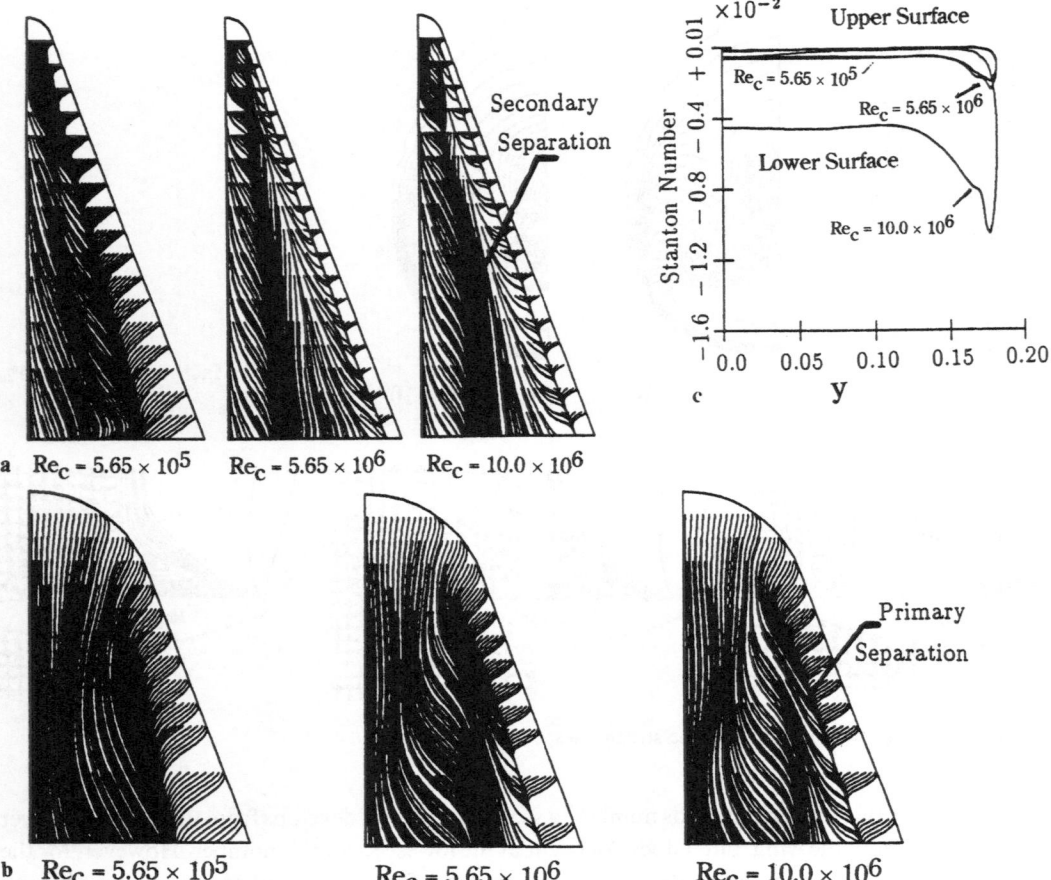

Fig. 8. a Upper surface skin friction lines, **b** zoom in nose region, **c** Stanton number distribution, 50% chord plane

bow shock, the highest heat transfer occurs around the leading edge of the wing. As the Reynolds number increases, the thermal boundary layer gets thinner, the temperature gradients get larger, and thus the Stanton number decreases, i.e. the heat transfer increases. Figure 8 c shows virtually no increase from the low to the intermediate Re case, but a rather large increase between the intermediate and the highest Reynolds number case. One conjecture to explain this may be that the thermal boundary layer thins much faster at higher Reynolds numbers.

5 Conclusions

Instead of the concentrated vortex usually found over a delta wing at transonic speed, the flow in hypersonic speed is dominated by a shear layer that separates just past the blunt leading edge and forms a more distributed vortical region over the wing. The behaviour of the shear layer variation with angle of attack and with the chord Reynolds number has been investigated. With increasing α the computations confirm the exptected trends, namely the peak heating at the leading edge increases with α, and that the secondary separation point moves outboard with increasing α. What was more unexpected was the small separation found near the symmetry plane at zero incidence. The Navier Stokes results agree qualitatively with the sparse experimental data that is available, but not quantitatively.

As the Reynolds number increases from 5.65×10^5 to 5.65×10^6, the shear layer separates closer to the leading edge and in turn gives rise to a secondary separation closer to the centre of the wing span. The shape of the wing apex influences the location where the separation begins. A further increase in the Reynolds number from 5.65×10^6 to 10.0×10^6 does not change the flow structure significantly on the leeside, but it does influence the Stanton number significantly on the windside. As the Reynolds number increases from 5.65×10^6 to 10.0×10^6, the Stanton number decreases dramatically on the lower surface. We have no clear explanation for this, but it may have to do with a nonlinear thinning of the thermal boundary layer at higher Reynolds numbers.

References

[1] Rizzi, A., Murman, E. M., Eliasson, P., Lee, M. K.: Calculation of hypersonic leeside vortices over blunt delta wings. In: Vortex Flow Aerodynamic., AGARD Conference Proceedings, CP-494, 1990.

[2] Srinivasan, S., Eliasson, P., Rizzi, A.: Navier-Stokes computations of hypersonic flow past a blunt edge delta wing at several angles of attack. AIAA Paper 91-1698 (1991).

[3] Srinivasan, S., Eliasson, P., Rizzi, A.: Hypersonic laminar flow computations over a blunt delta wing at three different chord Reynolds numbers. Proc. 9th GAMM Conf. Num. Meth. Fluid Mech. (Vos, J., Rizzi, A., Ryhming, I., eds.), pp. 152–160. Braunschweig: Vieweg 1992.

[4] Squire, L. C.: Regimes over delta wings at supersonic and hypersonic speeds. Aero Q. 12, 1–14 (1976).

[5] Miller, D. S., Wood, R. M.: Lee-side flow over delta wings at supersonic speeds. NASA TP 2430 (1985).

[6] Mueller, B., Rizzi, A.: Navier-Stokes calculations of transonic vortices over a round leading edge delta wing. Int. J. Num. Meth. Fluid Mech. 9, 943–962 (1989).

[7] Winzell, B.: Validation of a computational method for laminar hypersonic flow around blunt bodies and application to the full hermes spacecraft. In: Aerothermodynamics for Space Vehicles (Battrick, B., ed.), pp. 327–338. Noordwijk: ESA 1991.

[8] Desideri, J., Glowinsky, R., Periaux, J. (eds.): Proc. workshop hypersonic flows for reentry problems, Part I. Berlin Heidelberg New York Tokyo: Springer 1991.

[9] Linde, M.: Experimental test on a planar delta wing at high Mach number and high angle of attack. Step 2, FFA TN 1990-06, Stockholm 1990.

[10] Reshotko, E., Beckwith, I. E.: Compressible laminar boundary layer over a yawed infinite cylinder with heat transfer and arbitrary Prandtl number. NACA Report 1379, p. 15 (1958).

[11] Abgrall, R., Desideri, J., Glowinsky, R., Mallet, M., Periaux, J. (eds.): Hypersonic flows for reentry problems, Vol. 3. Berlin Heidelberg New York Tokyo: Springer 1992.

Authors' address: Professor Dr. A. Rizzi and Dr. P. Eliasson, FFA, The Aeronautical Research Institute of Sweden, S-16111 Bromma, Sweden

Acta Mechanica (1994) [Suppl] 4: 279—287

The computation of transonic airfoil and wing design

Z. Q. Zhu, Z. X. Xia, and L. Y. Wu, Beijing, People's Republic of China

Summary. An inverse computational method for transonic airfoil and wing design is presented. Improvements aimed at increasing abilities of the method and computational efficiency have been taken. For example, a Riegels type leading edge correction is introduced. An artificial viscosity term is added to the integral equation method and a smoothing-relaxation procedure is proposed. In 2D transonic flow case, a regularity condition in closed form to be satisfied by a target pressure distribution is used. A few given design results illustrate that the present method is an efficient tool for transonic airfoil and wing design.

1 Introduction

Recently, a number of design methods have been developed and used for design of transonic airfoils and wings. Sloff [1] reviewed these methods and divided them into three major categories: indirect, inverse and aerodynamic optimization. In [2] some authors made extensive survey on current activities of aerodynamic design in their own countries. The conventional inverse methods are the widely used ones in industry applications. The existing inverse methods for transonic airfoils and wings can be subdivided into two categories: (a) methods utilizing Dirichlet-type boundary conditions derived from the target pressure distribution; (b) methods utilizing Neumann-type boundary conditions with a geometry correction procedure (residual-correction method).

In the inverse problem, both Dirichlet- and Neumann-type boundary conditions must be satisfied on the airfoil contour which is to be determined. This leads to a nonlinear problem with an unknown boundary to be solved iteratively. In the first approach, the required target pressure distribution is imposed on an initial airfoil as a Dirichlet boundary condition, while geometry corrections are derived by integrating (either explicitly, or in some implicit manner) the transpiration mass flow over the initial geometry. The Neumann boundary condition is satisfied at the end of the iterations.

In the second approach, the pressure distribution on an initial geometry is determined by the use of an analysis code (Neumann boundary condition), and the residuals (i.e. differences between target pressure distribution and pressure distribution of the current iteration) are transformed to geometry corrections using some relatively simple approximate inverse methods. Here, the Dirichlet boundary condition is satisfied at the end of the iteration.

In the present paper a residual-correction type method is used, in which the transonic nonisentropic full potential method [3] is used as analysis code and the procedure to determine the geometry correction is similar to the one used by Takanashi [4]. In the airfoil design, a regularity condition in closed form to be satisfied by the target pressure distribution [5] is used.

2 Computational method of the inverse problem

In the residual-correction method, the solution $\varphi(x, y, z)$ of the full potential equation for an initial geometry $f_\pm(x, y)$ can be obtained by using an existing analysis code, the object here is to determine the amount of the geometry correction $\Delta f_\pm(x, y)$ corresponding to the pressure difference $\Delta C_{p\pm}(x, y)$ between the specified and the calculated pressure. If a small perturbation $\Delta\varphi(x, y, z)$ is further introduced, we can obtain the potential equation for $\Delta\varphi(x, y, z)$ according to the transonic small-disturbance theory

$$\Delta\varphi_{xx} + \Delta\varphi_{yy} + \Delta\varphi_{zz} = \frac{\partial}{\partial x}\left[\frac{1}{2}(\varphi_x + \Delta\varphi_x)^2 - \frac{1}{2}\varphi_x^2\right], \tag{1}$$

$$\Delta\varphi_x(x, y, \pm 0) = -\frac{k}{2\beta^2}\Delta C_{p\pm}\left(x, \frac{y}{\beta}\right), \tag{2}$$

$$\Delta f_\pm'(x, y) = \Delta\varphi_z(x, y, \pm 0), \tag{3}$$

where $\beta = \sqrt{1 - M_\infty^2}$, $k = (\gamma + 1)M_\infty^2$.

By applying Green's theory to Eq. (1) and introducing a decay function similar to that used by Nøstrud [6], the following integral equations can be obtained:

$$\Delta u_s(x, y) = -\frac{1}{2\pi}\iint\limits_{S_w}\Psi_x(x, y, 0; \xi, \eta, 0)\,\Delta w_s(\xi, \eta)\,d\xi\,d\eta + G_s(x, y)$$

$$+ \frac{1}{2\pi}\iint\limits_{S_w}[I_s(x, y; \xi, \eta, +0)\,G(\xi, \eta, +0) + I_s(x, y; \xi, \eta, -0)\,G(\xi, \eta, -0)]\,d\xi\,d\eta, \tag{4}$$

$$\Delta w_a(x, y) = \frac{1}{2\pi}\iint\limits_{S_w}\frac{\Delta u_a(\xi, \eta)}{(y - \eta)^2}\left[1 + \frac{x - \xi}{\sqrt{(x - \xi)^2 + (y - \eta)^2}}\right]d\xi\,d\eta$$

$$+ \frac{1}{2\pi}\iint\limits_{S_w}[I_a(x, y; \xi, \eta, +0)\,G(\xi, \eta, +0) - I_a(x, y; \xi, \eta, -0)\,G(\xi, \eta, -0)]\,d\xi\,d\eta, \tag{5}$$

where

$$\Delta u_s(x, y) = \Delta\varphi_x(x, y, +0) + \Delta\varphi_x(x, y, -0),$$

$$\Delta w_s(x, y) = \Delta\varphi_z(x, y, +0) - \Delta\varphi_z(x, y, -0),$$

$$\Delta u_a(x, y) = \Delta\varphi_x(x, y, +0) - \Delta\varphi_x(x, y, -0),$$

$$\Delta w_a(x, y) = \Delta\varphi_z(x, y, +0) + \Delta\varphi_z(x, y, -0),$$

$$\Psi(x, y, z, \xi, \eta, \zeta) = [(x - \xi)^2 + (y - \eta)^2 + (z - \zeta)^2]^{-1/2},$$

$$G(x, y, z) = \frac{1}{2}[(\varphi_x + \Delta\varphi_x)^2 - \varphi_x^2],$$

$$G_s(x, y) = G(x, y, +0) + G(x, y, -0),$$

$$I_s(x, y; \xi, \eta, \pm 0) = \int_0^\infty \Psi_{\xi x}(x, y, 0; \xi, \eta, \zeta)\exp[-2R_\pm(\xi, \eta)\,\zeta]\,d\zeta,$$

$$I_a(x, y; \xi, \eta, \pm 0) = \int_0^\infty \Psi_{\xi z}(x, y, 0; \xi, \eta, \zeta) \exp\left[-2R_\pm(\xi, \eta)\zeta\right] d\zeta,$$

$$R_\pm(x, y) = |f_\pm''(x, y)/\varphi_x(x, y, \pm 0)|.$$

The correction function $\Delta f_\pm(x, y)$ is also split into symmetric $\Delta f_s(x, y)$ and antisymmetric $\Delta f_a(x, y)$ parts:

$$\Delta f_s(x, y) = \Delta f_+(x, y) - \Delta f_-(x, y),$$

$$\Delta f_a(x, y) = \Delta f_+(x, y) + \Delta f_-(x, y).$$

Since $\Delta f_a'(x, y) = \Delta w_a(x, y)$, the antisymmetric part can be determined by direct evaluation of the right-hand side of Eq. (5). Since $\Delta f_s'(x, y) = \Delta w_s(x, y)$, the symmetric part and Eq. (4) must be solved implicitly. Consequently, the correction $\Delta f_\pm(x, y)$ is obtained by integrating $\Delta w_s(x, y)$ and $\Delta w_a(x, y)$ with respect to x.

For improving the convergence and increasing the application ability, the following modifications have been taken in this paper:

(A). In order to increase the ability of dealing with the shock, an artificial viscosity term is added to the present integral equation method. To obtain Eq. (1) the assumption of nonexistence of shock waves in the flow field has been made. Introducing the artificial viscosity concept to deal with shock wave has become a routine procedure in finite difference methods. In integral equation methods directly introducing the artificial viscosity term to the starting differential equation plays an efficient role in the analysis problem of supercritical flow with shocks [7]. A similar artificial viscosity concept is used in the present inverse problem of the integral equation method. To this end the following artificial viscosity is introduced into the right hand side of Eq. (1)

$$-\mu \frac{\partial}{\partial x}\left[(\varphi_{xx} + \Delta\varphi_{xx}) P(\varphi_{xx} + \Delta\varphi_x) - \varphi_{xx} P(\varphi_x)\right], \tag{6}$$

where

$$P(\varphi_x) = \begin{cases} \varphi_x - 1 & \text{if } \varphi_x > 1 \\ 0 & \text{if } \varphi_x \leqq 1 \end{cases}, \tag{7}$$

μ — coefficient of artificial viscosity.

After an analogous derivation, we obtain the equations with the same form as Eqs. (4, 5) except for the expression of $G(x, y, z)$:

$$G(x, y, z) = \frac{1}{2}\left[(\varphi_x + \Delta\varphi_x)^2 - \varphi_x^2\right]$$

$$-\mu\left[(\varphi_{xx} + \Delta\varphi_{xx}) P(\varphi_x + \Delta\varphi_x) - \varphi_{xx} P(\varphi_x)\right]. \tag{8}$$

(B). A Riegels type leading edge correction is applied in this paper to remove the singularity at the leading edge of the round-nosed airfoil. To solve the singularity problem, several efficient methods have been developed by using perturbation method or numerical iteration techniques. They are accurate, but complicated and computer-time consuming. For simplicity, as an approximation, a Riegels type modification factor is introduced into the integral equations, i.e., the first term of right hand side of Eqs. (4) and (5) is multiplied by factor

$$E(x, y) = \frac{0.5}{\sqrt{1 + [f_+'(x, y)]^2}} + \frac{0.5}{\sqrt{1 + [f_-'(x, y)]^2}}. \tag{9}$$

In incompressible flow introducing Riegels factor makes the solution uniformly valid up to the leading edge. It is strickly correct for airfoils with elliptic cross section. It gives a good approximation for arbitrary shapes up to a thickness-to-chord ratio of about 20% [8].

(C). A smoothing-relaxation procedure is proposed and used in this paper. Experience with numerical experiments indicates that the process of iteration might be diverging for flows with strong shock waves in spite of using above modifications. So, a smoothing-relaxation procedure is proposed in the process of iteration:

$$f_\pm^{n+1}(x_i^j, y_j) = f_\pm^n(x_i^j, y_j) + \delta\{\Delta f_\pm^{n+1}(x_i^j, y_j) + 0.5[\Delta f_\pm^{n+1}(x_{i+1}^j, y_j) + \Delta f_\pm^{n+1}(x_{i-1}^j, y_j)]\}, \quad (10)$$

where δ denotes the relaxation parameter, $0 \le \delta \le 0.5$. Numerical examples show that adopting such a smoothing-relaxation measure is efficient to speed up the convergence. In some cases of flows with strong shocks it makes the process convergent, while the process is divergent without this procedure.

The inverse (design) problem can be solved by using the iteration process:

(1). Use an existing analysis code to solve the flow field of an initial geometry $f_\pm(x, y)$. By comparing the calculated pressure distribution $C_{p\pm}(x, y)$ and the target pressure distribution $C_{ps\pm}(x, y)$, the residual of $\Delta C_{p\pm}(x, y) = C_{ps\pm}(x, y) - C_{p\pm}(x, y)$ can be obtained.

(2). Solve Eqs. (4) and (5) to determine the geometry correction $\Delta f_\pm(x, y)$. A new geometry is thus obtained from the smoothing-relaxation formula (10).

Repeat the above process until the calculated pressure distribution agrees with the prescribed one to a specified accuracy. A nonisentropic potential solver [3] is used in the computation as the analysis code.

3 Constraints for the inverse problem

It was demonstrated by Lighthill [9] that a unique and correct solution to the inverse problem of 2D, incompressible flow does not exist, unless a number of additional conditions in the form of certain integral constraints are satisfied, i.e.,

$$\int_0^{2\pi} \log\left|\frac{q_0}{q_\infty}\right| d\omega = 0, \tag{11}$$

$$\int_0^{2\pi} \log\left|\frac{q_0}{q_\infty}\right| \cos\omega \, d\omega = 0, \tag{12}$$

$$\int_0^{2\pi} \log\left|\frac{q_0}{q_\infty}\right| \sin\omega \, d\omega = 0, \tag{13}$$

where q_0 — prescribed velocity distribution on the airfoil surface, q_∞ — velocity at infinity, ω — polar angle in the transformed plane.

The first of these constraints, known as regularity condition, is a consequence of the fact that q_0 is required to be an analytical (i.e. non-singular) function. To satisfy the requirement of regularity condition a free parameter in the target pressure distribution must be introduced.

Equations (12) and (13), known as closure conditions, express the fact that the airfoil is closed and that the specified velocity distribution is consistent with the specified circulation. It is evident that similar constraints are also required for the compressible flow. Volpe and Melnik [10] mentioned that the inverse problem had never been properly addressed for 2D compressible flows.

For 3D flow the situation is still more unsatisfactory, even the 3D equivalents of Lighthill's constraints for incompressible flow have not been formulated. Apart from this, Sloff pointed out that the 3D inverse problem is ill-posed in the sense that small differences in the specified pressure distribution may lead to large differences in geometry [1]. So, the pressure distribution in 3D must be carefully specified by a trial and error method with experience to make sure that it exists physically.

In the present paper a regularity condition in integral form is proposed for 2D calculation based on transonic small perturbation theory and Nøstrud's assumption [6]:

$$\int_0^1 \Delta u_a(x)\, \omega_x\, dx = \int_0^1 [I_v(x, +0)\, G(x, +0) + I_v(x, -0)\, G(x, -0)]\, dx, \tag{14}$$

where $I_r(x, \pm 0) = \int_0^\infty \dfrac{rr_{xx} - r_x{}^2}{r} \exp\left[-2R_\pm(x)\, z\right] dz$, r, ω — the modulus and the polar angle of the complex variable $\chi = re^{i\omega}$ in the transformed plane. The above discussion indicates that the prescribed pressure distribution should contain an adjustable parameter σ to guarantee that the regularity condition be satisfied. When the target pressure distribution is not properly given, the code will adjust the target pressure distribution with the help of σ and design the airfoil corresponding to the modified (acceptable) pressure distribution [5].

The closure condition

$$\int_{x_{le}(y)}^{x_{te}(y)} \Delta w_s(x, y)\, dx = 0 \tag{15}$$

is used in present paper for both 2D and 3D calculations, where $x_{le}(y)$ and $x_{te}(y)$ are the leading and trailing edges at different spanwise stations (y). This closure condition is important for the solution. It assures that the trailing edge thickness of the designed airfoil is always kept equal to that of the initial airfoil.

4 Numerical aspects and results

To solve Eqs. (4) and (5) numerically, we have to specify the computational domain. To this end the plane of wing is divided into $I(2J + 1)$ rectangular subintervals (Fig. 1), where I, J are the numbers of chordwise and spanwise subintervals. In the spanwise direction it is uniform and in the chordwise direction the center coordinate of the subintervals can be determined by

$$\xi_{i+(1/2)} = \frac{\text{th}\,(A_1 i_0) + \text{th}\,(A_1 i)}{\text{th}\,(A_1 i_0) + \text{th}\,(A_1 I)} \qquad i = 0, 1, 2, \dots, I - 1, \tag{16}$$

where A_1 and i_0 are specified positive numbers.

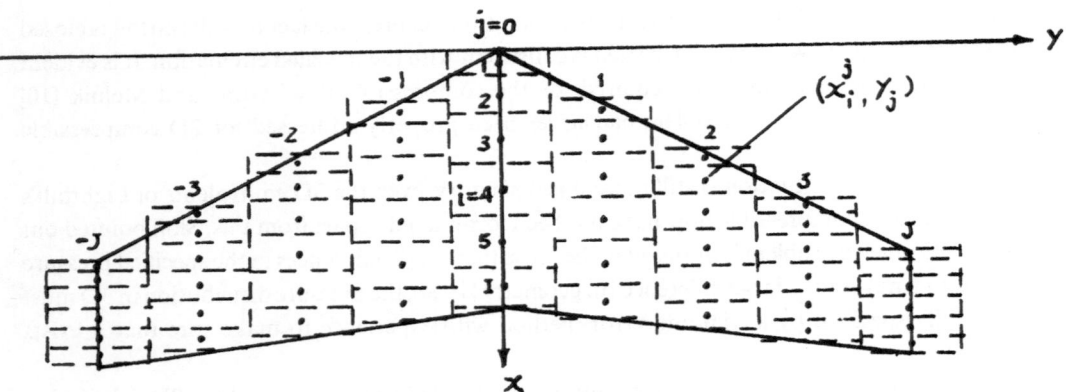

Fig. 1. Discretized panel on the surface of wing

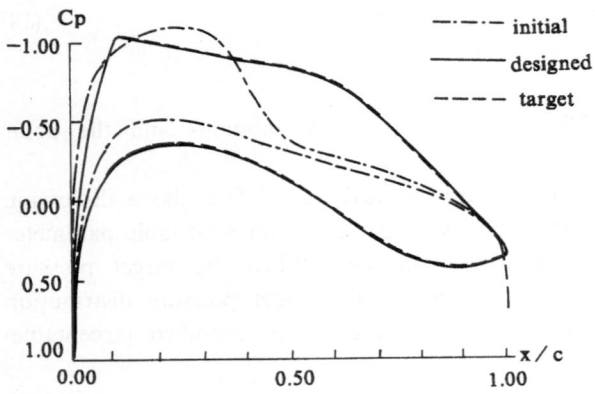

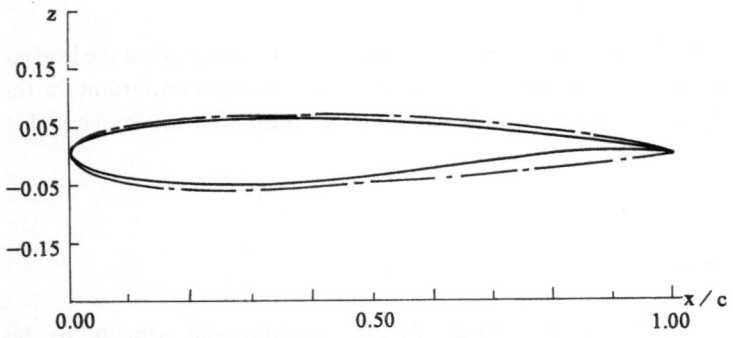

Fig. 2. Design of supercritical shock-free airfoil

In the ζ direction the space is divided into N subintervals, the coordinates of them can be determined by

$$\zeta_n = \frac{A[n/(N + 1)]}{\{1 - [n/(N + 1)]^2\}^2} \qquad n = 0, 1, 2, ..., N, \tag{17}$$

where A is a specified number, which controls the value of ζ_n.

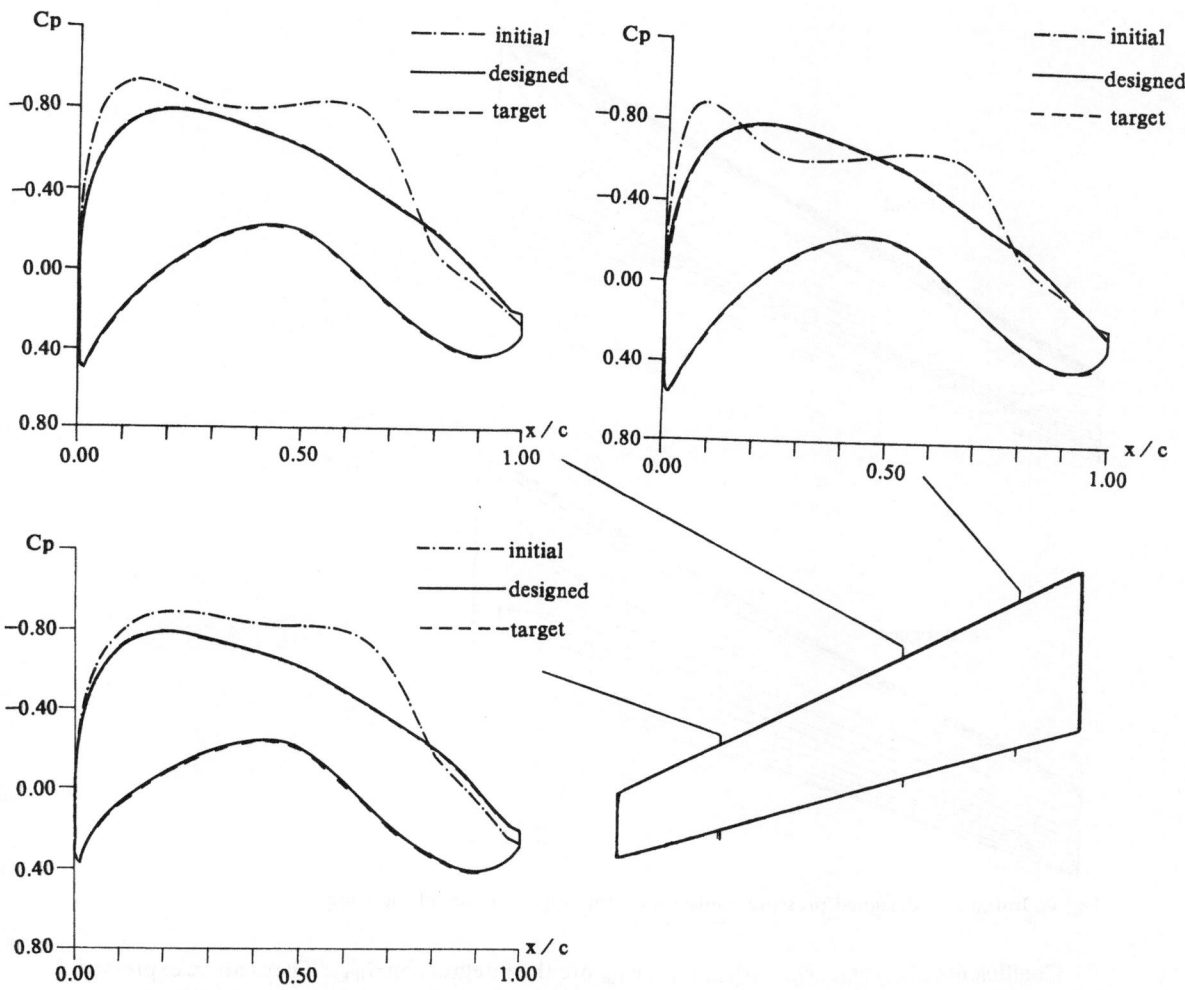

Fig. 3. Comparison of pressure distribution of wing B

On each ζ subinterval $[R_\pm(\xi, \eta)\, \zeta]$ is assumed to be constant.

On each of the rectangular subintervals on the surface of the wing $\Delta u_s(x, y)$, $\Delta u_a(x, y)$, $G(x, y, \pm 0)$, $R_\pm(x, y)$ and $\Delta w_a(x, y)$ are assumed to be constants, while $\Delta w_s(x, y)$ linearly varies with respect to x and is constant with respect to y. Based on these assumptions the final discretized form of Eqs. (4) and (5) can be expressed as

$$\Delta u_s(x_i{}^j, y_j) = \sum_{k=1}^{I} \sum_{m=0}^{J} \mu_{ijkm}^s \Delta w_s(x_{k-(1/2)}^m, y_m) + G_s(x_i{}^j, y_j)$$
$$+ \sum_{k=1}^{I} \sum_{m=0}^{J} [v_{ijkm}^s G(x_k{}^m, y_m, +0) + \hat{v}_{ijkm}^s G(x_k{}^m, y_m, -0)], \qquad (18)$$

$$\Delta w_a(x_i{}^j, y_j) = \sum_{k=1}^{I} \sum_{m=0}^{J} \mu_{ijkm}^a \Delta u_a(x_k{}^m, y_m)$$
$$+ \sum_{k=1}^{I} \sum_{m=0}^{J} [v_{ijkm}^a G(x_k{}^m, y_m, +0) + \hat{v}_{ijkm}^a G(x_k{}^m, y_m, -0)] \qquad (19)$$
$$(i = 1, 2, ..., I; \; j = 0, 1, 2, ..., J).$$

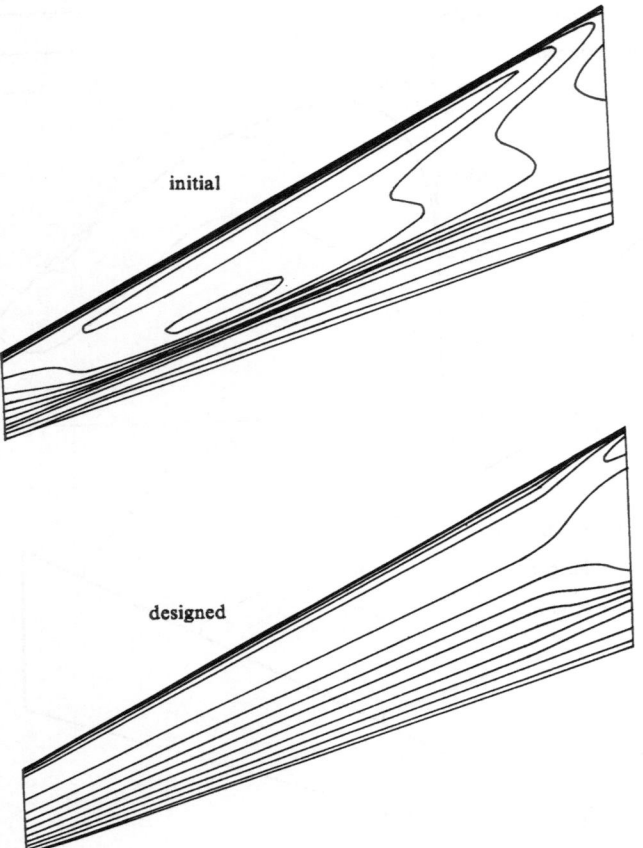

initial

designed

Fig. 4. Initial and designed pressure contours on the upper surface of the wing

Coefficients μ^s_{ijkm}, v^s_{ijkm}, $\hat{v}^s_{ijkm}$, μ^a_{ijkm}, v^a_{ijkm}, $\hat{v}^a_{ijkm}$ are the integrals on S_{Wkm}. They can be expressed analytically [11]. This not only reduces the calculation amount, but also eliminates the errors of numerical integration.

The discretized form of the closure condition (15) can be expressed in the form

$$\sum_{i=1}^{I} [\Delta w_s(x^j_{i-(1/2)}, y_j) + \Delta w_s(x^j_{i+(1/2)}, y_j)] \, (x^j_{i+(1/2)} - x^j_{i-(1/2)}) = 0 \quad (j = 0, 1, 2, ..., J). \qquad (20)$$

Equations (18) and (19) consist of $(I + 1)\,(J + 1)$ linear algebraical equations, from which Δw_s can be calculated.

Two numerical examples are given here to show the validity and applicability of the method.

The first example deals with the design of a supercritical shock free airfoil in 2D flow. The target pressure distribution is taken from one of supercritical shock free flow [8]. Initial airfoil is NACA0012, with $M_\infty = 0.75$, initial angle of attack $\alpha = 1.05°$. After 9 iterations the convergent solution is obtained. Figure 2 depicts the target, initial and designed pressure distributions. The geometry of the designed airfoil is also shown in the figure. The automatically adjusted velocity distribution which fulfils the regularity condition can be found in [5].

The second test case is a supercritical shock free wing design in 3D flow. The Lockhead-AFOSR Wing B [12] is taken as baseline wing. For simplicity, modifications are made on the upper surface only, i.e., a supercritical shock free pressure distribution is taken as the target pressure distribution on the upper surface at all sections along the span, while the pressure distribution on the lower surface is kept the same as the one on the original wing. For the sake of convergence the root section is not modified during the design process. After 7 iterations, the modification is almost completed, i.e., the target pressure distribution on the upper surface is almost realized except for the region near the root section. Figure 3 depicts the comparison of initial, target and designed pressure distribution at different spanwise positions. In Fig. 4, the pressure contours on the upper surface of the original and designed wing are plotted.

5 Conclusions

An improved residual-correction method for transonic airfoil and wing design is presented in this paper. All the improvements speed up the convergence, increase the application ability and computational efficiency. Several numerical examples show that the method is reliable and efficient.

References

[1] Sloff, J. W.: A survey of computational methods for subsonic and transonic aerodynamic design. Paper at ICIDES-I, Austin 1984.
[2] Inverse design concepts and optimization in engineering sciences. ICIDES-III, Washington D.C., 1991.
[3] Zhu, Z. Q., Bai, X. S.: Numerical computation of improved transonic potential method. J. Aircraft **29**, 180 – 184 (1992).
[4] Takanashi, S.: Iterative three-dimensional transonic wing design using integral equations. J. Aircraft **22**, 655 – 660 (1985).
[5] Zhu, Z. Q., Xia, Z. X., Wu, L. Y.: An inverse method with regularity condition for transonic airfoil design. Acta Mech. **95**, 59 – 68 (1992).
[6] Nøstrud, H.: High speed flow past wings. NASA CR-2246 (1973).
[7] Su, J. C., Wu, L. Y.: Calculation of transonic flow around airfoils by integral equation method (in Chinese). Acta Aeron. Astron. Sin. **8**, A 543 – A 552 (1987).
[8] Küchemann, D.: The aerodynamic design of aircraft. New York: Pergamon Press 1978.
[9] Lighthill, M. J.: A new method of two dimensional aerodynamic design. ARC R & M 2112 (1945).
[10] Volpe, G., Melnik, R. E.: The role of constraints in the inverse problem for transonic airfoils. AIAA 81 – 1233 (1981).
[11] Xia, Z. X.: Inverse design method of transonic airfoils and wings (in Chinese). Ph. D. Thesis, Beijing University of Aeronautics Astronautics 1991.
[12] Ohman, L. H.: Experimental data base for computers program assessment. AGARD-AR 138, Addendum 1984.

Authors' address: Professor Z. Q. Zhu, Z. X. Xia Ph. D., and Professor L. Y. Wu, Institute of Fluid Mechanics, Beijing University of Aeronautics and Astronautics, Beijing 100083, People's Republic of China

Acta Mechanica (1994) [Suppl] 4: 289–295
© Springer-Verlag 1994

Part 7: Miscellaneous Problems

Calculation of the Ranque effect in the vortex tube

A. A. Borissov, P. A. Kuibin, and **V. L. Okulov**, Novosibirsk, Russia

Summary. The theoretical model allowing to calculate correctly the observed temperature separation along the tube radius is presented here. The model is based on the original analytical solution of the general problem and calculation of hydrodynamics of complex spatial vortex flow in a tube. The model allows to derive three components of velocity analytically using the ideal incompressible liquid approximation and to obtain the complex spatial flow structure with helical vortex in a tube. Then the known hydrodynamic field of velocity is introduced into the equation of convective heat transfer. The solution of the obtained equation, in which only the convective transfer due to complex topology of hydrodynamic field is taken into account, permits to obtain the temperature field distribution. The region of decreased temperature is in the center, and that of increased temperature is placed on the periphery. Counter flow of liquid in the center is observed depending on the step of helix. The obtained solution for the velocity field allows to calculate the phenomenon of the vortex reverse effect too.

1 Introduction

The vortex effect or the Ranque effect is known long ago. It is realized on a simple device, the so-called vortex tube. The simplest vortex tube is a smooth cylindrical tube with tangential inlet, which provides an intensive swirling of a flow, and outlets on the opposite ends of the tube (the central annular outlet near inlet and the ring one at another end). Despite the construction simplicity, the process of energy separation in the tube (the Ranque effect) has no satisfactory explanation up to day. The authors consider that the attempts of preceding investigators failed, first of all, due to the choice of rather simplified hydrodynamic model of the flow in a tube. Really, even in the last studies [1] the Rankine vortex was used to approximate the flow in the vortex tube, i.e. the axial and radial motions were ignored. This excluded from the consideration a possibility of convective transfer in the process of energy separation. The choice of Rankine model was justified by two reasons. Firstly, the simplicity of the velocity field representation simplified significantly the studied equation of energy transfer. Secondly, the profile of averaged tangential flow velocity in a tube was similar to the velocity profile in the Rankine vortex.

Indeed, a flow in the vortex tube with strong swirl and non-zero axial component of velocity must have helical structure, which was not taken into consideration before.

New results on the determination of the velocity field induced by spiral vortex filament in unbounded space were obtained in [2]. However, it was difficult to use his approach based on the transformation of the Biot-Savart-law for the description of the flow induced by the vortex filament in the cylindrical tube.

In the given work the results were obtained allowing to analyze exactly a flow induced by vortex helical filaments in a tube and to estimate the effect of convective transfer in the process of energy separation in the Ranque vortex tube.

Hydrodynamical model

To describe the energy separation effect in swirling flow in vortex tube, we need to fill in the lack of knowledge, existing in the description of the velocity field induced by helical vortex flow in cylindrical tube of the radius R. The aim of the presented study is to obtain a principal answer to the question about the influence of convective transfer on the energy separation effect. Since this effect is observed at the Mach numbers $M = 0.4 \div 0.5$ and even less [1], and the terms of equations responsible for the effects of compressibility are proportional to M^2, we are limited to the consideration of the model of inviscid non-compressible liquid. In such a statement the velocity field $u \equiv (u_\varrho, u_\varphi, u_z)$ satisfies the continuity equation

$$\frac{1}{\varrho} \frac{\partial}{\partial \varrho} (\varrho u_\varrho) + \frac{1}{\varrho} \frac{\partial}{\partial \varphi} u_\varphi + \frac{\partial}{\partial z} u_z = 0;$$

ϱ, φ, z are cylindrical coordinates (oz-axis is directed along the tube axis). We assume that a liquid flow has a right helical symmetry with the step $h = 2\pi l$. Introducing new variables $x = z/l$, $\chi = \varphi - x$ and $r = \varrho/l$ the continuity equation can be rewritten in the form

$$\frac{1}{r} \frac{\partial}{\partial r} (r u_r) + \frac{1}{r} \frac{\partial}{\partial \chi} (u_\varphi - r u_x) = 0; \tag{1}$$

Equation (1) allows to introduce stream function $\Psi(r, \chi)$ and a new velocity component u_χ, which we determine by the following relationship:

$$u_r = \frac{\partial \psi}{\partial \chi}; \qquad u_\chi = u_\varphi - r u_x = -\frac{\partial (r\psi)}{\partial r}. \tag{2}$$

Let us obtain the equation for determination of stream function ψ, associated with the velocity field in the tube of radius R, induced by the given distribution of vorticity $\omega(r, \varphi, x) = \mathrm{rot}\, u$. Taking into account the assumption on helical symmetry of a flow, we set the vector ω as follows $\omega = \left(0, \omega r / \sqrt{r^2 + 1}, \omega / \sqrt{r^2 + 1}\right)$, where $\omega(r, \varphi, x)$ is the function determining the modulus of vector ω, which is directed along the tangent to the helical line passing through a point (r, φ, x). Taking into account definitions of stream function (Eq. 2) and ω, let us transform Eq. (1) to the equation

$$L\psi \equiv -\frac{\partial^2 \psi}{\partial r^2} + \left(1 + \frac{2}{r^2 + 1}\right) \frac{1}{r} \frac{\partial \psi}{\partial r} + \frac{1 - r^2}{(r^2 + 1) r^2} \psi + \frac{(r^2 + 1)}{r^2} \frac{\partial \psi}{\partial \chi} = \frac{2}{(r^2 + 1) r} f_0 - \omega l \frac{\sqrt{r^2 + 1}}{r}, \tag{3}$$

where $f_0 = \mathrm{const}$. From the non-penetration condition on the tube wall at $r = \mathbb{R} \equiv R/l$ the boundary conditions for the function ψ are the following:

$$u_r|_{r=\mathbb{R}} = \frac{\partial \psi}{\partial \chi}\bigg|_{r=\mathbb{R}} = 0.$$

To give concrete expression to ω, let us consider the simplest distribution of vorticity, which corresponds to infinitely thin vortex filament L of helical form, laying on the cylinder of radius $a < R$, which has the axis coaxial with the tube axis, step $h = 2\pi l$ and intensity Γ. Then

$$\omega(r, \varphi, x) = \frac{\Gamma}{l^2} \int_L \delta(r - b)\, \delta(x - x_0)\, \delta(\varphi - \varphi_0)\, \frac{ds}{r}, \tag{4}$$

where $\delta(x)$ is the Dirac function, ds is the element of the length of helical line passing through the point (b, φ_0, x_0); $b = a/l$. Integrating Eq. (4) over x $\left(ds = dx\sqrt{1 + r^2}\right)$ with a variable replaced by the $\varphi = \chi + x$ $(\varphi = \chi_0 + x_0)$ and introducing the result into Eq. (3), we obtain

$$L\psi = \frac{2}{(r^2 + 1)r} f_0 - \frac{\Gamma}{l} \delta(r - b)\, \delta(\chi - \chi_0)\, (r^2 + 1)/r^2 . \tag{5}$$

To simplify the calculations, let us consider the complex function Φ, satisfying Eq. (5). In this case the solution is $\Psi = \mathrm{Re}\,(\Phi)$. Taking into account the periodicity with respect to χ, one can represent the unknown solution in the form

$$\Phi(r, \chi) = \sum_{m = -\infty}^{\infty} e^{\mathrm{i}m(\chi - \chi_0)} p_m(r, b). \tag{6}$$

Substituting Eq. (6) into Eq. (5), let us multiply both parts by $\exp\{-in\chi\}$, $(n = 0, \pm 1, \ldots)$ and integrate with respect to χ from 0 to 2π. Using orthogonality of the function system $\{\exp(im\chi)\}$ at integration, we obtain

$$p_m'' + \left(1 + \frac{2}{r^2 + 1}\right)\frac{1}{r}\, p_m' + \left[\frac{1 - r^2}{(r^2 + 1)\,r^2} - \frac{m^2(r^2 + 1)}{r^2}\right] p_m$$

$$= \frac{2f_0 E_m}{(r^2 + 1)\,r} - \frac{\Gamma}{2\pi l} \delta(r - b)\frac{r^2 + 1}{r^2}, \tag{7}$$

where $E_m = 0$ for all m except $E_0 = 1$. Substituting solution of Eq. (7), satisfying the boundary non-penetration condition at $r = \mathbb{R}$ and regularity at $r = 0$, p_m into Eq. (6) and taking the real part from Φ, we obtain the expression for stream function

$$r\psi = c_1 + \frac{c_2 r^2}{2} + (c_2 - f_0)\ln r - \left\{\begin{matrix} c_3 \\ \dfrac{\Gamma}{4\pi l}\,(r^2 + \ln r^2) \end{matrix}\right\}$$

$$- \frac{\Gamma b r}{\pi l} \sum_{m = -\infty}^{\infty} \left\{\begin{matrix} I_m'(mr)\,[K_m'(mb) - \alpha_m I_m'(mb)] \\ I_m'(mb)\,[K_m'(mr) - \alpha_m I_m'(mr)] \end{matrix}\right\} \cos m(\chi - \chi_0),$$

where $I_m(x)$ and $K_m(x)$ are modified cylindrical functions; c_1, c_2 and c_3 are the constants of integration; $\alpha_m = K_m'(m\mathbb{R})/I_m'(m\mathbb{R})$. Here and further, the upper line in brackets corresponds to the case $r < b$, and the lower one — to the case $r \geq b$. From the continuity of solution at $r = b$, constant c_3 is determined as $c_3 = \Gamma(b^2 + \ln b^2)/4\pi l$. The requirement of the regularity of the solution at $r \to 0$ $(b \neq 0)$ yields $c_2 = f_0$. For convenience we shall introduce a new constant: $\beta = 2\pi l f_0/\Gamma$; the meaning and the conditions of its determination will be discussed below. The constant c_1 can have the arbitrary value. As a result, the solution is rewritten in the form

$$r\psi = c_1 + \frac{\Gamma}{4\pi} \left\{\begin{matrix} \beta r^2 - b^2 - \ln b^2 \\ (\beta - 1)\,r^2 - \ln r^2 \end{matrix}\right\} - \frac{\Gamma b r}{\pi l^2} \sum_{m=1}^{\infty} \left\{\begin{matrix} I_m'(mr)\,Z_m'(mb) \\ I_m'(mb)\,Z_m'(mr) \end{matrix}\right\} \cos m(\chi - \chi_0), \tag{8}$$

where $Z_m(x) = K_m(x) - \alpha_m I_m(x)$. Now introducing Eq. (8) into Eq. (2), the velocity vector components take the form

$$u_r = -\frac{\Gamma b}{\pi l} \sum_{m=1}^{\infty} m \left\{\begin{matrix} I_m'(mr)\,Z_m'(mb) \\ I_m'(mb)\,Z_m'(mr) \end{matrix}\right\} \sin m(\chi - \chi_0),$$

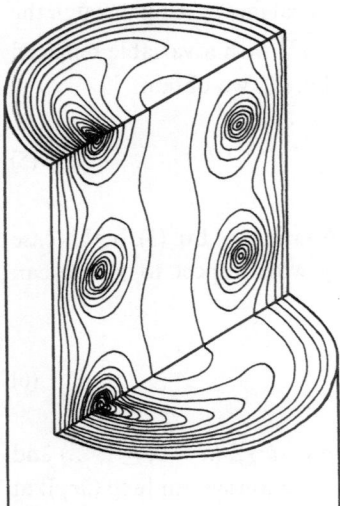

Fig. 1. Contours of $\psi = $ const

$$u_\varphi = \frac{\Gamma}{2\pi r l} \begin{Bmatrix} 0 \\ 1 \end{Bmatrix} + \frac{\Gamma b}{\pi r l} \sum_{m=1}^\infty \begin{Bmatrix} I_m(mr) \, Z_m'(mb) \\ I_m'(mb) \, Z_m(mr) \end{Bmatrix} \cos m(\chi - \chi_0),$$

$$u_x = -u_\varphi r + \beta\Gamma/2\pi l.$$

For vortex filament of the helical form in the unbounded space ($R \to \infty$), $\alpha_m \to \infty$ and we obtain a complete coincidence of the velocity vector components u_r, u_φ with the results of [2]. The value u_x differs only by the constant $\Gamma(\beta - 1)/2\pi l$, which represents a potential flow along the ox-axis. The difference in the results for the stream function at $\beta = 1$ and $r < b$ is explained by the fact, that in [2] integration constants at $r = b$ were not matched. To elucidate the physical sense of constant β and the conditions of its determination, let us find specific flow rate Q of a flow induced by helical vortex filament in an arbitrary cross-section $x = $ const of the tube

$$Q = l^2 \int_0^{\mathbb{R}} dr \int_0^{2\pi} u_x r \, d\varphi.$$

Having noted that inner integral of each term of series in the determination of u_x is equal to zero, the flow rate is

$$Q = \frac{\Gamma l}{2}\left[\mathbb{R}^2(\beta - 1) + b^2\right] \quad \text{and} \quad \beta = \frac{2Q}{\Gamma l \mathbb{R}^2} + \frac{\mathbb{R}^2 - b^2}{\mathbb{R}^2}.$$

Therefore, to determine the complete solution, the liquid flow rate across a tube should be set. The existence of stream function ψ for the given class of flows allows us to begin the study with the most visual representation through the level lines $\psi = $ const. In Fig. 1 they are given in the step cross-section of a tube for parameters $R = 1$; $a = 0.6$; $h = 1.0$, $\Gamma = \pi$; $\beta = 0.5$. The presence of intense counter flows on the tube axis and near its walls, which are induced by helical vortex structure is seen from the obtained data. This corresponds completely to a flow realized in the vortex tubes and gives the answer to the question on the mechanism of organization of counter flow in the centre of the tube and on its periphery.

However, two circumstances should be taken into account for more exact quantitative comparison of the obtained results with experimental data. First, the real vortex has a core of finite size, therefore the nearest velocity field of vortex differs from the above solution for infinitely thin filament. Secondly, the experimental data are time-averaged characteristiccs of a flow. In the obtained solution, taking into account precession of helical vortex, arising due to its self-induced motion, the time-averaging will correspond to the φ – angle coordinate averaging.

In accordance with this, let us consider a vortex of helical form with constant distribution of vorticity in a core of finite size, having the circle of ε – radius in cross-section, which is perpendicular to a helix. In this case, representations for velocity field, averaged over the angle coordinate φ will be written as integral

$$\langle \boldsymbol{u}_\varepsilon(r, \varphi, x) \rangle = \frac{1}{2\pi} \int\limits_0^{2\pi} d\varphi \, \frac{1}{\pi e^2} \int\limits_0^e ds \int\limits_0^{2\pi} \boldsymbol{u}(r, \varphi, x; r', \varphi', x') \, s \, d\theta, \tag{9}$$

where internal integration is performed over the circle of radius ε ($e = \varepsilon/l$) with center at the point $r' = b, \varphi' = \varphi_0, x' = 0$, besides, the local coordinates r, θ are connected with variables r', φ', x' by the expressions:

$$\begin{cases} r' \cos(\varphi' - \varphi_0) = b + s \cos\theta \\ r' \sin(\varphi' - \varphi_0) = s \sin(\theta)/\sqrt{b^2 + 1} \\ x' = -s \sin(\theta) \, b/\sqrt{b^2 + 1}. \end{cases}$$

Let us change the integration order in Eq. (9), taking into account that integral of series in determination of $\boldsymbol{u}$ over φ is equal to zero. Then the problem of finding the averaged values is reduced to the calculation of integrals included into the formulas

$$\langle u_{r\varepsilon} \rangle \equiv 0,$$

$$\langle u_{\varphi\varepsilon} \rangle = \frac{\Gamma}{2\pi r l} \begin{cases} 0, & r - b \leqq -e \\ \dfrac{1}{\pi e^2} \int\limits_0^e \int\limits_0^{2\pi} \begin{Bmatrix} 0, r < r' \\ 1, r \geqq r' \end{Bmatrix} r \, dr \, d\theta, & |r - b| < e \\ 1, & r - b \geqq e \end{cases} \tag{10}$$

$$\langle u_{x\varepsilon} \rangle = -\langle u_{\varphi\varepsilon} \rangle \, r + \beta \Gamma / 2\pi l.$$

Integral in Eq. (10) is evaluated through the area S of intersection of circle $r = e$ with ellipse given by the formula $(b + r \cos\theta)^2 + (r \sin\theta)^2/(b^2 + 1) = r^2$. Representation for averaged stream function at $|r - b| < e$ can be obtained to be missed after integration of the obvious relationship

$$\frac{\partial \langle r\psi_\varepsilon \rangle}{\partial r} = -\langle u_{\varphi\varepsilon} \rangle + r \langle u_{x\varepsilon} \rangle.$$

The obtained relationships for averaged velocities allow to compare calculated data with the experimental ones for mean cross-section of vortex tube [3]. The experimental and calculated profiles of radial distributions of averaged axial, transversal and radial velocities normalized by the maximal value of transversal velocity for dimensionless parameters $R = 1; a = 0.325; h = 2\pi; \varepsilon = 0.5$ are given in Fig. 2. Analyzing the obtained data, one can conclude that the chosen hydrodynamic model of a flow describes the distribution of peripheral velocity more exactly than the Rankine vortex (dotted line). But the main advantage of the given model is the correct description of the axial and radial velocities while it was not taken into consideration in any way in the Rankine vortex model.

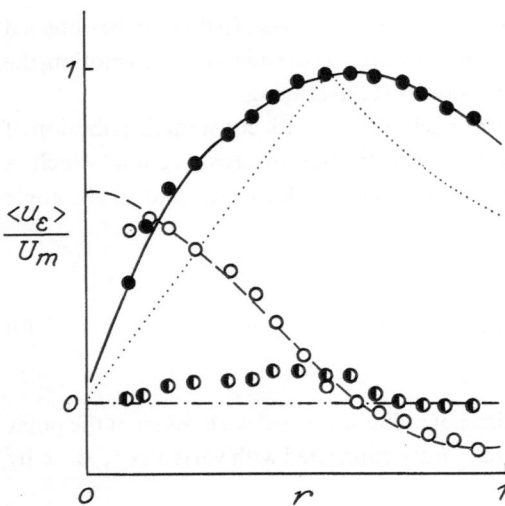

Fig. 2. Velocity profiles: lines − calculated; symbols − measured [3]; ——, ····· ● transversal velocity; − − − − ○ axial velocity; −·−·− ◐ radial velocity

Thermodynamical model

In the frame of the supposed model of a flow in vortex tube let us consider the energy transfer equation

$$c \frac{DT}{Dt} = \lambda V \, \Delta T + \mu VW, \tag{11}$$

where T is the temperature, c is the specific heat capacity, λ is the heat conduction, μ is the dynamic viscosity, V is the specific gas volume, W is the dissipative function.

As it was mentioned, in the preceding works [1, 3] due to unsuccessful choice of the flow model (Rankine vortex) the left-side part of Eq. (11) was not taken into consideration. The authors of these works attempted to explain the energy separation effect in the vortex tube completely by heat transfer and dissipation despite relatively small quantities of parameters λ and μ. The assumptions on a high intensity of the flow turbulence in vortex tubes were made, to increase their meaning, although this was insufficient, as the authors of the work [1] showed, for total description of the observed effect.

Neglecting the estimation of the role of turbulent energy transfer, we consider Eq. (11) in the frame of the model of a flow, which is obtained here. A special interest is in the possibility to estimate the role of convective transfer in the Ranque effect. We are restricted to the consideration of the left-side part of Eq. (11) alone but on the basis of our exact model of a flow.

For further simplification we assume that the temperature distribution has a time-steady character (to a certain extent, in the mean section of a tube). Then Eq. (11) (in variables r, χ) is written in the form

$$u_r \frac{\partial T}{\partial r} + u_\chi \frac{1}{r} \frac{\partial T}{\partial \chi} = 0. \tag{12}$$

Passing in Eq. (12) from velocities to stream function ψ, we get

$$\frac{\partial \psi}{\partial \chi} \frac{\partial T}{\partial r} = \frac{\partial (r\psi)}{\partial r} \frac{1}{r} \frac{\partial T}{\partial \chi}.$$

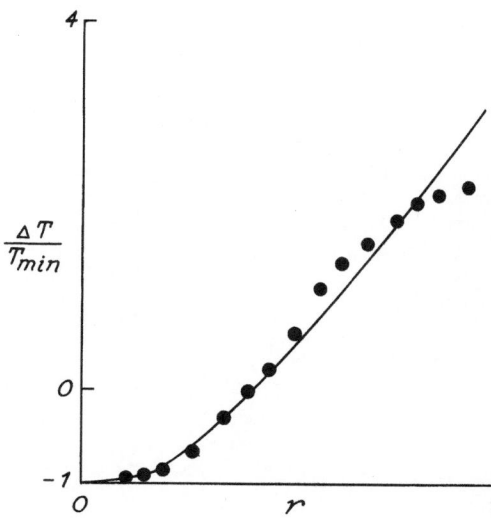

Fig. 3. Temperature profile: —— calculated; • measured [3]

Hence, distribution of the temperature in the case of purely convective transfer will be an arbitrary function of $r\psi$. We are limited to the consideration of the simplest case when temperature is a linear function of $r\psi$, i.e. $T = T_0 + Br\psi$ or for the excess temperature $\Delta T = T - T_0 = Br\psi$, where the constant B is determined from the boundary conditions, for instance, by the specification of the temperature flux at the tube wall.

Figure 3 presents the comparison between calculated φ — averaged excess temperature normalized to its minimal value and experimental data [3] corresponding to the earlier comparison of the velocity field (Fig. 2). Comparison yields a good qualitative coincidence with the observed effect of energy separation in vortex tubes. The obtained quantitative discrepancy is explained, first of all, by neglecting the flow compressibility in the pressure field deformed by the velocity distribution. Thus the determining role of convective transfer in the Ranque effect is revealed with the help of more exact modelling of the flow in a vortex.

References

[1] Gupta, A. K., Lilley, D. G., Syred, N.: Swirl flows, p. 402. New York: Abacus Press 1984.
[2] Hardin, J. C.: The velocity field induced by a helical vortex filament. Phys. Fluids **25**, 1949–1952 (1982).
[3] Shtym, A. N.: Airdynamics of cyclone-vortex chamber, p. 143. Vladivostok: Far-East Univ. 1985 (in Russian).

Authors' address: Professor Dr. A. A. Borissov, A. Kuibin, and V. L. Okulov, Institute of Thermophysics of the Sibirian Branch of the Academy of Sciences, Acad. Kutateladze Street 2, 630090 Novosibirsk, GUS, Russia

Acta Mechanica (1994) [Suppl] 4: 297–304

Topology of the flow resulting from vortex breakdown over a delta wing at subsonic speed

J. Délery and **P. Molton**, Châtillon, France

Summary. The breakdown of the primary vortex forming over a 70°-sweep angle delta wing has been investigated at a velocity of 24 m/s, the angle of incidence being 26°. Detailed velocity measurements in the breakdown region have been executed by means of a three component LDV system to allow an accurate definition of the mean flow structure in planes normal to the delta wing upper surface. The topology of these transverse fields is characterized by the existence of limit circles in the pattern of the lines of force of the projected vector field. The obtained results tend to show that, in spite of its spectacular aspect, the breakdown weakly affects the flow structure in the immediate vicinity of the wing surface.

1 Introduction

As it is well known, the setting of a delta wing at an angle of incidence leads to the formation of several vortical structures resulting from the rolling up of the rotational layers separating from the wing upper surface. The most intense structure comes from the so-called primary separation taking place along the wing sharp leading edge. The strength of this primary vortex, as measured by its swirling rate, steadily increases with the angle of incidence until a brutal disorganization of the structure occurs. This phenomenon, known as vortex breakdown, is characterized by a rapid deceleration of both the axial and swirl components of the mean velocity and an important dilatation of the vortex. At the same time, a spectacular amplification of the turbulence level in the vortex takes place resulting from the occurrence of large amplitude fluctuations. As a general rule, during the breakdown process, the mean axial velocity component rapidly decreases and becomes negative on the vortex axis. This behaviour corresponds to the existence of a stagnation point inside the flowfield, followed by a "recirculation" bubble, which most often resorbs itself with the progressive re-acceleration of the flow [1–5].

Vortex breakdown has already been thoroughly studied by considering the case of an isolated vortex submitted to an adverse pressure gradient [6–10]. These studies have demonstrated that the unsteady structure of the flow can adopt different organizations, the breakdown being of the double helix, the spiral or the bubble type, depending of factors difficult to apprehend. The aim of the present experimental analysis was to examine the conditions of vortex breakdown over a delta wing where the situation is more complex due to the interaction of the field induced by the wing.

2 Experimental arrangement and testing conditions

The tested delta wing has a sweep angle of 70° and a sharp leading edge bevelled at 15° on the pressure side, its chordlength being c = 950 mm. The tests were executed in he closed-loop F 2 subsonic wind tunnel. As shown in Fig. 1, the model was set a zero yaw angle by means of

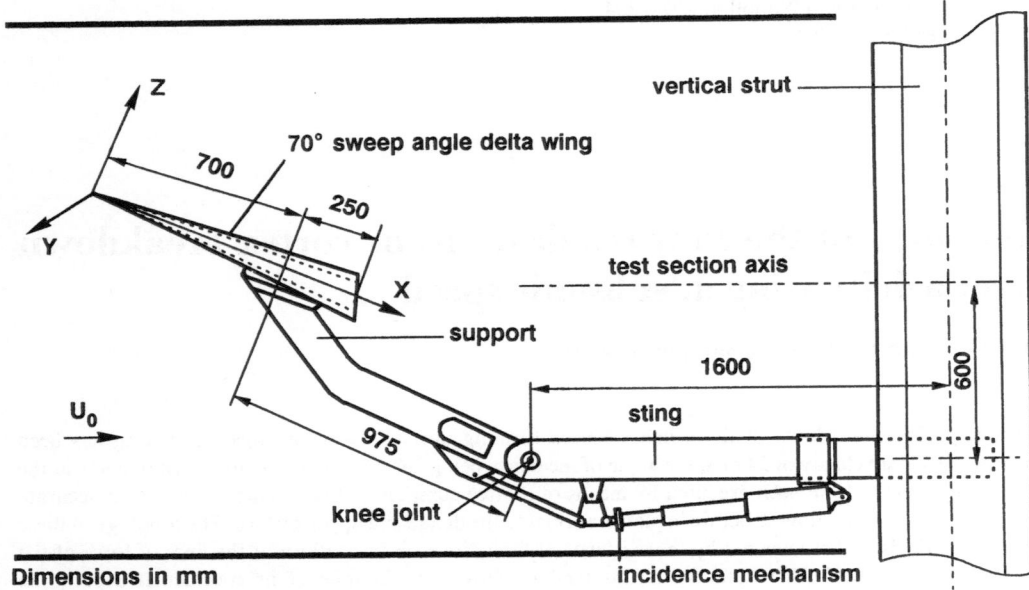

Fig. 1. Model installation in the F2 subsonic wind tunnel

a cylindrical support fixed to an horizontal sting through a knee joint. The sting itself was fixed to a vertical strut, traversing the test section, the fixation point being adjustable to allow the positioning of the model in the centre of the test section in order to minimize the perturbating effects caused by the tunnel walls. The wing could be set at a certain roll angle, without changing its incidence, to facilitate LDV measurements close to the model surface.

The upstream flow velocity U_0 was equal to 24 m/s, leading to a Reynolds number: $Re_c = 1.46 \cdot 10^6$; the wing was set at an angle of incidence $\alpha = 26°$.

The overall organization of the flow was first investigated by using the laser sheet technique which allowed an easy and fast detection of breakdown. The outer field in the breakdown region was then probed by means of a three-component LDV system working in the forward scattering mode of operation to take advantage of a higher signal/noise ratio [11]. In order to have a sufficiently high data rate, the flow was seeded with particles of incense smoke emitted downstream of the test section, thus avoiding perturbation of the incoming flow by the injection device. At each measurement point, the three instantaneous velocity components were acquired from a sample of 2000 particles.

3 Presentation of results

3.1 Overall field properties

A visualization of the flow for an incidence of 26° is shown in Fig. 2. In this case, the laser sheet is in a vertical plane containing the axis of the primary vortex developing on the model left side. The photograph clearly exhibits the vortical system which maintains a well organized structure between its birth, at the wing apex, and the breakdown point. The original vortex is the seat of very high swirl velocities creating intense centrifugal forces expelling the particles from the core of the structure which appears in black on the photograph. At the breakdown point, there is a rapid dilatation of the vortex core and an abrupt decrease of the centrifugal forces applied to the particles which can then penetrate in the centre of the structure, thus allowing its visualization. The picture of Fig. 2, as also

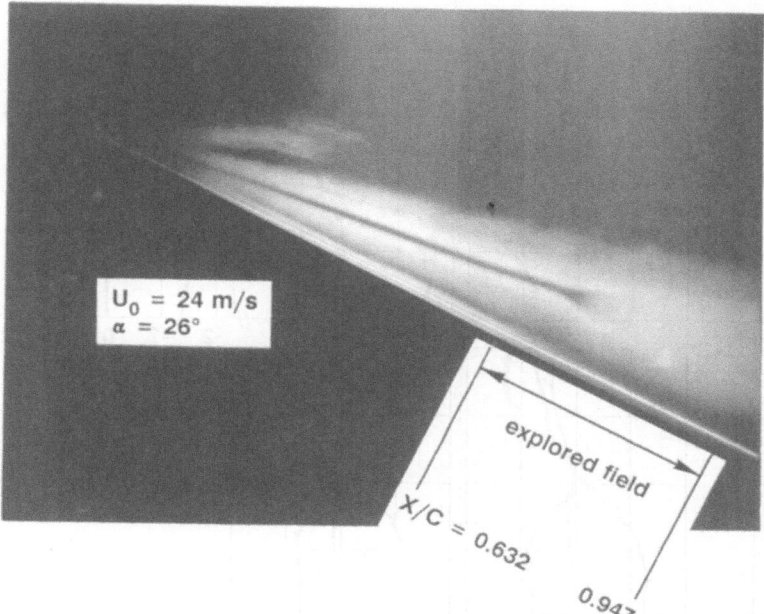

Fig. 2. Laser sheet visualization along a longitudinal plane

the LDV measurements, gives a time averaged representation of the flow field which is consistent with a modelling using the classical time-averaged version of the Navier-Stokes equations. In fact, the flow is highly fluctuating and can adopt a three-dimensional structure.

3.2 Detailed field organization

The flow above the wing has been explored with the LDV system in 7 planes perpendicular to the wing upper surface located at the following reduced distances from the apex: $X/c = 0.632, 0.684, 0.737, 0.789, 0.842, 0.894$, and 0.947. Vortex breakdown occurs approximately at $X/c = 0.684$.

Velocity profiles. Figure 3 shows the distributions of the mean velocity components U, V, W along axes Z passing through the vortex centre, for the seven X stations. These profiles reveal the following properties:

— Before breakdown, the streamwise component U reaches very high values on the axis of the structure, this behaviour being typical of the intense primary vortices generated by a delta wing. At station $X/c = 0.684$, the U-profile presents a dip in the vicinity of the axis, U becoming negative with a maximum amplitude U/U_0 equal to 0.2 This indicates the existence of a recirculation region which progressively broadens when the streamwise distance X increases, until it starts to diminish in size, U becoming everywhere positive again in the plane $X/c = 0.789$.

— The two components V and W, which can be respectively identified with the swirl V_θ and radial V_r components of a vortex which would be axisymmetric, have evolutions typical of a delta wing vortex. Upstream of the breakdown station ($X/c = 0.632$), the rotation rate $V_{\theta(max)}/U_0$ is close to 2, the viscous core having a diameter comprised between 3 and 4 mm. The breakdown of the structure has a modest effect on the modulus or the swirl component. On the other hand, the viscous core undergoes a strong dilatation, its diameter becoming equal to 70 mm at $X/c = 0.947$. The radial component V_r is everywhere small, which indicates that the considered profiles go practically through the vortex centre.

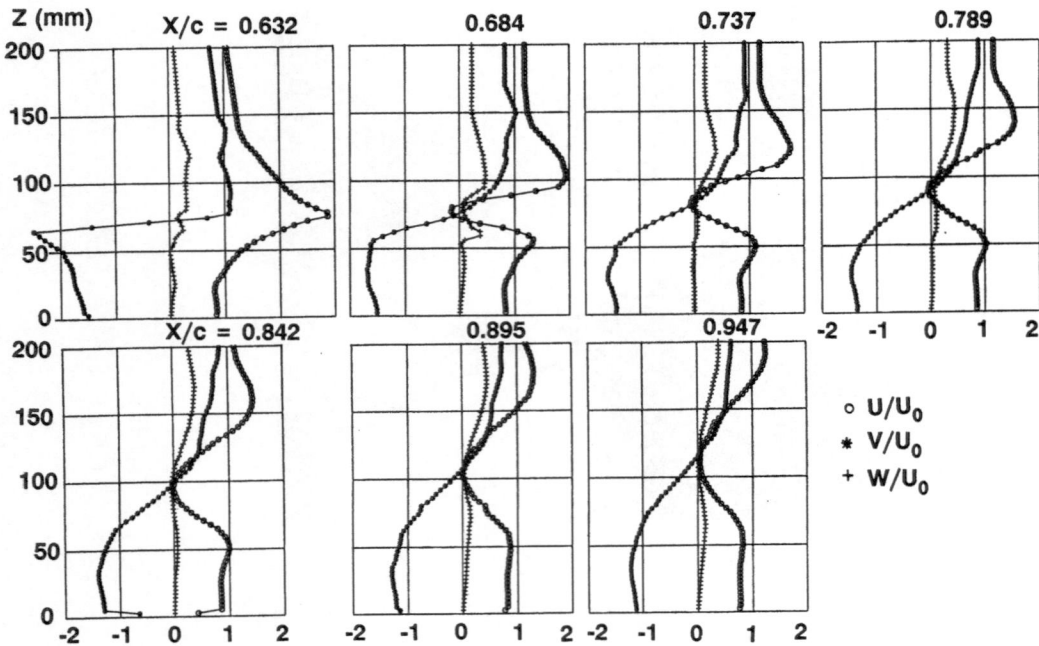

Fig. 3. Profiles of the mean velocity components through the vortex centre

Velocity field in transverse planes. The visual representation of a three-dimensional flow being difficult, it is usual to consider projections of the velocity field in planes adequately chosen. Then, one is led to interpret the structure of the projected field by tracing what are improperly called streamlines in the projection plane. This denomination is misleading, the picture given by this kind of representation being not objective, since it depends of the adopted projection plane. Thus, it is essential to use a more rigorous terminology by accurately defining the adopted two-dimensional mode of representation and by using the denomination lines of force of the projected vector field, rather than streamlines.

In the case of the flow over a delta wing, it is convenient to examine the field obtained by performing a *conical* projection in the considered plane (P), which will be here perpendicular to the wing upper surface. Then, at each measurement point, the mean velocity vector is decomposed into a component along the line coming from the wing apex and passing through the point and a component in (P).

The field projected in the plane located at $X/c = 0.632$ is represented in Fig. 4a. This tracing clearly shows the swirling motion associated with the primary vortex. On the other hand, the less intense secondary vortex is barely visible. The tracing of the lines of force, shown in Fig. 4b, reveals the following features:

— In the most outer part of the primary vortex, the lines of force, coming from the region located below the wing, roll up according to the expected way.

— A separator, difficult to capture by such a tracing, must ends at the leading edge to separate the lines of force going into the primary vortex from those constituting the secondary vortex.

— In the central part of the primary structure, the lines of force unroll from the *focus* type critical point F_1.

— The two families of lines of force constituting the primary vortex wind around a *limit circle* Γ_1, which means that the lines tending to Γ_1, some from outside, the other from inside, have no contact.

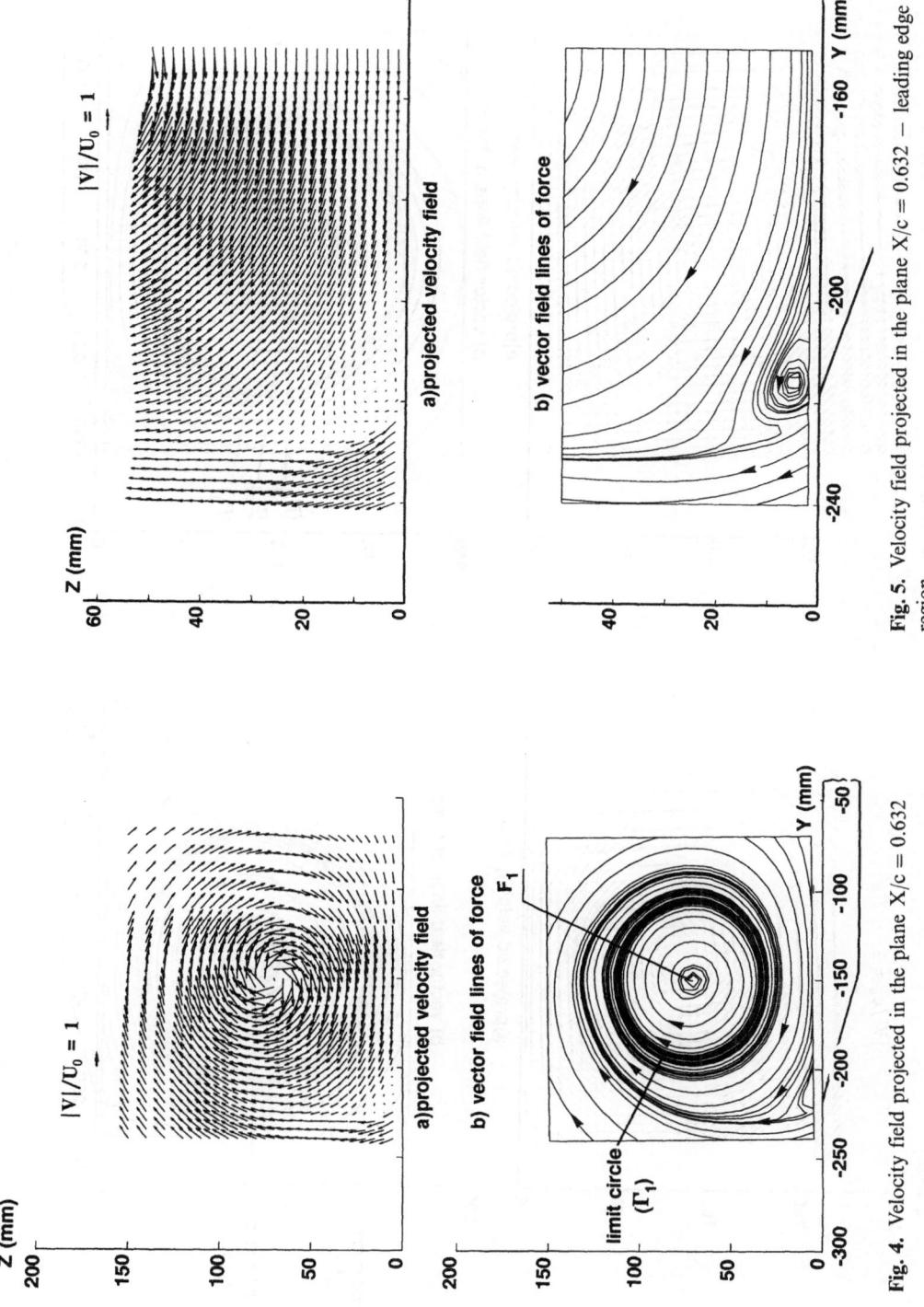

Fig. 4. Velocity field projected in the plane X/c = 0.632

Fig. 5. Velocity field projected in the plane X/c = 0.632 — leading edge region

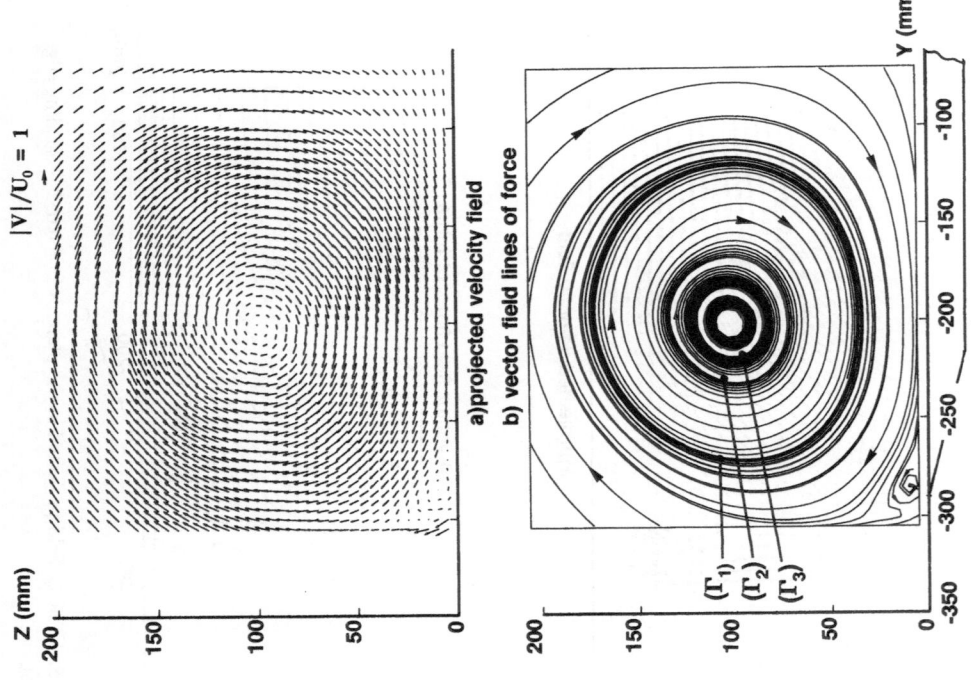

Fig. 7. Velocity field projected in the plane X/c = 0.842

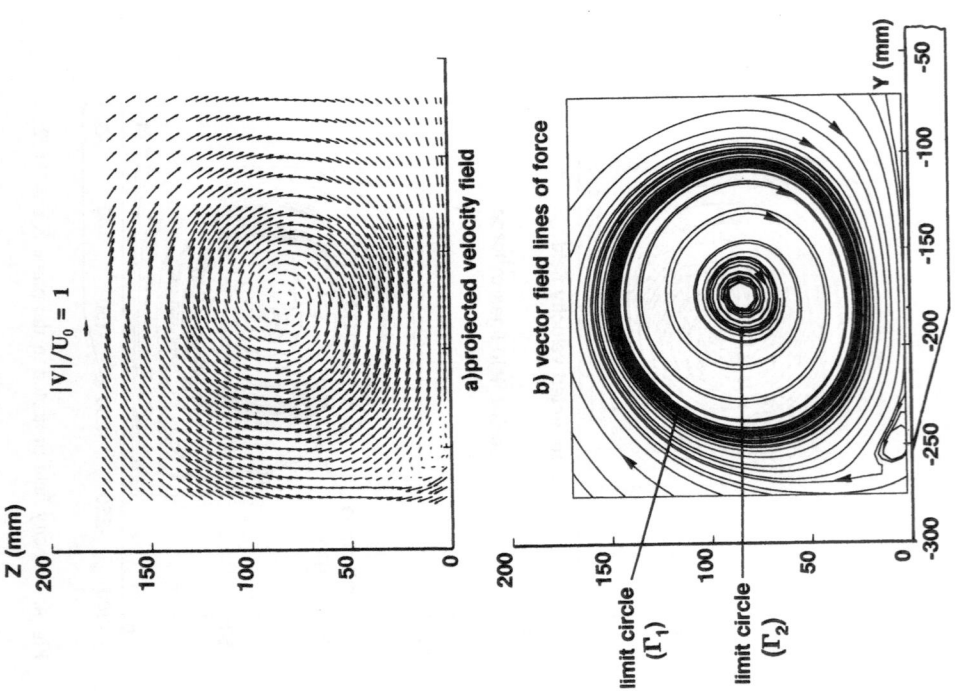

Fig. 6. Velocity field projected in the plane X/c = 0.737

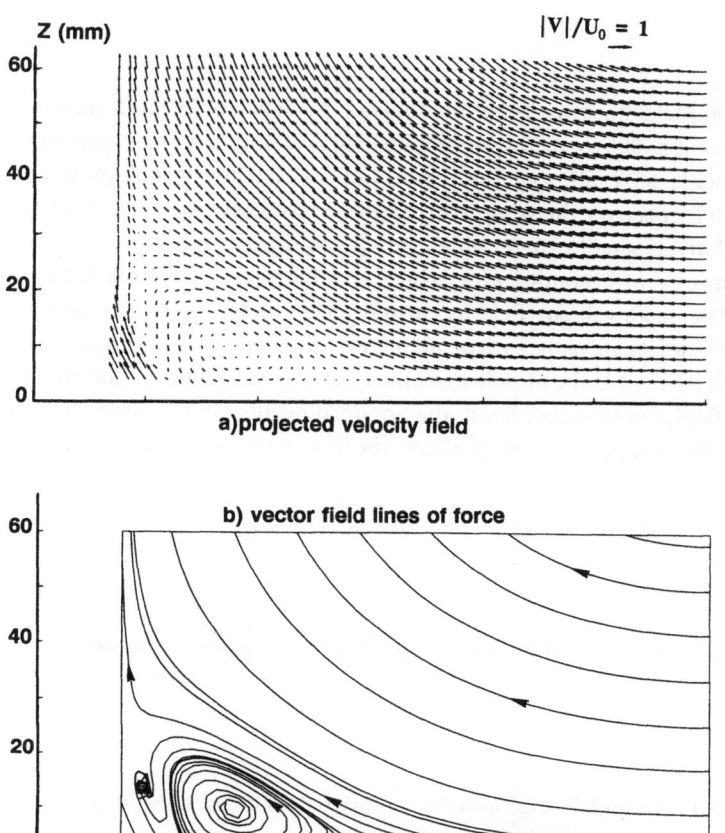

Fig. 8. Velocity field projected in the plane X/c = 0.842 — leading edge region

In the terminology of the dynamic system theory, a focus is said unstable when the lines of force emerge from it, since it is then the image of an unstable system in the plane of phase. However, one should avoid any analogy as concerns the stability of the vortex with respect to breakdown, since the picture is here established in the physical plane. A more detailed exploration of the plane X/c = 0.632 has been performed to clearly identify the flow organization in the vicinity of the wing leading edge. The corresponding vector field along with its lines of force are shown in Fig. 5, a 2 mm-wide band in contact with the wing being free of results because of the proximity of the surface which rendered measurements with the LDV system impossible.

The results relative to the plane at X/c = 0.737, located just downstream of the breakdown origin, are presented in Fig. 6. Now, one notes the existence of two limit circles, one being the continuation of the circle (Γ_1) observed at X/c = 0.632, the other — (Γ_2) — forming in the vicinity of the vortex centre. At the interior of (Γ_2), the lines of force wind around the core of the burst vortex, into which determination of these lines was not possible because of the extremely small values of the velocity. Between the two circles, the lines of force unroll from (Γ_2) and tend towards (Γ_1). Figure 7 shows the projected field in the plane located at X/c = 0.842 where a third limit circle (Γ_3) is detected. A close up of the leading edge region, given in Fig. 8, shows that the secondary vortex is practically not affected by the breakdown of the primary vortex.

4 Conclusion

The flow above a 70°-sweep angle delta wing, set at an angle of incidence of 26° in a subsonic stream, has been carefully investigated by means of detailed surveys with a three component LDV system. These experiments have allowed the description of the flow topology in the breakdown region which then forms over the wing rear part. The results show the existence of a reversed flow region, of relatively small extent, which rapidly contracts and disappears, the streamwise mean velocity component becoming everywhere positive again at a short distance from the breakdown point. The mean velocity field has also been characterized by considering the conical projection of the velocity in planes perpendicular to the wing upper surface. The topology of these transverse fields reveals the presence of limit circles in the pattern of the lines of force of the projected vector field. On the other hand, the obtained results tend so show that, in spite of its spectacular aspect, the breakdown weakly affects the flow in the immediate vicinity of the wing surface.

Acknowledgement

The authors are greatly indebted to the F 2 Wind Tunnel Group for performing the experiments.

References

[1] Wentz, W. H., Kohlmann, D. L.: Vortex breakdown on slender sharp edged wings. AIAA Paper 69-778 (1969).

[2] McKernan, J. F., Nelson, R. C.: An investigation of the breakdown of the leading edge vortices on a delta wing at high angle of attack. AIAA Paper 83-2114 (1983).

[3] Werlé, H.: Etude phénoménologique de la formation et de l'éclatement des tourbillons au tunnel hydrodynamique TH 2. ONERA RT N° 29/1147AY (1985).

[4] Payne, F. M., Ng, T. T., Nelson, R. C., Schiff, L. B.: Visualization and flow surveys of the leading edge vortex structure on delta wing planforms. AIAA Paper 86-390 (1986).

[5] Agrawal, S., Barnett, R. M., Robinson, B. A.: Investigation of vortex breakdown on a delta wing using Euler and Navier-Stokes equations. AGARD CP-494 (1990).

[6] Sarpkaya, T.: On stationary and travelling vortex breakdown. J. Fluid Mech. **45**, 545—559 (1971).

[7] Faler, J. H., Leibovich, S.: Disrupted states of vortex flow and vortex breakdown. Phys. Fluids **20**, 1385—1400 (1977).

[8] Escudier, M.: Vortex breakdown: observations and explanations. Prog. Aerospace Sci. **25**, 189—229 (1988).

[9] Délery, J., Horowitz, E., Leuchter, O., Solignac, J.-L.: Etudes fondamentales sur les écoulements tourbillonnaires. Rech. Aérospat. **2**, 81—104 (1984).

[10] Pagan, D.: Contribution à l'étude expérimentale et théorique de l'éclatement en air incompressible. Ph. D. Dissertation, Université Pierre et Marie Curie, Paris (1989).

[11] Afchain, D.: Mise au point du vélocimètre laser tridirectionnel de la soufflerie F 2. Etude de la veine vide. ONERA RT 4/3633 (1988).

Authors' address: Professor Dr. J. Délery and P. Molton, Aerodynamics Department Office National d'Etudes et de Recherches Aérospatiales ONERA, 29 Avenue de la Division Leclerc, B. P. 72, F-92322 Châtillon, France

Acta Mechanica (1994) [Suppl] 4: 305–311

The effect of the thermal conductivity on the propagation of small perturbations in viscous compressible fluid

C. Ferrari, Torino, Italy

Summary. The effect of thermal conductivity on the propagation of small perturbations in a viscous compressible fluid is investigated by adopting slow motion. The equations of motion are deduced with the constitutive equation between the stress and deformation rate tensors obtained by C. Ferrari in [1] and the equation of energy applying the constitutive equation between the heat flux density and the gradient of temperature obtained by C. Cattaneo in [2]. The initial value problem (Cauchy's problem) is represented by a system of five partial differential equations of first order which is totally hyperbolic. In each point of the flow field passes five characteristic lines, two corresponding to progressing waves in the direction of the flow, two corresponding to waves propagating in opposite direction, and one corresponding to a discontinuity line fixed to the fluid. The equations defining the variation law of any physical quantity along characteristics are deduced: the problem can be solved either numerically or iteratively. The propagation of the discontinuities of the initial data along these characteristics is then studied.

1 Introduction

In [1] I have proved that the paradox of the instantaneous propagation of any small perturbation in a viscous compressible flow is removed if one takes the relation between the stress (τ_{ik}) and deformation rate (γ_{ik}) tensors not that given by the classical theory, but that obtained taking into account the explicit dependence on the time of the distribution function of the molecular velocities. This result was obtained without considering the effect of the thermal conductivity on the phenomenon, and consequently considering it as isentropic: therefore, in this paper the problem is examined again assuming the constitutive equation relating τ_{ik} and γ_{ik} as obtained in [1] and the constitutive equation which gives the relation between the heat flux density and the gradient of the temperature obtained by C. Cattaneo in [2]. The problem is thus reduced to a system of five partial differential equations of the first order in two independent variables, $\underline{x}$ and $\underline{t}$, and it is proved that this system is totally hyperbolic. There are therefore five families of characteristic lines in this system, which are all real and different; they make it possible to calculate easily the numerical solution of the problem as well as to determine in the plane (x, t) the domain of dependence of any point P on the initial data, and the domain of influence of these data. If in the correspondent Cauchy's problem the initial data have discontinuities, they are propagated along the characteristics and the study of this propagation, leading to the kinematic and dynamic consistency conditions (T. Levi-Civita [3]), show that the majority of the kinematic discontinuity produced in a point $P_0(x_0, 0)$ propagates along that characteristic issuing from P_0 corresponding to the wave with a higher propagation speed, and along this characteristic the decay of the discontinuity is much less than that in the other.

2 Transfer equations of momentum and heat in slow motion of a compressible viscous fluid

We assume the constitutive equations of the stress tensor τ_{ik} and of the heat flux density vector q_k, respectively indicated in [1] by C. Ferrari and in [2] by C. Cattaneo; therefore we have

$$\tau_{ik} = \mu \Gamma_{ik} - \frac{\sigma^*}{\mu} \frac{\partial}{\partial t} \tau_{ik}; \quad q = -\chi \operatorname{grad} \vartheta - \frac{\sigma_\vartheta^*}{\chi} \frac{\partial q}{\partial t} \tag{1}$$

where

$$\tau_{ik} = \mu \Gamma_{ik} - \sigma^* \frac{\partial}{\partial t} \Gamma_{ik}; \quad \Gamma_{ik} = \gamma_{ik} - \frac{2}{3} \delta_{ik} \frac{\partial u_l}{\partial x_l}$$

$$\gamma_{ik} = \frac{\partial u_i}{\partial x_k} + \frac{\partial u_k}{\partial x_i}; \quad \delta_{ik} = \begin{array}{ll} 1 & \text{if} \quad i = k \\ 0 & \text{if} \quad i \neq k \end{array}$$

being: u_i the velocity component along the x_i axis; μ the viscosity coefficient; $\sigma^* = \frac{1}{2} \varrho \langle l_c^2 \rangle = \varrho \sigma$, where l_c^2 is the mean square value of the mean free-path of the molecules whose velocity is c; ϱ is the fluid density; χ the coefficient of the thermal conductivity; ϑ the temperature; $\sigma_\vartheta^* = \gamma \varrho c_v \sigma$; c_v the specific heat at constant volume. The first of Eqs. (1) in the case of slow non-stationary motion in the direction of the axis $x \equiv x_1$ leads to the equations

$$\sigma^* \left(\varrho_0 \frac{\partial^2 u}{\partial t^2} + \frac{\partial^2 P}{\partial x \, \partial t} \right) - \frac{4}{3} \mu^2 \frac{\partial^2 u}{\partial x^2} + \mu \varrho_0 \frac{\partial u}{\partial t} + \mu \frac{\partial P}{\partial x} = 0, \tag{2}$$

where P is the pressure, while the continuity equation can be written as

$$\frac{\partial \varrho}{\partial t} + \varrho_0 \frac{\partial u}{\partial x} = 0, \tag{3}$$

where the subscript zero stands for the unperturbed values of the various quantities.

Thus, the equation of the heat transfer is

$$\frac{\partial q}{\partial x} + \varrho_0 c_v \frac{\partial \vartheta}{\partial t} + P_0 \varrho_0 \frac{\partial}{\partial t} \left(\frac{1}{\varrho} \right) = 0. \tag{4}$$

By the elimination of q from Eqs. (1) and (4) one obtains

$$\sigma_\vartheta^* \left(\varrho_0 c_v \frac{\partial^2 \vartheta}{\partial t^2} - \frac{P_0}{\varrho_0} \frac{\partial^2 \varrho}{\partial t^2} \right) + \chi \left(\varrho_0 c_v \frac{\partial \vartheta}{\partial t} - \frac{P_0}{\varrho_0} \frac{\partial \varrho}{\partial t} \right) - \chi^2 \frac{\partial^2 \vartheta}{\partial x^2} = 0 \tag{5}$$

Let us say:

$$u = U c_s; \quad x = \mathscr{L} X; \quad t = \frac{\mathscr{L}}{c_s} T; \quad P = \varrho_0 c_s^2 \mathscr{P}; \quad \mathscr{L} = \frac{v}{u_0}; \quad \vartheta = \vartheta_0 \theta; \quad \varrho = \varrho_0 \Sigma, \tag{6}$$

where c_s is the sound speed in non viscous fluid, u_0 a reference velocity that characterizes the order of magnitude of u. Denoting besides

$$\frac{\partial U}{\partial T} = p_1; \quad \frac{\partial U}{\partial X} = q_1; \quad \frac{\partial \theta}{\partial T} = p_2; \quad \frac{\partial \theta}{\partial X} = q_2 \tag{7}$$

the system of Eqs. (2), (3), (5) can be written in matrix notation under the form

$$L(U) = AV_x + BV_T = C \tag{8}$$

where A and B are the matrices given by

$$
A = \begin{vmatrix}
0 & \dfrac{1}{\gamma} & -K & 0 & \dfrac{1}{\gamma} \\
\gamma - 1 & 0 & 0 & -H_\vartheta & 0 \\
-1 & 0 & 0 & 0 & 0 \\
0 & -1 & 0 & 0 & 0 \\
0 & 0 & 0 & 0 & 0
\end{vmatrix}, \quad
B = \begin{vmatrix}
1 & 0 & 0 & 0 & 0 \\
0 & 1 & 0 & 0 & 0 \\
0 & 0 & 1 & 0 & 0 \\
0 & 0 & 0 & 1 & 0 \\
0 & 0 & 0 & 0 & 1
\end{vmatrix} = I \tag{9}
$$

V is the matrix $V = [p_1, p_2, q_1, q_2, \Sigma]'$, using the notation []' to indicate the transpose of the matrix within the brackets, while $V_x = \left[\dfrac{\partial p_1}{\partial X}, \dfrac{\partial p_2}{\partial X}, \ldots \right]'$; $V_T = \left[\dfrac{\partial p_1}{\partial T}, \dfrac{\partial p_2}{\partial T}, \ldots \right]'$; $\underset{\sim}{C}$ is the vector whose components are

$$C^1 = -hp_1 - \frac{h}{\gamma} q_2; \quad C^2 = -h_\vartheta p_2 - (\gamma - 1) h_\vartheta q_1; \quad C^3 = 0; \quad C^4 = 0; \quad C^5 = -q_1. \tag{10}$$

The meaning of the other symbols is the following:

$$K = \frac{1}{\gamma} + \frac{4}{3} \frac{v^2}{\sigma c_s^2}; \quad h = \frac{v\mathscr{L}}{\sigma c_s}; \quad h_\vartheta = \frac{h}{\mathrm{Pr}}; \quad v_\vartheta = \frac{\chi}{\gamma \varrho_0 c_v}; \quad H_\vartheta = \gamma \frac{v_\vartheta^2}{\sigma c_s^2};$$

$$\mathrm{Pr} = \text{Prandtl number} = \frac{v}{v_\vartheta}; \quad v = \text{kinematic viscosity coefficient}; \tag{11}$$

γ = specific heat coefficient ratio.

3 Characteristic lines and variation laws of the quantities (p_i, q_i, Σ) along the characteristics

The characteristic lines of the system (8) are defined by the equation $\dfrac{dX}{dT} = \tau$, where τ are the eigenvalues of the matrix A, and therefore the roots of the characteristics equation

$$|A - \tau I| = 0. \tag{12}$$

Consequently one deduces

$$\tau \left[-\tau^4 + \left(H_\vartheta + K + \frac{\gamma - 1}{\gamma} \right) \tau^2 - K H_\vartheta \right] = 0 \tag{13}$$

from which one obtains

$$\tau_1 = \left\{ \frac{H_\vartheta + K_0}{2} + \left[\frac{(H_\vartheta + K_0)^2}{4} - K H_\vartheta \right]^{1/2} \right\}^{1/2}; \quad \tau_2 = -\tau_1;$$

$$\tau_3 = \left\{ \frac{H_\vartheta + K_0}{2} - \left[\frac{(H_\vartheta + K_0)^2}{4} - K H_\vartheta \right]^{1/2} \right\}^{1/2}; \quad \tau_4 = -\tau_3; \tag{14}$$

$$\tau_5 = 0$$

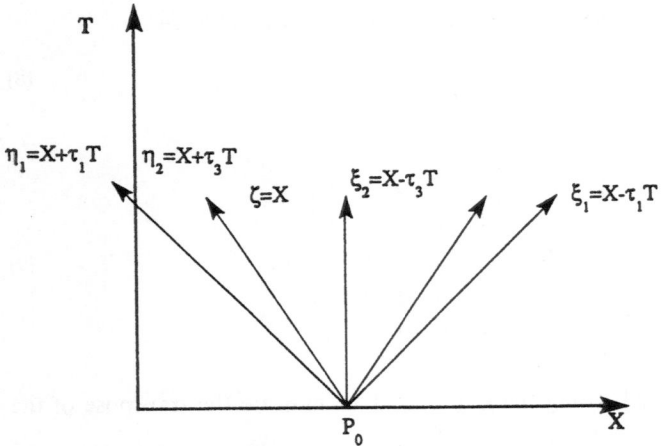

Fig. 1. System of characteristics issuing from a point P_0

where $K_0 = K + \dfrac{\gamma - 1}{\gamma}$. Since $K_0 > K$ the roots τ_i are always real, and therefore the differential system (8) is totally hyperbolic: there are five different families of characteristics, which in the plane (X, T) are straightlines whose equations are respectively

$$\xi_1 = X - \tau_1 T; \quad \eta_1 = X + \tau_1 T; \quad \xi_2 = X - \tau_3 T; \quad \eta_2 = X + \tau_3 T; \quad \zeta = X. \tag{15}$$

The last of Eqs. (15) defines a line of discontinuity fixed to the medium ($\tau_5 = 0$), while the others move in the medium with velocity τ_i and they are correspondent to real propagation waves of the perturbation produced in the medium (Fig. 1). It is easy to find the meaning of the constant K_0 and H_ϑ: the former is the square of the propagation velocity of the wave in viscous non conductive fluid, while the latter gives the square of the propagation velocity in the problem of the heat transfer in a non flowing medium ($U = 0$) (Cattaneo problem). From the inequality $K_0 > K$, one obtains $\tau_1 > K_0^{1/2}$, and thus the thermal conductivity increases the propagation velocity of the corresponding wave, and since it is $\tau_4{}^2 < H_\vartheta$ as a result of the flow, the propagation velocity of the corresponding wave decreses. On the other hand, one has to observe that in the problem now considered from any point P_0, in which a perturbation of the flow is produced, two waves start that propagate in the same direction of the flow with different velocities ($\tau_1 > \tau_3$) and share the kinematic and thermal discontinuities among them according to the law given later on.

The eigenvectors corresponding to the eigenvalues τ_i can be deduced by Eq. (16)

$$\underset{\sim}{l}A = \tau\underset{\sim}{l} \tag{16}$$

and therefore it turns out that

for $\quad \tau = \tau_i \quad (i = 1, 2, 3, 4)$

$$l_1 = \frac{\tau_i}{\gamma} \frac{l_1}{\tau_i{}^2 - H_\vartheta}; \quad l_2 = -\frac{K}{\tau_i} l_1; \quad l_3 = -\frac{1}{\gamma} \frac{H_\vartheta l_1}{\tau_i{}^2 - H_\vartheta}; \quad l_4 = \frac{h}{\gamma\tau_i} l_1 \tag{17}$$

for $\quad \tau = \tau_5 \quad l_i = 0.$

The variation laws of the physical quantities along the characteristics are then deduced by the relation

$$\underset{\sim}{l}(AV_X + IV_T) = l_i C^i \tag{18}$$

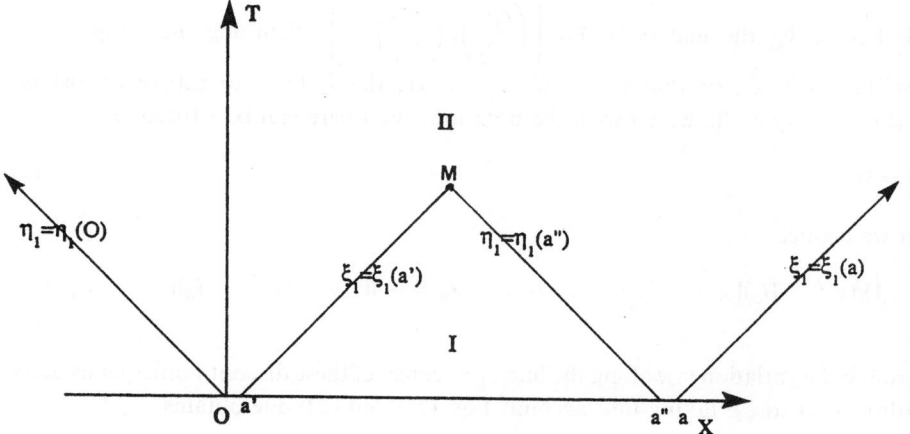

Fig. 2. *I* Domain of dependence of point M on the initial data; *II* domain of influence on the initial data on the segment $0a$

and therefore along the characteristics $\xi_1 = $ const. we have

$$\frac{\partial}{\partial \eta_1}\left[\left(p_1 - \frac{K}{\tau_1}q_1\right) + \frac{1}{\gamma}\frac{1}{\tau_1{}^2 - H_\vartheta}(\tau_1 p_2 - H_\vartheta q_2) + \frac{h}{\gamma \tau_1}\Sigma\right]$$

$$= \frac{1}{2}\left\{-\frac{h}{\tau_1}\left[p_1 + \left(\frac{1}{\gamma \tau_1} + \frac{\gamma - 1}{\gamma}\frac{\tau_1}{\tau_1{}^2 - H_\vartheta}\frac{h_\vartheta}{h}\right)q_1\right] - \frac{h}{\gamma \tau_1}\left(q_1 + \frac{h_\vartheta}{h}\frac{\tau_1}{\tau_1{}^2 - H_\vartheta}p_2\right)\right\} \quad (19)$$

The corresponding equation for the characteristic $\eta_1 = $ const. can be directly obtained by Eq. (19) substituting $\dfrac{\partial}{\partial \eta_1}$ with $\dfrac{\partial}{\partial \xi_1}$ and τ_1 with $(-\tau_1)$, while for the characteristics $\xi_2 = $ const. and $\eta_2 = $ const. the equations are completely analogous to those above indicated with τ_3 in place of τ_1. Finally for

$$\zeta = \text{const. it is } \frac{\partial \Sigma}{\partial T} = -q_1. \quad (20)$$

The equations now obtained are very suitable to solve any initial value problem: given the initial values of $U, \dfrac{\partial T}{\partial T}, \Sigma, \theta, \dfrac{\partial \theta}{\partial T}$ in the range $(X : 0 \doteq a)$ determine these functions in the domain in which they are defined. For any point M in the plane (X, T) the domain of dependence of M proves to be the segment $\sigma(a'a'')$ of X intercepted on this axis by the two outer characteristics whose slope are $(\tau_1, -\tau_1)$ (Fig. 2) while the domain of influence of the data in the segment $(X : 0 \doteq a)$ is the region of the half plane $T > 0$ lying between the two characteristics $\eta_1 = \eta_0; \xi_1 = \xi_a$ pointing from the end points of σ into the upper halfplane. Moreover, Eqs. (19) and similar have the form appropriate for obtaining either a numerical solution or a solution with the iterative method indicated by R. Courant [4], or with methods based on the Fourier or Laplace transformations.

4 Propagation of the discontinuities

Rewrite Eq. (8) under the form

$$A^i D_i V = \underset{\sim}{C} \quad (i = 0, 1) \quad (8')$$

Solutions of Eq. (8') can exist that are continuous, while their first derivates suffer jumps across the characteristics, so that for these solutions it is $A^i(D_i V) = 0$, if (G) denotes the jump of the quantity G. Let be V_{ξ_1} the matrix $(V_{\xi_1}) = \left[\left(\dfrac{\partial p_1}{\partial \xi_1}\right), \left(\dfrac{\partial p_2}{\partial \xi_1}\right) \ldots\right]'$; denoting the jumps of the elements of (V_{ξ_1}) with λ_1, so that $(V_{\xi_1}) = (\lambda_1, \lambda_2, \ldots \lambda_5)'$, the λ_i have to satisfy consistency conditions (Levi-Civity a [3]), which with the notations used here can be written as

$$(A - \tau_1 I)\, \lambda = 0 \tag{21}$$

from which we deduce

$$\lambda_2 = [\tau_1(\gamma - 1)/(\tau_1{}^2 - H_\vartheta)]\, \lambda_1; \quad \lambda_3 = -(\lambda_1/\tau_1); \quad \lambda_4 = -[(\gamma - 1)/(\tau_1{}^2 - H_\vartheta)]\, \lambda_1; \quad \lambda_5 = 0. \tag{22}$$

To determine the variation law along the line $\xi_1 = $ const. of these discontinuities let us derive Eq. (19) with respect to ξ_1: taking into account Eqs. (21) and (22) one obtains

$$\frac{\partial \lambda_1}{\partial \eta_1} = -\frac{\lambda_1}{4} \frac{[(h/\tau_1)(K_0 - 1)/(\tau_1{}^2)] + [(\gamma - 1)/(\gamma)](h_\vartheta/\tau_1)\, H_\vartheta/(\tau_1{}^2 - H_\vartheta)^2}{1 + [(\gamma - 1)/\gamma]\, H_\vartheta/(\tau_1{}^2 - H_\vartheta)^2} = -\alpha\lambda_1 \tag{23}$$

and therefore

$$\lambda_1 = (\lambda_1)_0 \exp(-\alpha\eta_1) \tag{24}$$

Considering now the characteristics $\xi_2 = $ const. in an analogous way one deduces the relations between the jumps of the physical quantities across this characteristics, and one obtains

$$\frac{\partial \mu_1}{\partial \eta_2} = -\frac{\mu_1}{4} \frac{[(h/\tau_3)(K_0 - 1)/(\tau_3{}^2)] + [(\gamma - 1)/(\gamma)](h_\vartheta/\tau_3)\, H_\vartheta/(\tau_3{}^2 - H_\vartheta)^2}{1 + [(\gamma - 1)/(\gamma)]\, H_\vartheta/(\tau_3{}^2 - H_\vartheta)^2} = -\beta\mu_1 \tag{25}$$

so that

$$\mu_1 = (\mu_1)_0 \exp(-\beta\eta_2) \tag{26}$$

Assuming for v and σ the values already indicated in [1] one obtains

$$v = 0.132\ \mathrm{cm}^2/\mathrm{sec}; \quad \sigma = 2.9 \times (2 \times 10^{-6})^2\ \mathrm{cm}^2; \quad P_r = 0.7; \quad \gamma = 1{,}4,$$

so that

$$H_\vartheta = 0.25; \quad K_0 = 1.12; \quad (h_\vartheta/h) = 1.43; \quad \tau_1{}^2 = 1.198; \quad \tau_3{}^2 = 0.172 \tag{27}$$

and thus $(\alpha/\beta) = 0.0188$: the discontinuity μ_1, across the line $\xi_2 = $ const., decay along it much more rapidly than λ_1 along the line $\xi_1 = $ const.

Let us now determine $(\lambda_1)_0, (\mu_1)_0$, i.e. how the discontinuity of the initial data in a point P is shared between the two characteristics issuing from the point P.

Assuming $\mathscr{A}$ and zero respectively the discontinuity of $\dfrac{\partial p_1}{\partial X}$ and $\dfrac{\partial p_2}{\partial X}$ in the initial data on the point $P \equiv 0$

$$\left(\frac{\partial p_1}{\partial X}\right)_{\xi_1 = 0} + \left(\frac{\partial p_1}{\partial X}\right)_{\xi_2 = 0} = \mathscr{A}; \quad \left(\frac{\partial p_2}{\partial X}\right)_{\xi_1 = 0} + \left(\frac{\partial p_2}{\partial X}\right)_{\xi_2 = 0} = 0 \tag{28}$$

from which one deduces

$$(\lambda_1)_0 + (\mu_1)_0 = \mathscr{A}; \quad 0.46(\lambda_1)_0 - 2.13(\mu_1)_0 = 0 \tag{29}$$

and therefore

$$(\lambda_1)_0 = 0.823 \,\mathscr{A}; \quad (\mu_1)_0 = 0.177 \,\mathscr{A}$$

so that it appears that the majority of the discontinuity in P is propagated along the line $\xi_1 = 0$.

5 Conclusions

The influence of the thermal conductivity on the propagation of small perturbations appears to be dependent on two parameters: γ and h_ϑ. The former regulates link between the two equations of the momentum and the heat transfer, the latter the decay of the perturbations. If $\gamma = 1$ the heat transfer is independent on the flow and the velocities of the waves propagating either the kinematic perturbations or the thermal perturbations have the same values $(K_0{}^{1/2}, H_0{}^{1/2})$ as though the influence of the thermal conductivity were null.

References

[1] Ferrari, C.: On the propagation of small perturbations in viscous compressible fluid. Acta Mech. [Suppl.] 3: 1–16 (1992).
[2] Cattaneo, C.: Sulla conduzione del calore. Seminario matematico e fisico. Università di Modena Società Tipografica Modenese 1948.
[3] Levi-Civita, T.: Caracteristiques des Systèmes differentiels et Propagation des ondes, pp. 56–61. Paris: Librairie Felix Akan 1932.
[4] Courant, R.: Partial differential equations, pp. 466–471. New York: London Interscience Publishers 1962.

Author's address: em. Professor Carlo Ferrari, Politecnico di Torino, Corso Galileo Ferraris 146, I-10129 Torino, Italy

Acta Mechanica (1994) [Suppl] 4: 313–324
© Springer-Verlag 1994

Imploding cylindrical temperature pulses in superfluid helium*

W. Fiszdon**, **T. Olszok**, **G. Stamm**, **B. Noack**, and **J. Piechna*****, Göttingen, Federal Republic of Germany

Summary. Unlike in normal fluids the flow and heat transfer of cylindrical imploding temperature pulses in superfluid helium can be fairly easily investigated. This possibility was taken advantage of to study the influence of quantized vorticity on the flow process with strong geometric constraints and to check the validity of the simplified theoretical model and the numerical simulation procedure used. At weak heat pulses, when the linear approximation can be used, the analytical solution and the numerical procedure compare very well with the experimental results. For medium heat pulses, when quantum turbulence and non linearity become important, the qualitative agreement of the numerical simulations with experimental data is also very satisfactory, reproducing all the flow features and showing the strong interaction between the geometric and vorticity effects. However the quantitative differences are of the order of 20%.

1 Introduction

A question occasionally asked by gas- and fluid-dynamicists is why bother to study superfluid flows when there still remain so many unsolved problems in classical fluids? Besides the general appeal to investigate new physical processes, there is some hope that they may be helpful, in the broad sense, to understand better some cases encountered in normal fluids particularly when their experimental study is in some respect easier or more precise in superfluids. We think that converging cylindrical waves, besides posing their own problems, have this character. Hopefully this can be of interest to researchers and students in the field of gas dynamics to which Professor Jürgen Zierep has contributed so much.

The study of non stationary flow processes in superfluid helium is also of fundamental interest of its own resulting from the unique properties of the medium as described e.g. in Puttermann [8]. The evolution of top hat axisymmetric perturbations as compared with planar configurations offers the possibility to investigate the inter-relation between geometric constraints and physical effects. Applying heat pulses in He II of very small to moderate amplitudes gives the possibility to investigate the propagation of so-called "second sound" shock waves in a laminar as well as in a turbulent environment. The second sound temperature entropy waves are analogous to "first sound" pressure density waves in gas dynamics. It should also be noted here that the established experimental techniques in He II allow to measure temperature differences of the order of μK (see v. Schwerdtner et al. [10]). The direct determination of the local vorticity by means of measuring

---

* Partially presented at the 1st ECFM, Cambridge, 1991 and ad the 18th ICTAM, Haifa, 1992.
** Permanent address: Polish Academy of Sciences and Technical University, Warszawa, Poland.
*** Permanent address: Technical University, Warszawa, Poland.

the attenuation of low intensity second sound waves as described by v. Schwerdtner [9] and Stamm [11] is an additional asset of superfluid experimentation.

The above considerations, we hope, justify the presentation of our research on this particular problem of a "quantum fluid" flow to a "newtonian fluid" dynamic environment.

2   Experimental setup

The test cavity is formed by a 6 mm high circular cylinder of 36 mm inner diameter made of lucite. Two endplates made of glass close this cylinder in the axial direction. A vapor deposited heating film of 100 Å chromium, 300 Å copper and 100 Å gold covers the inner surface of the cylinder. This film forms an electric heater with a resistance of approximately $50\,\Omega$ used to generate a second sound converging temperature or rather entropy wave. Typically heat pulses of a few hundreds of μs duration and a few W/cm^2 intensity are applied. On one of the cover plates a $\approx 2\,\text{mm}^2$, 300 Å thick chromium test pulse emitter and on the opposite plate a temperature sensor is located, directly facing the emitter. This sensor consists of 0.2 cm long, 20 μm wide and approximately 1000 Å thick gold-tin strip and works as a superconducting bolometer in the constant current mode. The position of the cover plates relative to the cylinder can be adjusted from outside the cryostat to obtain a space and time temperature and vortex line density distribution in radial direction by adequately phasing the release times of heat and test signals.

The vortex line density (VLD), in the assumed case of a homogeneous, isotropic distribution of the vortex lines, is obtained from the second sound amplitude attenuation as given by Tough [14]:

$$L = \frac{6c_{20}}{B\varkappa b} \ln\left(\frac{\tilde{T}}{\tilde{T}_0}\right),$$

where $\tilde{T}_0$ is the test pulse amplitude which crossed the undisturbed cavity height, $\tilde{T}$ the test pulse amplitude that crossed the cavity containing the quantized vortex lines produced by the main heating pulse, b is the cavity height, c_{20} is the second sound velocity and B the mutual friction parameter of order one (see Swanson et al. [13]). Details of the experimental arrangements can be found in the reports of Stamm et al. [12] and Olszok [7].

3   Approximate theoretical model

To assist in the interpretation of the experimental observations and gain a better understanding of the flow process a simplified theoretical description of the flow was used. The Tisza-Landau two-fluid model as given e.g. in Khalatnikov [3] or in Putterman [8] was supplemented by including the Gorter-Mellink terms [2] which take into account the dissipation due to the interaction of the flow field with the superfluid vorticity expressed in terms of the vortex line density L. To close the set of equations it was necessary to add Vinen's [15] vortex line density evolution equation with an additional term allowing for the drift of the quantum vortex tangle (see Nemirovskii and Lebedev [5]).

To simplify the rather cumbersome set of differential equations it was assumed that the total density ϱ, which is the sum of the normal and superfluid density fractions ϱ_n, ϱ_s, respectively, remains constant. It was also assumed that counterflow conditions prevail, i.e.

$$\varrho\underline{v} = \varrho_s\underline{v}_s + \varrho_n\underline{v}_n = 0, \tag{1}$$

where $\underline{v}_n$, $\underline{v}_s$ are the flow velocities of the normal and the superfluid components. The bulk viscosity terms were neglected. Keeping only terms up to second order in the variation of the dependent variables temperature T, counterflow velocity $\underline{w} = \underline{v}_n - \underline{v}_s$ and the vortex line density L leads to

$$\frac{\partial \underline{w}}{\partial t} + \frac{\varrho_s}{\varrho}\left((\underline{w}\nabla)\,\underline{w} + (\nabla \cdot \underline{w})\,\underline{w} + \frac{1}{2}\,\nabla(\underline{w} \cdot \underline{w})\right) + \frac{\varrho_s}{\varrho_n}\,\nabla T = \frac{\varrho_s}{\varrho\varrho_n}\,(\eta\varDelta\underline{w}) - \frac{1}{3}\,\varkappa BL\underline{w}, \tag{2}$$

$$\frac{\partial T}{\partial t} + \frac{\varrho_s}{\varrho}\,\underline{w}\cdot\nabla T + \frac{Ts\varrho_s}{c_p\varrho}\,\nabla\cdot\underline{w} = \frac{k}{\varrho c_p}\,\varDelta T + \frac{1}{3}\,\frac{\varrho_s\varrho_n}{\varrho^2 c_p}\,\varkappa BLw^2, \tag{3}$$

$$\frac{\partial L}{\partial t} + \nabla\cdot(\underline{v}_L L) = \chi_1\,\frac{B}{2}\,\frac{\varrho_n}{\varrho}\,|\underline{w}|\,L^{3/2} - \chi_2\varkappa L^2, \tag{4}$$

where η is the shear viscosity coefficient, c_p the specific heat, k the coefficient of thermal conductivity, and χ_1 and χ_2 are the VLD growth and decay terms given by Vinen (1957). The value of the tangle drift velocity $\underline{v}_L$ is not well known in the literature and we will take it proportional to $\underline{w}$.

Supplying, during a period t_H, a heat flux Q at the heated surface generates there the entropy s_0 which is entrained with the normal velocity $\underline{v}_n$. According to (1) this velocity can be expressed by the counterflow velocity $\underline{v}_n = \varrho_s\underline{w}/\varrho$. Hence the boundary condition at the heated surface is given by the relation

$$w_0(x = x_0, t) = \frac{Q}{\varrho_s s_0 T_0} \quad \text{for} \quad 0 \leq t \leq t_H, \quad \text{and} \quad w_0(x = x_0, t) = 0 \quad \text{for} \quad t > t_H. \tag{5}$$

The form of the above set of equations, disregarding the small second order derivative terms on the rhs. of Eqs. (3) and (4), indicates that the flow field is described by a set of hyperbolic equations and hence is of a wave like character.

The above set of equations is conveniently non-dimensionalized with respect to a length of one cm and the second sound velocity $c_{20}^2 = \dfrac{\varrho_s}{\varrho_n}\,\dfrac{s_0{}^2 T_0}{c_{p0}}$, where T_0, s_0, and c_{p0} are the initial values of temperature, entropy and specific heat respectively at the bath temperature.

The temperature increase $T' = T - T_0$ is non-dimensionalized with respect to $s_0\varrho/c_{20}^2\varrho_n$. The densities of the components are non-dimensionalized with respect to the total density ϱ.

The set of Eqs. (3), (4), and (5) written in divergence form for the cylindrical case becomes

$$\frac{\partial}{\partial t'}\begin{bmatrix} w' \\ T' - \gamma\,\dfrac{w'^2}{2} \\ L' \end{bmatrix} + \frac{\partial}{\partial x'}\begin{bmatrix} T' + \dfrac{3}{2}\,\varrho_s{'}w'^2 \\ w' + \varrho_s{'}\,\dfrac{s_0}{c_p}\,T'w' \\ L'v_L{'} \end{bmatrix}$$

$$= \begin{bmatrix} \eta'\varDelta_j'w' - b_1 L'w' - \varrho_s{'}\,\dfrac{w'^2}{x} \\ k'\varDelta_j'T' + \gamma\eta'w'\varDelta_j'w' + (b_2 - \gamma b_1)\,L'w'^2 - \left(\dfrac{w'}{x'} + \varrho_s{'}\,\dfrac{s_0}{c_p}\,\dfrac{T'w'}{x'}\right) \\ \chi_1{'}w'L'^{3/2} - \chi_2{'}L'^2 - \dfrac{L'v_L{'}}{x'} \end{bmatrix}, \tag{6}$$

where $\gamma = \varrho_s{'}(1 - s_0/c_p)$ and the primes denote non-dimensional quantities.

Using the Richtmyer two-step version of the Lax-Wendroff procedure a general numerical method of solution of the set of non-homogenous partial differential equations (6) was obtained (see Noack and Fiszdon [6]).

4 Small amplitude temperature pulses

The first experiments made using low intensity heat pulses to avoid the complications due to non-linearities and vorticity effects have exhibited some interesting features of the temperature evolution illustrated in Fig. 1. The upper diagramm shows the simple wave trajectories subject to

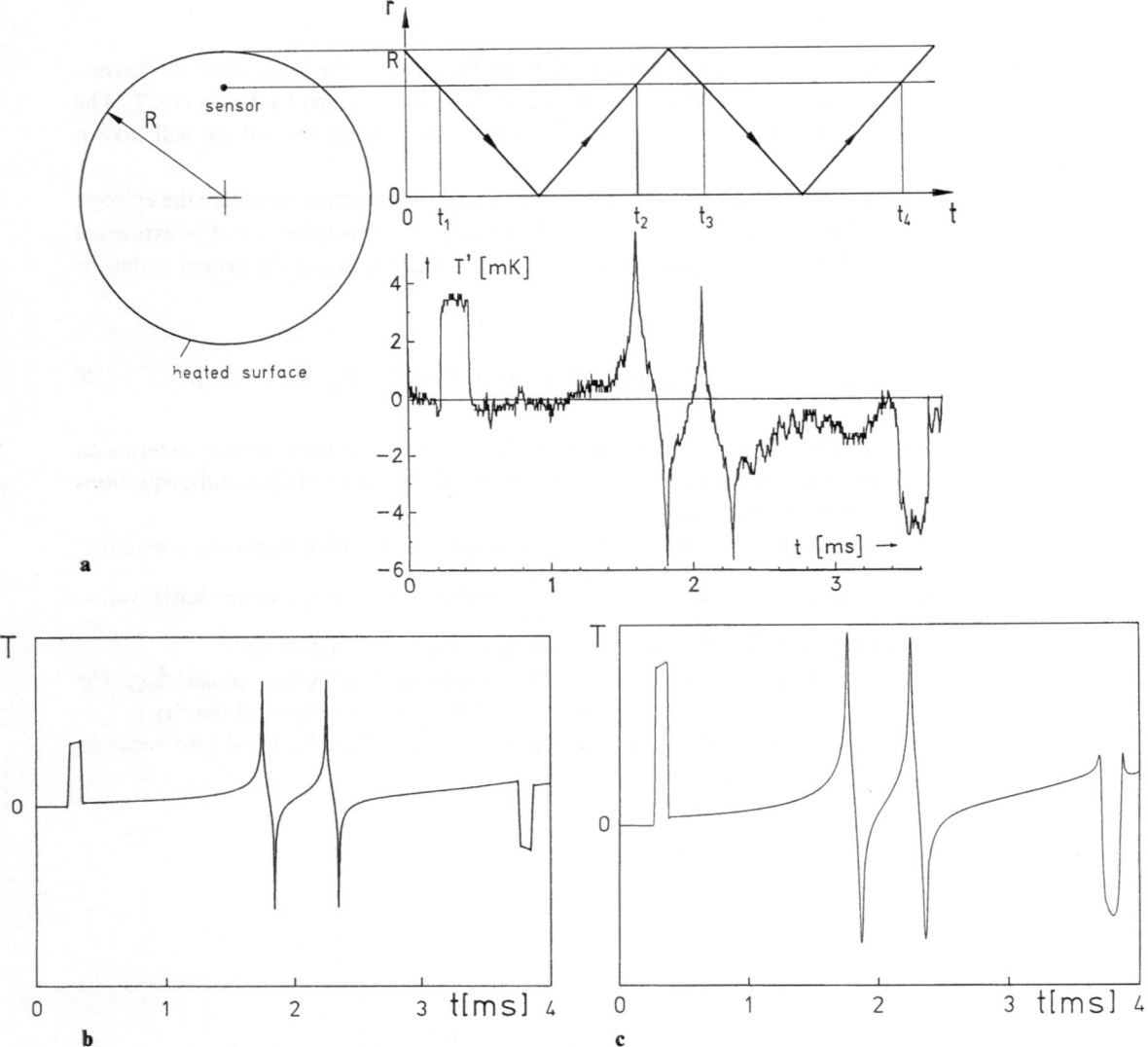

Fig. 1. Wave trajectory and temperature versus time curves after the release of a short rectangular heat pulse. **a** Time trace recorded experimentally by the sensor located at 0.45 cm = 1/4 · R ($Q = 1.0\,\text{W/cm}^2$, $t_H = 0.2\,\text{ms}$, $t_R = 2\,s$). **b** Corresponding analytical solution with the temperature in arbitrary units. **c** Corresponding numerical simulation

consecutive reflections at the axis and at the inner surface of the cylinder. At t_1 the first passage of the top hat pulse generated at the wall is recorded by the temperature sensor located at a distance of 1/4 of the cylinder radius R from the heated inner surface. The following signal received at the sensor at the time t_2 corresponds to the pulse reflected at the axis and has a "N-wave" like shape. This wave travels to the wall where it is reflected again and measured by the sensor at the time t_3, with the same "N-wave" like shape. The consecutive pulse recorded at the time t_4 by the sensor has recovered its top hat shape but inverted, i.e. with changed sign. A similar sequence, but with changed signs (not shown in Fig. 1) takes place during the next period which ends with a complete restoration of the original rectangular starting pulse. Hence the full repetition period corresponds to twice the number of reflections at the axis. This wave propagation and reflection process was confirmed and clarified by analysing the exact analytical solution of the linearized dissipationless vortex free problem to which the small amplitude case can be reduced. The set of Eqs. (6) in this approximation reduces to

$$\frac{\partial T}{\partial t} + V \cdot \underline{w} = 0, \quad \frac{\partial \underline{w}}{\partial t} + VT = 0, \tag{7}$$

and eliminating $\underline{w}$ the respond single second order wave equation is

$$\frac{\partial^2 T}{\partial t^2} - VT = 0, \tag{8}$$

with the boundary condition (5) at the inner cylinder surface taken as $x_0 = 1$ and $T(x, t < 0) = 0$. This case was solved analytically by Möhring and Fiszdon [4] using the Fourier integral method. They showed that the incoming rectangular pulse is reflected as a logarithmic singularity at the axis, reversed in sign at the reflection from the cylinder wall and the logarithmic singularity is then reflected at the axis restoring the rectangular pulse. Thereafter follows the second half of the whole period but with opposite sign.

The results of the analytical solution and numerical simulation showing the temperature variation with time at a selected distance from the cylinder axis are shown in Fig. 1b and c respectively for the same point as in Fig. 1a. In Fig. 2 are also shown for comparison the temperature versus time curves obtained experimentally and from the numerical solution of Eqs. (2) to (4) half way between the axis and the inner cylinder wall and close ($\varepsilon \cdot R$) to the cylinder axis. As can be seen, the qualitative agreement between the analytical, numerical, and experimental results is very satisfactory.

5 Medium amplitude temperature pulses

Pulses whose parameters, such as temperature amplitude and pulse duration, are large enough to produce flows involving superfluid vorticity but not as large as to provoke evaporation or even film boiling, are understood as medium pulses.

Experimentally only the temperature and vortex line density could be measured. The variation of the initial vorticity level L_0 was achieved in the experiments by varying the rest time t_R between consecutive heat pulses. The temperature and VLD signals shown in Figs. 3a and 5a were recorded after the dynamic equilibrium of the involved processes had settled, practically after four consecutive pulses (for details of this dynamic process see e.g. Fiszdon et al. [1]). Comparing the curves in Figs. 3a and 5a with those of Figs. 3b and 5b the influence of the

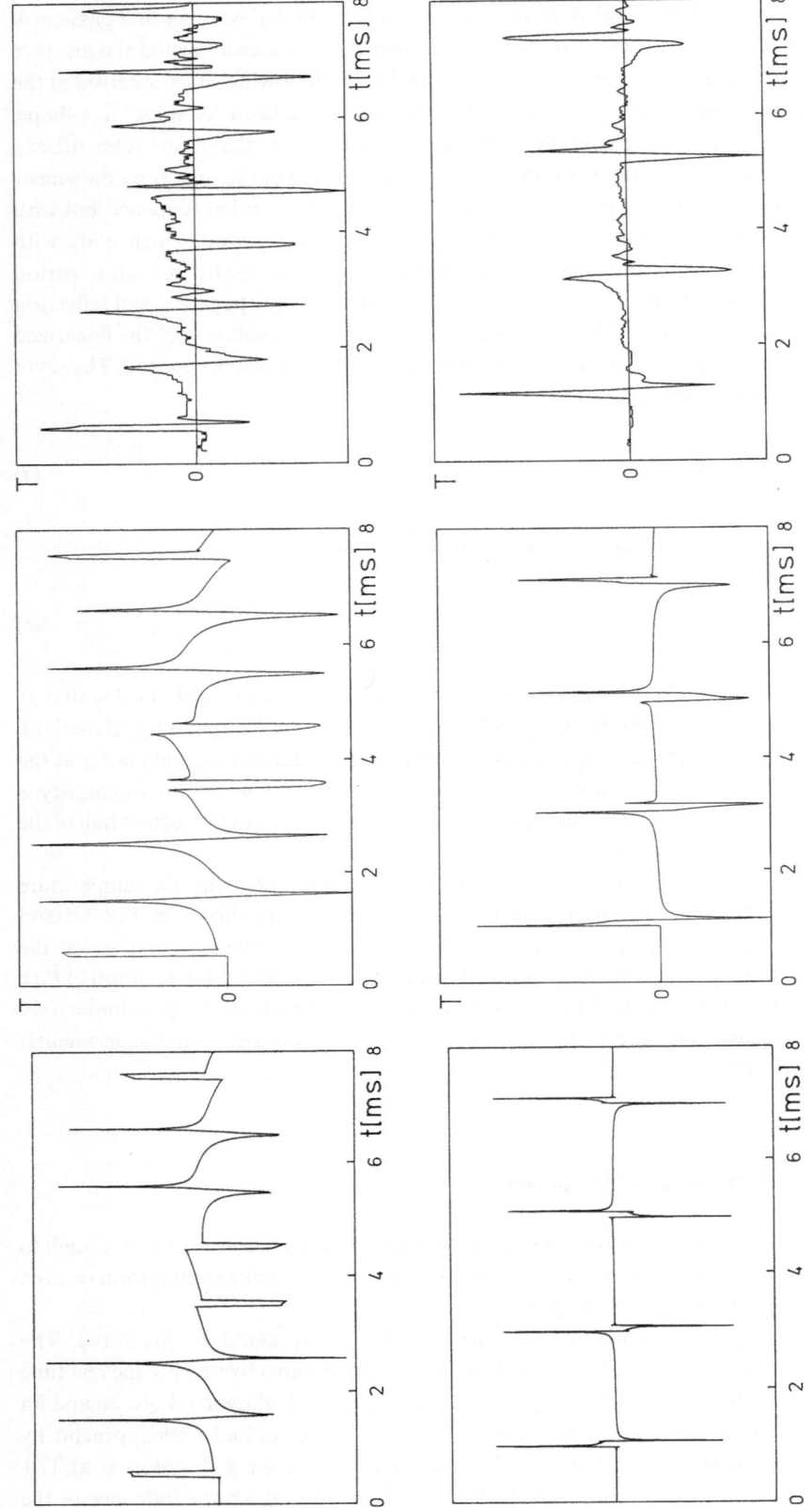

Fig. 2. Analytical (left column), experimental (middle) and numerical (right column) temparature evolutions at $1/2 \cdot R$ (R = radius of the cylinder) and at $\varepsilon \cdot R$ ($\varepsilon \ll 1$, which means very close to the cylinder axis)

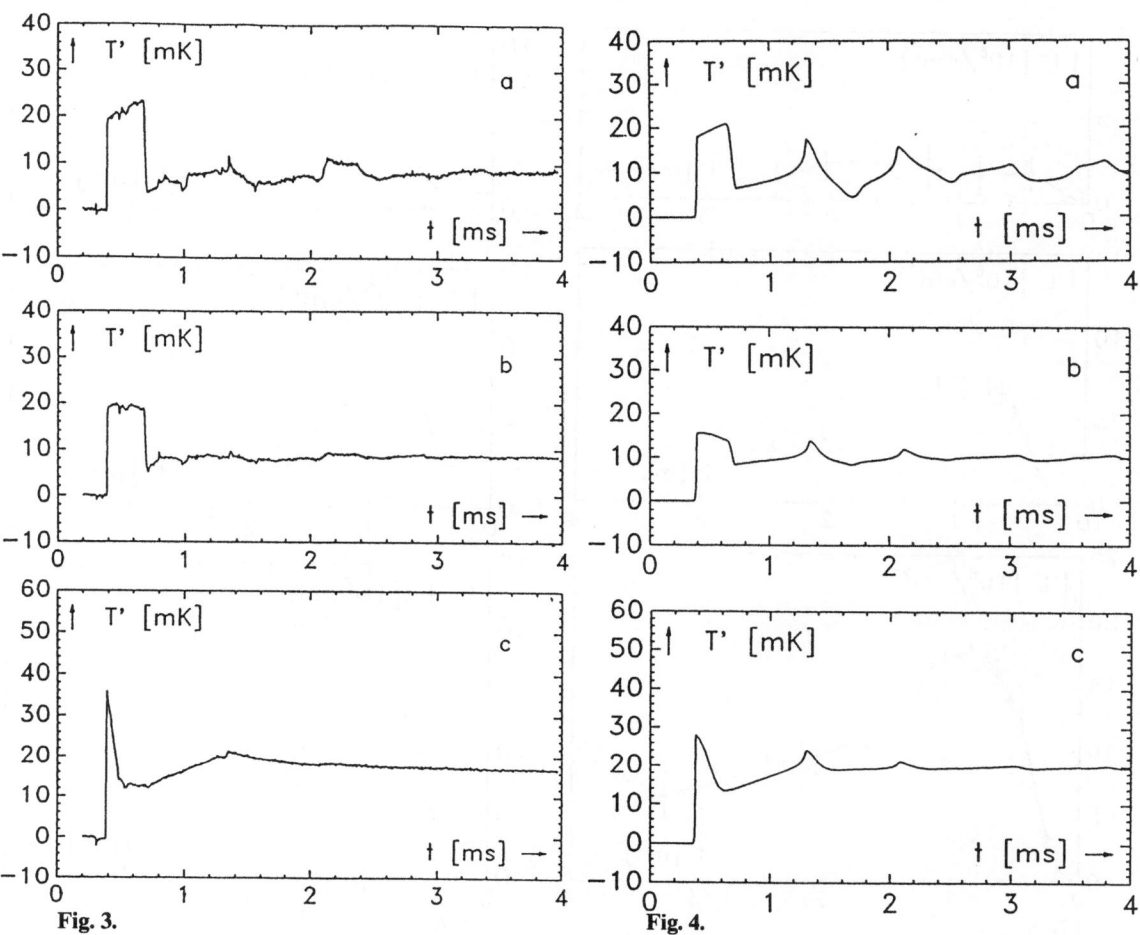

Fig. 3.   **Fig. 4.**

Fig. 3. Influence of the VLD on the evolution of initially rectangular heat pulses of 0.3 ms duration at a distance of 0.8 cm from the heated surface. **a** $Q = 4$ W/cm^2 and $t_R = 2$ s; **b** same as **a** but at a smaller rest time of $t_R = 0.25$ s; **c** same as **b** but at twice the heat input $Q = 8$ W/cm^2. **Fig. 4.** Numerical simulation of the temperature evolution with the same as parameters in Fig. 3 except that the different rest times t_R were simulated by different initial vortex line densities L_0. **a** $Q = 4$ W/cm^2 and $L_0 = 10^6$/cm^2; **b** same as **a** but for $L_0 = 3 \cdot 10^6$/cm^2; **c** same as **b** but at twice the heat input $Q = 8$ W/cm^2

vorticity level on the temperature and vortex line density evolution can be noticed. The results of the numerical simulations of the corresponding cases, fitting the different rest times of the experimental curves by using different initial vortex line densities L_0, can be seen in Figs. 4a, 4b, 6a, and 6b. The qualitative agreement is satisfactory although the damping of the pulses, particular after consecutive reflections at the center and at the wall, seems to be much weaker in the numerical cases where these reflected signals are much more pronounced compared with the experimental ones.

The influence of an increase of the heat input observed experimentally can be seen in Figs. 3c and 5c and the corresponding numerical calculations in Figs. 4c and 6c. It can be noticed that increasing the VLD at the lower heat input (which means decrsing the rest time from Fig. 3a to Fig. 3b) changed the inclination of the pulse plateau from a positive slope due to the convergent cylindrical symmetry to a negative one showing a strong influence of the prevailing VLD. Increasing the heat input while retaining the high initial vorticity level (from Fig. 3b to Fig. 3c) leads to a strong damping of the pulse directly behind the shock front, so that from the initially rectangular pulse only a small "delta"-shaped peak remains.

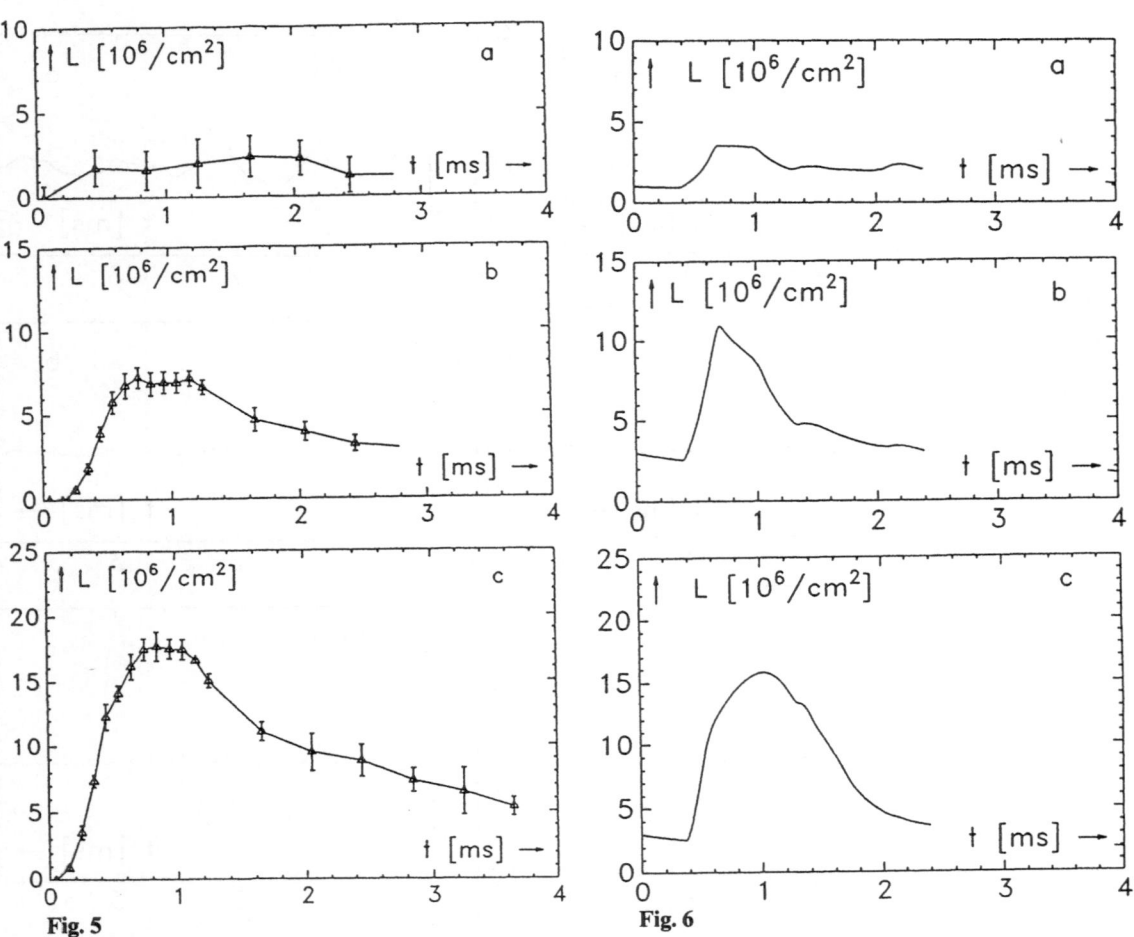

Fig. 5

Fig. 6

Fig. 5. Experimentally measured evolutions of the VLD for the same parameters as in Fig. 3
Fig. 6. Numerical simulations of the VLD evolution for the same parameters as in Fig. 4

Results of numerical calculations according to Eqs. (2) to (5) are shown in Fig. 7 illustrating typical evolutions in time of all three flow parameters: counterflow velocity $w = |\underline{w}|$, temperature T and vortex line density L. The influence of an increase in the initial VLD L_0 from $1 \cdot 10^6/cm^2$ (a) to $3 \cdot 10^6/cm^2$ (b) on the evolution of w (upper), T (middle), and L (lower) at distances of 0.1, 0.4, 0.8 and 1.6 cm from the heated surface is illustrated in Fig. 7. These numerical simulations show clearly the strong influence of the VLD on the temporal evolution of the velocity and the temperature. While at the lower initial VLD of $L_0 = 10^6/cm^2$ in Fig. 7a the counterflow as well as the temperature show the predominant focusing effect when approaching the axis of the cavity, they become strongly damped as illustrated in Fig. 7b only by increasing the initial vorticity level by about 30%.

An increase of the heating time t_H and heat intensities Q produces appreciable temperature overshoots like in the other geometric cases described in the previously mentioned paper by Fiszdon et al. [1]. The experimental measurements and numerical simulations for a distance of 0.8 cm (≈ 0.44 cylinder radius) from the heated surface illustrating these overshoots are shown in Fig. 8 for different heating times t_H while keeping the heat flux constant at $Q = 10\,W/cm^2$ and in Fig. 9 for different heat fluxes while keeping constant the pulse duration $t_H = 0.7$ ms. The simulated curves were obtained assuming an initial VLD of $L_0 = 6 \cdot 10^6/cm^2$ and a vortex tangle drift velocity of $v_L \approx 0.35 \cdot w$ to

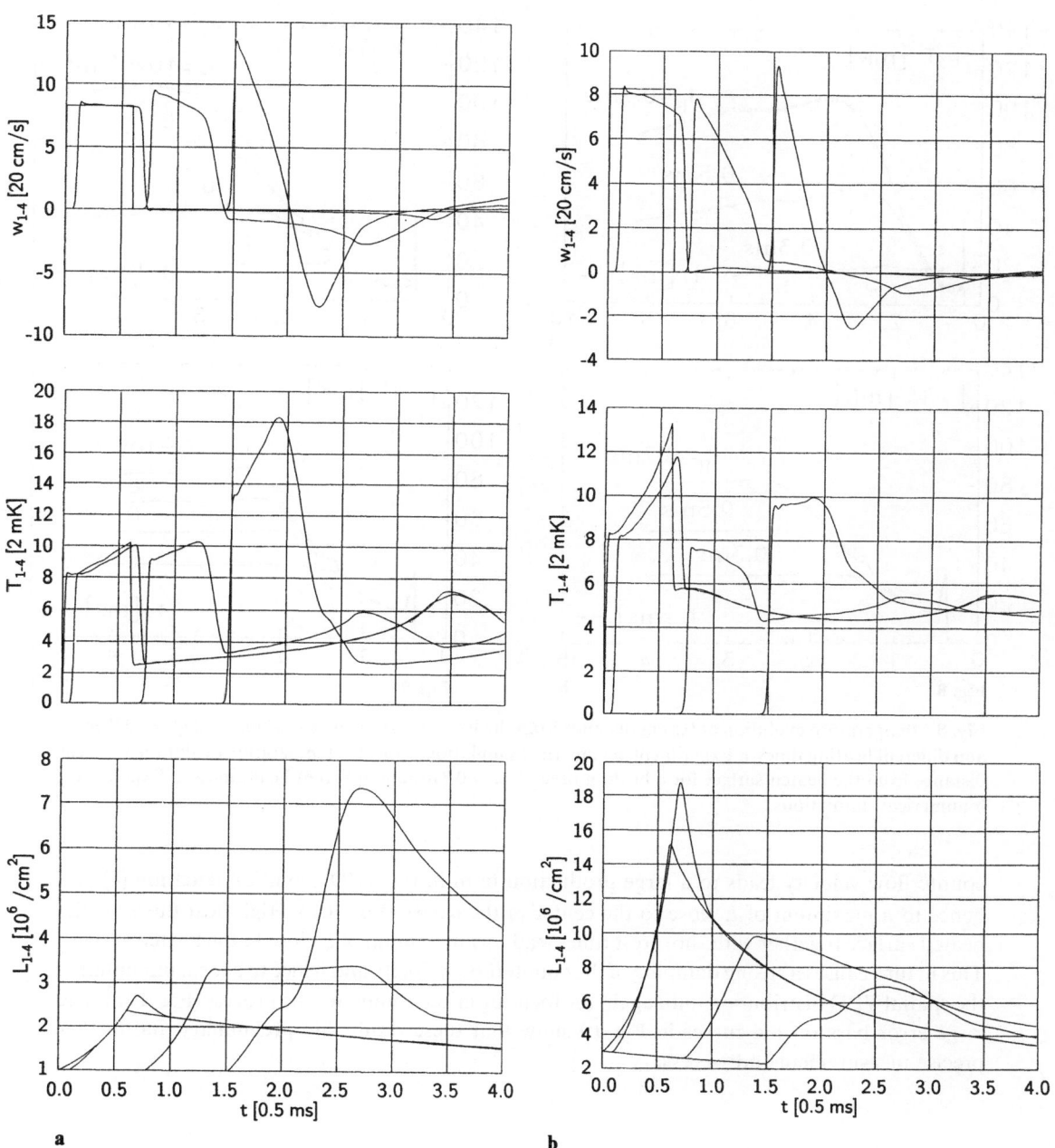

Fig. 7. Numerical simulations of the counterflow velocity w, the temperature T and the VLD L at distances of 0.1, 0.4, 0.8, and 1.6 cm from the heated surface for a heat input of $Q = 4\,\text{W/cm}^2$ and an initial VLD of $L_0 = 10^6/\text{cm}^2$ (**a**) and $L_0 = 3 \cdot 10^6/\text{cm}^2$ (**b**)

obtain the best fit. Although the general character of the time variations is quite similar the amplitudes are about 30% to low.

The competition between the geometric constraints imposed by the requirements of cylindrical convergence and the vorticity impeding the heat transport away from the heated surface by reducing the velocity of the normal component are clearly visible in Fig. 10. As can be noticed at smaller heat fluxes and longer rest times, hence smaller initial VLD, the focusing of the

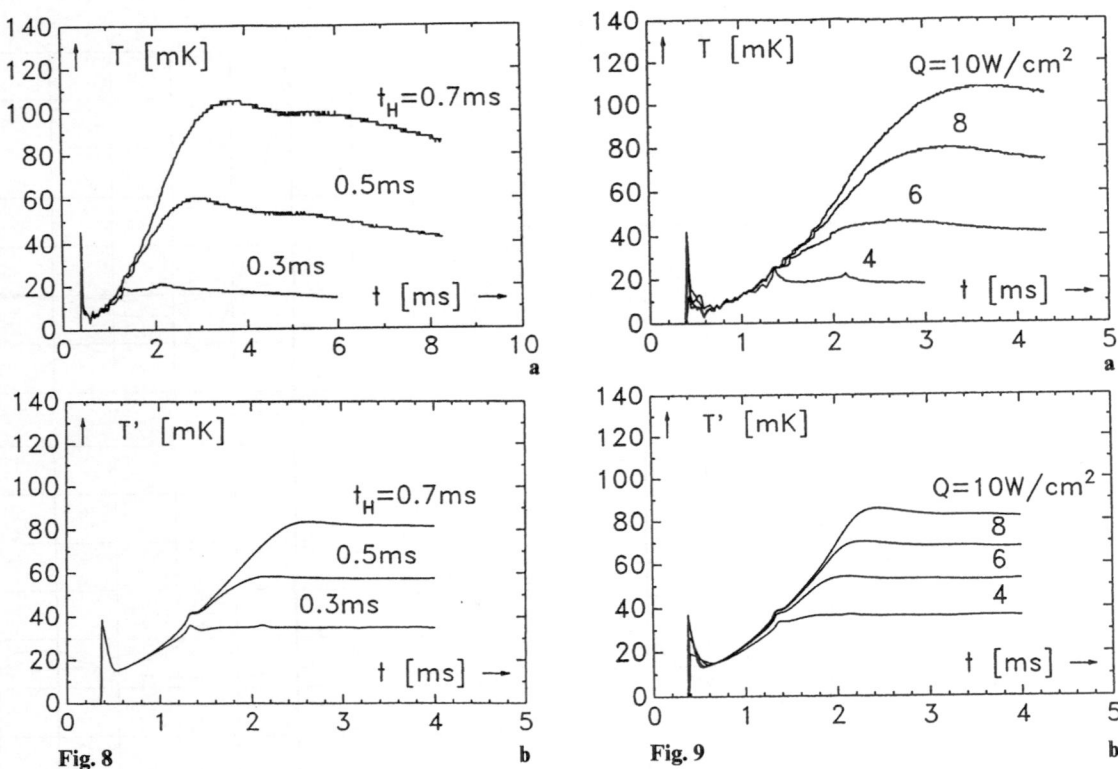

Fig. 8. Temperature evolution at 0.8 cm distance from the heated surface for a heat input of $Q = 10 \, \text{W/cm}^2$ and different heating times. **a** Experiment, **b** numerical simulations. **Fig. 9.** Temperature evolution at 0.8 cm distance from the heated surface for a heating time of $t_H = 0.7$ ms and different heat inputs. **a** Experiment, **b** numerical simulations

counterflow velocity leads to a large production term in the VLD evolution equation (4) and hence to a maximum of L close to the center of the cavity (Fig. 10b). High heat fluxes at the heated surface together with short rest times lead to a maximum of L close to the heated surface. This in turn causes a strong damping of the counterflow velocity and hence to a local minimum in the spatial VLD distribution until again the focusing effect dominates the process (Fig. 10a). The large error bars on the curves in Fig. 10 show that these results are preliminary and further precise measurements are necessary.

6 Concluding remarks

It should be noted that although the height of the cylindrical cavity was only 1/3 of its radius the comparison of the experimental results with the theoretical and numerical simulation results shows that, at least within the investigated parameter range, the influence of the boundary layer formed on both endplates is rather small. This is confirmed by some preliminary numerical calculations. Anyhow the qualitative picture remains unaltered except at temperatures very close to the superfluid transition temperature T_λ.

May be some theoretical solutions concerning problems of cylindrical symmetry of interest for normal fluids could be checked in superfluid helium. At larger heat inputs and longer heating times the effect of superfluid quantized vorticity becomes predominant but its relation to normal

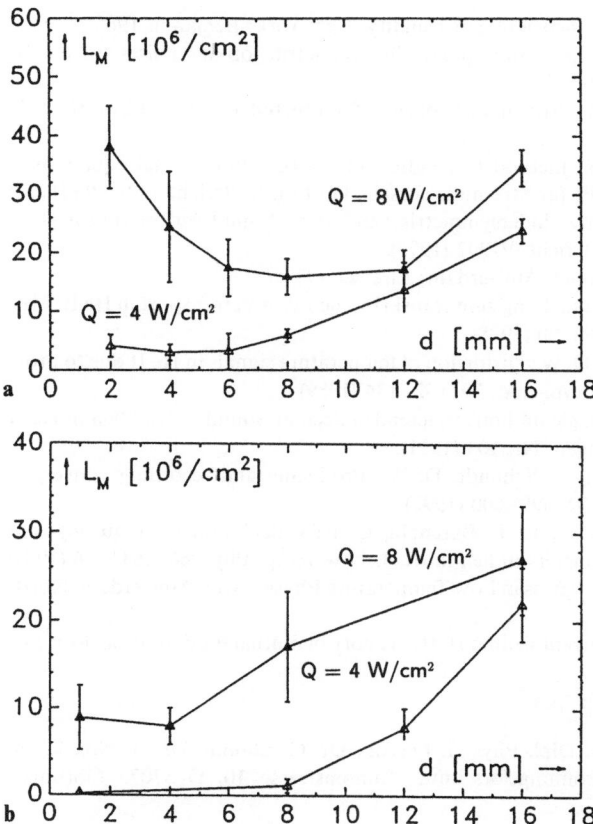

Fig. 10. Radial distribution of the maximum vortex line density L_M occurring during a heating time of $t_H = 0.3\,\text{ms}$, for two different heat inputs and rest times of $t_R = 0.25\,\text{s}$ **a** and $t_R = 20\,\text{s}$ **b**

turbulence is unknown. Another important point is that only its evolution is described by the phenomenological Vinen equation (4) but the production processes for quantum vorticity are still not known. Hence the problem of quantum turbulence and its relation with respect to normal turbulence remains still in many respects open.

However, the present work has shown that there is a strong interaction between geometrical constraints and the physical processes involved in superfluid flow, particularly on the evolution of the VLD and that by carefully matching the time and space parameters a closer insight into the flow evolution can be obtained.

Acknowledgements

The authors would like to express their thanks to professor Dr. E.-A. Müller and Dr. D. W. Schmidt for their advice, interest and support of this research project. We would also like to express our special thanks to Dr. M. v. Schwerdtner, H.-U. Vogel and F. Bielert for the valuable discussions and constant readiness to advice and help during different stages of this work. Thanks are also due to J. Krüger for careful typing and preparing the manuscript.

References

[1] Fiszdon, W., Schwerdtner, M. v., Stamm, G., Poppe, W.: Temperature overshoot due to quantum turbulence during the evolution of moderate heat pulses in He II. J. Fluid Mech. **212**, 663–684 (1990).
[2] Gorter, C. J., Mellink, J. H.: On the irreversible processes in liquid helium II. Physica **XV**, 285–304 (1949).

[3] Khalatnikov, I. M.: Introduction to the theory of superfluidity. New York: Benjamin 1965.

[4] Möhring, W., Fiszdon, W.: Converging axi- and spherically-symmetric top-hat pulses (in He II), accepted for publication in J. Sound Vibr.

[5] Nemirovskii, S. K., Lebedev, V. V.: The hydrodynamics of superfluid turbulence. Sov. Phys. JETP **57** (1983), 1009–1016.

[6] Noack, B. R., Fiszdon, W.: A numerical method of solution of the one-dimensional equations of turbulent second sound He II flows. MPI für Strömungsforsch., Göttingen, Bericht 102/1990 (1990).

[7] Olszok, T.: Experimente zur konvergenten zylindersymmetrischen Second-Sound-Ausbreitung in He II. MPI für Strömungsforsch., Göttingen, Bericht 9/1992 (1992).

[8] Putterman, S. J.: Superfluid hydrodynamics. Amsterdam: Elsevier 1974.

[9] Schwerdtner, M. v.: Experimentelle Untersuchung zum transkritischen Wärmetransport in He II. Mitt. MPI für Strömungsforsch., Göttingen, Nr. 90 (1988).

[10] Schwerdtner, M. v., Poppe, W., Schmidt, D. W.: Distortion of temperature signals in He II due to probe geometry, and a new improved probe. Cryogenics **29**, 132–134 (1989).

[11] Stamm, G.: Experimentelle Untersuchungen an konvergierenden Second-Sound-Stoßwellen in He II. Mitt. MPI für Strömungsforsch., Göttingen, Nr. 103 (1991).

[12] Stamm, G., Olszok, T., Schwerdtner, M. v., Schmidt, D. W.: Producing and recording converging second-sound shock waves, Cryogenics **32**, 598–600 (1992).

[13] Swanson, C. E., Wagner, W. T., Donnelly, R. J., Barenghi, C. F.: Calculation of frequency- and velocity-dependent mutual friction parameters in helium II, J. Low Temp. Phys. **66**, 263–276 (1987).

[14] Tough, J. T.: Superfluid Turbulence. In: Progress in Low Temperature Physics, vd **8**. Amsterdam: North-Holland 1982.

[15] Vinen, W. F.: Mutual friction in a heat current helium II. III. Theory of mutual friction. Proc. Roy. Soc. London Ser. A **242**, 493–515 (1957).

Authors' address: Professor Dr. W. Fiszdon, Dipl.-Phys. T. Olszok, Dr. G. Stamm, Dr. B. Noack, and Dr. J. Piechna, Max-Planck-Institut für Strömungsforschung, Bunsenstrasse 10, D-37073 Göttingen, Federal Republic of Germany

Acta Mechanica (1994) [Suppl] 4: 325—334

Self-similar solutions for two-dimensional slender-channel flows

K. Gersten and **B. Rocklage,** Bochum, Federal Republic of Germany

Summary. Laminar flows having self-similar velocity distributions in slender convergent or divergent channels are special cases of the well-known Jeffery-Hamel flows for the double limit that the Reynolds number tends to infinity and the angle of divergence (or convergence) tends to zero such that the product of Reynolds number and the divergence angle, called the slender-channel parameter, is kept finite. It is shown that equivalent self-similar solutions exist also for turbulent flows. If the slender-channel parameter tends to minus infinity, the solutions reduce to the well-known boundary-layer solutions for sink flows.

1 Introduction

Laminar flows in convergent or divergent two-dimensional channels are fully developed when the velocity distributions are self-similar. Figure 1 shows the geometry under consideration. Stars as superscripts refer to dimensional values. With the self-similarity condition

$$\frac{u^*(r^*, \varphi)}{u^*_{\max}(r^*)} =: F(\eta) \qquad \eta := \frac{\varphi}{\Phi} \tag{1}$$

the full Navier-Stokes equations reduce to the ordinary differential equation, see [1], page 113,

$$F''' + 2\Phi \, \mathrm{Re} \, FF' + 4\Phi^2 F' = 0 \tag{2}$$

with the boundary conditions

$$\eta = 0: \quad F = 1, \quad F' = 0$$

$$\eta = 1: \quad F = 0. \tag{3}$$

The Reynolds number is defined by

$$\mathrm{Re} := \frac{u^*_{\max} r^* \Phi}{v^*}. \tag{4}$$

It is always positive, whether divergent channels (diffusers, $\Phi > 0$, $u^*_{\max} > 0$) or convergent channels (nozzles, $\Phi < 0$, $u^*_{\max} < 0$) are considered.

At high Reynolds numbers two different limiting processes have to be distinguished:

(1) Boundary-Layer Theory: $\mathrm{Re} \to \infty$, $\Phi = O(1)$
   This is a singular perturbation problem. The naive solution, see [1], page 251, $F'(\eta) = 0$ or $F(\eta) = 1$ is not valid near the walls. Near the walls Eq. (2) reduces to

$$\ddot{F} - 2F\dot{F} = 0 \tag{5}$$

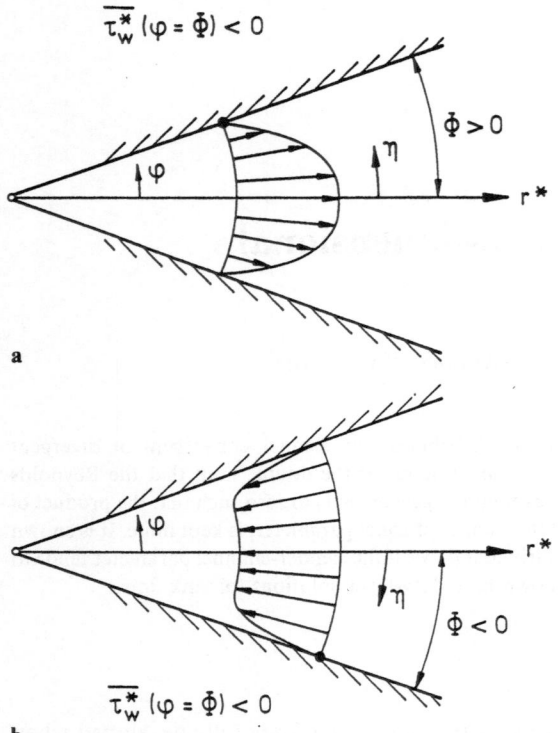

$\overline{\tau_w^*} \, (\varphi = \Phi) < 0$

$\Phi > 0$

φ

η

r^*

a

φ

r^*

η

$\Phi < 0$

$\overline{\tau_w^*} \, (\varphi = \Phi) < 0$

b

Fig. 1. Geometry of slender channels. **a** Divergent channels (diffusers, $\Phi > 0, u_{max}^* > 0$),   **b** convergent channels (nozzles, $\Phi < 0, u_{max}^* < 0$)

with

$$\xi = 0: \qquad F = 0$$

$$\xi \to \infty: \qquad F = 1, \qquad \dot{F} = 0, \tag{6}$$

where dots refer to differentiation with respect to the boundary-layer coordinate

$$\xi := (1 - \eta) \, \sqrt{-\Phi \, \mathrm{Re}}. \tag{7}$$

The solution of Eqs. (5) and (6) is

$$F(\xi) = 3 \tanh^2 \left(\frac{\xi}{\sqrt{2}} + \operatorname{arctanh} \sqrt{\frac{2}{3}} \right) - 2. \tag{8}$$

Only for convergent channels ($\Phi < 0$) such solutions are possible, where the flow field is divided into the inviscid core region and the two boundary layers at the walls.

(2) Slender-Channel Theory: $\mathrm{Re} \to \infty$, $\Phi \to 0$, $\Phi \, \mathrm{Re} = O(1)$
Since in these cases $\Phi \to 0$, Eq. (2) reduces to

$$F''' + 2\Phi \, \mathrm{Re} \, FF' = 0 \tag{9}$$

with unchanged boundary conditions, Eq. (3). This so-called slender-channel theory is valid for nozzles as well as for diffusers. It is worth mentioning that for slender channels a coupling between the Reynolds number and the geometry exists characterized by the slender-channel parameter Φ Re. Figure 2 shows the value $[F'(1)]^2$, representing the skin friction at the wall, as function of the parameter Φ Re. For Φ Re $\to -\infty$ the slender-channel theory reduces to the boundary-layer solution Eqs. (5) to (8).

The purpose of this investigation is to show that those two distinguished limiting solutions for high Reynolds numbers exist also in turbulent flows.

2 Boundary-layer theory for turbulent convergent channel flows

The velocity of the inviscid solution for the core region is $U^*(r^*) = u^*_{\max}(r^*) \sim -1/r^*$. This special distribution of the "outer" flow leads to a so-called equilibrium boundary layer, see [1], page 641 or [5], p. 189. In this case the well-known Clauser equilibrium parameter is $\beta = -(dp^*/dr^*)\,\delta_1^*/\overline{\tau_w^*} = -0.5$ (δ_1^* is the displacement thickness). Whereas in the laminar case $1/\sqrt{\text{Re}}$ was used as perturbation parameter, in turbulent flows the perturbation parameter is chosen as

$$\gamma := \frac{u_\tau^*(r^*)}{U^*(r^*)}, \tag{10}$$

where

$$u_\tau^* := -\sqrt{\frac{|\overline{\tau_w^*(r^*)}|}{\varrho^*}} \tag{11}$$

is the local skin-friction velocity. It can be shown a posteriori that the positive parameter γ tends to zero for the limit of high Reynolds numbers.

Using the perturbation equation for the velocity distribution

$$\frac{u^*(r^*, \varphi)}{U^*(r^*)} = 1 - \gamma \dot{F}(\xi) \tag{12}$$

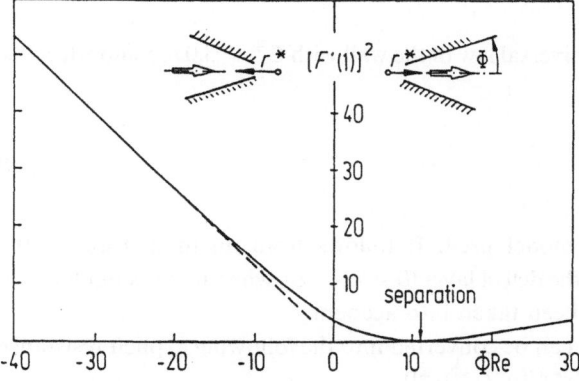

Fig. 2. Skin-friction coefficient c_f for laminar slender channels as function of the slender-channel parameter Φ Re. $c_f := 2|\tau_w^*|/(\varrho^* u_{\max}^{*2}) = 2F'(1)/\text{Re}$; $\;----\;$ $[F'(1)]^2 = -4\Phi\,\text{Re}/3$ (Boundary-layer theory); $\;c_f = 0$ for Φ Re $= 10.3$

and the equation

$$\frac{\tau_t^*(r^*, \varphi)}{\tau_w^*(r^*)} = S(\xi) \tag{13}$$

for the turbulent shear stress distribution, where

$$\xi := \frac{-r^*(\Phi - \varphi)}{\Delta^*(r^*)} = -\frac{r^*\Phi}{\Delta^*(r^*)}(1 - \eta) \tag{14}$$

is the boundary-layer coordinate, leads for $\gamma \to 0$ to the differential equation

$$\dot{F} + \dot{S} = 0 \tag{15}$$

with the boundary conditions

$$\xi = 0: \quad \ddot{F} = -\frac{1}{\varkappa \xi}, \quad S = 1$$

$$\xi \to \infty: \quad \dot{F} = 0, \quad S = 0. \tag{16}$$

Dots refer to differentiation with respect to ξ. The value $\varkappa$ is the well-known Karman constant $\varkappa = 0{,}41$. Further correlation between $F(\xi)$ and $S(\xi)$ must result from an appropriate turbulence model and will thus close the system of equations. The correlation between the perturbation parameter γ, representing the skin friction, and the Reynolds number $\text{Re}_r = |u_{\max}^*| r^*/v^*$ reads

$$\frac{1}{\gamma} = \frac{1}{\varkappa} \ln(\text{Re}_r \gamma^2) + C^+ + \tilde{C} - \frac{1}{\varkappa} \ln 2. \tag{17}$$

It is worth mentioning that this particular equilibrium boundary layer is additionally also self-similar, which means that the scale $\Delta^*(r^*)$ for the boundary-layer thickness is exactly proportional to r^*. The analysis, see [1], p. 639 or [5], p. 189, leads to

$$\Delta^*(r^*) = \frac{\gamma}{2} r^*. \tag{18}$$

The value C^+ in Eq. (17) stems from the universal law of the wall with $C^+ = 5{,}0$ for smooth walls. The value $\tilde{C}$ in Eq. (17) is defined by

$$\tilde{C} = \lim_{\xi \to 0} \left[\dot{F}(\xi) + \frac{1}{\varkappa} \ln \xi \right] \tag{19}$$

and hence depends on the turbulence model used. It follows from an integration of the dimensionless velocity gradient $\ddot{F}(\xi)$ over the defect layer $(0 < \xi < \infty)$, whereby the singularity of $\ddot{F}(\xi)$ for $\xi \to 0$ according to Eq. (16) has been taken into account.

The implicit relation $\gamma(\text{Re}_r)$ of Eq. (17) can be converted into the following explicit resistance law for the friction coefficient $c_f = 2|\overline{\tau_w^*}(r^*)|/(\varrho^* U^{*2}(r^*))$

$$c_f = 2 \left[\frac{\varkappa}{\ln \text{Re}_r} G(\Lambda; D) \right]^2. \tag{20}$$

The universal function $G(\Lambda; D)$ is defined by the implicit equation

$$\frac{\Lambda}{G} + 2 \ln \frac{\Lambda}{G} - D = \Lambda, \tag{21}$$

from which follows that $G(\Lambda; D) \to 1$ for $\Lambda \to \infty$, see [1], page 782. Here we have $\Lambda = \ln \mathrm{Re}_r$ and $D = 2 \ln \varkappa + \varkappa(C^+ + \tilde{C}) - \ln 2$.

3 Slender-channel theory for fully developed turbulent flows

Since the angle Φ of divergence is small and the viscosity effects in the core region can be neglected at high Reynolds numbers the radial momentum equation in polar coordinates reduces to

$$\varrho^* u^* \frac{\partial u^*}{\partial r^*} = -\frac{dp^*}{dr^*} + \frac{1}{r^*} \frac{\partial \tau_t^*}{\partial \varphi}. \tag{22}$$

The inviscid flow satisfies the equation

$$\varrho^* U_0^* \frac{dU_0^*}{dr^*} = -\frac{dp_0^*}{dr^*} \tag{23}$$

with the solution

$$U_0^* = \mathrm{sign}\left(\frac{dp_0^*}{dr^*}\right) \sqrt{\frac{r^{*3}}{\varrho^*} \left|\frac{dp_0^*}{dr^*}\right|} \frac{1}{r^*}. \tag{24}$$

This solution will be disturbed by an asymptotic expansion with the expansion parameter $\gamma := u_\tau^*(r^*)/U_0^*(r^*)$.

To keep the expansion parameter γ positive not only for diffusers $\left(\Phi > 0, U_0^*(r^*) > 0\right)$ but also for nozzles $\left(\Phi < 0, U_0^*(r^*) < 0\right)$, the friction velocity $u_\tau^*(r^*)$ is defined as

$$u_\tau^*(r^*) := \mathrm{sign}\,(\Phi) \sqrt{\frac{-\tau_w^*}{\varrho^*}},$$

where $\tau_w^*\,(\varphi = \Phi)$ is always a negative number.

The dimensionless values $F'(\eta)$, $S(\eta)$ and P are introduced as

$$\frac{u^*(r^*, \varphi)}{U_0^*(r^*)} = 1 - \gamma F'(\eta), \tag{25}$$

$$\frac{\tau_t^*(r^*, \varphi)}{\varrho^* U_0^{*2}} = \gamma^2 S(\eta), \tag{26}$$

$$\frac{dp^*}{dr^*} = \frac{dp_0^*}{dr^*} + \gamma^2 \frac{\varrho^* U_0^{*2}}{r^* \Phi} P, \tag{27}$$

where $\eta = \varphi/\Phi$ varies between 0 and 1, if the symmetry condition is used. The values $F'(\eta)$, $S(\eta)$ and P represent the velocity, shear stress, and pressure gradient of the perturbation describing the deviation from the inviscid solution.

As in the laminar case, the slender-channel theory for fully developed turbulent flows is a two-parameter perturbation problem. In the turbulent case the two perturbation parameters are γ, representing the Reynolds number effect (Re $\to \infty$ for $\gamma \to 0$), and Φ.

The limiting process is characterized by $\gamma \to 0$ and $\Phi \to 0$, but such that the new similarity parameter

$$\alpha = \frac{\Phi}{\gamma} = O(1) \tag{28}$$

is kept constant. For this so-called distinguished limit the momentum equation leads to the following differential equation, when Eqs. (22) to (27) are applied,

$$-2\alpha F'(\eta) = S'(\eta) + P \tag{29}$$

with the boundary conditions:

$$\eta = 0: \quad F' = 0, \quad F'' = 0, \quad S = 0$$
$$\eta \to 1: \quad F'' = \frac{1}{\varkappa(1 - \eta)}, \quad S = 1. \tag{30}$$

Integration of Eq. (29) over the half cross-section yields

$$P = -1 - 2\alpha \lim_{\eta \to 1} \int_0^\eta F'(\eta)\, d\eta. \tag{31}$$

In order to close the system of equations for the functions $F'(\eta)$ and $S(\eta)$, an additional equation resulting from a turbulence model is needed.

The solutions depend on the similarity parameter α. The special case $\alpha = 0$ corresponds to the well-known solution for the two-dimensional channel of constant cross-section. Solutions for positive as well as for negative values of α are possible.

The limit $\alpha \to +\infty$ is supposed to represent cases where the wall shear stress is equal to zero. However, these cases cannot be covered by this analysis, since a non-zero wall shear stress has been assumed. In the limit $\alpha \to -\infty$, the slender-channel solution tends to the already mentioned boundary-layer solution.

The skin-friction formula for slender channels follows from matching the solutions of the core region with the wall regions and reads

$$\frac{1}{\gamma} = \frac{1}{\varkappa} \ln (\mathrm{Re}\, \gamma) + C^+ + \bar{C}, \tag{32}$$

where

$$\mathrm{Re} = \frac{U_0^* r^* \Phi}{\nu^*} > 0 \tag{33}$$

and

$$\bar{C}(\alpha) = \lim_{\eta \to 1} \left[F'(\eta) + \frac{1}{\varkappa} \ln (1 - \eta) \right]. \tag{34}$$

The volume flow rate can be determined by integrating the velocity over the cross-section. It follows:

$$\sqrt{\frac{2}{c_{fm}}} = \frac{u_m^*}{u_\tau^*} = \lim_{\eta \to 1} \int_0^\eta \frac{u^*(\eta)}{u_\tau^*} \, d\eta = \frac{1}{\gamma} + \bar{C}, \tag{35}$$

where

$$\bar{\bar{C}}(\alpha) = -\lim_{\eta \to 1} \int_0^\eta F'(\eta) \, d\eta = (P + 1)/(2\alpha). \tag{36}$$

Combining Eqs. (32) and (35) leads after inversion to the explicit skin-friction formula

$$c_{fm} = \frac{2|\overline{\tau_w^*}|}{\varrho^* u_m^{*2}} = 2 \left[\frac{\varkappa}{\ln \mathrm{Re}_D} G(\Lambda; D) \right]^2 \tag{37}$$

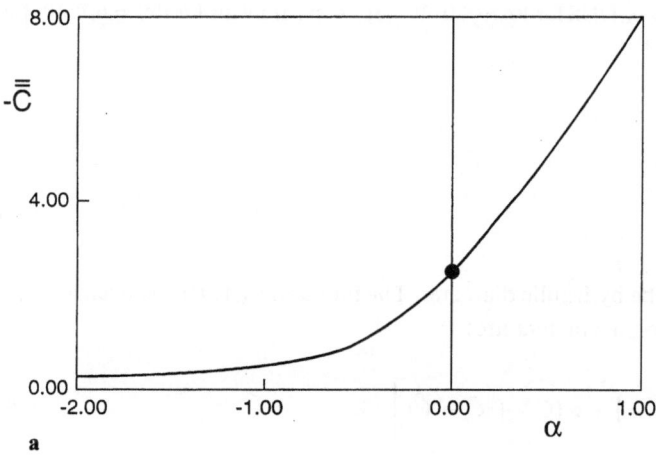

a

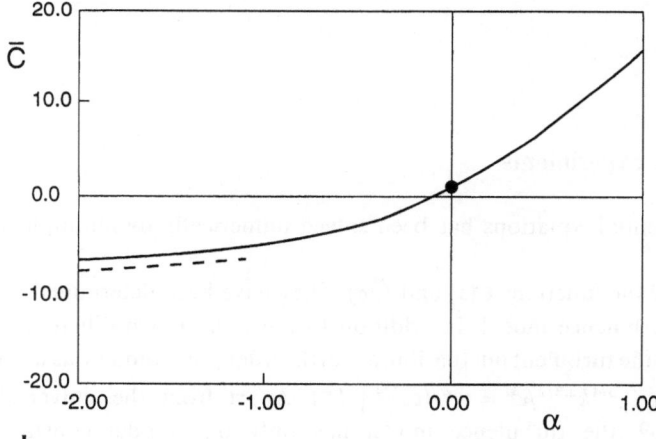

b

Fig. 3. Constants $\bar{C}$ (b) and $\bar{\bar{C}}$ (a) of the skin-friction formulas as function of the similarity parameter α

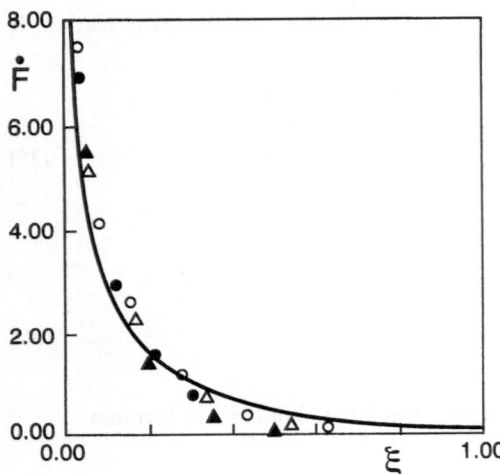

Fig. 4. Distribution of the velocity defect in a sink flow. Comparison between theory, direct numerical simulations (DNS) by Spalart [3], and experiments by Jones and Launder [2]. —— Theory according to Eq. (12), ○ DNS for $Re_r = 6.7 \cdot 10^5$ [3], △ DNS for $Re_r = 4.0 \cdot 10^5$ [3], ● experiments for $Re_r = 6.7 \cdot 10^5$ [2], ▲ experiments for $Re_r = 4.0 \cdot 10^5$ [2]

where

$$Re_D = \frac{4u_m^* r^* \Phi}{\nu^*} \tag{38}$$

is the Reynolds number based on the hydraulic diameter. The function $G(\Lambda; D)$ has been already defined in Eq. (21), where here the parameters are:

$$\Lambda = \ln(Re_D^2), \quad D(\alpha) = 2\left[\ln\left(\frac{1}{2}\varkappa\right) + \varkappa(C^+ + \bar{C} + \bar{\bar{C}})\right]. \tag{39}$$

By using the given formulas, the well-known blockage factor can be determined as follows:

$$B = 1 - u_m^*/U_0^* = -\bar{\bar{C}}\sqrt{c_{fm}/2}. \tag{40}$$

4 Results and comparison with experiments

The system of the ordinary differential equations has been solved numerically by an implicit central-difference scheme.

Figure 3 shows as main results the functions $\bar{C}(\alpha)$ and $\bar{\bar{C}}(\alpha)$. They have been determined by using a so-called one-equation turbulence model. In addition to the well-known differential equation for the kinetic energy k of the turbulent fluctuation, a fourth-order polynomial was used for the turbulent length $L^*(\eta) = c_\mu^{3/4} k^{*3/2}/\varepsilon^* = \nu_t^*/(c_\mu^{1/4}\sqrt{k^*})$. Apart from the universal constants $\varkappa = 0.41$ and $c_\mu = 0.09$, the turbulence model has only one model constant $L_{max}^*/(r^*|\Phi|) = 0.20$, which was chosen such that full agreement with experimental results was achieved for $\alpha = 0$, see [1], p. 593.

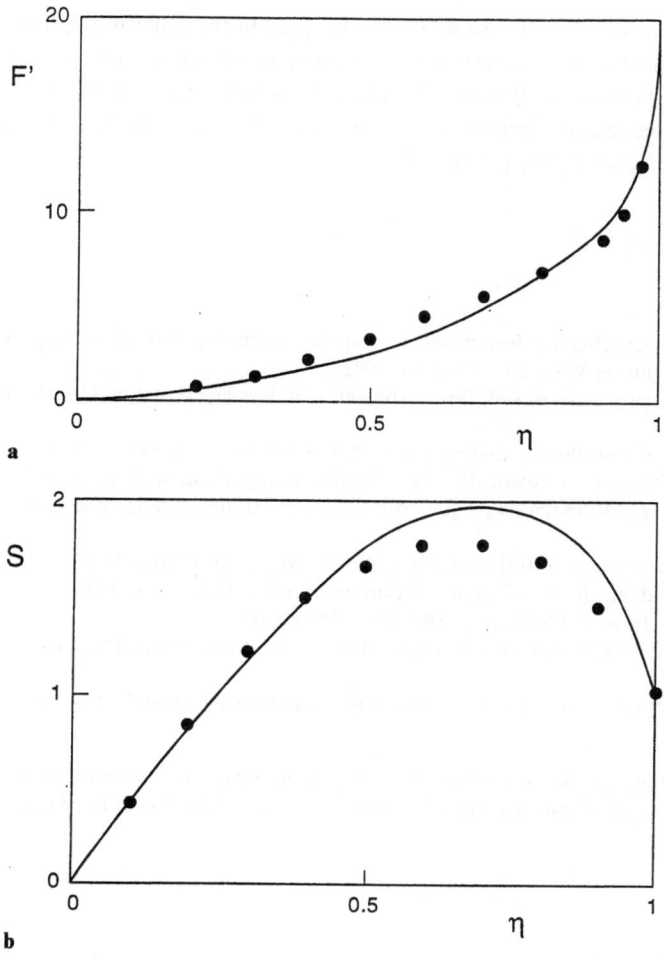

Fig. 5. Velocity and shear-stress profiles for a diffuser of $\Phi = 1°$. Comparison between the theory and experiment by Ruetenik and Corrsin [4], $\alpha_{EXP} = 0{,}48$. **a** Defect velocity $F'(\eta)$, **b** shear stress $S(\eta)$

For the limit $\alpha \rightarrow -\infty$, the turbulence model is free from empirical constants since then $L^* = \varkappa r^*(\varphi - \Phi)$ is simply proportional to the wall distance, see [1], p. 621. It follows

$$\lim_{\alpha \rightarrow -\infty} \bar{C} = \tilde{C} - \frac{1}{\varkappa} \ln(-2\alpha), \qquad \lim_{\alpha \rightarrow -\infty} \bar{\bar{C}} = 0, \qquad \lim_{\alpha \rightarrow -\infty} P = 0. \tag{41}$$

The numerical calculations led to the constant $\tilde{C} = -4.4$. This is in good agreement with *direct numerical simulations* (DNS) of the full flow equations according to [3]. As an example, Fig. 4 shows the distribution of the defect velocity in comparison with data of DNS in [3] and with experimental results in [2].

In Fig. 5 the functions $F'(\eta)$ and $S(\eta)$ for $\alpha = 0.48$ are compared with expeeriments by Ruetenik and Corrsin [4]. The agreement is also quite satisfying.

The range of validity of Eq. (37) could be determined only by a higher-order slender-channel theory or by experiments, which unfortunately are not available. But certainly the Reynolds numbers must be higher than the critical Reynolds numbers Re_c to ensure turbulent flow. The critical Reynolds number, which is $Re_c = 1.2 \cdot 10^3$ for $\alpha = 0$, see [6], will increase due to acceleration of the flow up to about $Re_c = 3 \cdot 10^5$ for $\alpha \to -\infty$, see [7]. On the other hand, there will be an upper bound for the angle Φ at a given Reynolds number. These bounds are supposed to be quite large for accelerated flows ($\alpha < 0$), since the angle Φ is unbounded in the limit $\alpha \to -\infty$. For decelerated flows, however, the limit $\alpha \to +\infty$ (zero wall shear stress) is reached at rather small angles of the order of $\Phi = 4°$, see [8], p. 176.

References

[1] Gersten, K., Herwig, H.: Strömungsmechanik, Grundlagen der Impuls-, Wärme- und Stoffübertragung aus asymptotischer Sicht. Braunschweig Wiesbaden: Vieweg 1992.
[2] Jones, W. P., Launder, B. E.: Some properties of sink-flow turbulent boundary layers. J. Fluid Mech **56**, 337−351 (1972).
[3] Spalart, Ph. R.: Numerical study of sink-flow boundary layers. J. Fluid Mech. **172**, 307−328 (1986).
[4] Ruetenik, J. R., Corrsin, S.: Equilibrium turbulent flow in a slightly divergent channel. In: 50 Jahre Grenzschichtforschung (Görtler, H., Tollmien, W., eds.), pp. 446−459. Braunschweig Wiesbaden: Vieweg 1955.
[5] Tennekes, H., Lumley, J. L.: A first course in turbulence. Cambridge Mass: MIT Press 1972.
[6] Dean, R. B.: Reynolds number dependence of skin friction and other bulk flow variables in two-dimensional rectangular duct flow. J. Fluids Eng. **100**, 215−223 (1978).
[7] Narasimha, R., Sreenivasan, K. R.: Relaminarization of fluid flows. Adv. Appl. Mech. **19**, 221−309 (1979).
[8] Townsend, A. A.: The structure of turbulent shear flow. Cambridge: Cambridge University Press 1976.

Authors' address: Prof. Dr.-Ing. Dr.-Ing. E. h. K. Gersten and Dipl.-Ing. B. Rocklage, Institut für Thermo- und Fluiddynamik, Ruhr-Universität Bochum, Universitätsstrasse 150, D-44801 Bochum, Federal Republic of Germany

Acta Mechanica (1994) [Suppl] 4: 335–349
© Springer-Verlag 1994

Interacting laminar boundary layers of dense gases

A. Kluwick, Vienna, Austria

Summary. The concept of triple deck theory is applied to study laminar interacting boundary layers of dense gases in external subsonic, supersonic and transonic flow. If the flow outside the boundary layer is either purely subsonic or purely supersonic the unusual thermodynamic properties of dense gases do not enter the description of interaction processes to leading order thus leaving the basic scaling laws of standard triple deck theory unchanged. This is no longer true in the transonic flow regime where the streamwise extent and, more important, the magnitude of the induced pressure disturbances are seen to depend on the size of the fundamental derivative Γ. The size of Γ also strongly influences the distance from the wall at which nonlinear cumulative effects lead to a significant distortion of outgoing pressure waves generated by supersonic interacting boundary layers.

1 Introduction

The theory of viscous-inviscid interactions is one of the corner stones of modern boundary layer theory and plays an important role for the understanding of external and internal aerodynamic flows. Studies dealing with the outer inviscid part of the problem have for many years focused primarily on ideal working fluids such as incompressible fluids and perfect gases. Real fluid effects of interest included transport phenomena reflecting deviations from thermodynamic equilibrium and were restricted, in general, to the boundary layer regime.

In recent years, however, there has been a rapidly growing interest in a different type of real fluid effects which may be of significance also in situations where dissipation plays an insignificant role. These effects come into play if the fluid under consideration consists of relatively complex molecules and they are associated with the magnitude and sign of the socalled fundamental derivative (Duhem [1])

$$\Gamma = \frac{1}{\tilde{c}} \left(\frac{\partial \tilde{\varrho} \tilde{c}}{\partial \tilde{\varrho}} \right)_{\tilde{s}} \qquad (1.1)$$

which characterizes the qualitative behaviour of any single phase gas, Hayes [2], Tompson [3]. Here $\tilde{c} = \sqrt{(\partial \tilde{p}/\partial \tilde{\varrho})_{\tilde{s}}}$ is the thermodynamic sound speed and $\tilde{\varrho}$, $\tilde{p}$, $\tilde{s}$ denote the density, the pressure and the entropy.

The fundamental derivative of dilute (perfect) gases can be expressed in terms of the ratio γ of the specific heats $\Gamma = (\gamma + 1)/2$ and, therefore, is seen to satisfy the condition $\Gamma > 1$. In contrast, dense gases may have $\Gamma < 1$ and even $\Gamma < 0$ if the specific heats are sufficiently high as, for example, in the case of high molecular hydrocarbons and fluorocarbons. Following Cramer [4] we refer to fluids having the property that Γ changes sign in the single phase gas region as BZT fluids thus stressing the pioneering character of studies by Bethe [5], Zel'dovich [6], Thompson [3].

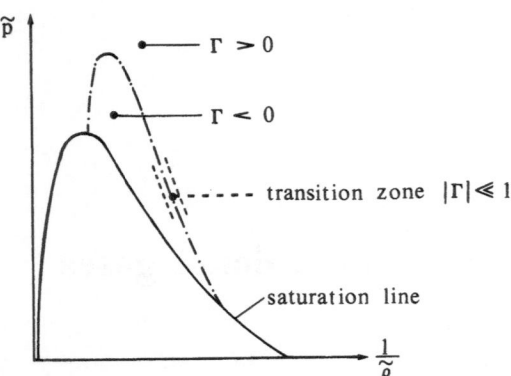

Fig. 1. Schematic of dense gas region of Bethe-Zel'dovich-Thompson fluid in $\tilde{p}$, $1/\tilde{\varrho}$-plane

Meanwhile, the interesting and often antiintuitive properties of waves in BZT fluids with special emphasis on states in the transition zone where Γ is small, Fig. 1 have been investigated in a number of papers, Cramer and Kluwick [7], Cramer, Kluwick, Watson and Pelz [8], Kluwick and Czemetschka [9], Cramer and Sen [10], Cramer and Crickenberger [11]. For a summary of the results which include new phenomena such as expansion shocks, sonic shocks, collisions between compression and expansion shocks, novel shock structures etc. the reader is referred to review articles by Cramer [4] and Kluwick [12]. In addition, internal as well as external steady flows have been studied also, Cramer and Best [13], Chandrasekar and Prasad [14], Cramer [4], Cramer and Crickenberger [15], Cramer and Tarkenton [16], Kluwick [17]. The results suggest that BZT fluids may prove beneficial in a number of practical applications, for example, as working fluids in organic Rankine cycle power systems. Before a final conclusion can be drawn it will be necessary, however, to take into account viscous effects and to investigate in particular how inviscid flows are affected by and interact with boundary layers. The present paper is intended to provide a first step towards this goal.

2 Unperturbed boundary layer

As mentioned earlier, the properties of boundary layers in the dense gas regime have received scant attention so far. Investigations of laminar boundary layers on flat plates without and with heat transfer are currently carried out by Whitlock [18] and Zieher [19]. Owing to the complicated dependence of the transport coefficients on the thermodynamic state variables these studies are premarily numerical in nature. In addition, some analytical progress seems possible, however, by exploiting the fact that the substances under consideration have large specific heats.

It is convenient to introduce Cartesian coordinates $\tilde{x}$ and $\tilde{y}$ as indicated in Fig. 2. The corresponding velocity components are $\tilde{u}$ and $\tilde{v}$. Furthermore, $\tilde{p}$, $\tilde{\varrho}$, $\tilde{h}$, $\tilde{T}$, $\tilde{c}$, $\tilde{\mu}$, $\tilde{\lambda}$ and $\tilde{L}$ denote the pressure, the density, the specific enthalpy, the temperature, the speed of sound, the dynamic viscosity, the thermal diffusivity and a reference length. Subscripts ∞ and w are used to characterize free stream and wall values of the various field variables.

Introducing the nondimensional quantities

$$x = \frac{\tilde{x}}{\tilde{L}}, \quad y = \frac{\tilde{y}}{\tilde{L}}, \quad u = \frac{\tilde{u}}{\tilde{u}_\infty}, \quad v = \frac{\tilde{v}}{\tilde{u}_\infty}, \quad p = \frac{\tilde{p} - \tilde{p}_\infty}{\tilde{\varrho}_\infty \tilde{u}_\infty^2}, \quad \varrho = \frac{\tilde{\varrho}}{\tilde{\varrho}_\infty}, \quad h = \frac{\tilde{h}}{\tilde{h}_\infty},$$

$$T = \frac{\tilde{T}}{\tilde{T}_\infty}, \quad c = \frac{\tilde{c}}{\tilde{c}_\infty}, \quad \mu = \frac{\tilde{\mu}}{\tilde{\mu}_\infty}, \quad \lambda = \frac{\tilde{\lambda}}{\tilde{\lambda}_\infty}$$

$$\tag{2.1}$$

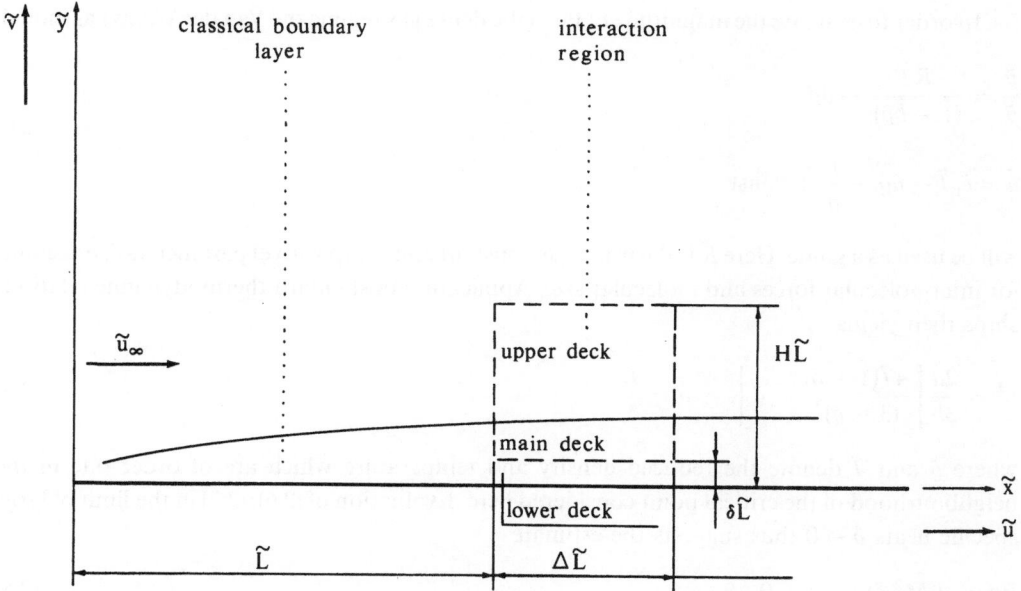

Fig. 2. Laminar boundary layer on flat plate encountering a surface mounted obstacle

and

$$u = f'(\eta), \quad h = g(\eta), \quad \eta = \left(\frac{\text{Re}}{2x}\right)^{1/2} \int_0^y \varrho \, dy \tag{2.2}$$

the boundary layer equations are written in classical self similar form

$$(Cf'')' + ff'' = 0, \tag{2.3}$$

$$\left(\frac{C}{\text{Pr}} g'\right)' + fg' = -\text{Ec}\,C(f'')^2. \tag{2.4}$$

Here

$$C = \varrho\mu, \quad \text{Re} = \frac{\tilde{u}_\infty \tilde{\varrho}_\infty \tilde{L}}{\tilde{\mu}_\infty}, \quad \text{Pr} = \frac{\tilde{\mu}_\infty \tilde{c}_{p\infty}}{\tilde{\lambda}_\infty}, \quad \text{Ec} = \frac{\tilde{u}_\infty^2}{\tilde{h}_\infty} \tag{2.4}$$

are the Chapman Rubesin parameter, the Reynolds number, the Prandtl number and the Eckert number.

Equations (2.3) have to be solved subjected to the boundary conditions

$$f(0) = f'(0) = 0, \quad f'(\infty) = 1,$$
$$g(0) = g_w, \quad g(\infty) = 1. \tag{2.5}$$

In the perfect gas case Ec and the freestream value M_∞ of the Mach number satisfy the wellknown relationship $\text{Ec} = (\gamma - 1) M_\infty^2$. Since the ratio of the specific heats γ is of order one $\text{Ec} = O(M_\infty^2)$ and, consequently, dissipative effects and compressibility effects are seen to be of equal importance.

In order to estimate the magnitude of Ec in the dense gas regime the Van der Waals gas model

$$\frac{\tilde{p}}{\tilde{\rho}} = \frac{\tilde{R}\tilde{T}}{(1 - \tilde{b}\tilde{\varrho})} - \tilde{a}\tilde{\varrho},$$

$$\tilde{h} = \tilde{c}_v\tilde{T} - \tilde{a}\tilde{\varrho} + \frac{\tilde{p}}{\tilde{\varrho}} + \text{const}$$

(2.6)

will be used as a guide. Here $\tilde{R}$ is the usual gas constant and the (positive) parameters $\tilde{a}, \tilde{b}$ account for intermolecular forces and molecular size. Application of standard thermodynamic relationships then yields

$$\tilde{c}^2 = \frac{2\tilde{a}}{3\tilde{b}}\left[\frac{4\tilde{T}(1 + \delta)}{(3 - \bar{\varrho})^2} - \bar{\varrho}\right], \quad \delta = \frac{\tilde{R}}{\tilde{c}_v}$$

(2.7)

where $\bar{\varrho}$ and $\bar{T}$ denote the reduced density and temperature which are of order one in the neighbourhood of the critical point considered here. Evaluation of (2.6), (2.7) in the limit of large specific heats $\delta \to 0$ thus suggests the estimate

$$\text{Ec} = O(M_\infty^2 \delta), \quad \delta \to 0.$$

(2.8)

In marked contrast to the perfect gas case the limit $\text{Ec} \to 0$, therefore, no longer requires $M_\infty^2 \to 0$. In the dense gas regime, the dissipation term entering the energy equation (2.3) may play an insignificant role for high subsonic or even supersonic external flows provided the specific heats take on sufficiently large values.

Precise measurements of the thermal conductivity in the dense gas regime are extremely difficult. As a result, accurate Prandtl number data for BZT fluids seem not to be available at present and one has to resort to empirical correlations. Evaluation of the most simple correlation proposed by Eucken suggests that Pr tends to 1 asymptotically in the limit of large specific heats $\delta \to 0$. Preliminary calculations based on the more sophisticated methods by Chung, Ajlan, Lee and Starling (e.g. Reid, Prausnitz and Poling [20], chaps. 9–10) which have been carried out for PP 11 (Zieher [19]) seem to support the result that the values of Pr are reasonable close to 1 if $\delta \ll 1$. This may warrant the use of the Pr = 1 approximation which has been applied successfully in classical boundary layer theory. Since many of the assumptions implicit in the correlations for the viscosity and the thermal conductivity are open to questions – especially when applied to dense gases – this approximation has to be taken with caution, however.

3 Local interaction theory for external subsonic or supersonic flow

The results summarized in the previous section indicate that boundary layers of dense gases may exhibit interesting new properties owing to the unconventional relationships holding between the nondimensional groups which enter the governing equations. Nevertheless, they represent standard boundary layers insofar as they obey the scaling laws of classical theory. It thus follows that the description of local interaction processes follows the classical concept developed by Stewartson [21], Neiland [22], Messiter [23] provided that the nonlinearity of the governing equations in the external inviscid flow region can be neglected to leading order, e.g. if the external flow is purely subsonic or supersonic.

As in standard triple deck theory the length Δ and lateral extent H of the local interaction region are of $O(\text{Re}^{-3/8})$. Inside this region three layers with different physical properties have to be distinguished, Fig. 2. Outside the boundary layer, in the upper deck region, the perturbations of the field quantities are governed by the linearized equations of inviscid theory. The role of the main deck which comprises most of the boundary layer is essentially a passive one, to transfer displacement effects exerted by the viscous near-wall region – termed lower deck – to the upper deck and to transfer the resulting pressure disturbances back to the lower deck. Here the flow is governed by the boundary layer equations for an incompressible fluid. Supplemented with appropriate boundary conditions and matching conditions following from the analytical solutions holding in the upper and main deck, respectively, these equations constitute a closed problem which completely determines the properties of the interaction process. Introducing the scaled quantities

$$X = k^{5/4} \mu_W^{1/4} \varrho_W^{1/2} |M_\infty^2 - 1|^{3/8} \operatorname{Re}^{3/8}(x - 1),$$

$$Y = k^{3/4} \mu_W^{-1/4} \varrho_W^{1/2} |M_\infty^2 - 1|^{1/8} \operatorname{Re}^{5/8} y,$$

$$U = k^{-1/4} \mu_W^{-1/4} \varrho_W^{1/2} |M_\infty^2 - 1|^{1/8} \operatorname{Re}^{1/8} u,$$

$$V = k^{-3/4} \mu_W^{-3/4} \varrho_W^{1/2} |M_\infty^2 - 1|^{-1/8} \operatorname{Re}^{3/8} v,$$

$$P = k^{-1/2} \mu_W^{-1/2} |M_\infty^2 - 1|^{1/4} \operatorname{Re}^{1/4} p,$$

$$F = k^{3/4} \mu_W^{-1/4} \varrho_W^{1/2} |M_\infty^2 - 1|^{1/8} \operatorname{Re}^{5/8} f$$

(3.1)

the fundamental lower deck problem can be written in parameter free from

$$\frac{\partial U}{\partial X} + \frac{\partial V}{\partial Y} = 0,$$

$$U \frac{\partial U}{\partial X} + V \frac{\partial U}{\partial Y} = -\frac{dP}{dX} + \frac{\partial^2 U}{\partial Y^2},$$

$$U = V = 0 \quad \text{or} \quad U = \frac{\partial U}{\partial Y} = 0 \quad \text{on} \quad Y = F(X),$$

(3.2)

$$U = Y + A(X) \quad \text{for} \quad Y \to \infty \quad \text{all} \quad X,$$

$$P(X) = -\frac{1}{\pi} \int\limits_{-\infty}^{\infty} \frac{A'(\xi)}{\xi - X} d\xi \quad \text{subsonic flow}$$

or

$$P(X) = -A'(X) \quad \text{supersonic flow}.$$

Here $k = \dfrac{\partial u}{\partial y}(1, 0)$, μ_W and ϱ_W denote the slope of the velocity profile, the viscosity and the density at the wall of the unperturbed boundary layer for $x = 1$ while $f(x)$ characterizes the shape of the contour on which the boundary conditions are imposed.

Equation (3.1) expresses the similarity law of laminar locally interacting boundary layers in a general fluid. It generalizes the transformation originally derived by Stewartson [24] which is recovered by considering the flow of a perfect gas which obeys the Chapman viscosity law past a thermally isolated flat plate

$$\varrho_W = T_W^{-1}, \quad \mu_W = CT_W, \quad k = 0.332 \ldots C^{-1/2} T_W^{-1}. \tag{3.3}$$

The triple deck equations (3.2) have been studied extensively in the past, both analytically and numerically. As a result, solutions applying to a large variety of physical situations including trailing edge flows, flows past surface mounted obstacles, shock boundary layer interactions are available at present. For a discussion of these results in the context of classical gasdynamics the reader is referred to the review articles by Stewartson [24], Smith [25], Messiter [26] and Kluwick [27], [28]. The considerations leading to Eqs. (3.1) and (3.2), however, show that triple deck solutions have a much wider range of applicability. Indeed, the thermodynamic properties of the fluid enter the theory trough the distribution of the field quantities in the unperturbed boundary layer only and can thus be accounted for by a transformation of the dependent and independent variables.

Since expansion shocks cannot form in dilute gases the problem how such shocks interact with laminar boundary layers has not been adressed explicitly in triple deck studies carried out so far. It is easily confirmed, however, that the solution to this problem does not require additional computations but is already contained in the accumulated body of results. As shown by Ruban [29] and Burggraf [30] the triple deck solutions for the flow past a compression-ramp and for the compression-shock boundary layer interaction are closely related. Most important, the distributions of the pressure and the wall shear stress are identical provided the ramp angle is twice the flow deflection angle characterizing the impinging shock. Since the argument leading to this conclusion rests on an invariance property of the boundary layer equations for incompressible flow (Prandtl transposition theorem, e.g. Kluwick [31]) it remains intact if the notations compression-ramp and compression-shock are replaced by the notations expansion-ramp and expansion-shock. As a result, the basic properties of expansion-shock boundary layer interactions can simply be extracted from the solutions to the expansion-ramp problem. For example, it is well known that a supersonic laminar boundary layer encountering an expansion ramp remains attached for arbitrary large ramp angles. It thus follows that even strong oblique expansion shocks will not separate a laminar boundary layer.

4 Supersonic far field

Owing to the linearity of the governing equations in the upper deck the treatment of locally interacting laminar boundary layers with purely subsonic or supersonic external flow is simplified considerably. If the external flow is supersonic the resulting solution, however, ceases to be valid at large distances from the wall where nonlinear cumulative effects have to be accounted for. Adopting the assumption that the fluid under consideration is a perfect gas, the associated distortion of the pressure waves leaving the interaction region and leading to the formation of characteristic wave patterns which can be made visible, for example, by means of a Schlieren technique have been calculated first by Kluwick [31].

Here we are concerned with the modification of these results caused by dense effects. Specifically, it will be assumed that the thermodynamic state characterizing the unperturbed flow is in the vicinity of one of the high or low pressure zeros of the fundamental derivative. Then Γ is of the same order of magnitude as the pressure disturbances induced by the interaction process while its derivative with respect to ϱ at constant entropy s is of order one:

$$\Gamma = O(p), \quad \Lambda = \left.\frac{\partial \Gamma}{\partial \varrho}\right|_s = O(1). \tag{4.1}$$

For simplicity, the calculations will be restricted to the case of simple outgoing waves. Examples of practical importance include the flow past compression and expansion ramps and the flow near the trailing edge of a flat plate without and with incidence. As pointed out by Kluwick [31], however, effects of incoming pressure disturbances can be incorporated rather easily provided the width of the impinging wave is of order one in the triple deck scaling.

Owing to the simple wave assumption the pressure disturbances caused by the interaction process are constant on outgoing characteristics $\xi = $ const. Taking into account the order of magnitude estimate (4.1) the slope of these characteristics in the x, y-plane can be written in the form (Cramer [4])

$$\frac{dy}{dx}\bigg|_{\xi} = \frac{1}{(M_\infty^2 - 1)^{1/2}} + \frac{M_\infty^4}{(M_\infty^2 - 1)^2}\left[\Gamma_\infty\Theta + \frac{M_\infty^2}{(M_\infty^2 - 1)^{1/2}}\frac{\Lambda_\infty}{2}\Theta^2\right]. \tag{4.2}$$

Here Θ denotes the flow angle in the inviscid outer flow region which satisfies the Ackeret relationship

$$\Theta = (M_\infty^2 - 1)^{1/2}p \tag{4.3}$$

to leading order. Since the pressure disturbances do not vary across the boundary layer in a first approximation, Eq. (4.2) has to be solved subject to the boundary conditions

$$y = 0: p = p(x) \tag{4.4}$$

where $p(x)$ is determined by the pressure distribution following from Eqs. (3.2).

It is convenient to eliminate the parameters Re, M_∞, Γ_∞ and Λ_∞ entering Eqs. (4.2), (4.3) and (4.4) by introducing the scaled quantities

$$\hat{\Gamma}_\infty = \text{Re}^{1/4}\Gamma_\infty,$$

$$\hat{\eta} = \hat{\Gamma}_\infty^{-1}\Lambda_\infty M_\infty^{-2}(M_\infty^2 - 1)^{3/2}\,\text{Re}^{3/8}[x - 1 - y(M_\infty^2 - 1)^{1/2}],$$

$$\hat{y} = \text{Re}^{1/8}y, \tag{4.5}$$

$$\hat{p} = \hat{\Gamma}_\infty^{-1}\Lambda_\infty M_\infty^2\text{Re}^{1/4}p.$$

Solutions to Eqs. (4.3), (4.4) exhibit, in general, regions of multivaluedness which have to be eliminated by the insertion of shock fronts in order to obtain physically meaningful single valued results. The evolution equation for the pressure disturbances in the far field is, therefore, written in conservation form which properly accounts for the formation of shock discontinuities. One then obtains

$$\frac{\partial\hat{p}}{\partial\hat{y}} + \frac{\partial\hat{j}}{\partial\hat{\eta}} = 0, \quad \hat{j} = -\frac{\hat{p}^2}{2} - \frac{\hat{p}^3}{6}, $$

$$\hat{y} = 0: \hat{p} = \hat{p}(\hat{\eta}). \tag{4.6}$$

A formal solution to the kinematic wave equation (4.6) is given by:

$$\hat{p} = \text{const on characteristics} \quad \zeta(\hat{\eta}, \hat{y}) = \text{const}, \quad \frac{d\hat{\eta}}{d\hat{y}}\bigg|_{\xi} = \frac{d\hat{j}}{d\hat{p}} = -\hat{p} - \frac{\hat{p}^2}{2}. \tag{4.7}$$

If characteristics intersect each other this solution has to be supplemented with shock discontinuities $S(\hat{\eta}, \hat{y}) = $ const satisfying the slope conditions

$$\left.\frac{d\hat{\eta}}{d\hat{y}}\right|_s = \frac{[\hat{j}]}{[\hat{p}]} = -\frac{1}{6}\frac{[3\hat{p}^2 + \hat{p}^3]}{[\hat{p}]}. \tag{4.8}$$

Here $[Q] = Q_b - Q_a$ denotes the jump of the quantity Q, e.g. the difference between the values of Q before and after the shock.

Equations (4.6), (4.7), and (4.8) which fully determine the properties of the far field outside the local interaction region once the wall pressure distribution has been obtained are identical to the set of equations derived by Cramer and Kluwick [7] in a paper dealing with the propagation of unidirectional planar waves in media having mixed nonlinearity. All the results summarized in this study of unsteady wave phenomena including the criterion for admissible shocks, the occurrence of sonic shocks, the collision of expansion and compression shocks, etc. can, therefore, be used unchanged to analyse wave fields associated with laminar interacting boundary layers in dense gases. Owing to the complexity of these results and the limitation of space an exhaustive discussion of possible wave patterns is not attempted here, however. Rather we conclude this section by adding a few general remarks.

It is an interesting feature of many interaction processes that they lead to the occurrence of shock fronts. The correct description of these shocks appears to be especially important in investigations of internal flows. For example, shocks emanating from one blade of a cascade may impinge on adjacent blades thus severely influencing the boundary layer development there.

In the limit of large distances from the wall, $\hat{y} \to \infty$, the wave field predicted by purely inviscid theory is recovered from the far field equations (4.6) involving the induced wall pressure disturbances. Closer to the wall, however, the wave pattern may be considerable more complicated and, even more important, may include shocks of substantially higher strength than anticipated on the basis of inviscid theory. A prominent example is provided by the flow near the trailing edge of an aligned flat plate which remains unperturbed if viscous effects are ignored completely but which causes shocks to form if the rapid variation of the boundary layer displacement thickness is properly accounted for.

To determine the location and strength of interaction induced shocks it is necessary to solve Eqs. (4.6) numerically, in general. The shock formation distance $\hat{y}_s$, however, can easily be calculated analytically. One obtains

$$\hat{y}_s = \frac{1}{d\hat{p}/d\hat{\eta}[1 + \hat{p}]} = O(1) \tag{4.9}$$

where the maximum of the absolute value of the denominator is to be taken. Combination of Eqs. (3.1), (4.5) and (4.9) show that shocks form at distances

$$y_s = O(\mathrm{Re}^{1/8}) \tag{4.10}$$

from the wall, Fig. 3, e.g. at much larger distances than in the perfect gas case where $y_s = O(\mathrm{Re}^{-1/8})$, Kluwick [31]. Of course, this is a direct consequence of the assumption adopted here that the free stream value of the fundamental derivative Γ_∞ is small rather than of order one. Furthermore, Eq. (4.10) indicates that the shock formation distance increases with increasing values of the Reynolds number. Again, this is an immediate consequence of the scaling leading to Eq. (4.6) which require $\Gamma_\infty \to 0$ as $\mathrm{Re} \to \infty$.

In order to derive results holding for $|\Gamma_\infty| = O(1)$ we first observe from Eqs. (4.5) that $\hat{p} \to 0$ as $\hat{\Gamma}_\infty \to \infty$ so that the cubic term in the expression (4.6) for $\hat{j}$ can be neglected to leading order.

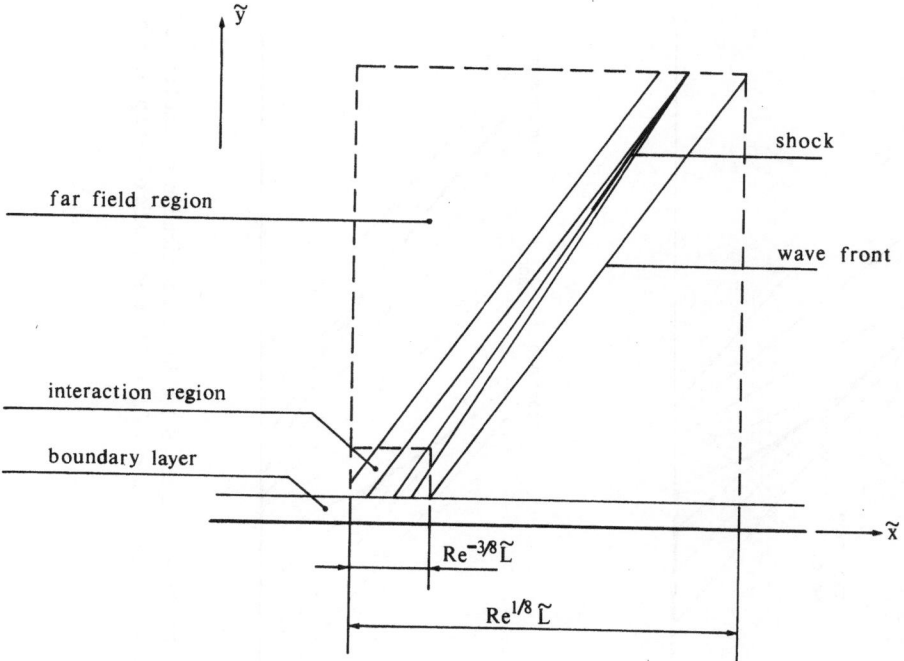

Fig. 3. Local interaction region and far field region of a supersonic boundary layer: $|\Gamma_\infty| = O(\text{Re}^{-1/4})$.

Physically this means that the parameter Λ_∞ does not enter the description of the flow at this level of approximation. As a consequence, it is possible to keep the triple deck scalings (3.1) for the pressure and the coordinate in the streamwise direction which occurs in the definition of the linear phase variable. Here, however, we prefer a representation similar to that when Γ_∞ is small:

$$\bar{\eta} = \Gamma_\infty^{-1} M_\infty^{-4}(M_\infty^2 - 1)^{3/2} \text{Re}^{3/8}[x - 1 - y(M_\infty^2 - 1)^{1/2}],$$

$$\bar{y} = \text{Re}^{1/8} y, \tag{4.11}$$

$$\bar{p} = \text{Re}^{1/4} p.$$

$\bar{p}(\bar{\eta}, \bar{y})$ then satisfies the simplified set of equations

$$\frac{\partial \bar{p}}{\partial \bar{y}} + \frac{\partial \hat{j}}{\partial \bar{\eta}} = 0, \quad \bar{j} = -\frac{\bar{p}^2}{2}, \tag{4.12}$$

$$\bar{y} = 0: \bar{p} = \bar{p}(\bar{\eta}).$$

Equations (4.11) and (4.12) generalize the results holding for perfect gases given by Kluwick [31] which are recovered by setting $\Gamma_\infty = (\gamma + 1)/2$.

As a typical example of wave patterns generated by interaction processes we briefly consider the flow near the trailing edge of an aligned flat plate in a supersonic stream. The triple deck solution to the trailing edge local interaction problem has been calculated by Daniels [32] and the distribution of the induced pressure disturbances at the wall and the wake centerline which determine the wave field is depicted in Fig. 4a. It is seen that the pressure drops monotonically as the trailing edge is approached from upstream. In the wake region the pressure returns – again monotonically – to the free stream level.

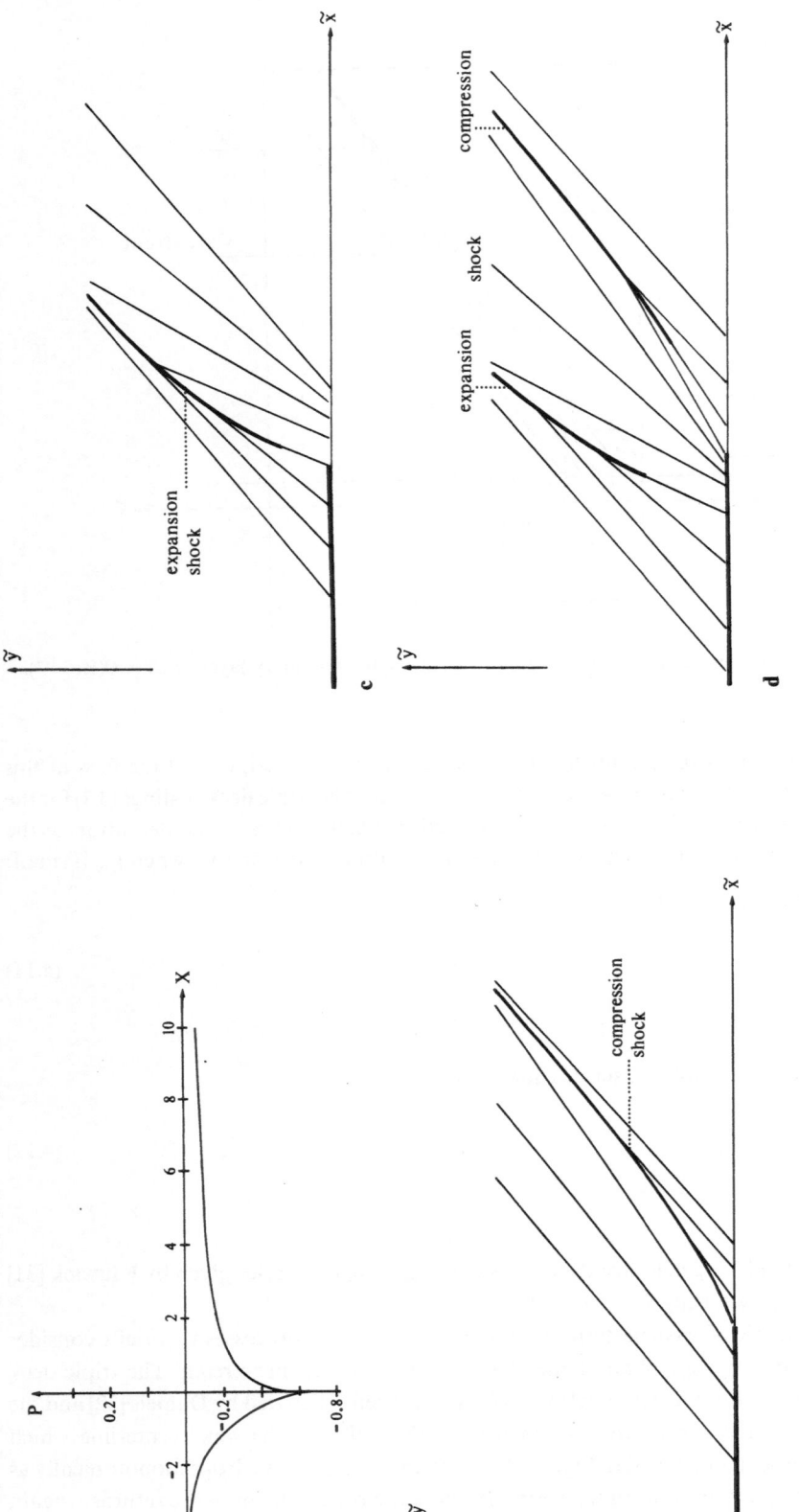

Fig. 4. Supersonic flow near the trailing edge of flat plate at zero incidence (waves in the upper half plane $\tilde{y} \geqq 0$ are displayed only). **a** Pressure distribution predicted by local interaction theory, Daniels [32]; **b** Schematic of wave pattern for $\Gamma_\infty > 0, \Gamma_\infty = O(1)$; **c** Schematic of wave pattern for $\Gamma_\infty < 0, \Gamma_\infty = O(1)$; **d** Schematic of wave pattern for $|\Gamma_\infty| = O(Re^{-1/4})$

If Γ_∞ is a positive order one constant, compression waves emanating from the wake steepen to form a compression shock, Fig. 4b. Moreover, since $(dp/dx)(x, 0) \to \infty$ as $x - 1 \to 0 +$, the origin of this shock coincides with the location of the trailing edge when expressed in terms of the scaled variables (4.11): $\bar{\eta} = \bar{y} = 0$, Kluwick [31]. In the case of strictly negative values of Γ_∞, however, the compressive part of the outgoing pressure waves flattens out while expansion waves originating in the interaction zone upstream of the trailing edge focus. This mechanism leads to the occurrence of an expansion shock which forms at $\bar{\eta} = 0$, $\bar{y} = \bar{y}_s > 0$, Fig. 4c. The complexity of the wave pattern generated by the trailing edge increases significantly if Γ_∞ is small and if the pressure reduction associated with the interaction process causes the local value of Γ to change sign twice. In general, one compression and one expansion shock will then be generated, Fig. 4d. The compression and expansion shock must merge eventually since only compression/expansion shocks can exist at arbitrarily large distances from the line of symmetry if Γ_∞ is positive/negative.

5 Local interaction theory for external transonic flow

If the flow outside the boundary layer is purely subsonic or supersonic the nonlinearity of the governing equations does not affect the interaction process in the upper deck region of the triple deck structure to leading order. Although nonlinear effects have to be taken into account at large distances from the wall in the case of external supersonic flow they do not, however, feed back into the interaction region and are thus passive in this sense. As a result, the fundamental derivative does not occur explicitly in the description of interacting laminar subsonic and supersonic boundary layers.

This is no longer true if the free stream Mach number M_∞ differs only slightly from its critical value 1. As in the perfect gas case the linear versions of the upper deck equations have then to be replaced by the transonic small perturbation equations (Messiter, Feo and Melnik [34], Bodonyi and Kluwick [35]). However, as shown by Cramer [33] and Kluwick [17] the proper form of the nonlinear terms in these equations depends crucially on the magnitude of Γ_∞ and Λ_∞. It is via these terms that dense gas effects may enter the description of transonic interaction processes. Anticipating that the local interaction region will again exhibit a three tiered structure the scales involved may be derived from an order of magnitude argument as follows, Fig. 2.

(i) Owing to the passive nature of the main deck region the disturbances of u there and in the lower deck zone are of equal order of magnitude. Similarly, the v-component of the velocity in the main deck and in the upper deck are found to be comparable in magnitude. (ii) In order to allow for a fully nonlinear response of the boundary layer the u-disturbances in the lower deck must be of the same order of magnitude as the velocities in the unperturbed flow: $u_l = O(\delta \text{Re}^{1/2})$. (iii) Evaluation of the continuity equation in the main deck then leads to the estimate $v_u = O(\delta \Delta^{-1})$ for the induced lateral velocity components in the upper deck. (iv) Assuming that the thermodynamic state of the fluid at free stream conditions is in the neighbourhood of one of the high or low pressure zeros of the fundamental derivative, (Fig. 1), Γ_∞ satisfies the order of magnitude relationship $\Gamma_\infty = O(p_u)$. As shown by Cramer [33] then the estimates $p_u = O(u_u) = O(v_u^{1/2})$, $H = O(u_u^{-1}\Delta)$ hold in the transonic flow regime. (v) Equating the inertia, pressure gradient and viscous terms in the lower deck taking into account $p_u = O(p_l)$ yields the estimates

$$\Delta = O(\text{Re}^{-1/4}), \quad \delta = O(\text{Re}^{-7/12}), \quad H = O(\text{Re}^{-1/12}),$$

$$u_l = O(\text{Re}^{-1/12}), \quad v_l = O(\text{Re}^{-5/12}),$$

$$u_m = = O(\text{Re}^{-1/12}), \quad v_m = O(\text{Re}^{-4/12}), \tag{5.1}$$

$$u_u = O(\text{Re}^{-2/12}), \quad v_u = O(\text{Re}^{-4/12}),$$

$$p_l = O(p_m) = O(p_u) = O(\text{Re}^{-1/6})$$

which determine the length scales of the interaction zone and the magnitude of the disturbances of the field quantities in the upper deck, main deck and lower deck, respectively.

Based on (5.1) it is possible to formulate an asymptotic theory for transonic interacting flows holding in the limit

$$|\Gamma_\infty| = O(\text{Re}^{-1/6}). \tag{5.2}$$

Here we briefly summarize the final result. Introducing suitably scaled quantities

$$x - 1 = k^{-7/5} \mu_W^{-2/5} \varrho_W^{-1/5} \text{Re}^{-3/12} x_T,$$

$$y = k^{-4/5} \mu_W^{1/5} \varrho_W^{-1/5} \text{Re}^{-7/12} y_T,$$

$$u = k^{1/5} \mu_W^{1/5} \varrho_W^{-1/2} \text{Re}^{-1/12} u_T, \tag{5.3}$$

$$v = k^{4/5} \mu_W^{4/5} \varrho_W^{-1/2} \text{Re}^{-5/12} v_T,$$

$$p = k^{2/5} \mu_W^{2/5} \text{Re}^{-2/12} p_T$$

for the lower deck region and

$$y = k^{-8/5} \mu_W^{-3/5} \varrho_W^{-1/5} \text{Re}^{-1/12} z,$$

$$u - 1 = k^{2/5} \mu_W^{2/5} \text{Re}^{-2/12} \bar{u}_T, \tag{5.4}$$

$$\hat{K}_O = k^{-2/5} \mu_W^{-2/5} (M_\infty^2 - 1) \, \text{Re}^{4/12},$$

$$\hat{\Gamma}_\infty = \text{Re}^{1/6} \Gamma_\infty,$$

$$\hat{\Lambda}_\infty = k^{2/5} \mu_W^{2/5} \Lambda_\infty$$

in the upper deck region the fundamental problem can be cast into the form

$$u_T \frac{\partial u_T}{\partial x_T} + v_T \frac{\partial u_T}{\partial y_T} = -\frac{dp_T}{dx_T} + \frac{\partial^2 u_T}{\partial y_T^2},$$

$$\frac{\partial u_T}{\partial x_T} + \frac{\partial v_T}{\partial y_T} = 0. \tag{5.5}$$

$$u_T = v_T = 0 \quad \text{on} \quad y_T = F(x_T),$$

$$u_T = y_T \quad \text{for} \quad x_T \to -\infty, \quad \text{all} \quad y_T, \tag{5.6}$$

$$u_T = y_T + A(x_T) \quad \text{for} \quad y_T \to \infty \quad \text{all} \quad x_T.$$

$$(\hat{K}_O + 2\hat{\Gamma}_\infty \bar{u}_T - \hat{\Lambda}_\infty \bar{u}_T{}^2) \frac{\partial \bar{u}_T}{\partial x_T} + \frac{\partial v_T}{\partial z} = 0,$$

$$\frac{\partial \bar{u}_T}{\partial z} - \frac{\partial v_T}{\partial x_T} = 0.$$

(5.7)

$$v_T(x_T, z = 0) = \frac{dA}{dx_T},$$

$$\bar{u}_T{}^2 + v_T{}^2 \to 0 \quad \text{for} \quad x_T{}^2 + z^2 \to \infty.$$

(5.8)

$$p_T(x_T) = -\bar{u}_T(x_T, z = 0).$$

(5.9)

Here Eqs. (5.5) and (5.7) describe the fluid motion in the lower and upper deck regions, respectively. Equation (5.6) expresses the initial condition at the start of the interaction zone, the boundary conditions at the wall and at the base of the main deck. Similarly, Eqs. (5.8) state the boundary conditions to be satisfied by the modified transonic small perturbation equations (5.7). The set of fundamental equations is closed by means of the coupling condition (5.9) which defines the pressure driving the boundary layer in terms of the velocity disturbances at the inner edge of the upper deck.

Equations (5.5) to (5.9) which apply to interaction processes caused by surface mounted obstacles can easily be adapted to other types of problems including shock boundary layer interactions and trailing edge flows. In any of these cases, however, the solution of the interaction equations poses a formidable numerical problem which has not been attacked yet.

The estimates (5.1) include the key results of this section, namely that the pressure rise needed to separate an interacting transonic laminar boundary layer in the dense gas regime is of order $Re^{-1/6}$ provided Γ_∞ is small and of the order $Re^{-1/6}$ also. In contrast, studies dealing with transonic flows of perfect gases (Messiter, Feo and Melnik [34], Bodonyi and Kluwick [35]) which are representative also for gases with $|\Gamma_\infty| = O(1)$ have shown that the pressure increase required to separate the boundary layer is of order $Re^{-1/5}$. We thus conclude that dense gases having small values of Γ_∞ are (slightly) less susceptible to boundary layer separation than gases with $|\Gamma_\infty|$ being of order 1.

6 Concluding remarks

The present work extends previous studies dealing with laminar interacting boundary layers of perfect gases to the dense gas regime in which the fundamental derivative is small rather than of order one and may even assume negative values. To this end it is assumed that the interaction processes under consideration are local and that they can be treated by means of triple deck theory.

In the case of external subsonic or supersonic flow the nonlinearity of the governing equations in the region outside of the boundary layer has a small effect on the interaction process and can thus be neglected to leading order. As a result, the explicit Reynolds number and Mach number dependence of the disturbances inside the local interaction zone is of exactly the same form as in the standard triple deck theory of perfect gases. Specifically, it is found that the pressure rise leading to separation is of order $Re^{-1/4}$. Dense gas effects enter the scaling laws via the values of the shear stress, the density and the viscosity at the wall of the unperturbed boundary layer which, however, may be significantly different from the predictions of classical boundary layer theory for perfect gases.

Although nonlinear effects are small in the external flow region which actively takes part in the interaction process if the flow there is purely supersonic these effects accumulate with increasing distance from the wall thus leading to an order one distortion of the outgoing pressure signal eventually. In general, this mechanism forces the occurence of shocks which are found to form at distances of order $Re^{1/8}$. In contrast, Kluwick [31] has shown that shocks form at distances of the order $Re^{-1/8}$ if the flow medium is a perfect gas. The vast increase of the shock formation distance observed in the dense gas regime – which may be especially important in studies of internal flows – is, of course, a direct consequence of the assumption adopted here that the free stream value Γ_∞ of the fundamental derivative is small rather than of order one.

While the thermodynamic properties of the fluid do not influence the basic scaling laws of laminar interaction processes if the external flow is purely subsonic or supersonic this is not true in the transonic flow regime. As pointed out first by Cramer [33] and Kluwick [17] the nonlinear terms of the transonic small perturbation equations have to be modified if Γ_∞ is small. Since it is through these terms that transonic effects make themselves felt in the interaction process the scaling laws of transonic viscous-inviscid interactions have to be modified too. Specifically it is found that decreasing values of Γ_∞ cause the interaction zone to expand in the streamwise direction. This in turn makes the boundary layer less susceptible to separation by raising the required pressure increase from its classical value of $O(Re^{-1/5})$ to the higher $O(Re^{-1/6})$ level.

Acknowledgement

This work was supported in part by the Fonds zur Förderung der wissenschaftlichen Forschung in Österreich under Contract P8320-TEC and NSF Grant 427246.

References

[1] Duhem, P.: Sur la propagation des ondes de choc au sein des fluides. Z. Phys. Chem. **69**, 169–186 (1909).

[2] Hayes, W. D.: Gasdynamic discontinuities. Princeton Series on High Speed Aerodynamics and Jet Propulsion. Princeton: Princeton University Press (1960).

[3] Thompson, P. A.: A fundamental derivative in gasdynamics. Phys. Fluids **14**, 1843–1849 (1971).

[4] Cramer, M. S.: Nonclassical dynamics of classical gases. In: Nonlinear waves in real fluids (Kluwick, A. ed.), pp. 91–145. Wien, New York: Springer 1991.

[5] Bethe, H. A.: The theory of shock waves for an arbitrary equation of state. Technical report, Office Sci. Res. Dev. Rep. **545** (1942).

[6] Zel'dovich, Ya. B.: On the possibility of rarefaction shock waves. Zh. Eksp. Teor. Fiz. **4**, 363–364 (1946).

[7] Cramer, M. S., Kluwick, A.: On the propagation of waves exhibiting both positive and negative nonlinearity. J. Fluid Mech. **142**, 9–37 (1984).

[8] Cramer, M. S., Kluwick, A., Watson, L. T., Pelz, W.: Dissipative waves in fluids having both positive and negative nonlinearity. J. Fluid Mech. **169**, 329–336 (1986).

[9] Kluwick, A., Czemetschka, E.: Kugel- und Zylinderwellen in Medien mit positiver und negativer Nichtlinearität. ZAMM **70**, 207–208 (1990).

[10] Cramer, M. S., Sen, R.: Exact solutions for sonic shocks in van der Waals gases. Phys. Fluids **30** 377–385 (1987).

[11] Cramer, M. S., Crickenberger, A. B.: The dissipative structure of shock waves in dense gases. J. Fluid Mech. **223**, 325–355 (1991).

[12] Kluwick, A.: Small-amplitude finite-rate waves in fluids having both positive and negative nonlinearity. In: Nonlinear waves in real-fluids (Kluwick, A. ed.), pp. 1–43. Wien, New York: Springer 1991.

[13] Cramer, M. S., Best, L. M.: Steady isentropic flow of dense gases. Phys. Fluids A 3, 219–226 (1991).

[14] Chandrasekar, D., Prasad, Ph.: Transonic flow of a fluid with positive and negative nonlinearity through a nozzle. Phys. Fluids A 3, 427–438 (1991).

[15] Cramer, M. S., Crickenberger, A. B.: Prandtl-Meyer function for dense gases. AIAA J. 30, 561–564 (1992).

[16] Cramer, M. S., Tarkenton, G. M.: Transonic flows of Bethe-Zel'dovich-Thompson fluids. J. Fluid Mech. 240, 197–228 (1992).

[17] Kluwick, A.: Transonic nozzle flow of dense gases. J. Fluid Mech 247, 661–688 (1993).

[18] Whitlock, S. T.: Compressible flows of dense gases in boundary layers. Master's thesis, Department for Engineering Science and Mechanics, Virginia Polytechnic Institute and State University, 1992.

[19] Zieher, F.: Laminare Grenzschichten in schweren Gasen. Master's thesis, Institute for Fluid Dynamics and Heat Transfer, Technical University, Vienna 1993.

[20] Reid, R. C., Prausnitz, J. M., Poling, B. E.: The properties of gases and liquids, 4th edn. New York: Wiley 1987.

[21] Stewartson, K.: On the flow near the trailing edge of a flat plate II. Mathematika 16, 106–121 (1969).

[22] Neiland, V. Ya.: Towards a theory of separation of the laminar boundary layer in a supersonic stream, Izv. Akad. Nauk. SSSR, Mekh. Zhidk. Gaza 4, (1969).

[23] Messiter, A. F.: Boundary layer flow near the trailing edge of a flat plate. SIAM J. Appl. Math. 18, 241–257 (1970).

[24] Stewartson, K.: Multistructured boundary layers on flat plates and related bodies. Adv. Appl. Mech. 14, 145–239 (1974).

[25] Smith, F. T.: On the high Reynolds number theory of laminar flows. IMA J. Appl. Math. 28, 207–281 (1982).

[26] Messiter, A. F.: Boundary-layer interaction theory. ASME J. Appl. Mech. 50, 1104–1113 (1983).

[27] Kluwick, A.: Stationäre, laminare wechselwirkende Reibungsschichten. Z. Flugwiss. Weltraumforsch. 3, 157–174 (1979).

[28] Kluwick, A.: Interacting boundary layers. Z. Angew. Math. Mech. 67, T3–13 (1987).

[29] Ruban, A. I.: Akademia Nauk SSSR 18, 1253–1265, (1978).

[30] Burggraf, O.: Private communication (1980).

[31] Kluwick, A.: On the nonlinear distortion of waves generated by interacting supersonic boundary layers. Acta Mech. 55, 177–189 (1985).

[32] Daniels, P. G.: Numerical and asymptotic solutions for the supersonic flow near the trailing edge of a flat plate. Q. J. Mech. Appl. Math. 27, 175–191 (1974).

[33] Cramer, M. S.: Transonic flows of BZT fluids. Proceedings of the 13th World Congress on Computation and Applied Mathematics (IMACS 91), 570–571 (1991).

[34] Messiter, A. F., Feo, A., Melnik, R. E.: Shock-wave strength for separation of a laminar boundary layer at transonic speeds. AIAA J. 9, 1197–1198 (1971).

[35] Bodonyi, R. J., Kluwick, A.: Freely interacting transonic boundary layers. Phys. Fluids 20, 1432–1437 (1977).

Author's address: o. Univ. Professor Dipl.-Ing. Dr. techn. A. Kluwick, Institut für Strömungslehre und Wärmeübertragung, Technische Universität Wien, Wiedner Hauptstrasse 7, A-1040 Wien, Austria

Acta Mechanica (1994) [Suppl] 4: 351—357
© Springer-Verlag 1994

Base pressure of a rotating axially symmetrical body

V. M. Kovalenko, Sumy, Ukraine

Summary. The base pressure is studied on axially symmetrical finned and finless bodies. The revolutions are either forced or selfsustained (in the case of finned bodies). It is shown that the revolution of the axisymmetric body results in an essentially reduced base pressure and hence an increased base drag. The structure of the flow in the base region is considerably modified by the rotation. The presence of tail fins leads to an additional pressure reduction at the base of a not rotating body. This effect turns out to be more pronounced in the case of the forced rotation. The pressure along the base radius varies almost uniformly. By varying the fin setting angle it is possible to control not only the magnitude but also the sign of the pressure gradient.

List of symbols

L, D, R	model length, diameter and radius, respectively, [m]
Ri	distance from the base center to a pressure hole, [m]
$\bar{R} = Ri/R$	relative radius
$\lambda = L/D$	model length-to-diameter ratio
κ	number of tail fins
S_F	tail fin area, [m²]
S_R	ring area produced by the rotating tail fin, [m²]
$\sigma = \kappa \cdot \dfrac{S_F}{S_R}$	coefficient of filling
φ	fin setting angle, i.e. the angle between the tail fin chord and the body axis, deg.
α	angle between the tail fin chord/body axis and the free-stream flow direction, i.e. incidence, deg.
U_∞	flow velocity, [m/s]
ϱ	air density, [kg/m³]
v	kinematic coefficient of viscosity, [m²/s]
$\mathrm{Re} = \dfrac{U_\infty \cdot L}{v}$	Reynolds number
ω	angular revolution velocity, [s⁻¹]
n	revolution frequency, [turns/min]
$\Theta = \dfrac{\omega \cdot R}{U_\infty}$	revolution parameter
P_B	base pressure, [Pa]
P_∞	pressure at the entrance to the test section [Pa]
$\bar{P}_B = \dfrac{P_B - P_\infty}{\dfrac{1}{2}\varrho U_\infty{}^2}$	base pressure coefficient.

1 Introduction

The problem of the base pressure of an axially-symmetrical body is of great practical interest and has been studied rather extensively. The least investigated factor which influences the base pressure of an axisymmetric body is its revolutions, especially if the body has tail fins. Today there are very few published papers devoted to the influence of the body revolutions on the base pressure. C. Wieselsberger [1] found experimentally how the total drag of three bodies of various shape is changed during the turning at low subsonic velocities. Wieselsberger's experiments permitted estimates to be made of the base drag contribution due to the body rotation. Some results related to that problem are obtained for supersonic flows in [2] and [3], for low subsonic velocities — in [4] and [5]. In the experiments [4, 5] the models are supported by a sting, mounted on the base of the body. The pressure probing tubes went through the hollow model and the sting. The presence of the sting as well as static pressure tubes in the first model [4] could have distorted the flow structure in the base region.

The rarefaction in the base region of an axisymmetric body is determined not only by the flow regime but also by the curling of the flow separating from the body. Regarding the curling flow, the drag depends on the frequency of rotation and the presence of tail fins. Those were the factors which had to be examined.

2 Test object and measurement technique

The model was an axially symmetrical cylindric body with $\lambda = 6.65$ which was either finless or fitted out with four tail fins (Fig. 1). The central part of the model casing was fixed to the side support, s.c. the base of models have been absolutely free of obstacles. Inside the model an air motor was mounted which rotated simultaneously the front and the back parts of the model surface. The model rotation may be realized also by means of inclined tail fins.

The tail fins (flat plates 1 mm thick with a sweepback angle $\chi = 30°$) were placed at the back part of the model. The fin setting angle was $\varphi = 0, \pm 15, \pm 30°$ (see Fig. 1), the coefficient of filling was $\sigma = 0,45$.

As shown in the experiments [4], the base rotation does not affect the base pressure significantly. That allowed us to simplify the measurements of the base pressure by probing it through pressure holes (0.3 mm in diameter and 25 in number) in the not rotating base.

Because of the side support the central part of the model casing did not move, while the front and the back parts rotated. Considering a short length of the motionless part, its retardation effect on the circumferential velocity component of the flow could be neglected [6].

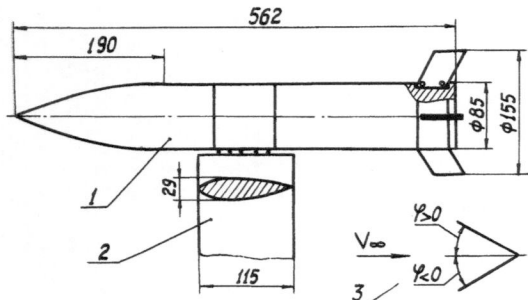

Fig. 1. A schematic view of the model. *1* Model; *2* support; *3* direction of reading the tail fin incidence

The experiments were done in wind tunnels with open or closed test section. The flow velocity U_∞ was varied from 6.5 to 50 m/s, with a Reynolds number of $Re = 2.5 \cdot 10^5 \cdots 1.9 \cdot 10^6$. The rotation frequency n reached 5000 turns/min for the finned model and about 10000 turns/min for the finless one. The clockwise direction of the model revolution as viewed from the base was assumed as positive.

Based on the test data, the $\bar{P}_{Bj} = f(\bar{R})$ curves were constructed where $\bar{P}_{Bj}$ is the average base pressure coefficient determined in four directions from the base center on the fixed radius. From the $\bar{P}_{Bj} = f(\bar{R})$ distribution the base pressure coefficient $\bar{P}_B$ was calculated which was averaged over the entire base area (taking into account the fraction of the corresponding ring area).

3 Finless model in forced rotation

As follows from the analysis of measurements being performed, at the non-rotating model the base pressure exhibits considerable fluctuations. The minimum pressure variations are observed in the vicinity of the outer base edge, while in the region where $\bar{R} \approx 0.5$ the fluctuations are the largest. The pressure is also somewhat nonuniform in different directions on the base surface especially in the experiments [4] where the incidence angle α is varied. As the relative rotating velocity is increased, the initial base pressure distribution asymmetry (at the base top and bottom) smooths down, and the original pressure nonuniformity along the radius vanishes, the base pressure fluctuations are damped, and the pressure in the bottom areas adjacent to the lee and windward sides of the body, become uniform at $\alpha > 0$.

The main result obtained is that the rarefaction in the base region increases appreciably during the rotation. The increase, being monotonous if the state of the boundary layer on the

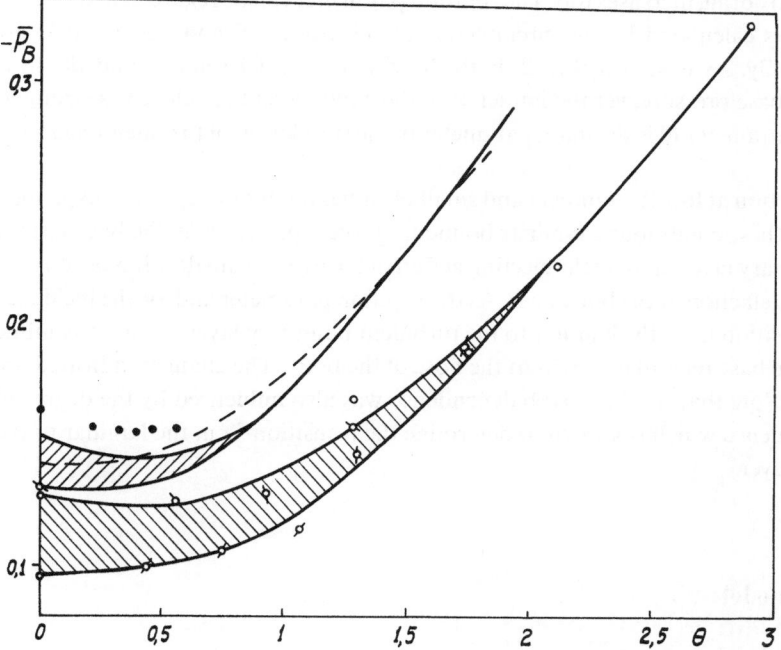

Fig. 2. Effect of revolutions on the base pressure coefficient of the finless model. \\\\\ $Re = 5.7 \cdot 10^5 \cdots 1.6 \cdot 10^6$; ⎯ $Re = 1.3 \cdot 10^6$; ○ $Re = 5.7 \cdot 10^5$ ⎯ $Re = 1.6 \cdot 10^6$. Comparison with another experiments: ---- $Re = 1.4 \cdot 10^6$, [4]; ///// $Re = 7 \cdot 10^5 \cdots 2.5 \cdot 10^6$, [5]; ● $Re = 1.9 \cdot 10^6$ [5]

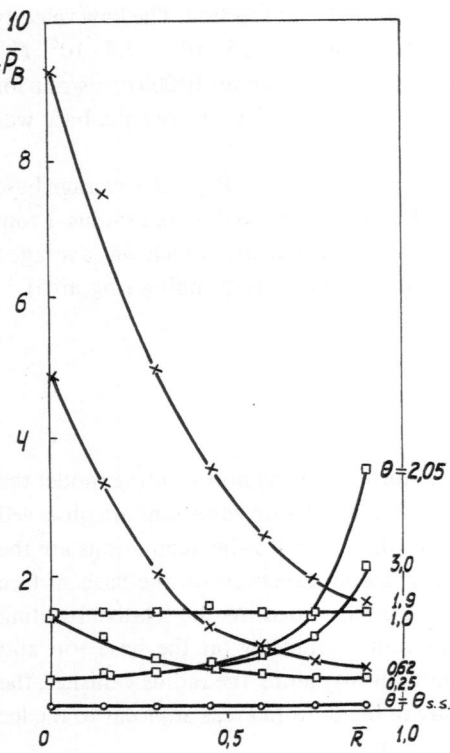

Fig. 3. Base pressure coefficient distribution along the base radius of the finned model at varied Θ and φ. x, $\varphi = 30°$; $\square$, $\varphi = -15°$

model surface is not changed. Naturally, in this situation the base drag which is an appreciable fraction of the total body drag, increases too. The rotation parameter can be regarded as a ratio of two Reynolds numbers calculated by the circumferential velocity $\omega \cdot R$ and free-stream flow velocity U_∞, respectively. As is seen in Fig. 2, both the circumferential velocity and the flow velocity influence the base pressure, yet the influence of the circumferential velocity is stronger. Thus, the revolution parameter Θ is the main parameter in the simulation of the phenomenon in question.

The greater rarefaction at low Re numbers and small Θ values (or at $\Theta = 0$) can be explained by the fact that under these conditions a laminar boundary layer is preserved in the base surface region. The thin boundary layer hinders the ejecting action of the free-stream flow less effectively, and results in larger rarefaction at the body base. As the rotation parameter and/or the incidence are increased, the transition from the laminar to the turbulent boundary layer on the streamline surface (and also in the base region) is shifted to the nose of the body. The changes indicated are clearly seen in Fig. 2. Note that the $\bar{P}_B = f(\Theta)$ dependence was also influenced by the degree of free-stream flow turbulence which is known to determine the transition from the laminar to the turbulent boundary layer.

4 Rotating finned model

The experiments were carried out in the regions of self-sustained ($\Theta = \Theta_{s.s.}$) and forced revolutions. The initial experimental data are plotted as $\bar{P}_B = f(\bar{R})$ curves for varied Θ at fixed φ (Fig. 3). The model performs a self-sustained revolutions with the aid of tail fins ($\varphi \gtrless 0$), driven by

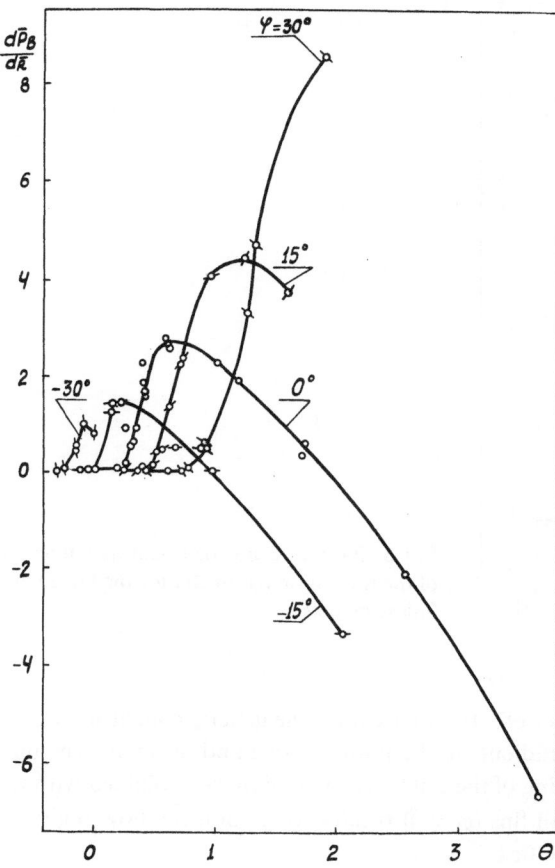

Fig. 4. Average base pressure gradient versus the revolution parameter at varied tail fin incidence

the free-stream flow. In this case the distribution of the base pressure coefficient almost coincides with that one for the finless model at the corresponding Θ value.

Of practical interest is the question of how the base pressure and average $\bar{P}_B$ values vary during forced revolutions of the finned model. For sufficiently large Θ's the pressure distribution along the base radius is essentially nonuniform. As a criterion for estimation of the pressure variation rate we take the quantity $\Delta \bar{P}_B / \Delta \bar{R}$, where $\Delta \bar{P}_B = \bar{P}_B|_{\bar{R}=0} - \bar{P}_B|_{\bar{R}=1}$, $\Delta \bar{R} = 1$. This quantity is in fact the pressure gradient on the base surface and, roughly speaking, can be termed so. The analysis of the $\Delta \bar{P}_B / \Delta \bar{R}$ curves (Fig. 4) demonstrates the following:

— to each φ corresponds a certain Θ value, and until this Θ value is reached, the base pressure gradient is zero or increases insignificantly;

— as the indicated Θ value is exceeded, the pressure gradient increases rapidly, the greatest rarefaction being observed in the center of the base surface (the gradient sign is assumed positive);

— for each φ there is a limiting Θ value and as soon as it is reached, the pressure gradient decreases as Θ increases further;

— for $\varphi = 0$ and $-15°$ the pressure gradient $\Delta \bar{P}_B / \Delta \bar{R}$ is found to decrease so that at large Θ's the gradient becomes negative (the greatest rarefaction is achieved at the base periphery);

— the maximum pressure gradient is different for different φ, increasing as φ increases.

The behaviour of the curves is qualitatively reproduced for all the φ values investigated. This suggests that for $\varphi = 15°$ and $30°$ the pressure gradient variations would be similar to those

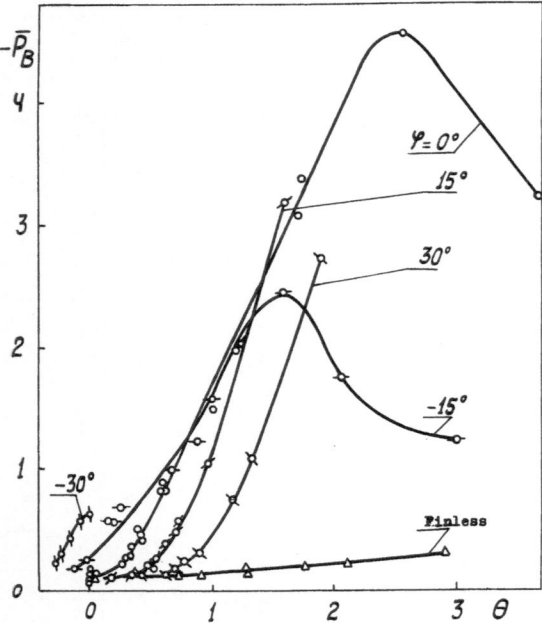

Fig. 5. Base pressure coefficient as a function of the revolution parameter for the finned and finless models

discussed above, only shifted towards greater Θ's. In other words, the general conclusion is that for $\varphi < 0$, $\varphi = 0$ and $\varphi > 0$ the pressure gradient can be both positive and negative. The flow structure in the base region due to the turning of the tail fins can possibly be explained with the following assumption: as the body with tail fins ($\varphi < 0$) rotates, very soon the flow regime is reached, where the flow separates from the fins.

Average $\bar{P}_B = f(\Theta)$ values corresponding to the forced revolution of the finned bodies with varying tail fin setting angles are presented in Fig. 5 for $-0.28 < \Theta < 3.67$. It can be seen that with increasing Θ the base pressure coefficient and the pressure gradient vary in a similar way, i.e. for each φ there is a corresponding Θ value and as soon as that Θ is exceeded, the rarefaction in the base region increases up to a certain maximum value, passes through a vertex and then starts to decrease. $\bar{P}_B$ may be rather large in magnitude, e.g. for $\varphi = 0$ the coefficient $\bar{P}_B = -4.6$ is reached at $\Theta \approx 2.5$. From the behaviour of the $\bar{P}_B = f(\Theta)$ curves it might be deduced that maximum $\bar{P}_B$ values for $\varphi > 0$ would be larger with increasing φ's.

The greater rarefaction in the base region is due to the larger flow twisting, which is determined by the fin setting relative to the free-stream flow direction, i.e. by the fin setting angle.

Thus, it may be concluded that the tail fins permit both the magnitude and the rate of variation of the base pressure along the base radius to be controlled on the revolving axisymmetric body. This result is belived to be promising for a practical use.

References

[1] Wieselsberger, C.: Über den Luftwiderstand bei gleichzeitiger Rotation des Versuchskörpers. Physik. Zeitschr. **28** (1927).

[2] Kavanau, L., Lenert, P., Hastings, C.: The result of the experimental investigation of the base drag at average value of the Reynolds numbers and the influence of the rotation of models. Woprosy raketnoy techniki **1** (1955) (in Russian, original in English).

[3] Kovalenko, V. M., Machnin, A. M.: The aerodynamic characteristics of a rotating axially symmetrical body with perforation of the back parts by number M = 2. Izwestiya SO AN SSSR **13**, 75−82 (1981) (in Russian).

[4] Kovalenko, V. M., Kisel, G. A.: Base drag of a rotating axisymmetric body. Izwestiya SO AN SSSR **16**, 58−66 (1985) (in Russian).

[5] Kovalenko, V. M., Voronov, G. M.: Base pressure of a rotating axisymmetrical finned body. Sbornik "Aerodynamika letatelnych apparatow i ich system", Kuybischew 1987 (in Russian).

[6] Higuchi, H., Rubesin, M. W.: An experimental and computational investigation of the transport of Reynolds stress in an axisymmetric swirling boundary layer. AIAA Paper 416 (1981).

Author's address: Professor Dr. Sc. V. M. Kovalenko, Academy of the Ukraine, Sumy Institute of Physics and Technology, 2 R-Korsakov St., 244007 Sumy 7, Ukraine

Acta Mechanica (1994) [Suppl] 4: 359–365
© Springer-Verlag 1994

A comparison of metastable flows of condensation in Laval nozzles and vaporization in capillary tubes

S. Lin, Montreal, Quebec

Summary. One-dimensional metastable flows with condensation of moist air in Laval nozzles and vaporization of Refrigerant-12 in capillary tubes are considered. The locations of the inception of condensation in Laval nozzles and vaporization in capillary tubes can be predicted. A steady "condensation" shock occurs in the supersonic part of a Laval nozzle. The effect of moisture condensation on the flow properties is in the downstream of the shock. For an unsteady and periodic condensation process, the effect of moisture condensation on the flow properties may be either in the downstream of the shock, or in both directions, downstream and upstream of the shock, dependent on the location of the shock. For vaporization of refrigerant in capillary tubes, the back pressure has an effect on the pressure distribution along the whole capillary tube, when the flow is at a non-critical flow condition. For a critical flow condition, the back pressure has no effect on the pressure distribution upstream of the location of the inception of vaporization. However, it does have an effect on the pressure distribution downstream of the location of the inception of vaporization. It shows that in the two-phase flow with vaporization, the critical flow rate is not controlled by the sonic speed at the exit of the capillary tube.

1 Introduction

When moist air flows through a Laval nozzle, due to decreasing temperature and pressure, condensation phenomena take place in the supersonic part of the nozzle, despite that the thermodynamic saturation state of the moisture occurs at the upstream of the throat. Figure 1 shows the temperature-entropy diagram in which vapor moisture flows isentropically from the stagnation state, o, passes the saturation state, s, and reaches the position c where the condensation begins. In the region below the saturated-vapor line, vapor is supersaturated and in a metastable state. The undercooling, $T_s - T_c$, is a characteristic quantity for moisture condensation in the Laval nozzle.

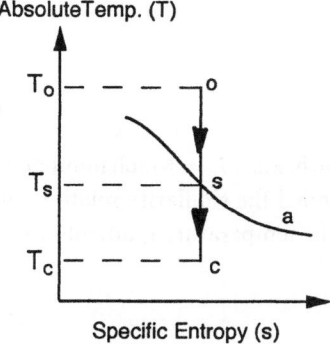

Fig. 1. Isentropic flow of moist air in a Laval nozzle in a temperature-entropy diagram: a saturated vapor line, o stagnation state with temperature T_o, s saturation state with temperature T_s, c inception of condensation at temperature T_c

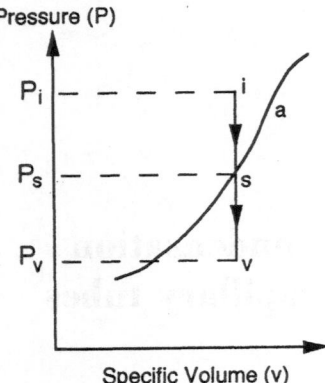

Fig. 2. Isochoric flow of liquid refrigerant in a capillary tube in pressure-volume diagram: a saturated liquide line, i inlet state with pressure p_i, s saturation state with pressure p_s, v inception of vaporization at pressure p_v

Consider a liquid refrigerant flowing through an insulated capillary tube having a diameter of about 1 mm. Due to friction losses, the pressure drop along the capillary tube causes the change of the refrigerant state from liquid to a vapor-liquid mixture. In such a process, there exists also a metastable phenomenon in the flow. This means that the inception of vaporization does not take place at the location of the thermodynamic saturation state, s, with a pressure p_s, but takes place at a location v with a pressure P_v downstream from location s as shown in Fig. 2. The pressure difference, $P_s - P_v$, is a characteristic quantity for the metastable flow, which is designated as the Underpressure of Vaporization.

In the present paper, a comparison of the features is made between the metastable flows of moisture condensation in Laval nozzles and refrigerant vaporization in capillary tubes.

2 Determination of the inception of condensation in Laval nozzles and vaporization in capillary tubes

The moist air flow through Laval nozzles and the refrigerant flow through capillary tubes are considered to be one-dimensional. For the determination of the inception of moisture condensation in Laval nozzles, Zierep and Lin [1] applied a dimensional analysis and used gas dynamic relations and a geometrical characteristic quantity (y^*/R^* = throat diameter/streamwise radius of curvature at the throat) to obtain the following similarity relation:

$$\phi_o{}^a = [(\varkappa + 1)/2]/[1 + (\varkappa - 1)\, M_c{}^2/2] \tag{1}$$

with

$$a = 0.208[\, y^*/(\varkappa + 1)\, R^*]^{0.295}, \tag{2}$$

where ϕ_o = stagnation relative humidity, $\varkappa = c_p/c_v$, ratio of specific heats, M_c = Mach number at the begin of condensation. Schnerr [2] and Schnerr et al. [3] extended the similarity relation for condensation of moist air transonic flow over profiles by using the temperature gradient at the nozzle throat, $(dT/dx)^*$, as follows:

$$a = 0.049\,8(-dT\,[\mathrm{K}]/dx\,[\mathrm{cm}])^{*0.3010} \tag{3}$$

It can be seen that an increase of the nozzle curvature or an increase of the temperature gradient at the throat results in an increase of the condensation Mach number.

For the determination of the inception of vaporization of Refrigerant-12 in capillary tubes, Chen et al. [4] used the classic nucleation theory in combination with dimensional analysis and 238 experimental sets of data to obtain the following correlation:

$$\Delta P_{sv} = 0.679\ \mathrm{Re}^{0.914}\ (\Delta T_{sc}/T_c)^{-0.208}\ (D/D')^{-3.18} \tag{4}$$

with

$$\Delta P_{SV} = (P_s - P_v)\ \sqrt{kT_s/\sigma^{3/2}}, \tag{5}$$

$$D' = \sqrt{kT_s/\sigma} \times 10^4, \tag{6}$$

where k = Boltzmann's constant ($= 1.380\,662 \times 10^{-23}$ J/K), T_s, T_c = temperatures at saturation and critical states, respectively, (K), σ = surface tension (N/m), $\mathrm{Re} = uD/v$ Reynolds number, u = mean velocity at the inlet of capillary tube (m/s), ΔT_{sc} = degree of subcooling at the inlet of capillary tube (K), D = diameter of capillary tube (m). Equation (4) applies in the following range:

$$0.464 \times 10^4 \lesssim \mathrm{Re} \lesssim 3.76 \times 10^4,$$

$$0 \lesssim \Delta T_{sc} \lesssim 17\,^{\circ}\mathrm{C},$$

$$0.66\ \mathrm{mm} \lesssim D \lesssim 1.17\ \mathrm{mm}.$$

It can be seen that the dimensionless underpressure of vaporization, ΔP_{sv}, is mostly affected by the size of the tube diameter, D. The smaller the diameter of the capillary tube, the larger is the value of ΔP_{sv}. Also, ΔP_{sv} is nearly proportional to Re, and inversely proportional to $(\Delta T_{sc})^{0.208}$.

3 Location of condensation shock in a Laval nozzle

The release of the latent heat of moisture condensation represents heat addition to the air stream in the Laval nozzle. Zierep [5] indicated that heat addition has an effect like that of cross-section contraction with a single exception that the temperature gradient changes its sign for heat addition with Mach number, $M \leq 1/\varkappa$. The value of the ratio of specific heats, $\varkappa$, is always larger than one. The Mach number, $M \leq 1/\varkappa$, is in the subsonic region of the nozzle. Thus heat addition caused by moisture condensation in the supersonic part of the nozzle always has an effect like that of cross-section contraction on flow properties.

Figure 3 shows a schematic diagram of moisture condensation in a Laval nozzle. The dashed line represents the effective streamtube cross-section resulting from the release of the latent heat of moisture condensation. Depending on the magnitude of the relative humidity, three different forms of flow have been observed [5]: (a) the flow exhibits a steady and continuous behaviour; (b) a steady shock occurs at the start of the condensation region; and (c) the flow becomes unsteady.

In general, for heat addition, the effective streamtube cross-section area, A_e, may be equal to, smaller or larger than the nozzle throat area, A^*. Zierep [5] discussed the possibility of the shock occuring in the nozzle. For heat addition resulting from moisture condensation in a Laval nozzle, and for the case $A_e > A^*$ (Case a in Fig. 3), a steady shock due to condensation occurs only at the

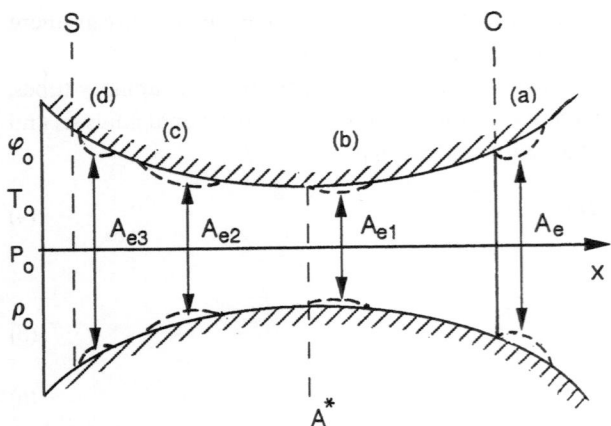

Fig. 3. Schematic diagram of moisture condensation in a Laval nozzle (the dashed line represents the effective streamtube cross-section in the condensation zone): S, C respectively the positions of thermodynamic saturation state and inception of condensation, a steady state condensation, $A_e > A^*$, b, c, d unsteady and periodic condensation ($b\ A_{e1} < A^*$, $c\ A_{e2} = A^*$, $d\ A_{e3} > A^*$), φ_o, T_o, p_o, ϱ_o respectively the relative humidity, temperature, pressure, and density at the stagnation state

beginning of condensation, section c, because a shock occurring between sections c and A_e is unstable. Therefore the location of a steady "condensation" shock represents the location of the inception of condensation.

If heat addition, caused by moisture condensation, increases significantly, the condensation process taking place in the Laval nozzle becomes unsteady and periodic. These phenomena were discovered by Schmidt [6], discussed in detail by Barschdorff [7] and Wegener and Cagliostro [8]. During such an unsteady process, the "condensation" shock moves upstream. When it reaches in the vicinity of the nozzle throat, the effective streamtube cross-section A_{e1} may become smaller than the throat cross-section, A^* (Case b in Fig. 3). For such a case, the passage through sonic condition at A^* can no longer be maintained and critical conditions set in only in the region of heat addition at A_{e1} [5]. The flow upstream of A_{e1} is now subsonic. A "condensation" shock can then appear only downstream of A_{e1}. The "condensation" shock then becomes gradually weaker when it moves towards the thermodynamic saturation point, s, because reduction of condensation takes place in the direction towards point s. (Note that no condensation can take place in the upstream of the saturation point, s.) Finally, the shock caused by condensation disappears in the Laval nozzle when either one of the following two conditions is satisfied:

(1) The "condensation" shock reaches the thermodynamic saturation point, s.

(2) The effective streamtube cross-section, A_{e2}, becomes equal to (Case c in Fig. 3) or A_{e3}, becomes larger than (Case d in Fig. 3) the nozzle throat cross-section, A^*, because for such a case, it is not possible for a shock occuring upstream of A^*.

It can be seen that, at the instance when A_{e1} in the supersonic part of the nozzle becomes smaller than A^*, there is a sudden change of critical conditions and a sudden change of the location of the shock due to condensation which moves from the upstream to the downstream of A_{e1}. These changes cause disturbances in the downstream of A_{e1}, which favours the formation of moisture condensation at the downstream of A_{e1}. Thus the condensation process is periodically starting up again. The frequency of the periodic flow can be predicted by using a similarity relation obtained by Zierep and Lin [9].

4 Effect of back pressure on metastable flow in capillary tubes

When a flow at the exit of a capillary tube is subsonic, the back pressure has an important effect on flow properties inside the capillary tube. Experimental results conducted by Li et al. [10] showed that the metastable flow of Refrigerant-12 through capillary tubes has different characteristics in comparison with that of moist air through Laval nozzles.

Figure 4 shows the mass flux, G, and the underpressure of vaporization, $P_s - P_v$, as functions of the back pressure, P_b. It can be seen that the variation of G and $p_s - p_v$ is separated by the back pressure p_c into two regions: In region (1), $p_b > p_c$, a decrease of p_b results in an increase of G and $p_s - p_v$. In region (2), $p_b < p_c$, a decrease of p_b has no effect on G and $p_s - p_v$. This means that G and $p_s - p_v$ remain constant in region (2) for $p_b < p_c$. This separation back pressure, P_c, is defined as the critical pressure, and the corresponding flow at this condition is defined as the critical flow.

Figure 5 shows the pressure distribution measured along the capillary tube with the back pressure as parameter for the Re-number in the range: $550 < \text{Re} < 600$. The flows represented by curves 1, 2 and 3 do not reach the critical condition. At the non-critical flow condition, the back pressure has an effect on the pressure distribution along the whole capillary tube. The flows represented by curves 4, 5, 6 and 7 have reached the critical condition. For a critical flow, the back pressure has no effect on the pressure distribution upstream of the location of the inception of vaporization. However, it does have an effect on the pressure distribution downstream of the location of the inception of vaporization, as shown in Fig. 5. This phenomenon of two-phase flow is significantly different from that of single phase flow.

In a single phase flow, the flow speed reaches its sonic speed at the exit under the critical flow condition. For such a case, a drop of the back pressure has no effect on the flow properties in the upper stream. Therefore the pressure may have discontinuity at the exit plane.

In a two-phase flow with vaporization, the critical flow condition is different. Before the flow reaches the critical state, a decrease of the back pressure causes an increase of the flow-rate and of

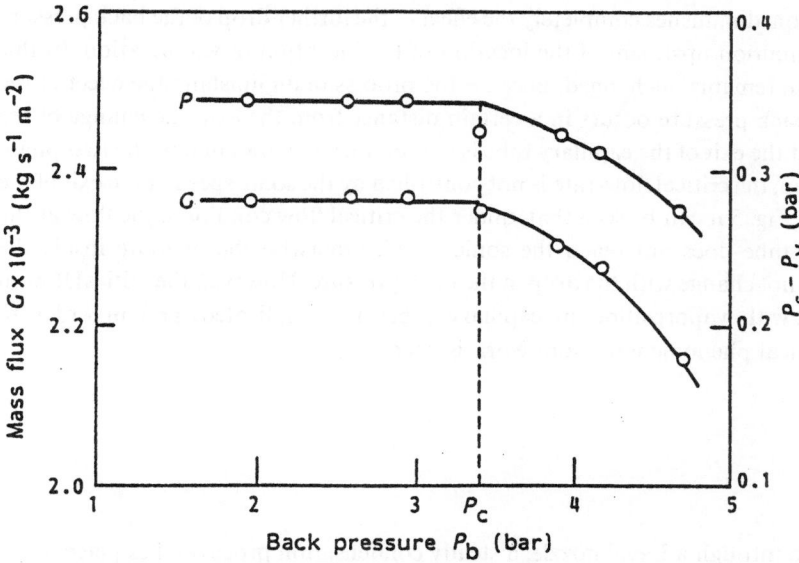

Fig. 4. Relationships between mass flux and back pressure, and between underpressure of vaporization and back pressure for flow of Refrigerant-12 through a capillary tube. $D = 0.66$ mm, $p_i = 7.17$ bar, $T_i = 23.2\,°\text{C}$

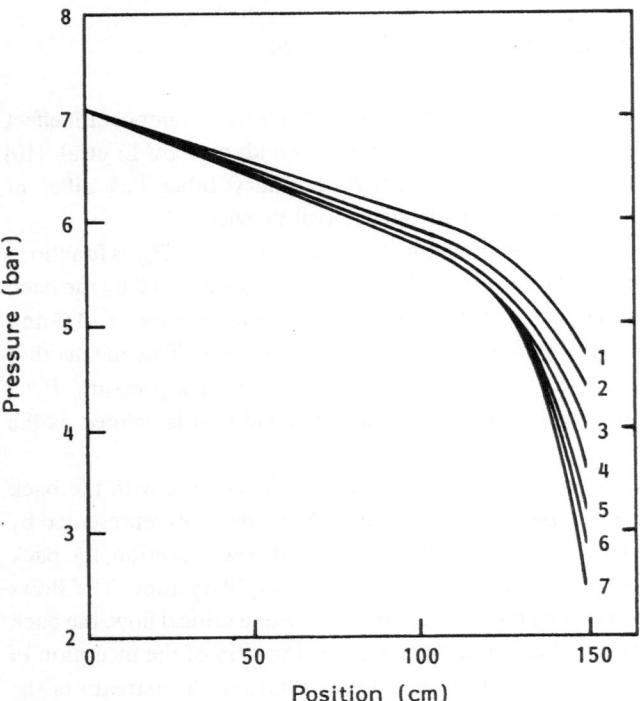

Fig. 5. Pressure distributions of Refrigerant-12 along a capillary tube with back pressure as parameter. $D = 0.66$ mm, $p_i = 7.17$ bar, $T_i = 23.2\,°C$, $p_b = 4.69$ (*1*), 4.21 (*2*), 3.91 (*3*), 3.43 (*4*), 3.00 (*5*), 2.60 (*6*), 1.96 (*7*) bar, $p_c = 3.43$ bar

vaporization too. The increase of vaporization results in increases in the frictional pressure drop and the accelerational pressure drop. When the back pressure drops to the critical pressure, the increase of the frictional pressure drop and the accelerational pressure drop, due to vaporization in the two-phase region, diminishes completely the effect of the further drop of the back pressure on the pressure distribution upstream of the location of the inception of vaporization. In this instance, the flow-rate remains unchanged. Because the process of diminishing the effect of the further drop of the back pressure occurs in a certain distance from the exit, the change of the pressure inside and at the exit of the capillary tube is continuous. It shows that in the two-phase flow with vaporization, the critical flow-rate is not controlled by the sonic speed at the exit of the capillary tube. From Fig. 5 it can be seen that, under the critical flow condition, the flow at the exit of the capillary tube does not reach the sonic speed, otherwise the pressure inside the capillary tube would not change with the drop in the back pressure. However, the critical flow in the two-phase flow with vaporization in capillary tubes is complicated, and in order to understand the physical phenomenon, more work is needed.

5 Conclusions

When moist air flows through a Laval nozzle, a steady condensation process takes place in the supersonic part of the nozzle. Equation (1) predicts the location of the inception of condensation in Laval nozzles. The "condensation" shock occurs at the beginning of the condensation region.

The effect of moisture condensation on the flow properties occurs downstream of the shock. For an unsteady and periodic condensation process, the condensation zone moves upstream into the vicinity of the nozzle throat. If the effective streamtube cross-section, A_{e1}, becomes equal to or smaller than the throat cross-section, A^*, the "condensation" shock has to change its location from the upstream to the downstream side of A_{e1}. For this specified case, because a part of the condensation zone is located in the subsonic part of the nozzle, the effect of moisture condensation on the flow properties occures downstream and upstream of the shock.

For liquid Refrigerant-12 flowing through a capillary tube, the location of the inception of vaporization can be predicted by Eq. (4). At a non-critical flow condition, the back pressure has an effect on the pressure distribution along the whole capillary tube. For a critical flow, the back pressure has no effect on the pressure distribution upstream of the location of the inception of vaporization. However, it does have an effect on the pressure distribution downstream of the location of the inception of vaporization. It shows that in the two-phase flow with vaporization, the critical flow rate is not controlled by the sonic speed at the exit of the capillary tube.

References

[1] Zierep, J., Lin, S.: Bestimmung des Kondensationsbeginns bei der Entspannung feuchter Luft in Überschalldüsen. Forsch. Ingenieurwes. **33**, 169−172 (1967).
[2] Schnerr, G.: Homogene Kondensation in stationären transsonischen Strömungen durch Lavaldüsen und um Profile. Habilitationsschrift Universität (TH) Karlsruhe, Fakultät für Maschinenbau 1986.
[3] Schnerr, G., Dohrmann, U., Jantzen, H.-A., Huber, R. R.: Transsonische Strömungen mit Relaxation und Energiezufuhr durch Wasserdampfkondensation. Strömungsmech. Strömungsmasch. **40**, 39−79 (1989).
[4] Chen, Z.-H., Li, R.-Y., Lin, S., Chen, Z.-Y.: A correlation for metastable flow of refrigerant-12 through capillary tubes. ASHRAE Trans. **96**, 550−554 (1990).
[5] Zierep, J.: Theory of flows in compressible media with heat addition. AGARD-AG **191** (1974).
[6] Schmidt, B.: Beobachtungen über das Verhalten der durch Wasserdampf-Kondensation ausgelösten Störungen in einer Überschall-Windkanaldüse. Dissertation Universität Karlsruhe; also: Jahrbuch der WGLR, 160−167 (1962).
[7] Barschdorff, D.: Kurzzeitfeuchtemessung und ihre Anwendung bei Kondensationserscheinungen in Lavaldüsen. Dissertation Universität Karlsruhe; also: Mitt. Inst. Strömungsl. Strömungsmasch. Univ. Karlsruhe **6**, 18−39 (1967).
[8] Wegener, P. P., Cagliostro, D. J.: Periodic nozzle flow with heat addition. Combust. Sci. Technol. **6**, 269−277 (1973).
[9] Zierep, J., Lin, S.: Ein Ähnlichkeitsgesetz für instationäre Kondensationsvorgänge in Lavaldüsen. Forsch. Ingenieurwes. **34**, 97−99 (1968).
[10] Li, R. Y., Lin, S., Chen, Z. Y., Chen, Z. H.: Metastable flow of R-12 through capillary tubes. Int. J. Refig. **13**, 181−186 (1990).

Author's address: Professor Dr.-Ing. S. Lin, Department of Mechanical Engineering, Concordia University, 1455 de Maisonneuve Blvd. West, Montreal, Quebec, Canada H3G 1M8

Acta Mechanica (1994) [Suppl] 4: 367–376
© Springer-Verlag 1994

Liquid metal magnetohydrodynamics in complex pipe geometries

U. Müller and **L. Bühler**, Karlsruhe, Federal Republic of Germany

Summary. The paper outlines the principal physical effects of liquid metal channel flow in strong magnetic fields. The magnetohydrodynamic equations and the boundary conditions for channels of thin conducting walls are presented. In particular, the inertialess core flow model is discussed. Results from calculations are compared with experimental data.

1 Introduction

In nuclear power engineering liquid metal is often used to cool high heat flux surfaces. For self-cooled blankets surrounding the magnetically confined plasma of future fusion reactors liquid lithium or lithium-lead alloy have been proposed as a coolant. A typical self-cooled blanket design has been developed at the Nuclear Research Center Karlsruhe [1]. Blankets are the crucial components in which the high neutron and γ-radiation energy from the plasma is transformed to utilizable thermal energy. Blankets have to be designed as a compact channel system to assure uniform heat removal from the structural material of the plasma chamber and to shield the plasma controlling magnets from nuclear irradiation. Therefore, blankets have to be placed within the range of very strong magnetic fields of fusion machines. Strong electromagnetic forces acting on the liquid metal flow may give rise to large pressure losses. Furthermore, they may cause velocity distributions unfavorable for heat removal. An accurate prediction of the pressure losses and the velocity distributions in the cooling channel systems is of paramount importance for the design of self-cooled blankets. Computational MHD-models are being developed and experiments are being conducted in order to demonstrate the technical feasibility of a self-cooled blanket. Some basic ideas as well as some theoretical and experimental results are outlined in this article.

2 Fundamental physical effects

We consider channel flow of a conducting fluid under the influence of a strong external transverse magnetic field of induction **B**. The shape of the cross-section may be arbitrary (see Fig. 1). In the fluid flowing with a velocity **v**, due to the interaction with the magnetic field an electric field $\mathbf{E}^* = \mathbf{v} \times \mathbf{B}$ is induced. According to Ohm's law the electric field drives currents of density **j**. These currents together with the magnetic field generate a force density $F = \mathbf{j} \times \mathbf{B}$, called Lorentz force. In the center of the channel this force points opposite to the flow direction and acts against the forcing external pressure gradient, thus causing electromagnetic pressure losses in addition to

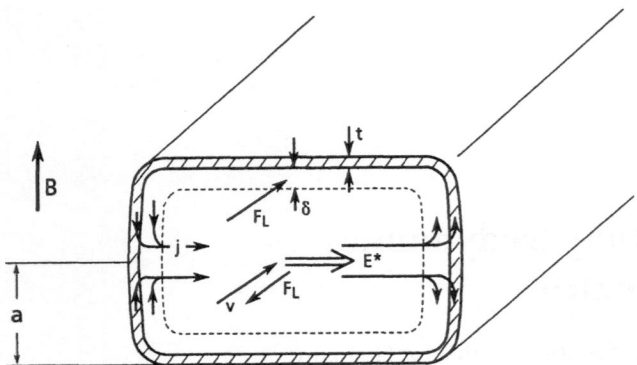

Fig. 1. Principle sketch of MHD-pipe flow

hydraulic ones. The conservation of electric charge requires closed current paths. In the channel system the currents find closed circuits along areas of very small or zero velocity. Such areas are the thin viscous boundary layers near the channel walls or the conducting walls themselves. It is obvious that the intensity of the electrical currents, and thus the flow opposing force density, is limited by both the resistance of returning currents in the walls and the resistance of the fluid.

For strong magnetic fields the whole flow region may be divided into distinct subregions. One is the core, where viscous and inertia effects are unimportant and the driving pressure gradient is mainly balanced by the Lorentz force. The core is surrounded by boundary layers of two types, depending on the orientation of the magnetic field to the wall normal direction. At walls where a high normal magnetic field component exists we find so-called Hartmann layers, where the viscous force is mainly balanced by the Lorentz force of returning current and the inviscid core is matched to the no-slip boundary condition. At walls almost tangential to the *B*-field Lorentz forces are of minor importance. Along these walls thin, high velocity jets may occur, carrying a significant part of the total volumetric flux. In addition interior layers may be present, which spread through the fluid along magnetic field lines, starting from discontinuities at channel walls caused by abrupt changes of wall slopes, wall conductivity, wall thickness or by sudden changes in the applied magnetic field. These layers are formed around a current sheet and match the core variables at both sides [2].

3 Equations and boundary conditions

Commonly the physical quantities velocity $\mathbf{v}$, magnetic induction $\mathbf{B}$, pressure p, current density $\mathbf{j}$ and the electric field $\mathbf{E}$ are scaled by the control parameters mean velocity U_0, the external magnetic induction B_0 and the composite quantities $a\sigma U_0 B_0^2$, $\sigma U_0 B_0$, $U_0 B_0$. The geometry is scaled by a characteristic dimension a of the channel cross-section.

Liquid-metal MHD flow is governed by the following set of dimensionless mechanical and electromagnetic equations:

Conservation of mass: $\qquad \nabla \cdot \mathbf{v} = 0,$ $\qquad\qquad\qquad\qquad\qquad\qquad$ (3.1)

Conservation of momentum: $\quad 1/N[\partial_t \mathbf{v} + (\mathbf{v} \cdot \nabla)\,\mathbf{v}] = -\nabla p + \mathbf{j} \times \mathbf{B} + 1/M^2 \nabla^2 \mathbf{v},$ $\qquad$ (3.2)

Ampère's law: $\qquad\qquad \nabla \times \mathbf{B} = R_m \mathbf{j},$ $\qquad\qquad\qquad\qquad\qquad$ (3.3)

Faraday's law: $\qquad\qquad \nabla \times \mathbf{E} = -\partial_t \mathbf{B},$ (3.4)

Ohm's law: $\qquad\qquad \mathbf{j} = \mathbf{E} + \mathbf{v} \times \mathbf{B}.$ (3.5)

Here $N = \sigma a B_0^2/(\varrho U_0)$ is the Interaction parameter describing the ratio of electromagnetic and inertial forces. The square of the Hartmann number $M = aB_0(\sigma/(\varrho \cdot v))^{1/2}$ ratios electrodynamic and viscous forces. The magnetic Reynolds number $R_m = \mu \sigma a U_0$ is a measure for the ratio of the induced magnetic induction to the applied induction B_0. Here μ is the magnetic permeability, σ is the specific electric conductivity, v the kinematic viscosity and ϱ the liquid density. From Ampère's and Faraday's law we conclude, that $\mathbf{j}$ and $\mathbf{B}$ are solenoidal.

$$\nabla \cdot \mathbf{j} = 0, \quad \nabla \cdot \mathbf{B} = 0.$$ (3.6), (3.7)

Furthermore, for design calculations we consider steady flows only. Faraday's law, therefore, reduces to $\nabla \times \mathbf{E} = 0$ and the electric field can be expressed by the gradient of the scalar electric potential Φ.

$$\mathbf{E} = -\nabla \Phi.$$ (3.8)

In particular for non-moving conductors the conservation of charge gives

$$\nabla^2 \Phi = 0.$$ (3.9)

For the above partial differential equations boundary conditions must be specified. At the channel walls the no-slip condition

$$\mathbf{v}|_{\text{wall}} = 0$$ (3.10)

holds for viscous fluids. The conservation of electric charge requires continuous normal current density components at the fluid-wall interface. This condition can be transformed into a relation for the electrical potential Φ. Using Ohm's law we obtain

$$\mathbf{j} \cdot n|_{\text{fluid}} = -\sigma_w/\sigma \cdot \partial_n \Phi|_{\text{wall}},$$ (3.11)

where ∂_n denotes the wall-normal derivative and σ_w/σ the ratio of wall and fluid conductivity. Furthermore, the potential across the wall-fluid interface can be assumed continuous, since contact resistance is excluded:

$$\Phi|_{\text{fluid}} = \Phi|_{\text{wall}}.$$ (3.12)

If the channel walls are insulating relation (3.11) reduces to

$$\mathbf{j} \cdot \mathbf{n}|_{\text{fluid}} = 0.$$ (3.13)

An equivalent relation holds at the outer surface of conducting channel walls towards an insulating surrounding.

To solve the full set of partial differential equations $(3.1 - 3.9)$ together with the boundary condition equations $(3.10 - 3.13)$ in general is a formidable task. Attempts to calculate MHD-flow numerically based on the above equations are limited to relatively small values of the Hartmann number and to simple rectangular duct geometries. A direct numerical treatment of the problem for high parameter values and complex geometries is currently prohibitive because of the limitations in computational capacity.

4 Flow in the core

The magnetic field for the plasma confinement of fusion reactors has an induction intensity of the order of $5-10$ Tesla. Taking into account the blanket dimensions, the thermohydraulic requirements and material properties, the previously defined dimensionless groups become typically of the order $M \approx 10^4$, $N \approx 10^2 - 10^5$ and $R_m \sim 0.07$. The extremely high values of Hartmann number and Interaction parameter, and the low value of magnetic Reynolds number permit remarkable simplification of the governing equations under the assumption: $M \gg 1$, $N \gg 1$, $\mathrm{Re}_m \ll 1$. These relations suggest to search for a solution in form of a power series expansion in the small parameters $1/N$, $1/M$, Re_m. To leading order the set of Eqs. $(3.1-3.7)$ is then simplified to the following set of linear equations:

$$\nabla \cdot \mathbf{v} = 0, \qquad -\nabla p + \mathbf{j} \times \mathbf{B}_0 = 0, \tag{4.1}, (4.2)$$

$$\nabla \cdot \mathbf{j} = 0, \qquad \mathbf{j} = -\nabla \Phi + \mathbf{v} \times \mathbf{B}_0. \tag{4.3}, (4.4)$$

Within leading order in Eqs. $(4.1-4.4)$ viscous and inertia effects are neglected and pressure forces are balanced by Lorentz forces only. $\mathbf{B}_0$ is the imposed external magnetic induction since induced magnetic fields are also neglected. The boundary condition equation (3.10) has to be replaced by the non viscous condition $\mathbf{v} \cdot \mathbf{n}|_{\mathrm{wall}} = 0$. This so-called core flow approximation, first proposed by Kulikovskii [3], has frequently been applied. In general the determination of all flow variables requires to solve Eqs. $(4.1-4.4)$, for the fluid and the Laplace equation (3.9) for Φ in the solid walls coupled to the fluid domain by the boundary condition equation (3.11). For thin channel walls of thickness $t \ll a$ Walker [4] has proposed an approximation to replace Eq. (3.9) and boundary condition equation (3.11) at the fluid-wall interface. He obtains for the thin wall sheet the following condition for the potential at conducting channel walls

$$\mathbf{j} \cdot \mathbf{n}|_{\mathrm{fluid}} = -c \nabla_t^2 \Phi, \tag{4.5}$$

where $c = \sigma_w t/(\sigma a)$ is the wall conductance ratio and ∇_t^2 the tangential Laplacian at the wall. This relation has been generalized later to account for variable wall thickness or variable wall conductivity [5, 6].

5 Boundary layers

The previously outlined approximation suffers from neglecting boundary layers. As in ordinary fluid mechanics boundary layers can be taken into account by solving boundary layer equations and matching their solution to the external flow solution. A correction is important especially if the case of insulating channel walls has to be treated, since returning currents can close their circuit only through the boundary layers. The local form of the equations for the Hartmann layer can be easily derived from Eqs. $(3.1-3.7)$ by assuming $N \gg M \gg 1$ and by an appropriate coordinate stretching procedure. In the following we introduce a wall-fitted coordinate system (t_1, t_2, n) where $t_{1,2}$ are coordinates tangential to the wall scaled by the channel dimension a and n is the inward normal coordinate scaled by the Hartmann layer thickness δ. Furthermore, let t_1 be the coordinate in the direction of the core velocity (see Section 4) at the wall, and M_L the local Hartmann number based on the external B-field component normal to the wall. For $N \to \infty$ we get from the Eq. (3.2) the order of magnitude relation $\delta \sim M_L^{-1}$. The mass and current conservation equations (3.1, 3.6) give for the velocity and current density components perpendicular to the wall $v_n \sim M_L^{-1}$, $j_n \sim M_L^{-1}$. We conclude from Ohm's law (Eq. (4.4)) that $\partial_n \Phi \sim M_L^{-2}$ holds for the normal derivative of the potential. Therefore, the potential is constant across the Hartmann boundary layer to the order M_L^{-1}. An equivalent relation is obtained for the pressure gradient normal to the wall, namely $\partial_n p \sim M_L^{-2}$. The governing boundary layer equation for the velocity is derived from the momentum equation in t_1 direction using Ohm's law and the above relations. Defining the boundary layer length scale by $\delta = M_L^{-1}$ we get

$$\partial_n^2 \mathbf{v} - \mathbf{v} = -\mathbf{v}_c. \tag{5.1}$$

Obviously the velocity distribution in the boundary layer is determined by the ordinary differential equation, whose inhomogenious term, the negative core velocity $-\mathbf{v}_c$, is defined by Eqs. (4.1−4.4), describing the flow in the core. The solution of Eq. (5.1) which satisfies the no-slip boundary condition at the wall for $n = 0$ and which matches the inviscid core solution as $n \to \infty$ reads:

$$\mathbf{v} = \mathbf{v}_c[1 - \exp(-n)]. \tag{5.2}$$

The Hartmann layer has implications on the boundary condition for the core current. The current $\mathbf{j}$ entering the boundary layer from the core is deflected tangentially to the wall; it flows partly in the Hartmann layer and partly in the electrically conducting thin wall. This reasoning leads to the following modified condition for the normal component of the core current density at the edge of the Hartmann layer [6]

$$\mathbf{j} \cdot \mathbf{n}|_{\text{core}} = -\nabla_t[(\delta + c)\,\nabla_t \Phi]. \tag{5.3}$$

Here $\nabla_t = (\partial_{t1}, \partial_{t2}, 0)$, is the tangential gradient. The set of Eqs. (4.1−4.5; 5.1−5.3) forms a basis for calculating inertialess MHD flow in complex channel geometries, if an appropriate numerical scheme is employed.

6 Numerical procedure

Instead of solving Eqs. (4.1−4.4) and Eq. (5.3) numerically we initially perform some basic analytical treatment. From Eq. (4.2) it is obvious that the pressure p as well as the current components $\mathbf{j}_\perp$ perpendicular to $\mathbf{B}_0$ do not vary along magnetic field lines, since $\mathbf{B}_0 \cdot \nabla p = 0$. This implies immediately, that the B-field parallel current component $j_\parallel$ may only vary linearly due to the charge conservation equation (4.3) and that the potential Φ may vary at least as a quadratic function along field lines. These characteristic properties of inertialess MHD core flow allow for an analytical integration of Eqs. (4.1−4.4). This reduces the general 3-D problem to a 2-D problem which finally is treated by numerical methods without loosing any 3-D information. The procedure is as follows: Currents $\mathbf{j}_\perp$ are calculated as a function of the 2-D pressure p using Eq. (4.2) and eliminated in Eq. (4.3). $j_\parallel$ is given by an integration along B-field lines, introduced in Eq. (4.4) to give after additional integration the potential variation in the $\mathbf{B}_0$-direction. Potentials Φ_t and Φ_b at both channel walls intersecting the considered B-field line are introduced as integration constants. From Eq. (4.4) we get the velocity components $\mathbf{v}_\perp$ perpendicular to $\mathbf{B}_0$ as a function of Φ_t, Φ_b and p. They are introduced in Eq. (4.1) to give after an integration along $\mathbf{B}_0$ a second order partial differential equation for the 2-D pressure if Φ_t and Φ_b are known. In general they are unknown 2-D functions which may be determined by the charge conservation

Eq. (5.3) in Hartmann layers and conducting channel walls. Thus the 3-D problem is essentially reduced to a 2-D problem for the unknown pressure p, coupled to the potentials Φ_t and Φ_b at the channel walls. Once the pressure and surface potentials are known we are able to reconstruct all flow variables inside the fluid domain. To obtain a unique solution procedure applicable for general geometries we introduced curvilinear coordinates. One coordinate (u^3) is chosen in $\mathbf{B}_0$-direction to provide easy analytical treatment, the others are chosen to be tangential to the channel walls. Every arbitrary geometry is transformed to a standard volume $0 \leq u^1, u^2 \leq 1, -1 \leq u^3 \leq 1$. The channel wall on which the equation for potentials Φ_t and Φ_b has to be solved reads simply $u^3 = +1$ and $u^3 = -1$ respectively. The 2-D equations for wall potentials and for pressure are solved numerically using a staggered grid for discrete potential and pressure values. A direct linear solution algorithm is used, since the convergence criteria for fast iterative solvers are not satisfied for all possible general applications (for details see [6]).

7 Experimental investigations

If the core flow model described in Sections 4 – 6 is to be used as a computational design tool for blanket development, a validation of its predictive capability is required. Therefore MHD-experiments were carried out using a superconducting magnet of an intensity up to 3.5 Tesla generating a homogeneous magnetic field of 45 cm length in a working cross section of 40 cm [7]. A stainless steel tube of rectangular cross-section forming a Z-shaped double bend was placed in the homogeneous range of the B-field. A schematic sketch of the experimental arrangement is shown in Fig. 2. Along the test tube the static pressure was measured at different locations of the perimeter. Furthermore, thin electrodes were welded to the tube surface to measure the local potential. Positions for pressure and potential measurements are indicated in Fig. 3 a.

8 Results and discussion

The considered 90° bend is a basic element of the self-cooled blanket concept [1]. If inertial effects play a significant role at all for high values of N this should be most expressed in the geometry of a sharpe edge. The inlet and outlet connecting pipes of the bend are tilted by an angle of 15° to the magnetic field lines (see Fig. 3 a). This arrangement was chosen to provide fully developed MHD-flow in each part of the bend and to avoid entrance length problems occurring in pipe flow

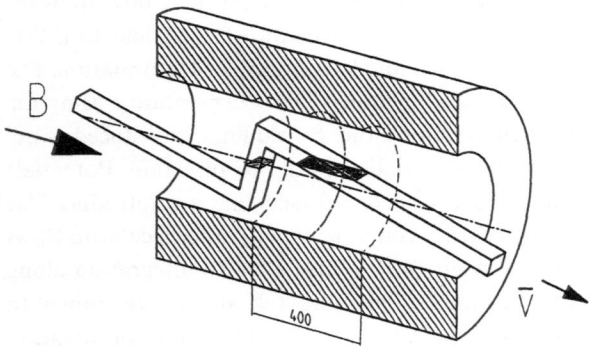

Fig. 2. Principle arrangement of the z-bend in the solenoidal magnet

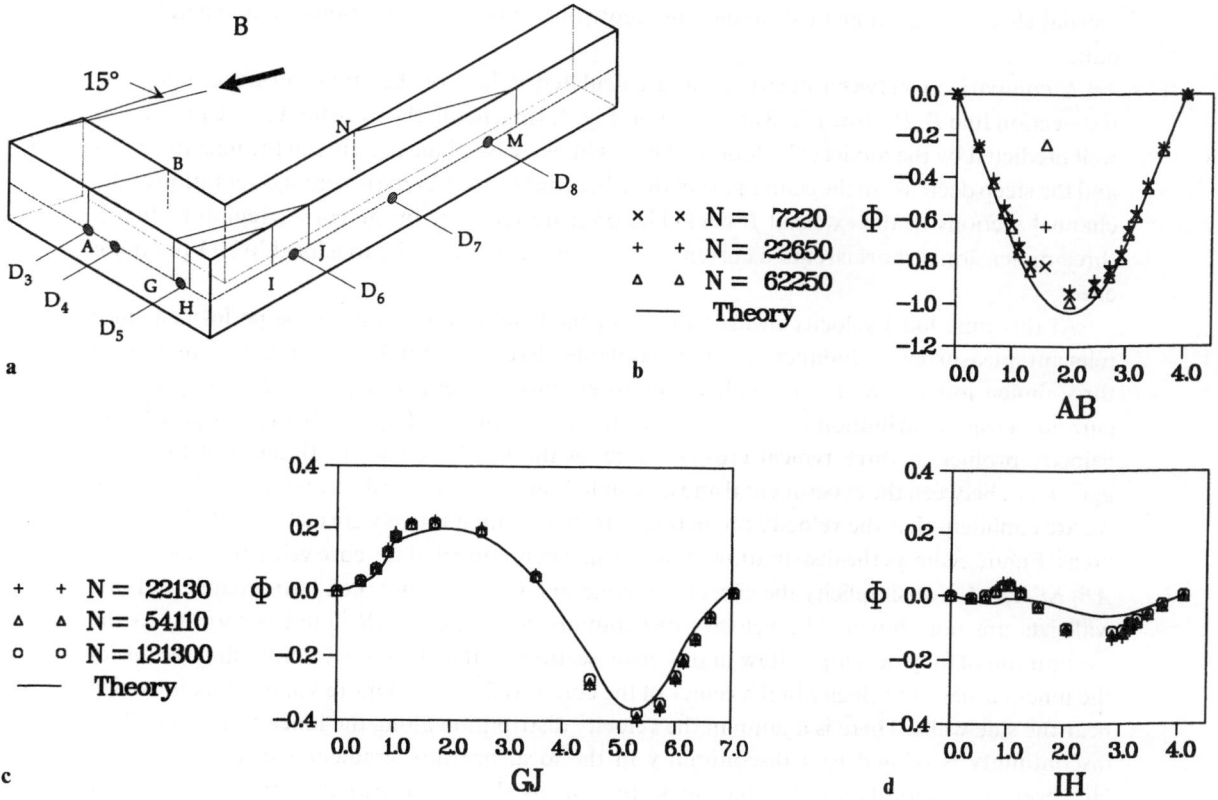

Fig. 3. a Principle sketch of one half of the z-bend showing the positions of pressure taps $D_3 - D_8$ and the characteristic surface lines AB, GJ, HI, NM where potentials were measured and calculated. **b – d** Measured and calculated values of surface potentials at AB, GJ, IH

aligned with the magnetic field lines. The calculations were performed under the assumption of fully developed flow in the inlet plane NM and a symmetry plane AB both aligned to the magnetic field. In the numerical procedure the potential Φ and the pressure p are the leading variables to be determined from coupled linear second order, elliptic differential equations. All other variables can be derived thereof by analytical relations. An appropriate validation of the core flow model is thus achieved by comparing calculated and measured data of these quantities. Indeed, p and Φ are currently the only quantities, which can be measured reliably. In Figs. 3 b – d the calculated and measured surface potential is shown along the channel surface lines AB, GJ, HI (see Fig. 3 a).

There is good agreement between the calculated and measured values in Figs. 3 b and 3 c and fairly good agreement in Fig. 3 d. Figure 3 b demonstrates that the assumption of fully developed flow in the symmetry plane of the z-bend is justified. A similar good agreement was obtained for the surface line MN at the pipe inlet not shown here. Figure 3 c gives the potential at an intermediate perimeter line GJ. It reflects the strong three-dimensional distortions of the potential due to three-dimensional electrical effects, described adequately by the inertialess model. The inertialess assumption at the position GJ is confirmed by the experimental data, which show even for the lowest considered interaction parameter of $N = 22130$ almost asymptotic conditions for wall potential. Measurements of wall potential at the position IH closer to the outer corner also do not indicate a significant influence of inertial effects. Thus numerical predictions show only moderate deviations from experimental data.

A comparison between measured and calculated values of the static pressure along the test-section line $D_3 D_8$, (see Fig. 3 a) is given in Fig. 4. In general, the measured static pressure is well predicted by the model calculations. The slight decrease of the pressure in the inlet (near D_8) and the steep decrease in the center part of the z-bend (at D_3) reflect the different orientation of the channel sections in the external B-field. The pressure recovery in the corner region is due to three-dimensional short circuited currents in the fluid which give rise to a local MHD-pumping effect.

At this time local velocity measurements in the bend interior can not be performed since relevant measuring techniques are not available. Even potential measurements in typical three-dimensional flows are difficult to interpret since influences of the probe prong on the current density distribution is not well understood. Therefore in Fig. 5 only the computed core velocity profiles in three typical cross-sections of the bend are shown. Because of the good agreement between the experimental and calculated surface potential data and the static pressure we are confident that the velocity predictions are at least qualitatively correct even in the corner areas. Figure 5 shows the distribution of the normal component of the core velocity in the planes AB, EFK, GJ. For simplicity the effect of Hartmann layers was not taken into account and side wall jets are not shown. The velocity distribution in the plane AB is flat according to the assumption of fully developed flow in this cross-section. In the plane EFK containing the line of the inner corner the velocity in the center of the core is reduced and more volume flux is carried near the side walls. There is a jump in the velocity distribution along the inner corner line. This discontinuity is related to a discontinuity in the local pressure gradient indicated in Fig. 4. However, it is smoothed out immediately by viscous shear in Hartmann layers at the inner channel walls, which are not shown in Fig. 5. The physical interpretation of the velocity distribution presented in the plane GJ is the same as in EFK except that it is continuous everywhere.

Finally we mention here that the characteristic dimension a of one cross section may be not appropriate to account for inertial effects. If the interaction parameter is based

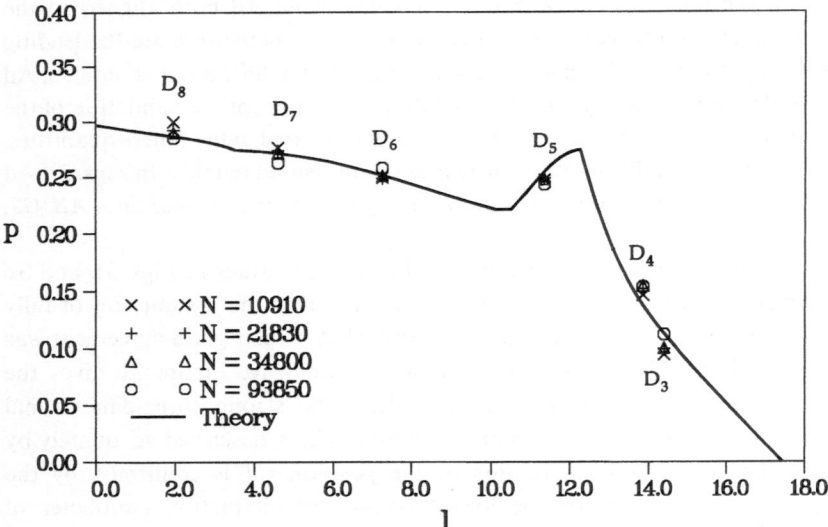

Fig. 4. Pressure variation along the surface line $D_3 D_8$

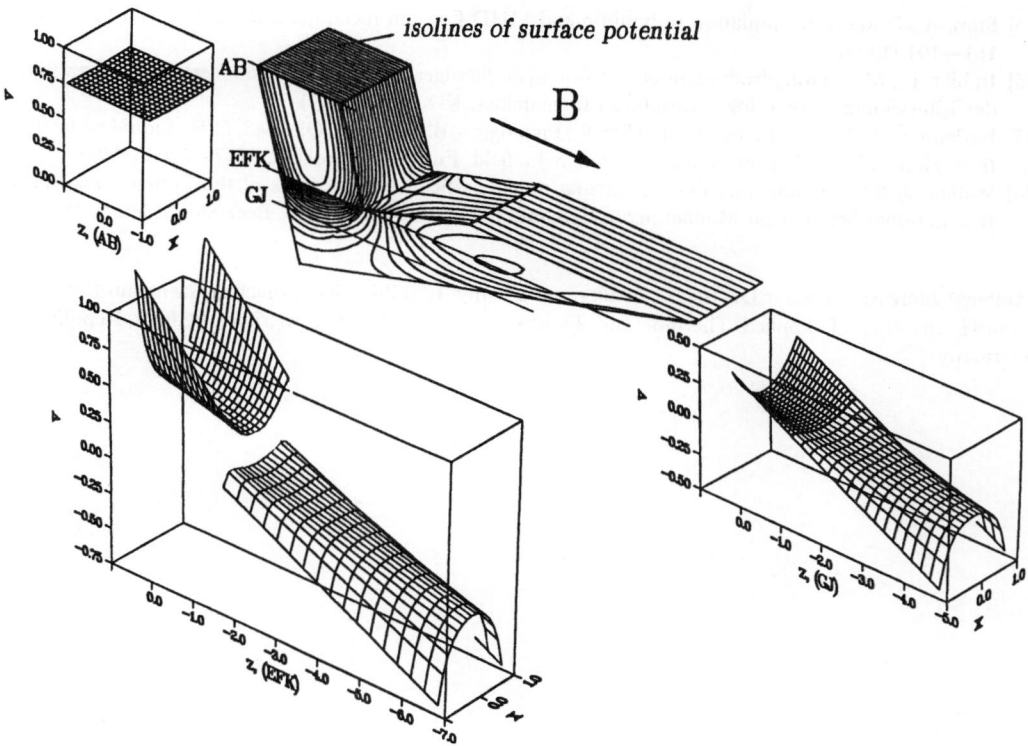

Fig. 5. Velocity distribution in different cross-sections of the *z*-bend, lines of constant potentials on the surface of the bend

on this scale the precise condition for inertialess flow is $N \gg M^{3/2} \gg 1$ [8]. The experimental results, however, indicate that the measured quantities reach their asymptotic limit even before.

Acknowledgements

The authors would like to thank their colleagues L. Barleon and R. Stieglitz for providing the experimental data for the comparison and, moreover, for many stimulating discussions during the preparation of this article.

References

[1] Malang, S., Reimann, J., Sebening, H.: Status report KfK. Contribution to the development of DEMO-relevant test blankets for NET/ITER. KfK 4907 (1991).

[2] Hunt, J. C. R., Shercliff, J. A.: Magnetohydrodynamics at high Hartmann number. Annu. Rev. Fluid Mech. **3**, 37−60 (1971).

[3] Kulikovskii, A. G.: Slow steady flows of a conducting fluid at large Hartmann numbers. Fluid Dyn. **3**, 1−5 (1968).

[4] Walker, J. S.: Magnetohydrodynamic flows in rectangular ducts with thin conducting walls. Part 1. J. Mec. **20**, 79−112 (1981).

[5] Sterl, A.: Numerical simulation of liquid-metal MHD flows in rectangular ducts. J. Fluid Mech. **216**, 161—191 (1990).

[6] Bühler, L.: Magnetohydrodynamische Strömungen flüssiger Metalle in allgemeinen Geometrien unter der Einwirkung starker, lokal variabler Magnetfelder. KfK 5095 (1993).

[7] Barleon, L., Bühler, L., Mack, R., Stieglitz, R., Picologlou, B. F., Hua, T. Q., Reed, C. B.: Liquid metal flow through a right angle bend in a strong magnetic field. Fusion Technol. **21**, 2197—2203 (1992).

[8] Walker, J. S.: Laminar duct flows in strong magnetic fields. Proceedings of the Fourth Beer-Sheva International Seminar on Magnetohydrodynamic Flows and Turbulence, Beer-Sheva, Israel, 1984.

Authors' address: Professor Dr.-Ing. U. Müller and Dr.-Ing. L. Bühler, Kernforschungszentrum Karlsruhe GmbH, Institute of Applied Thermo- and Fluiddynamics, D-76021 Karlsruhe, Federal Republic of Germany

Acta Mechanica (1994) [Suppl] 4: 377–388

Optimum regularity of Navier-Stokes solutions at time $t = 0$ and applications

R. Rautmann, Paderborn, Federal Republic of Germany

Summary. Having in our mind the general rule from numerical mathematics: "more regular solutions allow approximations of a higher convergence rate also in stronger norms" first of all we point out the optimum regularity of a Navier-Stokes solution at its initial time. Then by means of our result we state H^2-convergence of a sequence of linearizations and of Rothe's scheme to the Navier-Stokes initial-boundary value problem. By a numerical realization of this general approach, W. Borchers and our Paderborn group have calculated 3-dimensional viscous incompressible flows past a sphere (without symmetry assumptions) at Reynolds numbers 20 000 [5].

1 Introduction

Numerical solutions to the Navier-Stokes initial-boundary value problem have special interest in practice. However in 3-dimensional cases even if we use the new generation of computing machinery our numerical approach is restricted by the limited operational speed and storage capacity. E.g. taking 100 grid points (say on 10 length units) in one dimension we can guarantee high numerical accuracy in a few cases only. But in 3 dimensions even on such a rough grid at each time step we have to compute 10^6 vector values from complicated systems of equations. In order to make the calculations as efficient as possible, (at least) first (or higher) order approximation schemes would be required.

In addition there is another argument. From L. Prandtl's work [19] we know the decisive role of the vorticity and its boundary values for the development in time of a viscous incompressible flow. Also with a view of efficient numerical approximations Lighthill has renewed this picture strikingly in [15]. Therefore, since the calculated approximations to the flow velocity in any case should reflect this essential structure of the flow, we must ensure the convergence of their spatial first order partial derivatives even on the boundary, which would result e.g. from the H^2-convergence of the numerical approximations (i.e. from the convergence in the L^2-sense of the approximations and their spatial partial derivatives up to the second order).

Since, as a rule, more regularity of the solutions in question leads to a higher order convergence of their approximations also in stronger norms, our first question will be: *How smooth can a Navier-Stokes solution u be on a compact time interval* $[0, T]$, *the initial value u(o) being arbitrarily prescribed (in a suitable function space)?* For the answer, which will be given in Sections 3 and 4 below, in the next Section we will give a precise formulation of our problem. The applications of the results will be stated in Sections 5 and 6. Section 7 presents a numerical result (due to W. Borchers [5]) of this general approach.

2 Notations. Rothe's scheme and a sequence of linearizations to the Navier-Stokes equations

We consider a viscous incompressible flow at time $t \geqq 0$ in a bounded open set Ω of the Euclidean (x_1, x_2, x_3)-space $\mathbb{R}^3$, the boundary $\partial\Omega$ being a compact 2-dimensional C^3-submanifold of $\mathbb{R}^3$. The velocity vector $u(t, x) = (u_1, u_2, u_3)$ and the kinematic pressure $p(t, x) \geqq 0$ of the flow fulfil the Navier-Stokes equations

$$\frac{\partial}{\partial t} u - \Delta u + u \cdot \nabla u + \nabla p = F, \quad \nabla \cdot u = 0 \quad \text{for } t > 0, \quad x \in \Omega,$$

$$u|_{\partial\Omega} = 0, \quad u(0, x) = u_o \tag{2.1}$$

with the prescribed density $F = F(t, x)$ of the outer forces if we assume the constant mass density and the kinematic viscosity constant to equal 1. Δ denotes the Laplacean, ∇ the gradient operator in $\mathbb{R}^3$.

We will investigate solutions $u(t)$ of (2.1) with values in one of the Hilbert spaces H^m (with norms $\|\cdot\|_{H^m}$) of measurable real vector functions, which are square integrable on Ω together with their spatial derivatives $\partial_x^\alpha u$ up to the order $m \geqq |\alpha| = \alpha_1 + \alpha_2 + \alpha_3$, $\alpha_j = 0, 1, \ldots, m$. The incompressibility condition $\nabla \cdot u = 0$ and the condition of adherence $u|_{\partial\Omega} = 0$ at the boundary $\partial\Omega$ are defined in the usual generalized sense that u belongs to the closure H or V with respect to the space $H^0 = L^2(\Omega)$ or H^1, respectively, of the space D of real test functions on Ω. The latter functions are divergence free, have spatial derivatives of any order and compact support in Ω. Using H. Weyl's orthogonal projection $P: H^0 \to H$, which sends into zero exactly the generalized gradients $\nabla q \in H^0$, we get from (2.1) the initial value problem of the evolution Navier-Stokes equation

$$\partial_t u + Au + Pu \cdot \nabla u = PF, \quad t > 0,$$

$$u(0) = u_0 \tag{2.2}$$

for the function $u: [0, T] \to D_A$ which takes its values in the domain of definition $D_A (= V \cap H^2)$ of the Stokes operator $A = -P\Delta$. We note that our assumption $u(t) \in D_A$ includes that u is divergence-free and fulfils the condition of adherence $u|_{\partial\Omega} = 0$. In the following we will always assume $PF = 0$. This implies that the density of the outer force in the Navier-Stokes equation is the gradient of a scalar function. We use this unessential restriction only in order to simplify the notations below.

The Hilbert space H^0 is equipped with the bilinearform

$$\langle f, g \rangle = \int_\Omega f(x) \cdot g(x) \, dx \quad \text{for } f, g \in H^0.$$

We will always write $\tag{2.3}$

$$\|v\| = \langle v, v \rangle^{1/2} \quad \text{or} \quad \|\nabla v\| = \langle \nabla v, \nabla v \rangle^{1/2}$$

for the $L^2(\Omega)$-norm of the vector function v or ∇v, respectively. Let J be a real interval, B a Banach space. Then by $C^0(J, B)$ we denote the set of all uniformly bounded and continuous functions f defined on J with values in B. By $c, c_0, \ldots$ we will denote positive constants which may have different values at different places below.

Let us now assume that the function $u \in C^0([0, T], D_A)$ is a solution of (2.2) with $PF = 0$.

Example 1. We will approximate u by means of the Rothe scheme: We divide the interval $J = [0, T]$ by the grid points $t_k = k\varepsilon$, $k = 0, ..., K$, where $\varepsilon = \dfrac{T}{K}$ denotes the length of the time step. The approximations

$$v_k{}^\varepsilon \approx u(t_k)$$

to the values of the solution u of (2.2) will be calculated with the help of the scheme

$$\frac{v_{k+1}^\varepsilon - v_k{}^\varepsilon}{\varepsilon} + C_k v_{k+1}^\varepsilon = 0, \tag{2.4}$$

$k = 0, ..., K - 1$, starting from the given initial value $v_0{}^\varepsilon \in D_A$, where

$$C_k = A + P v_k{}^\varepsilon \cdot \nabla. \tag{2.5}$$

The unique solution

$$v_{k+1}^\varepsilon = (1 + \varepsilon C_k)^{-1} v_k{}^\varepsilon \in D_A, \tag{2.6}$$

of (2.4) exists, since the positive reals belong to the resolvent set of the operator $-C_k$. In addition, the linear operator C_k with given $v_k{}^\varepsilon \in D_A$ is m-sectorial, and $-C_k$ generates the holomorphic semigroup e^{-tC_k} [21, 18, 22].

In order to estimate the error in the approximation scheme (2.4) we integrate (2.2) from t_k to t_{k+1} and divide by ε. After some rearrangements on both sides of the equation we find

$$\frac{u(t_{k+1}) - u(t_k)}{\varepsilon} + A u(t_{k+1}) + P u(t_k) \cdot \nabla u(t_{k+1})$$

$$= \frac{1}{\varepsilon} \int\limits_{t_k}^{t_{k+1}} \left\{ A\big(u(t_{k+1}) - u(\tau)\big) + P\big(u(t_k) \cdot \nabla u(t_{k+1}) - u(\tau) \, \nabla u(\tau)\big) \right\} d\tau \equiv E_R, \tag{2.7}$$

where the right hand side describes the error at each single time step.

Example 2. In order to approximate u by the solutions of a sequence of linear initial value problems, again we divide the interval $(0, T]$ in K parts $J_k = (t_k, t_{k+1}]$ of length $\varepsilon = \dfrac{T}{K}$, $k = 0, ..., K - 1$. On each J_k we consider the linear initial value problem

$$\partial_t u^\varepsilon + A u^\varepsilon + P u_k^{\varepsilon *} \cdot \nabla u^\varepsilon = 0, \quad t_k < t \leq t_{k+1},$$

$$u^\varepsilon(t_k) = u_k{}^\varepsilon, \tag{2.8}$$

where

$$u_k^{\varepsilon *} = (1 + rA)^{-1} u_k{}^\varepsilon \in D_A \tag{2.9}$$

denotes the Yosida-approximation to the initial value $u_k{}^\varepsilon$, $r > 0$, and compute

$$u_{k+1}^\varepsilon = u^\varepsilon(t_{k+1}), \tag{2.10}$$

$k = 0, ..., K - 1$, one after the other from the given initial value $u_0{}^\varepsilon \in D_A$.

Because of the linearity of problem (2.8) its solution will be represented by the semigroup $e^{-(t-t_k)C_k{}^*}$ generated by the operator $-C_k{}^*$, where

$$C_k{}^* = A + P u_k^{\varepsilon *} \cdot \nabla, \tag{2.11}$$

[21, 18, 22]. We get an explicit expression for the error E_L in the scheme (2.8) at each single time step by writing (2.2) (with $PF = 0$) in the equivalent form

$$\partial_t u + Au + Pu(t_k) \cdot \nabla u = P(u(t_k) - u) \cdot \nabla u \equiv E_L. \tag{2.12}$$

Useful estimates to the expressions steming from $Pu \cdot \nabla u$ hold in terms of the fractional powers A^α with domains $D_{A^\alpha} \subset H$ which are defined for any real α by means of the spectral representation of A [9, p. 281; 25, p. 10, p. 44, Theorem 2.3.2]. For $\alpha < \beta$ the imbedding $D_{A^\beta} \hookrightarrow D_{A^\alpha}$ is compact, D_{A^β} being dense in D_{A^α}. On $V = D_{A^{1/2}}$ we have $\|\nabla f\| = \|A^{1/2}f\|$ [9, p. 270]. Thus

$$\|f\|_{H^1} \leqq c \, \|A^{1/2}f\| \quad \text{for } f \in V \tag{2.13}$$

follows from Poincarés inequality. By Cattabriga and Solonnikov's general estimates for the Stokes operator in [6, 24] we have

$$\|f\|_{H^{m+2}} \leqq c \, \|Af\|_{H^m} \quad \text{for } f \in D_A, \quad Af \in H \cap H^m, \quad m = 0, 1, \dots \tag{2.14}$$

The inequality (2.14) results from the a priori estimate

$$\|f\|_{H^{m+2}} + \|\nabla p\|_{H^m} \leqq c_o \, \|g\|_{H^m}$$

for any solution f of the Stokes boundary value problem

$$-\Delta f + \nabla p = g \in H \cap H^m, \quad \nabla \cdot f = 0, \quad f|_{\partial\Omega} = 0, \tag{2.15}$$

if we write the first Eq. (2.15) in the equivalent form

$$-\Delta f + \nabla \tilde{p} = Pg, \tag{2.16}$$

where $\nabla \tilde{p} = \nabla p - (1 - P)\, g$, the projection $(1 - P)\, g$ (by definition of Weyl's projection P) being a generalized gradient.

Proposition 1.1. *Assume* $0 \leqq \delta < \dfrac{5}{4}$. *Then*

$$\|A^{-\delta}Pu \cdot \nabla v\| \leqq c \, \|A^\theta u\| \cdot \|A^\varrho v\|$$

holds for any $u \in D_{A^\theta}$, $v \in D_{A^\varrho}$ *with some constant* $c = c(\delta, \theta, \varrho)$, *provided that* $\dfrac{5}{4} \leqq \delta + \theta + \varrho$,

$0 < \theta, \varrho, \dfrac{1}{2} < \delta + \varrho.$

This proposition is a special case of Giga and Miyakawa's Lemma 2.2 in [11, p. 270−271]. It suggests, to measure the smoothness of a Navier-Stokes solution u by the exponents $\alpha \geqq 0$ and β, for which $u \in C^o([0, T], D_{A^\alpha})$ and $\partial_t u \in C^o([0, T], D_{A^\beta})$ holds. Thus we reformulate our question above: *How large can α and β become in case of a given initial value $u(0) \in D_{A^\alpha}$?*

3 The compatibility condition $t = 0$

Let u denote a solution of (2.2). We try to assume $\alpha = 1$, $\beta = \dfrac{1}{2}$, $u \in C^o([0, T], D_A)$, $\partial_t u \in C^o([o, T], D_{A^{1/2}})$. Then because of (2.13) the time derivative $\partial_t u$ of the solution u of (2.2) is strongly H^1-continuous. Since $D_A = V \cap H^2$, using the explicit expression for $\nabla(u \cdot \nabla u) = \nabla\left(\sum\limits_{i=1}^{3} u_i \dfrac{\partial}{\partial x_i} u\right)$, Hölders inequality and Sobolev's imbedding theorem we find

$u \cdot \nabla u \in H^1$ and

$$\|u \cdot \nabla u\|_{H^1} \leqq c(\|u\|_{H^2})^2, \tag{3.1}$$

thus $Pu \cdot \nabla u \in H^1 \cap H$, and the estimate

$$\|Pu \cdot \nabla u\|_{H^1} \leqq c_1(\|u\|_{H^2})^2 \tag{3.2}$$

holds for all $t \in [o, T]$, since the projection P is bounded in H^1 [26, p. 18]. Similarly we see $u \cdot \nabla u \in C^o([0, T], H^1)$ and $Pu \cdot \nabla u \in C^o([0, T], H^1 \cap H)$. Consequently by (2.2) we have $Au \in C^o([0, T], H^1 \cap H)$, thus $u \in C^o([0, T], H^3 \cap V)$ is fulfilled because of (2.14). Now in (1.1) all terms different from ∇p being H^1-continuous, ∇p must be strongly H^1-continuous, too.

Then the trace theorem in Sobolev spaces shows the strong continuity on $[0, T]$ of each term in (2.1) on $\partial\Omega$. In the limit $t \downarrow 0$ we find

$$\Delta u(0)|_{\partial\Omega} = \nabla p(0) \tag{3.3}$$

since $\dfrac{\partial}{\partial t} u|_{\partial\Omega} = u \cdot \nabla u|_{\partial\Omega} = 0$. However in general the system (3.3) of 3 scalar equations contradicts the fact that at any time $t \in [0, T]$ the pressure p is up to a constant $c = c(t)$ uniquely determined by the Neumann problem

$$\Delta p = -\nabla \cdot (u \cdot \nabla u) \quad \text{in } \Omega,$$

$$N \cdot \nabla p = N \cdot \Delta u \quad \text{on } \partial\Omega, \tag{3.4}$$

N denoting the outer unit normal at any point of $\partial\Omega$. (We note that (3.4) follows from (2.1) by taking the divergence or the normal components of the boundary values, respectively). Namely on the one side we can take u_o in $D_A \cap H^3$ arbitrary, but on the other side together with p it should fulfil (3.3) and the nonlinear boundary value problem (3.4). This compatibility condition, which is known from [24, 14], has recently been studied in [13, 17, 20, 27, 29]. Because of its nonlinear and non-local nature it would be "virtually uncheckable for given data" [13, p. 277].

4 Optimum regularity at time $t = 0$

We will see that a certain amount of regularity of the time derivative, namely $\partial_t u \in C^o([0, T], D_{A^\zeta})$ together with $u \in C^o([0, T], D_{A^{1+\zeta}})$ will be achieved, if

$$u \in C^o([0, T], D_A) \tag{4.1}$$

holds and we take the initial value

$$u(o) \in D_A^{1+\zeta} \tag{4.2}$$

for arbitrary $\zeta \in \left(0, \dfrac{1}{4}\right)$. The assumption (4.2) holds e.g. with any $u_o \in V \cap H^3$ (see Corollary 4.2 below), and it is well-known that assumption (4.1) can be fulfilled in case $u(0) \in D_A$ at least locally in time, with $T > 0$ sufficiently small [9, 12].

Theorem 4.1 [20, 29]. *Assume* $v, w \in D_A$. *Then (1)* $Pv \cdot \nabla w \in D_{A^\zeta}$ *and (2)* $\|A^\zeta Pv \cdot \nabla w\|$ $\leqq c_\zeta \|Av\| \|Aw\|$ *hold for all* $\zeta \in \left[0, \dfrac{1}{4}\right)$ *with a constant* $c = c(\zeta)$.

Sketch of proof: The same arguments which led us to (3.1) and (3.2) show $v \cdot \nabla w \in H^1$, $Pv \cdot \nabla w \in H^1 \cap H$ and

$$\|Pv \cdot \nabla w\|_{H^1} \leqq \|v \cdot \nabla w\|_{H^1} \leqq c_1 \|Av\| \cdot \|Aw\|, \tag{4.3}$$

if we use (2.14) again. For any $s \in [0, 1]$, H^1 is continuously imbedded in the fractional order Sobolev space H^s [1, p. 204−214]. In case $s \in \left[0, \frac{1}{2}\right]$ we have $H^s = \overset{0}{H}{}^s$, $\overset{0}{H}{}^s$ denoting the closure of C_c^∞ in the norm of H^s, where C_c^∞ denotes the linear space of all vector functions defined on Ω which have partial derivatives of any order and compact support in Ω [16, p. 55, Theorem 11.1]. If $s \neq \frac{1}{2}$ the space $\overset{0}{H}{}^s$ (with equivalent norms) equals the interpolation space $[\overset{0}{H}{}^1, H^0]_{1-s} = D_{B^{s/2}}$, B denoting the closure of the Laplacean in H^o with domain $D_B = \overset{0}{H}{}^1 \cap H^2$ [16, p. 64, Theorem 11.6]. Both theorems cited from [16] hold also under our assumption $\partial\Omega \in C^3$. Finally Fujita and Morimoto [10] have proved $D_{B^{s/2}} \cap H = D_{A^{s/2}}$ (with equivalent norms) for any $s \in [0, 1]$. Combining the arguments above we see $Pv \cdot \nabla w \in H^1 \cap H \hookrightarrow D_{A^{s/2}}$ which proves our claim for any $\zeta = \frac{s}{2} \in \left[0, \frac{1}{4}\right)$.

An immediate consequence of Theorem 4.1 is

Corollary 4.1. *Assume* $v, w \in C^o([0, T], D_A)$. *Then* $Pv \cdot \nabla w \in C^o([0, T], D_{A^\zeta})$ *holds for all* $\zeta \in \left[0, \frac{1}{4}\right)$.

Corollary 4.2. *For all* $\zeta \in \left[0, \frac{1}{4}\right)$ *the continuous imbedding* $V \cap H^3 \hookrightarrow D_{A^{1+\zeta}}$ *holds.*

Proof. By definition of $A \sim -P\Delta$ with the projection P being bounded in H^1, the assumption $v \in V \cap H^3$ gives $Av \in H^1$. From the proof of Theorem 4.1 we see that the latter implies $Av \in H^s = \overset{0}{H}{}^s = D_{A^{s/2}}$ (with equivalent norms) for all $\frac{s}{2} = \zeta \in \left[0, \frac{1}{4}\right)$, which proves our statement since it ensures the existence of $A^{s/2}Av = A^{1+(s/2)}v \in H$.

Theorem 4.2. *Assume* $u \in C^o([0, T], D_A)$ *fulfils* (2.2) *and* $u(0) \in D_{A^{1+\zeta}}$ *holds for some* $\zeta \in \left(0, \frac{1}{4}\right)$. *Then we have* $u \in C^o([0, T], D_A{}^{1+\zeta})$.

Proof. With some $\xi \in \left(\zeta, \frac{1}{4}\right)$ writing the identity $1 = A^{-\xi}A^\xi$ of D_{A^ξ} into the semigroup representation [9, p. 272, Eq. (1.11)] and using that any fractional power A^α commutes with the exponential e^{-tA} on D_{A^α} from [9, p. 280, Lemma 2.12.] we find

$$A^{1+\zeta}u(t) = e^{-tA}A^{1+\zeta}u(0) - \int_0^t A^{1+\zeta-\xi}e^{-(t-s)A}A^\xi Pu(s) \cdot \nabla u(s) \, ds \tag{4.4}$$

which proves our claim because of the weak singularity $\|A^{1+\zeta-\xi}e^{-(t-s)A}\| \leqq \dfrac{c}{|t-s|^{1+\zeta-\xi}}, \zeta < \xi$ [9, p. 280, Lemma 2.10] and the Lemma mentioned before.

Remark 4.1. As we have shown in [20], any smoothness $u \in C^o([0, T], D_{A^{1+\zeta}})$ with $\zeta > \frac{1}{4}$ would require the compatibility condition (3.3), (3.4) (in the sense of a ζ-dependent space of functions defined on $\partial\Omega$). Related results have been proved in [13, p. 281, Proposition 2.1 and Corollary 2.1].

5 Application to the Navier-Stokes linearization and to Rothe's scheme

The above results lead to nice estimates of the error terms E_L and E_R, which allow convergence- and error-estimates even in H^2. For this we have to subtract (2.12) from (2.8) or (2.7) from (2.4), respectively, and to estimate the differences $u^\varepsilon(t) - u(t)$, $u_k^\varepsilon - u(t_k)$ by energy- and semigroup methods [22, 23].

Theorem 5.1 (H^2-convergence of the linearizations [22]). *Assume $u \in C^o([0, T], D_A)$ denotes a solution of the Navier-Stokes initial value problem (2.2) with right hand side $PF = 0$, and the initial values u_o, $u_o^\varepsilon \in D_A$ fulfil*

$$\|u_o^\varepsilon - u_o\| \leq c_o \varepsilon \quad and \quad \|A(u_o^\varepsilon - u_o)\| \leq c\varepsilon^{1/2} \tag{5.1}$$

with constants $c_o, c \geq 0$. Then the solutions u^ε of (2.8), (2.9), (2.10) converge to u in D_{A^α} with $\varepsilon \to 0$, $0 < r \leq \varepsilon$ for all $\alpha \in [0, 1]$. For $\varepsilon \in (0, \varepsilon_o]$, ε_o being sufficiently small, the error estimate

$$\|A^\alpha(u^\varepsilon(t) - u(t))\| \leq a_\alpha \varepsilon^{1-(\alpha/2)} \tag{5.2}$$

holds on $[0, T]$ with the constant a_α depending on α, T, c_o, c, and $\sup\limits_{t \in [0, T]} \|Au(t)\|$.

The *proof* is given in [22]. In virtue of the a priori estimate (2.14) the theorem shows that the linearization scheme (2.8) is H^2-stable in so far, as any sufficiently small error at the initial time $t = 0$ can be controlled by error estimates in $H^2(\Omega)$ on the whole interval $[0, T]$.

Theorem 5.2 (H^2-convergence of Rothe's scheme [23]). *For some $\zeta \in \left(0, \dfrac{1}{4}\right)$ let $u \in C^0([0, T], D_{A^{1+\zeta}})$ denote a solution of the Navier-Stokes initial value problem (2.2) with right hand side $PF = 0$. Assume the initial values u_o, $v_o^\varepsilon \in D_{A^{1+\zeta}}$ satisfy*

$$\|v_o^\varepsilon - u_o\| \leq c\varepsilon \quad and \quad \|A^\zeta(v_o^\varepsilon - u_o)\| \leq c'\varepsilon^{1-\xi} \tag{5.3}$$

with some $\xi \in \left[0, \dfrac{1}{4}\right]$, ξ being sufficiently small, and constants $c, c' \geq 0$. Then we have $v_k^\varepsilon \in D_{A^{1+\zeta}}$, and for any $\varepsilon \in (0, \varepsilon_o]$, ε_o being sufficiently small, the error estimates

$$\|v_k^\varepsilon - u(t_k)\| \leq b_o \varepsilon, \tag{5.4}$$

$$\|A^\zeta(v_k^\varepsilon - u(t_k))\| \leq b_\zeta \varepsilon^{1-\xi}, \tag{5.5}$$

$$\|A^\beta(v_k^\varepsilon - u(t_k))\| \leq b_\beta \varepsilon^{1+\zeta'-\beta} \tag{5.6}$$

for $\beta \in \left[\dfrac{1}{2}, 1\right]$ and $\zeta' \in [0, \zeta)$ hold uniformly in $k = 1, \ldots, K \geq \dfrac{T}{\varepsilon_o}$. The constants b_o, b_ζ, b_β depend on ζ, $\sup\limits_{t \in [0, T]} \|A^{1+\zeta}u(t)\|$, c, T, the constants b_ζ, b_β additionally on ξ, c', and $\|A^{1+\zeta}v_o^\varepsilon\|$ (on the latter in a non-decreasing way), finally b_β on ζ and β, too.

For the *proof* see [23]. As a consequence we have

Corollary 5.1. *With the assumptions of Theorem 5.2, Rothe's scheme (2.4), (2.5) to the Navier-Stokes initial value problem (2.2) has the rate of convergence 1, or $\dfrac{1}{2} + \zeta$, or ζ' for all $\zeta' \in (0, \zeta)$ in the spaces H, or H^1, or H^2, respectively, if in addition the initial values v_o^ε are uniformly bounded in $D_{A^{1+\zeta}}$.*

In virtue of (2.14) Corollary 5.1 follows immediately from the error estimates (5.4), (5.6) in Theorem 5.2.

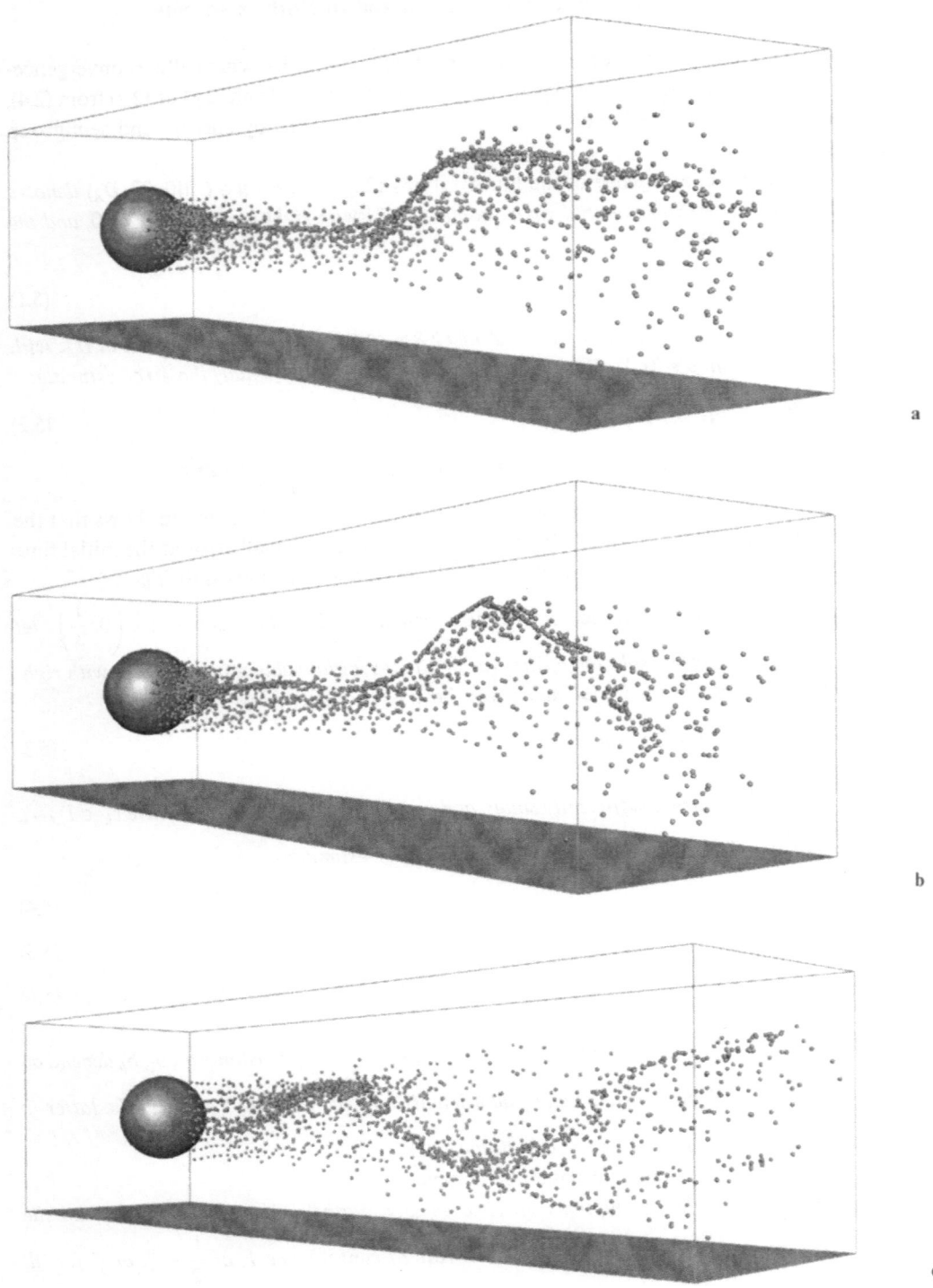

Fig. 1. Particles in a separated viscous flow behind a sphere at Reynolds numbers Re $= 2 \cdot 10^4$ and Re $= 10^3$ (time step length 10^{-2} sec; n number of time steps). **a** Re: 20000, $n = 45$; **b** Re: 20000, $n = 73$; **c** Re: 20000, $n = 86$; **d** Re: 1000, $n = 52$; **e** Re: 1000, $n = 56$; **f** Re: 1000, $n = 143$

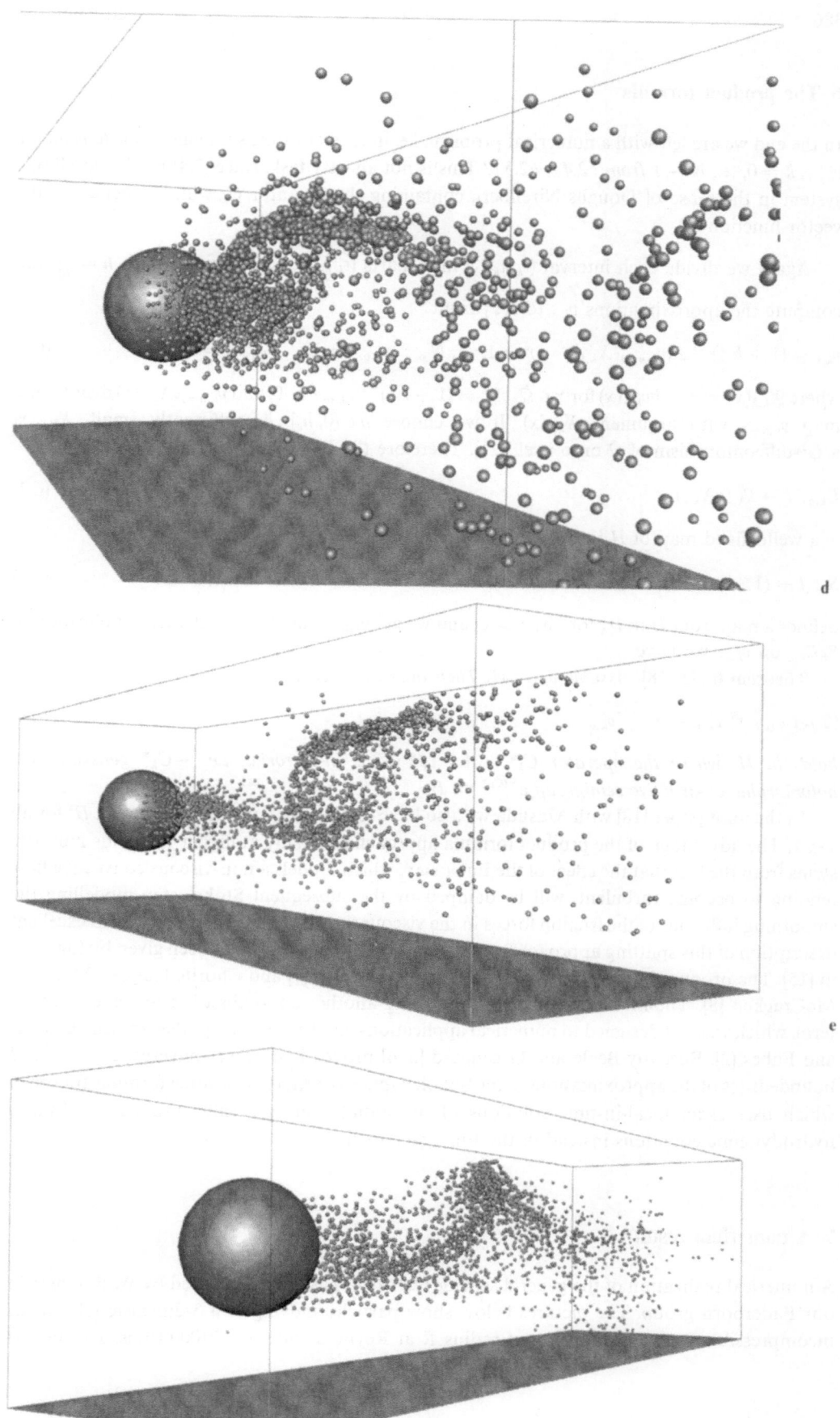

d

e

f

6 The product formula

In the end we are left with a numerical problem, i.e. in case of Rothe's scheme: *How to compute* v_{k+1}^ε, $k = 0, ..., N - 1$ *from (2.4), (2.5)?* This is not an easy task, since (2.4) is a linear elliptic system in the sense of Douglis-Nirenberg containing the operator C_k which depends on the vector function v_k^ε.

Again we divide each interval $(t_k, t_{k+1}]$ in N parts $(t_{k,n}, t_{k,n+1}]$ of equal length $h = \dfrac{\varepsilon}{N}$ and compute the approximations $v_{k,n}$ to v_{k+1}^ε from

$$v_{k,n} = (1 + hA)^{-1} P(v_{k,n-1} \circ X_{k,h}), \qquad n = 1, ..., N, \qquad v_{0,0} = v_0^\varepsilon, \qquad v_{k+1,0} = v_{k,N}, \tag{6.1}$$

where $X_{k,h}(x) = x - h v_{k,0}^{**}(x)$ for $x \in \bar\Omega$, $v_{k,0}^{**} = (1 + rA)^{-2} v_{k,0}$, $r > 0$, and $v_{k,n-1} \circ X_{k,h}(x)$ denotes the map $v_{k,n-1}$ with argument $X_{k,h}(x)$. If we choose $h \in (0, h_0]$, h_0 sufficiently small, $X_{k,h}$ is a C^1-diffeomorphism of $\bar\Omega$ onto itself [21]. Therefore the Euler step

$$E_{k,h}: f \to P(f \circ X_{k,h}) \tag{6.2}$$

is a welldefined map of H into H. In addition the Stokes step

$$S_h: f \to (1 + hA)^{-1} f \tag{6.3}$$

defines a map from H in D_A for any $h > 0$, and we get $v_{k,N}$ by an N-times iterated application of $S_h E_{k,h}$ on $v_{k,0}$. We have

Theorem 6 [21, 18]. *Assume $v_{k,0} \in H$. Then the convergence*

$$(S_{\varepsilon/N} E_{k,(\varepsilon/N)})^N v_{k,0} \to e^{-\varepsilon C_k^*} v_{k,0} \tag{6.4}$$

holds in H, where the operator $C_k^* = A + v_{k,0}^{**} \cdot \nabla$ is m-sectorial, i.e. $-C_k^*$ generates the holomorphic contractive semigroup $e^{-t C_k^*}$ on H.

In the joint paper [18] with Masuda we also have established convergence rates in H^s for all $s < 2$. The advantage of the product formula approach especially at higher Reynolds numbers stems from the fact that the effect of the Euler step, which models a non-viscous convective flow tending to become turbulent, will be damped by the subsequent Stokes step modelling the smoothing influence of the friction forces in the viscous flow in a quite natural way. An excellent description of this splitting approach from the physical point of view has been given by Lighthill in [15]. The product formula approach goes back to Chorin [7] and Chorin, Hughes, Marsden, McCracken [8]. The first convergence proof, using another approximation to the convective term, which was not designed to numerical applications, has been given by Alessandrini, Douglis and Fabes [2]. Recently Beale and Greengard [3, 4] proved first order convergence in L^p and boundedness of the approximations in the Sobolev space $W^{2,p}(\Omega)$ to a product formula approach which uses exact local-in-time solutions of the initial-boundary value problem to Euler's hydrodynamic equations instead of the Euler step (6.2).

7 A numerical result

A numerical realization of the general scheme (2.4), (6.4) has been developed by W. Borchers in our Paderborn group. The pictures below show particles moving in a 3-dimensional viscous incompressible flow past a sphere of radius R at Reynolds number 20000 (Figs. 1 a – c) and

Reynolds number 1000 (Figs. 1 d−f). The exterior domain of the sphere (centre at (0, 0, 0)) is approximated by a quadratic channel of length $32R$ in x_1-direction and of width $4R$ in x_2- and x_3-direction. The inflow at $x_1 = -2R$ is (u_∞, o, o), on the side-walls $x_2 = \pm 2R$ or $x_3 = \pm 2R$ we have periodic boundary conditions with respect to x_2 or x_3, respectively. The outflow condition is $\frac{\partial}{\partial x_1} u = 0$ at $x_1 = 30R$. The Reynolds number is defined in terms of u_∞ and R. The separated vortex layer is oscillating behind the sphere [5].

The helical particles' pathes are oscillating and spiraling away from the separation circle at backside of the sphere. With increasing Reynolds number the separation line tends to the equator.

Because the computational scheme is 3-dimensional and of first order, in regions of high flow velocity the numerical viscosity will be about twenty times higher than the physical one. A second order scheme is under development.

References

[1] Adams, R. A.: Sobolev spaces. New York: Academic Press 1975.

[2] Alessandrini, G., Douglis, A., Fabes, E.: An approximate layering method for the Navier-Stokes equations in bounded cylinders. Ann. Math. Pure Appl. **135**, 329−347 (1983).

[3] Beale, J. T.: The approximation of the Navier-Stokes equations by fractional time steps. Conference on the Navier-Stokes equations, theory and numerical methods, Oberwolfach 18.−24. 8. 1991 (conference lecture).

[4] Beale, J. T., Greengard, C.: Convergence of Euler-Stokes splitting of the Navier-Stokes equations. Research Report RC 18072 (79337) 6/11/92 Mathematics 30, IBM Research Division Almaden 1992.

[5] Borchers, W.: A Fourier-spectral method for flows past obstacles. In: Finite approximations in fluid mechanics (Hirschel, E. H., ed.), pp. 223−248. Notes on Numerical Fluid Mechanics, 1992/93.

[6] Cattabriga, L.: Su un problema al contorno relativo al sistema di equazioni di Stokes. Rend Mat. Sem. Univ. Padova **31**, 308−340 (1961).

[7] Chorin, A. J.: Numerical study of slightly viscous flow. J. Fluid Mech. **57**, 785−796 (1973).

[8] Chorin, A. J., Hughes, T. J. R., McCracken, M. F., Marsden, J. E.: Product formulas and numerical algorithms. Comm. Pure Appl. Math. **31**, 205−256 (1978).

[9] Fujita, H., Kato, T.: On the Navier-Stokes initial value problem. I. Arch. Rational Mech. Anal. **16**, 269−315 (1964).

[10] Fujita, H., Morimoto, H.: On fractional powers of the Stokes operator. Proc. Japan. Acad. **46**, 1141−1143 (1970).

[11] Giga, Y., Miyakawa, T.: Solutions in L_r of the Navier-Stokes initial value problem. Arch. Rational Mech. Anal. **89**, 267−281 (1985).

[12] Heywood, J. G.: The Navier-Stokes equations: on the existence, regularity and decay of solutions. Indiana Univ. Math. J. **29**, 639−681 (1980).

[13] Heywood, J. G., Rannacher, R.: Finite element approximation of the nonstationary Navier-Stokes problem I. Siam. J. Numer. Anal. **19**, 275−311 (1982).

[14] Ladyzhenskaya, O. A.: The mathematical theory of viscous incompressible flow, 2nd ed. New York: Gordon and Breach 1969.

[15] Lighthill, M. J.: Introduction: boundary layer theory. In: Laminar boundary layers (Rosenhead, L., ed.), pp. 43−113. Oxford: University Press 1963.

[16] Lions, J. L., Magenes, E.: Non-homogeneous boundary value problems and applications, vol. 1. Berlin Heidelberg New York: Springer 1972.

[17] Masuda, K.: Remarks on compatibility conditions for solutions of Navier-Stokes equations. J. Fac. Sci. Univ. Tokyo Sect. IA Math. **34**, 155−164 (1987).

[18] Masuda, K., Rautmann, R.: Convergence rates for product formula approximations to Navier-Stokes problems (to appear).

[19] Prandtl, L.: Tragflügeltheorie. I. Mitteilung, Nachrichten der K. Gesellschaft der Wissenschaften zu Göttingen. Mathematisch-physikalische Klasse 451−477 (1918).

[20] Rautmann, R.: On optimum regularity of Navier-Stokes solutions at time $t = 0$. Math. Z. **184**, 141–149 (1983).

[21] Rautmann, R.: Eine konvergente Produktformel für linearisierte Navier-Stokes-Probleme. Z. Angew. Math. Mech. **69**, 181–183 (1989).

[22] Rautmann, R.: H^2-convergent linearizations to the Navier-Stokes initial value problem. In: Proc. Intern. Conf. on New Developments in Partial Differential Equations and Applications to Mathematical Physics (Buttazo, G., Galdi, G. P., Zanghirati, L., eds.), pp. 135–156. Ferrara 14–18 October 1991. New York: Plenum Press 1992.

[23] Rautmann, R.: H^2-convergence of Rothe's scheme to the Navier-Stokes equations. J. Nonlin. Anal. (to appear).

[24] Solonnikov, V. A.: On differential properties of the solutions of the first boundary-value problem for nonstationary systems of Navier-Stokes equations. Trudy Mat. Inst. Steklov **73**, 221–291 (1964) (Transl.: British Library Lending Div., RTS 5211).

[25] Tanabe, H.: Equations of evolution. London: Pitman 1979.

[26] Temam, R.: Navier-Stokes equations. Amsterdam: North-Holland 1979.

[27] Temam, R.: Behaviour at time $t = 0$ of the solutions of semi-linear evolution equations. MRC Technical Summary Report 2162, University of Wisconsin, Madison 1980.

[28] Varnhorn, W.: Time stepping procedures for the nonstationary Stokes equations. Preprint 1353 (1991).

[29] Wahl, W. von: The equations of Navier-Stokes and abstract parabolic equations. Braunschweig: Vieweg 1985.

Author's address: Professor Dr. rer. nat. R. Rautmann, Fachbereich Mathematik—Informatik, Universität GHS Paderborn, Warburger Strasse 100, D-33098 Paderborn, Federal Republic of Germany

Acta Mechanica (1994) [Suppl] 4: 389–399

Generalised similarity solutions for 3-D laminar compressible boundary layer flows on swept profiled cylinders

V. Saljnikov, Belgrade, **Z. Boričić**, and **D. Nikodijević**, Niš, Yugoslavia

Summary. The method of generalised similarity allows us to derive solutions for the laminar steady, compressible three-dimensional boundary layer equations for flows over swept wings. The model configuration is that of an infinitely long swept profiled cylinder in an ideal gas flow whose dynamic viscosity is a linear function of temperature. A universal mathematical model is derived by successively introducing three transformations: Stewartson's, Saljnikov's Generalized Similarity [2] and Loitsianskii's Group of Parameters [7]. It can be integrated numerically by the finite difference iterative method — Tridiagonal Algorithm (TDA) — with an arbitrary choice of pressure distribution, Pr and Ma numbers, as well as of the angle of sweep and the temperature boundary condition. In the article, however, an interesting practical problem of transonic swept wing with air flowing around it is considered for two cases of: a) isothermal and b) adiabatic wall. The obtained results are presented graphically and analysed.

1 Introduction

From the standpoint of contemporary fluid mechanics research the Boundary Layer Theory remains highly actual. Recently it has been playing an important role in investigations based upon the numerical integration of Navier-Stokes equations. A conclusion was drawn, namely that the solutions of these equations should be tested against the solutions of the corresponding boundary layer equations, and not vice versa as one might expect because of the approximate nature of the boundary layer theory. This follows from the fact that, by their structure they are typical for boundary layer at high Re-numbers. Thus, one should check if the code for the numerical integration of Navier-Stokes equations, with respect to its asymptotic nature, is correct i.e. do the obtained results match the solutions of the boundary layer equations [1]. Special value is attributed to the so-called "similar" solutions for their use for testing results in flow stability investigations. They are therefore often used for developing semi-empirical criteria of the flow regime transition. Thus, the generalised similarity method [2] has driven a special attention of researchers, lately. The corresponding solutions are in their accuracy and generality, compared to classical "similar" solutions of the Falkner-Skan type, more appropriate for this purpose. It should be stressed that this, from the theoretical standpoint, fundamental analytical-numerical method, previously successfully applied for solving various two-dimensional boundary layer problems, matches fully the contemporary trends of fluid mechanics development [3]. Accordingly, the generalised similarity solutions, apart from their previously mentioned virtues, enable:

— before, their application to boundary layer computation for special pressure distribution cases, conducting in general form the analysis of influence of relevant parameters;

— in a particular section of boundary layer, computing the appropriate characteristic quantities, without the need of performing each time the integration starting again from the initial cross-section; pre-history of the boundary layer development being, namely — for the analysed section-included by the means of generalised similarity transformations in the corresponding universal solutions, taken from the ready-for-use tables;

— time and financial savings achieved by much shorter computer time and using the ready equations when applying the universal solutions for the cases of special pressure distributions.

2 Physical model and the corresponding system of starting equations

In this article generalised similarity solutions are determined and analysed for the case of three-dimensional laminar stationary boundary layers on a swept wing in a compressible fluid flow.

As a physical model we adopt an infinitely long swept airfoiled cylinder (see Fig. 1a).

One assumes that in the considered range of values of Ma-number of the incident uniform flow (M_∞) the assumptions of the ideal state of gas and the linear relationship between dynamic viscosity and temperature are apt to be made.

The system describing the adopted physical model, with nomenclature usual in the boundary layer theory: $u; v; w$ — velocity components in $x; y; z$ directions (see Fig. 1), ϱ — density, p — pressure, μ — dynamic viscosity coefficient, Pr — Prandtl number, h — enthalpy, c — speed of sound comprises the following equations [5]

— momentum in the chordwise direction (x)

$$\varrho u u_x + \varrho v u_y = \varrho u_e u_e' + (\mu u_y)_y \tag{1}$$

and in the spanwise direction (z)

$$\varrho u w_x + \varrho v w_y = (\mu w_y)_y \tag{2}$$

— continuity

$$(\varrho u)_x + (\varrho v)_y = 0 \tag{3}$$

— energy

$$\varrho u (h_0)_x + \varrho v (h_0)_y = [(\mu/P_r)\,(h_0)_y]_y - \{[\mu(1 - P_r)/P_r]\,[(u^2 + w^2)/2]_y\}_y \tag{4}$$

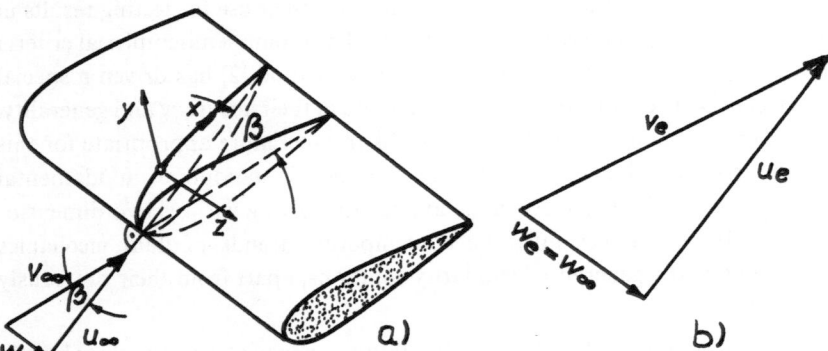

Fig. 1. a Physical model of the problem; **b** velocity components at the boundary layer outer edge

with the total or stagnation enthalpy

$$h_0 = h + (u^2 + w^2)/2 \tag{5}$$

− state

$$p = \varrho RT \tag{6}$$

− linear viscosity law

$$\mu = (\mu_w/T_w) \; T \tag{7}$$

− and the corresponding boundary conditions

$$y = 0: \quad u = v = w = 0; \quad h_0 = h_{0w} \quad \text{or} \quad (h_0)_y = 0$$

$$y = \infty: \quad u = u_e; \quad w = w_e; \quad h_0 = h_{0e} \tag{8}$$

Subscript "0" denotes "total", subscript "w" − "at wall" and subscript "e" denotes "outer edge of the boundary layer", whereas subscripts "x" and "y" represent partial derivatives over the corresponding coordinates.

One should note that the cases of: a) isothermal and b) adiabatic wall differ in the system of Eqs. (1)−(8) only by the thermal boundary condition at the wall; that is

in case a) for $y = 0$: $h_0 = h_{0w}$ and in case b) for $y = 0$: $(h_0)_y = 0$.

3 Transformations of the mathematical model

In order to solve the basic system of boundary layer equations (1)−(8), in the paper, the method of generalised similarity is used. It is necessary to perform three successive transformations (I, II, III) to achieve the final universal equations. Due to shortcoming of space and having in mind that the complete derivation for a similar problem is given in [4], here only the transformations and procedure without the interphase forms of the equation system which are the results of first two transformations will be shown.

I. First, by means of modified Stewartson's transformations [4]

− for coordinates

$$X = \int\limits_0^x (\mu_w T_0/\mu_0 T_w) \; (c_e/c_0) \; (p_e/p_0) \; dx; \quad Y = (c_e/c_0) \int\limits_0^y (\varrho/\varrho_0) \; dy \tag{9}$$

and components of velocity

$$U = u(c_0/c_e) = \partial\psi/\partial Y; \quad V = v(c_0/c_e) \; (p_0/p_e) \; (\varrho/\varrho_e) = -\partial\psi/\partial X \tag{10}$$

the system (1)−(8) is transformed into the mathematical model of the analogous incompressible boundary layer problem.

Here, one assumes that the free-stream flow is adiabatic, with the corresponding energy equation

$$c_0{}^2 = c_e{}^2 + [(\varkappa - 1)/2] \; (u_e{}^2 + w_e{}^2). \tag{11}$$

Introduced are also dimensionless velocity in spanwise direction

$$G = w/W_\infty \tag{12}$$

and dimensionless enthalpies

— in case a) (isothermal wall)

$$\theta = (h_0 - h_{0w})/(h_{0e} - h_{0w}) \tag{13}$$

— and in case b) (adiabatic wall)

$$\vartheta = (T - T_e)/[(u_e^2 + w_e^2)/2c_p] \tag{14}$$

II. Then, by applying Saljnikov's Generalised Similarity Transformation [2] on the equations obtained by the first transformation for the stream function $\psi(X, Y)$, velocity $G(X, Y)$, enthalpy $\theta(X, Y)$ that is $\vartheta(X, Y)$

$$X \equiv X; \quad \eta = U_e^{b/2} \left(a_0 v_0 \int_0^x U_e^{b-1} \, dX\right)^{-1/2} Y; \quad \psi = U_e^{1-\frac{b}{2}} \left(a_0 v_0 \int_0^x U_e^{b-1} \, dX\right)^{1/2} \Phi(X, \eta), \tag{15}$$

where: η — transformed dimensionless transversal coordinate, Φ — dimensionless stream function, a_0, b — presently arbitrary coordinates, one obtains the corresponding equations of generalised similarity.

One should note that in a special case for $a_0 = b = 2$ relation (15) yields first the Görtler's transformation [6],

$$\eta = U_e \left(2v_0 \int_0^x U_e \, dX\right)^{-1/2} Y; \quad \psi = \left(2v_0 \int_0^x U_e \, dX\right)^{1/2} \Phi(X, \eta), \tag{16}$$

which, for distribution of free stream velocities on wedge-formed airfoils $U_e = CX^m$ reduces to the well-known Falkner-Skan coordinates of "similar" solutions

$$\eta = [(m + 1)/2]^{1/2} (U_e/v_0 X)^{1/2} Y; \quad \psi = [2v_0 U_e X/(m + 1)]^{1/2} \Phi(\eta). \tag{17}$$

III. The generalised similarity equations, obtained in such a way for $\Phi(X, \eta)$, $G(X, \eta)$, $\theta(X, \eta)$ that is $\vartheta(X, \eta)$ finally, after introducing Loitsianskii's group of generalised similarity parameters [7]:

$$f_0 = (u_e^2/2h_0) + (w_e^2/2h_0) = f_{0x} + f_{0z};$$

$$f_k = U_e^{k-1}(d^k U_e/dX^k) \, [f_1/(dU_e/dX)]^k \quad (k = 1, 2, ..., \infty) \tag{18}$$

by the means of differential operator

$$(U_e f_1/U_e') \, \partial/\partial X = \sum_{k=1}^\infty E_k(\partial/\partial f_k) \tag{19}$$

where

$$E_0 = 2f_0(1 - f_0) \, f_1; \quad E_k = [(k - 1) \, f_1 + kF_x] \, f_k + f_{k+1} \quad (k = 1, 2, ..., \infty);$$

$$F_x = 2[\zeta_x - (2 + H_x) \, f_1]; \quad \zeta_x = B_x(\partial^2 \Phi/\partial \eta^2)_{\eta=0}; \quad H_x = A/B_x;$$

$$A = \int_0^\infty \left[1 + \frac{f_{0z}}{1 - f_{0z}} \, (1 - G^2) + \frac{(T_w/T_0) - 1}{1 - f_{0z}} \, (1 - \theta) - \Phi_\eta^2\right] d\eta, \tag{20}$$

transform into final systems of equations of boundary layer. The first parameter f_1 of the group (18) represents the integral of the corresponding momentum equation − in which the characteristic function $F_x(f_k)$ and the dimensionless decay of momentum thickness B_x play an important role.

$$f_1 = a_0 B_x^2 U_e' U_e^{-b} \int_0^x U_e^{b-1} \, dX, \qquad B_x = \int_0^\infty \Phi_\eta (1 - \Phi_\eta) \, d\eta \tag{21}$$

4 The universal mathematical model of generalised similarity [5]

The final system of equations for the case a) (isothermal wall) comprises the following universal equations:

− momentum in the chordwise direction

$$\Phi_{\eta\eta\eta} + \frac{1}{2B_x^2} [a_0 B_x^2 + f_1(2 - b)] \Phi\Phi_{\eta\eta} + \frac{f_1}{B_x^2} \left[1 + \frac{f_{0z}}{1 - f_{0z}} (1 - G^2) + \frac{(T_w/T_0) - 1}{1 - f_{0z}} (1 - \theta) - \Phi_\eta^2 \right]$$

$$= \frac{1}{B_x^2} \sum_{k=0}^\infty E_k (\Phi_{\eta f_k} \Phi_\eta - \Phi_{\eta\eta} \Phi_{f_k}) \tag{22}$$

and in the spanwise direction

$$G_{\eta\eta} + \frac{1}{2B_x^2} [a_0 B_x^2 + f_1(2 - b)] \Phi G_\eta = \frac{1}{B_x^2} \sum_{k=0}^\infty E_k (G_{f_k} \Phi_\eta - G_\eta \Phi_{f_k}) \tag{23}$$

− energy

$$\theta_{\eta\eta} + \frac{P_r}{2B_x^2} [a_0 B_x^2 + f_1(2 - b)] \Phi\theta_\eta - \frac{1 - P_r}{1 - (T_w/T_0)} [f_{0x}(\Phi_\eta^2)_{\eta\eta} + f_{0z}(G^2)_{\eta\eta}]$$

$$= \frac{P_r}{B_x^2} \sum_{k=0}^\infty E_k (\theta_{f_k} \Phi_\eta - \theta_\eta \Phi_{f_k}) \tag{24}$$

− and the corresponding boundary conditions

$$\eta = 0: \quad \Phi = \Phi_\eta = G = \theta = 0; \qquad \eta = \infty: \quad \Phi_\eta = G = \theta = 1;$$

$$f_1 = f_2 = \cdots = 0: \quad \Phi = \Phi_0; \qquad G = G_0; \qquad \theta = \theta_0. \tag{25}$$

In case b) (adiabatic wall) one should establish the relationship between the dimensionless enthalpies θ and ϑ. That relation is, by means of the corresponding definitions (13) and (14), obtained as

$$\theta = 1 + \frac{(G^2 + \vartheta - 1) f_{0z} + (\Phi_\eta^2 + \vartheta - 1) f_{0x}}{1 - (T_w/T_0)}. \tag{26}$$

By introducing relation (26) in the system of Eqs. (22)−(25) one obtains the universal mathematical model of generalised similarity for the case b) (adiabatic wall) which comprises the following equations:

— momentum in the chordwise direction

$$\Phi_{\eta\eta\eta} + \frac{1}{2B_x^2}\left[a_0 B_x^2 + f_1(2-b)\right]\Phi\Phi_{\eta\eta} + \frac{f_1}{B_x^2}\left[1 - \Phi_\eta^2 + \frac{\vartheta f_{0z} + (\Phi_\eta^2 + \vartheta - 1)\, f_{0x}}{1 - f_{0z}}\right]$$

$$= \frac{1}{B_x^2}\sum_{k=0}^{\infty} E_k(\Phi_{\eta f_k}\Phi_\eta\vartheta V - \Phi_{\eta\eta}\Phi_{f_k}); \tag{27}$$

and in the spanwise direction

$$G_{\eta\eta} + \frac{1}{2B_x^2}\left[a_0 B_x^2 + f_1(2-b)\right]\Phi G_\eta = \frac{1}{B_x^2}\sum_{k=0}^{\infty} E_k(G_{f_k}\Phi_\eta - G_\eta\Phi_{f_k}) \tag{28}$$

— energy

$$\vartheta_{\eta\eta} + \frac{P_r}{2B_x^2}\left[a_0 B_x^2 + f_1(2-b)\right]\Phi\vartheta_\eta + \frac{2P_r}{1 + (f_{0x}/f_{0z})}\left\{\frac{f_{0x}}{f_{0z}}\,\Phi_{\eta\eta}^2 + G_\eta^2\right.$$

$$\left. - \frac{f_{0x}}{f_{0z}}\frac{f_1}{B_x^2}\Phi_\eta\left[1 - \Phi_\eta^2 + \frac{\vartheta f_{0z} + (\Phi_\eta^2 + \vartheta - 1)\, f_{0x}}{1 - f_{0z}}\right]\right\} = \frac{P_r}{B_x^2}\sum_{k=0}^{\infty} E_k(\Phi_\eta\vartheta_{f_k} - \vartheta_\eta\Phi_{f_k}) \tag{29}$$

and the corresponding boundary conditions

$$\eta = 0: \quad \Phi = \Phi_\eta = G = \vartheta_\eta = 0; \qquad y = \infty: \quad \Phi_\eta = G = 1; \qquad \vartheta = 0;$$

$$f_1 = f_2 = \cdots = 0: \quad \Phi = \Phi_0; \qquad G = G_0; \qquad \vartheta = \vartheta_0. \tag{30}$$

Since the systems of Eqs. (22)−(25) and (27)−(30) do not contain the free stream velocity distribution $U_e(X)$, which defines each particular case of the considered flow, these mathematical models are considered to be universal. Still remaining are the following parameters: in cases a) and b) the constants a and b, explained later on, the Pr-number and the compressibility parameters f_{0x} and f_{0z} and (only in case a)) the dimensionless temperature at wall (T_w/T_0), the values of which are adopted when performing the numerical integration.

The "similar" solution equations of the considered boundary layer problem derived by Reshotko and Beckwith [8] yield from the system (22)−(25) i.e. (27)−(30) for values of the constants $a_0 = b = 2$, after retaining the values of parameters f_{0x}, f_{0z}, f_1 and localising at the parameters f_0 and f_1

$$f_0 = f_{0x} + f_{0z} \neq 0, \quad f_1 \neq 0, \quad f_2 = f_3 = \cdots = 0, \quad \partial/\partial f_0 = \partial/\partial f_1 = 0.$$

One should note that constants a_0, b determine the location of the tangent to the curve $F_x(f_1)$ in the starting point of the numerical integration: $f_1 = 0$ (see Fig. 2). By adopting values of a_0, b defined in this way which are being determined as a result of gradual approximation during the numerical integration one attains, i.e., according to existing experience in applying the universal solutions to concrete flow problems, an optimal accuracy of results compared to exact solutions. This answers the question of choice of constants a_0, b which are thus being determined automatically during the numerical integration.

Regarding the compressibility parameters, in the chordwise direction $-f_{0x}$ and in the spanwise direction $-f_{0z}$, they are expressed by the means of Ma-number of the incident flow M_∞ and the wing yaw angle β as:

$$f_{0x} = \bar{u}_e^2 (f_{0x})_\infty = \bar{u}_e^2\frac{[(\varkappa - 1)/2]\, M_\infty^2 \cos^2\beta}{1 + [(\varkappa - 1)/2]\, M_\infty^2}, \quad f_{0z} = (f_{0z})_\infty = \frac{[(\varkappa - 1)/2]\, M_\infty^2 \sin^2\beta}{1 + [(\varkappa - 1)/2]\, M_\infty^2} \tag{31}$$

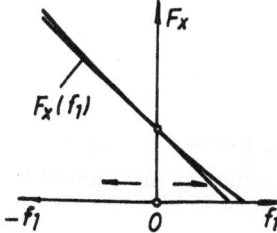

Fig. 2. Schematic representation of the characteristic boundary layer function $F_x(f_1)$

where: $(f_{0x})_\infty$, $(f_{0z})_\infty$ — are compressibility parameters of the incident flow, $\bar{u}_e = u_e/U_\infty$ — dimensionless velocity of the outer edge of the boundary layer, $\varkappa$ — power of the adiabate.

Therefore one can choose arbitrarily: — in case a) (isothermal wall) — Pr-number, M_∞, β, distribution of pressure i.e. free stream velocity $\bar{u}_e(x)$ and dimensionless temperature at wall (T_w/T_0); — in case b) (adiabatic wall) all listed quantities except T_w/T_0.

5 Range of variation of compressibility parameters f_{0x}, f_{0z} and of the corresponding velocity distribution $\bar{u}_e$

Before starting the numerical integration it is useful to define the range in which values of compressibility parameters f_{0x}, f_{0z} can be varied.

Range of variation of the parameter f_{0z} depends, according to (31), solely upon the quantities M_∞ and β, the values of which are adapted from the corresponding ranges relevant for practical application. So, in accordance with the contemporary problem of transsonic air flow over swept wings of aircrafts, one adopts the following values here: Pr = 0.72; $M_\infty = 0.82$; $\beta = 0 \div 40$.

However, for the parameter f_{0x}, which according to (31) encompasses, besides M_∞ and β, also the distribution of dimensionless free-stream velocity $\bar{u}_e$, one should perform a more detailed analysis.

Since, from the definition of the parameter f_{0x} (18), one concludes that $f_{0x} \geqq 0$, then by means of (32)

$$f_{0x} = \frac{u_e^2/2}{h_0} = \frac{\varkappa - 1}{2}\left(\frac{u_e}{c_0}\right)^2 = \frac{[(\varkappa - 1)/2]\,M_e^2\cos^2\beta_e}{1 + [(\varkappa - 1)/2]\,M_e^2} \tag{32}$$

where $M_e = v_e/c_e$, (see Fig. 1 b), one obtains first for the lower bound of the range

i.e. $f_{x0} = 0 \Rightarrow \begin{cases} \text{according to (32)} & u_e = 0 \text{ stagnation line} \\ \text{according to (32)} & M_e\cos\beta_e = 0 \Rightarrow \begin{cases} M_e = 0 & \Rightarrow c_e = \infty \\ \cos\beta_e = 0 & \Rightarrow \beta_e = 90°. \end{cases} \end{cases}$

The upper bound of the range follows from the condition: $M_e \to \infty$. In that case $c_e \to 0$, and with respect to the energy equation written as

$$f_{0x} + f_{0z} + (c_e/c_0)^2 = 1 \tag{33}$$

one obtains $(f_{0x})_{M_e \to \infty} = 1 - (f_{0z})_\infty$, i.e. because of $(c_e/c_0)^2 \geqq 0 \Rightarrow$ from (33) one obtains the upper bound

$$(f_{0x})_{\max} \leqq 1 - (f_{0z})_\infty, \tag{34}$$

where

$$(f_{0z})_\infty = \frac{\varkappa - 1}{2} \left(\frac{W_\infty}{c_0} \right)^2 = \text{const} \geqq 0. \tag{35}$$

In the special case of an unswept wing ($\beta = 0 \Rightarrow W_\infty = 0$) from (35) it follows that $(f_{0z})_\infty = 0$, therefore the corresponding upper bound $(f_{0x})_{\max} \leqq 1$. The searched range of variation of the parameter f_{0x} is, hence

$$0 \leqq f_{0x} \leqq 1 - (f_{0z})_\infty \tag{36}$$

i.e. for the unswept wing ($\beta = 0$)

$$0 \leqq f_{0x} \leqq 1. \tag{37}$$

By the means of (31) the range (36) is given as

$$0 \leq \bar{u}_e^2 (f_{0x})_\infty \leqq 1 - (f_{0z})_\infty \quad \text{i.e.} \quad 0 \leq \bar{u}_e \leqq \left[\frac{1 - (f_{0z})_\infty}{(f_{0x})_\infty} \right]^{1/2}. \tag{38}$$

After introducing expressions (31) in (38) one obtains the range for the corresponding free-stream velocity

$$0 \leq \bar{u}_e \leqq \left\{ \frac{1 + [(\varkappa - 1)/2] \, M_\infty^2 \cos^2 \beta}{[(\varkappa - 1)/2] \, M_\infty^2 \cos^2 \beta} \right\}^{1/2} \quad \text{where} \quad \bar{u}_e = \left\{ f_{0x} \frac{1 + [(\varkappa - 1)/2] \, M_\infty^2}{[(\varkappa - 1)/2] \, M_\infty^2 \cos^2 \beta} \right\}^{1/2}. \tag{39}$$

6 Determination of dimensionless temperature T_w/T_∞ and enthalpy θ

Whereas in case a) (isothermal wall) the non-dimensional temperature at wall T_w/T_0 is being adopted in advance, in case b) (adiabatic wall) it is being determined on the basis of the computed, by numerical integration, dimensionless enthalpy ϑ.

Definition of ϑ (14), namely, yields the distribution

$$T/T_0 = \vartheta[f_{0x} + (f_{0z})_\infty] - [f_{0x} + (f_{0z})_\infty] + 1 \tag{40}$$

which at the wall surface becomes

$$T_w/T_0 = (\vartheta)_{\eta=0} \, [f_{0x} + (f_{0z})_\infty] - [f_{0x} + (f_{0z})_\infty] + 1. \tag{41}$$

By introducing (41) into the relation (26) one obtains the formula for determining the dimensionless enthalpy

$$\theta = 1 + \frac{(\Phi_\eta^2 + \vartheta - 1) \, f_{0x} + (G^2 + \vartheta - 1) \, (f_{0z})_\infty}{[f_{0x} + (f_{0z})_\infty] - (\vartheta)_{\eta=0} \, [f_{0x} + (f_{0z})_\infty]}, \tag{42}$$

which can be, after differentiation by the variable η, used for determining the corresponding gradient $\partial\theta/\partial\eta$.

7 Numerical integration and analysis of the obtained results

"Independence principle", which could have been used in the case of the analogous problem of the incompressible boundary layer, is not valid for the integration of the system of equations (22)−(25) i.e. (27)−(30). They are, indeed, interdependent and therefore they should be integrated simultaneously.

For numerical integration the finite difference method with the implicit scheme [4] is used, known in the western literature as *"Tridiagonal algorithm"* (TDA) and in the Russian as *"Progonka"*.

Numerical integration is performed with the bi-parametrical approximation

$$f_0 = f_{0x} + f_{0z} \neq 0; \quad f_1 \neq 0; \quad f_2 = f_3 = \cdots = 0.$$

Accordingly, respecting the experience acquired with computation of characteristic quantities for special flow cases of the considered problems of incompressible boundary layer [2], satisfactory accuracy of the results compared to the exact solutions, even in the very vicinity of the singular separation point, is obtained even when keeping only the first parameter of the form f_1 of the set f_k (18).

For the case of the analogous incompressible boundary layer problems, it is however necessary, because of the more complicated mathematical model, besides the parameter of the form f_1 to account for compressibility by the means of the corresponding parameter $f_0 = f_{0x} + f_{0z}$.

So, the systems of universal equations, in the mentioned bi-parametrical approximation ($f_0 \neq 0$; $f_1 \neq 0$) undergo simultaneous numerical integration by the finite difference method (TDA).

From numerous solutions, computed for different combinations of values of free — when speaking of choice — parameters: Pr, M_∞, β, T_w/T_0, and those tabulated, chosen are the most interesting i.e. characteristic results. Based upon their graphical presentation an analysis was performed of the influence of particular parameters on the boundary layer development.

The variations of the dimensionless velocities Φ_η and G, for the case of adiabatic and isothermal wall, as functions of compressibility parameter f_{0x}, are much smaller than the variations of the thermodynamic quantities of the boundary layer. In this paper, due to limitations of space only the dimensionless temperatures are shown, while the character of the variations of the dimensionless velocities one can see in the paper [4] treating the similar problem.

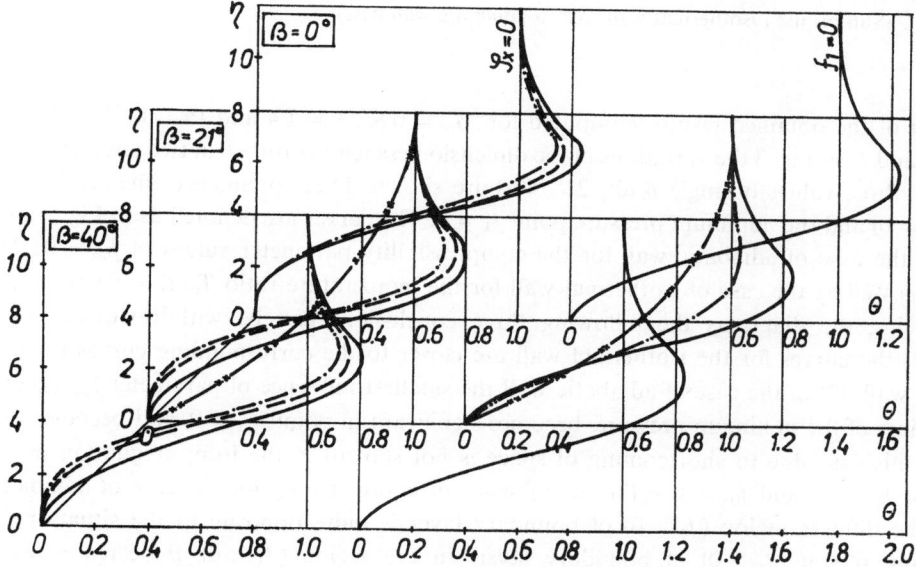

Fig. 3. Variations of dimensionless temperature θ in two characteristic points ($\zeta_x = 0$; $f_1 = 0$) for three values of wing yaw angle β (0°, 21°, 40°)

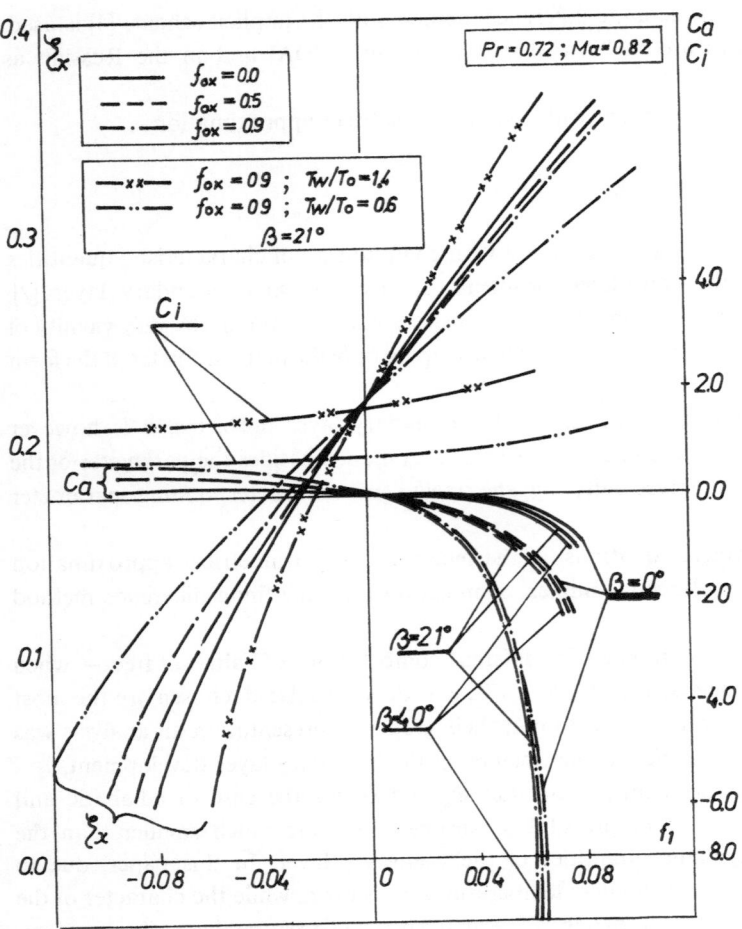

Fig. 4. Variations of both the dimensionless friction factor at wall ζ_x and the integral thermodynamic parameter C (Subscripts: i isothermal wall case, a adiabatic wall case)

A part of the obtained results, computed for $M_\infty = 0.82$, $\varkappa = 1.4$ and $\mathrm{Pr} = 0.72$ is shown in Figs. 3 and 4. In Fig. 3 the variations of the dimensionless temperature θ in two characteristic points for three values of angle β ($0°$, $21°$, $40°$) are shown. These points are: the separation point ($\zeta_x = 0$) and the minimum pressure point $f_1 = 0$. The curves are denoted as in Fig. 3 and represent the case of adiabatic wall for the compressibility parameter values of $f_{0x} = 0$; 0.5 and 0.9, as well as the case of isothermal wall for the temperature ratio $T_w/T_0 = 0.6$ and 1.4. By analysing the diagrams the following three conclusions can be withdrawn: (1) with $T_w/T_0 \to 1$, the curves for the isothermal wall are closer to the corresponding curves for the adiabatic wall; (2) in the case of adiabatic wall the smallest influence of parameter f_{0x} occurs in the points of the minimum pressure, becomes significant in points $\zeta_x = 0$, and becomes the greatest (this case due to shortcoming of space is not shown) in the front stagnation point $F_x = 0$ for $\beta = 21°$ and $f_{0x} = 0.9$; (3) the influence of parameter f_{0x} for the case of adiabatic wall in the diffuser region ($f_1 > 0$) of boundary layer is quite opposite to the situation in accelerating region ($f_1 > 0$) of boundary layer. In the region ($f_1 < 0$), θ decreases with increasing f_{0x}, while in the region ($f_1 > 0$) θ increases and has maximum value 4.8 in point $F_x = 0$ ($\beta = 21°$ and $f_{0x} = 0.9$).

In Fig. 4 the variations of both the dimensionless friction factor at wall ζ_x and one of the integral thermodynamic parameter of boundary layer $C = \int\limits_0^\infty \Phi_\eta(1 - \theta)\,dy$ are given, where subscripts (i) and (a) correspond to the cases of isothermal and adiabatic wall, respectively. By analysing the curves for ζ_x, one can notice that the most unconvenient flow for the case of isothermal wall is for $T_w/T_0 = 1.4$, since the separation of boundary layer occurs earlier in upstrem direction ($\zeta_x = 0$), and that the most convenient flow is for $T_w/T_0 = 0.6$. As far as adiabatic wall case is considered, increasing of parameter f_{0x} moves the separation point in downstream direction. Analysis of the thermodynamic quantity C indicates the significant differences between the shapes of curves for the cases (a) and (i). For the case (i), quantity C decreases in downstream direction, while for the case (a), C increases with increasing of parameter f_1. The quantity C states as follows: $C < 0$ in accelerating region and $C > 0$ in diffuser region of boundary layer.

References

[1] Gersten, K.: Die Bedeutung der Prandtlschen Grenzschichttheorie nach 85 Jahren. Z. Flugwiss. Weltraumforsch. **13**, 209−218 (1989).

[2] Saljnikov, V. N.: A contribution to universal solutions of the boundary layer theory. Teor. i prim. Meh. **4**, 139−163 (1978).

[3] Schneider, W.: Über die Bedeutung analytischer Methoden für die Strömungs-Mechanik im Zeitalter des Computers. Vortrag am Festkolloquium, 60jähriges Jubiläum von Prof. Zierep, Karlsruhe 1989.

[4] Saljnikov, V. N., Dallmann, U.: Verallgemeinerte Ähnlichkeitslösungen für 3-D lamin. stat. kompress. Grenzschichtströmungen an schiebenden profilierten Zylindern. DLR-FB **34** (1989).

[5] Saljnikov, V. N., Boričić, Z., Nikodijević, D.: Lösungen verallgemeinerter Ähnlichkeit für 3-D lamin. kompress. Flügelgrenzschichten. ZAMM **70**, T462−T465 (1990).

[6] Görtler, H.: A new series for the calculation of steady laminar boundary layer flows. J. Math. Mech. **6**, 1−66 (1957).

[7] Loitsianski, L. G.: Die universellen Gleichungen und die parametrigen Näherungen in der Theorie der laminaren Grenzschicht (in Russian). PMM **29**, 70−87 (1965).

[8] Reshotko, E., Beckwith, I.: Compressible laminar boundary layer over a yawed infinite cylinder with heat transfer and arbitrary Prandtl number. NASA Report **1379** (1958).

Authors' addresses: Professor Dr. V. Saljnikov, Nevesinjska 17/V, YU-11000 Belgrade, Professor Dr. Z. Boričić, T. Roksandića 3a, YU-18000 Niš, and Prof. Dr. D. Nikodijević, S. Mladenovića 138/21, YU-18000 Niš, Yugoslavia

Franz Ziegler

Mechanics of Solids and Fluids

1991. 352 figures. XXI, 735 pages.
Cloth DM 148,–, öS 1040,–
ISBN 3-211-97529-2

Prices are subject to change without notice

This book offers a unified presentation of the concepts and most of the practical principles common to all branches of solid and fluid mechanics. Its design should be appealing to advanced undergraduate students in engineering sciences and should also enhance the insight of both graduate students and practitioners. A profound knowledge of applied mechanics as understood in this book may help to meet the needs for necessary versatility which the engineering community has to face in the modern world of high-technology.
Although the book grew out of lecture notes for a three-semester course for advanced undergraduate students taught by the author and several colleagues during the past 20 years, it contains sufficient material for a subsequent two-semester graduate course. The only prerequisites are the basic algebra and analysis as usually taught in the first year of an undergraduate engineering curriculum. Advanced mathematics as it is required in the progress of mechanics teaching may be taught in parallel classes, but also introduction into the art of design should be offered at that stage. The book is divided into thirteen chapters which are arranged in such a way as to preserve a natural sequence of thoughts and reflections. Within the single chapter, however, the presentation is in general done such as to proceed from undergraduate level to the intermediate level and eventually to the graduate level.

German edition also available:

Technische Mechanik der festen und flüssigen Körper

Zweite, verbesserte Auflage.
1992. 333 Abbildungen, 93 Aufgaben mit Lösungen und einem dreiteiligen Anhang.
564 Seiten. Broschiert DM 85,–, öS 595,–
ISBN 3-211-82335-2

Springer-Verlag Wien New York

Sachsenplatz 4–6, P.O.Box 89, A-1201 Wien · 175 Fifth Avenue, New York, NY 10010, USA
Heidelberger Platz 3, D-14197 Berlin · 37-3, Hongo 3-chome, Bunkyo-ku, Tokyo 113, Japan